A Guide to the Scientific Career

A Guide to the Scientific Career

Virtues, Communication, Research, and Academic Writing

Edited by

Mohammadali M. Shoja, MD
Tuberculosis and Lung Disease Research Center, Tabriz University of Medical Sciences, Tabriz, Iran
Division of General Surgery, University of Illinois at Chicago Metropolitan Group Hospitals,
Chicago, IL, USA

Anastasia Arynchyna, MPH, CCRP
Pediatric Neurosurgery, Children's of Alabama, Birmingham, AL, USA

Marios Loukas, MD, PhD
Department of Anatomical Sciences, St. George's University School of Medicine, St. Georg's
Grenada, West Indies

Anthony V. D'Antoni, MS, DC, PhD
Division of Anatomy, Department of Radiology, Weill Cornell Medicine, New York, NY, USA

Sandra M. Buerger, PhD
Department of Natural Sciences and Mathematics, College of General Studies, Boston University,
Boston, MA, USA

Marion Karl, MFA, MS
Lure Animations, Reno, NV, USA

R. Shane Tubbs, Ph.D., PA-C, MSc
Seattle Science Foundation, Seattle, WA, USA

Registered Offices
John Wiley & Sons, Inc., 111 River Street, Hoboken, NJ 07030, USA

Editorial Office
The Atrium, Southern Gate, Chichester, West Sussex, PO19 8SQ, UK

For details of our global editorial offices, customer services, and more information about Wiley products visit us at www.wiley.com.

Wiley also publishes its books in a variety of electronic formats and by print-on-demand. Some content that appears in standard print versions of this book may not be available in other formats.

Library of Congress Cataloging-in-Publication data applied for

ISBN: 9781118907429 (Hardback)

Cover Design: Wiley
Cover Image: © David Perry/Getty Images

Set in 10/12pt WarnockPro by SPi Global, Chennai, India

Printed and bound in Singapore by Markono Print Media Pte Ltd

10 9 8 7 6 5 4 3 2 1

Contents

List of Contributors

Paul S. Agutter
Theoretical Medicine and Biology Group
Glossop, Derbyshire
United Kingdom

Katherine G. Akers
Shiffman Medical Library
Wayne State University
Detroit, MI
USA

Eyyub S. M. Al-Beyati
School of Medicine, Department of
Anatomy
Ankara University
Ankara
Turkey

and

School of Medicine, Department of
Neurosurgery
Ankara University
Ankara
Turkey

Naomi Andall
Department of Anatomical Sciences
St. George's University School of
Medicine
Grenada
West Indies

Rebecca Andall
Department of Anatomical Sciences
St. George's University School of
Medicine
Grenada
West Indies

Nihal Apaydin
Department of Anatomy and Brain
Research Center, Faculty of Medicine
Ankara University
Ankara
Tur

Anastasia A. Arynchyna
Department of Neurosurgery, Children's
of Alabama
University of Alabama at Birmingham
Birmingham, AL
USA

Arindam Basu
School of Health Sciences
University of Canterbury
Christchurch
New Zealand

Jeffrey Beall
Retired from Jeffrey Beall Auraria Library
University of Colorado Denver
Denver, CO
USA

Frederic J. Bertino
Department of Radiology and Imaging Sciences
Emory University School of Medicine
Atlanta, GA
USA

and

Department of Anatomical Sciences
St. George's University School of Medicine
Grenada
West Indies

Bharti Bhusnurmath
Department of Anatomical Sciences
St. George's University School of Medicine
Grenada
West Indies

Shivayogi Bhusnurmath
Department of Anatomical Sciences
St. George's University School of Medicine
Grenada
West Indies

Anand N. Bosmia
Department of Psychiatry, LSU Health Sciences Center
Shreveport, LA
USA

Philip R. Brauer
Department of Clinical Anatomy
Kansas City University of Medicine & Biosciences
Kansas City, MO
USA

Sandra M. Buerger
Department of Natural Sciences and Mathematics
Boston University, College of General Studies
Boston, MA
USA

Gernot Buerger
Department of Natural Sciences and Mathematics
Boston University, College of General Studies
Boston, MA
USA

Tracy E. Bunting-Early
Bunting Medical Communications
Newark, DE
USA

Jacopo Buti
School of Dentistry
The University of Manchester
Manchester
UK

Marcoen J.T.F. Cabbolet
Department of Philosophy
Free University of Brussels
Brussels
Belgium

Loretta Cacace
Division of Anatomy, Department of Radiology
Weill Cornell Medicine
New York, NY
USA

Stephen W. Carmichael
Department of Anatomy, Mayo Clinic, Emeritus Center
Rochester, MN
USA

Ayhan Cömert
School of Medicine, Department of
Anatomy
Ankara University
Ankara
Turkey

Arthur C. Croft
Spine Research Institute of San Diego
San Diego, CA
USA

Mariana Cuceu
Program on Medicine and Religion
The University of Chicago
Chicago, IL
USA

and

Grigore T. Popa University of Medicine
and Pharmacy
Iasi
Romania

Marcel Cuceu
Romanian Catholic Byzantine Diocese of
Saint George
Canton, OH
USA

Anthony V. D'Antoni
Division of Anatomy, Department of
Radiology
Weill Cornell Medicine
New York, NY
USA

Ganesh N. Dakhale
Department of Pharmacology
Government Medical College
Nagpur, Maharashtra
India

Michael Dieter
College of Applied Health Sciences,
Department of Biomedical and Health
Information Sciences
University of Illinois at Chicago
Chicago, IL
USA

Jamie Dow
Department of Neurosurgery
Vanderbilt University Medical School
Nashville, TN
USA

Leslie A. Duncan
Dartmouth College
Hanover, NH
USA

Jean Anderson Eloy
Center for Skull Base and Pituitary
Surgery
Rutgers New Jersey Medical School
Newark, NJ
USA

Edzard Ernst
Department of Complementary
Medicine, Peninsula Medical School
University of Exeter
Exeter
UK

Joseph Fernandez
Houston Methodist Hospital,
Department of Surgery
Houston, TX
USA

Adam J. Folbe
Department of Neurological Surgery
Rutgers New Jersey Medical School
Newark, NJ
USA

Jerzy Gielecki
Department of Anatomical Sciences
St. George's University School of
Medicine
Grenada
West Indies

Juan Guillermo Gormaz
Molecular and Clinical Pharmacology
Program, Institute of Biomedical
Sciences, Faculty of Medicine
University of Chile
Santiago
Chile

Maria J. Grant
Liverpool John Moores University
Liverpool
UK

Ray Greek
Americans for Medical Advancement
Goleta, CA
USA

Christoph J. Griessenauer
Department of Neurosurgery
University of Alabama at Birmingham
Birmingham, AL
USA

Department of Neurosurgery
Geisinger,
Danville, PA
USA

and

Research Institute of Neurointervention
Paracelsus Medical University
Salzburg
Austria

James Hartley
School of Psychology
Keele University, Staffordshire
UK

Daniel Hasson
Molecular and Clinical Pharmacology
Program, Institute of Biomedical
Sciences, Faculty of Medicine
University of Chile
Santiago
Chile

Mark P. Henderson
Theoretical Medicine and Biology Group
Glossop, Derbyshire
UK

Philipp Hendrix
Department of Neurosurgery,
Saarland University Hospital
Homburg, Saar
Germany

Sachin K. Hiware
Department of Pharmacology
Government Medical College
Nagpur, Maharashtra
India

Barbara J. Hoogenboom
Department of Physical Therapy
Grand Valley State University
Grand Rapids, MI
USA

David C. Howell
Department of Psychological Sciences
University of Vermont
Burlington, VT
USA

R. Peter Iafrate
Health Center IRB
University of Florida
Gainesville, FL
USA

Beatrice Gabriela Ioan
Faculty of Medicine
University of Medicine and Pharmacy
"Gr. T. Popa"
Romania

Talal A. Kaiser
Department of Internal Medicine
University of Connecticut School of
Medicine
Hartford, CT
USA

Kyle E. Karches
Internal Medicine Residency Program
The University of Chicago Medicine
Chicago, IL
USA

Marion Karl
Lure Animations
Reno, NV
USA

John R.W. Kestle
Division of Pediatric Neurosurgery,
Primary Children's Hospital
University of Utah
Salt Lake City, UT
USA

Sung Deuk Kim
Department of Anatomical Sciences
St. George's University School of
Medicine
Grenada
West Indies

Kristin L. Kraus
Division of Pediatric Neurosurgery,
Primary Children's Hospital
University of Utah
Salt Lake City, UT
USA

José Florencio F. Lapeña
Department of Otorhinolaryngology,
Philippine General Hospital
University of the Philippines Manila
Manila
Philippines

Matías Libuy
Molecular and Clinical Pharmacology
Program, Institute of Biomedical
Sciences, Faculty of Medicine
University of Chile
Santiago
Chile

Marios Loukas
Department of Anatomical Sciences
St. George's University School of
Medicine
Grenada
West Indies

Thomas F. Lüscher
Heart Division
Royal Brompton and Harefield Hospitals
and Imperial College
London
UK

and

Center for Molecular Cardiology
University Zürich
Zürich
Switzerland

and

European Heart Journal Editorial Office
Zurich Heart House
Zürich
Switzerland

Mohini S. Mahatme
Department of Pharmacology
Government Medical College
Nagpur, Maharashtra
India

Chrisovalantis Malesios
Department of Environment
University of the Aegean Mytilene
Greece

James M. Markert
Department of Neurosurgery
University of Alabama at Birmingham
Birmingham, AL
USA

Izet Masic
Faculty of Medicine
University of Sarajevo
Sarajevo
Bosnia and Herzegovina

Petru Matusz
Department of Anatomical Sciences
St. George's University School of
Medicine
Grenada
West Indies

S. Jane Millward-Sadler
Newcastle University
Manchester
UK

Haley J. Moon
Department of Anatomy and Cell Biology
Indiana University School of Medicine
Fort Wayne, IN
USA

Dan O'Brien
Rhazes Publishing
Birmingham, AL
USA

Aasim I. Padela
MacLean Center for Clinical Medical
Ethics
The University of Chicago
USA

John Panaretos
Department of Statistics
Athens University of Economics and
Business
Athens
Greece

Toral R. Patel
Department of Cardiology
University of Virginia
Charlottesville, VA
USA

Daxa M. Patel
Department of Neurosurgery
University of Alabama at Birmingham
Birmingham, AL
USA

Sanjay Patel
Department of Anatomical Sciences
St. George's University School of
Medicine
Grenada
West Indies

Marco Pautasso
Forest Pathology and Dendrology
Institute for Integrative Biology
ETH Zurich, Zurich
Switzerland

and

Animal and Plant Health Unit
European Food Safety Authority (EFSA)
Parma
Italy

Wilfred C.G. Peh
Department of Diagnostic Radiology
Khoo Teck Puat Hospital
Singapore
Yong Loo Lin School of Medicine
The National University of Singapore
Singapore

Matusz Petru
Department of Anatomical Sciences
St. George's University School of
Medicine
Grenada
West Indies

Sonali A. Pimpalkhute
Department of Pharmacology
Government Medical College
Nagpur, Maharashtra
India

Paul Posadzki
The Centre for Public Health
Liverpool John Moores University
Liverpool
UK

Sarah B. Putney
Administrative Compliance, Internal
Audit Division
Emory University
Atlanta, GA
USA

Vijay M. Ravindra
Division of Pediatric Neurosurgery,
Primary Children's Hospital
University of Utah
Salt Lake City, UT
USA

Jay K. Riva-Cambrin
Division of Pediatric Neurosurgery,
Primary Children's Hospital
University of Utah
Salt Lake City, UT
USA

Michelle K. Roach
Department of Obstetrics and
Gynecology
Vanderbilt University
Nashville, TN
USA

Ramón Rodrigo
Molecular and Clinical Pharmacology
Program, Institute of Biomedical
Sciences, Faculty of Medicine
University of Chile
Santiago
Chile

Priyanka Shah
Department of Otolaryngology–Head
and Neck Surgery
Rutgers New Jersey Medical School
Newark, NJ
USA

Chevis Shannon
Department of Neurosurgery
Vanderbilt University Medical School
Nashville, TN
USA

Mohammadali M. Shoja
Division of General Surgery
University of Illinois at Chicago
Metropolitan Group Hospitals
Chicago, IL
USA

George K. Simon
Character Development
Forensic, and Clinical Psychology
Little Rock, AR
USA

Walavan Sivakumar
Division of Pediatric Neurosurgery,
Primary Children's Hospital
University of Utah
Salt Lake City, UT
USA

Cristinel Stefanescu
Faculty of Medicine
University of Medicine and Pharmacy
"Gr. T. Popa"
Romania

Peter F. Svider
Department of Otolaryngology–Head
and Neck Surgery
Wayne State University School of
Medicine
Detroit, MI
USA

Paul Tremblay
Sidney Druskin Memorial Library
New York College of Podiatric Medicine
New York, NY
USA

R. Shane Tubbs
Seattle Science Foundation
Seattle, WA
USA

Avinash V. Turankar
Department of Pharmacology
Government Medical College
Nagpur, Maharashtra
India

Dirk T. Ubbink
Department of Surgery
Amsterdam University's Academic
Medical Center
Amsterdam
The Netherlands

Paige E. Vargo
Department of Internal Medicine
Northeast Ohio Medical University
Rootstown, OH
USA

Joel A. Vilensky
Department of Anatomy and Cell Biology
Indiana University School of Medicine
Fort Wayne, IN
USA

Elizabeth Wager
SideView
Princes Risborough, Buckinghamshire
UK

Thomas P. Walker
Holy Spirit Library
Cabrini University
Radnor, PA
USA

Beverly C. Walters
Department of Neurosurgery
University of Alabama at Birmingham
Birmingham, AL
USA

Peter J. Ward
West Virginia School of Osteopathic
Medicine
Lewisburg, WV
USA

Koichi Watanabe
Department of Anatomy
Kurume University School of Medicine
Fukuoka
Japan

Bradley K. Weiner
Weill Cornell Medical College, Houston
Methodist Hospital
The Methodist Hospital Research
Institute
Houston, TX
USA

Bulent Yalcin
Department of Anatomy, Gulhane
Medical Faculty
University of Health Sciences
Ankara
Turkey

Marilyn Michael Yurk
Department of Neurological Surgery
Indiana University School of Medicine
Indianapolis, IN
USA

Faizan Zaheer
School of Dentistry
The University of Manchester
Manchester
UK

Genevieve Pinto Zipp
School of Health and Medical Sciences,
Department of Graduate Programs in
Health Sciences
Seton Hall University
South Orange, NJ
USA

Preface

"And those who were seen dancing were thought to be insane by those who could not hear the music." — Friedrich Nietzsche

Too often in academia, students and professionals are expected to be knowledgeable of the various practical nuances of research that are not readily available or are only learned by experience. Herein, we strive to present a well-defined and comprehensive textbook that provides the academician with the most commonly encountered topics in higher education and research.

This textbook is an easy-to-read source of essential tips and skills for a scientific career. The topics have been chosen to be pragmatic and to enhance a career in academia, whether focused on didactics, basic science, or clinical research. There has seldom been any effort in the past to comprehensively and systematically address the academic lifestyle. *A Guide to the Scientific Career* fills a gap in this arena. It inspires and motivates research activity among a new generation of researchers all around the world.

Section I

Successful Career

1

Defining and Re-Defining Success

Mohammadali M. Shoja[1], R. Shane Tubbs[2] and Dan O'Brien[3]

[1] *Division of General Surgery, University of Illinois at Chicago Metropolitan Group Hospitals, Chicago, IL, USA*
[2] *Seattle Science Foundation, Seattle, USA*
[3] *Rhazes Publishing, Birmingham, AL, USA*

1.1 Introduction

The life of a human being in the modern era is centered on a struggle to succeed and attain whatever is initially thought near-impossible or very difficult. Each and every day we are moving farther apart from our traditional striving for personal fulfillment and peaceful content. For some, these are two sides of a same coin, labeled as prosperity or fortune. I do not intend to ignite a philosophical debate here. My intention is to encourage you to rethink success in all meaningful dimensions at different stages of your life, as it needs continual redefinition or refinement. The potential impact of this rethinking is profound, as it will naturally influence your decision-making.

Success and failure are not dichotomous, black and white, yes or no phenomena. They are two ends of a spectrum; we are born and live somewhere in between them. Success is a state of feeling content with who we are, and perceiving that the images of our self and past, current or future status correspond to what we long to be.

In this chapter, we identify five core elements that determine one's success: mindsets, prerequisites, methods, enhancers, and inhibitors. Mindsets are the sets of attitudes or core beliefs necessary for envisioning and establishing a successful career. Prerequisites are internal factors or personal qualities required to become successful. Methods are conscious actions or a plan of actions one should take to pave the road to success. Enhancers are external factors that increase one's chance of success. Inhibitors are internal or external factors that diminish one's chance of success.

1.2 Success Mindsets

1.2.1 Success Is a State of Mind

There are three characteristics that can drive success: communication efficacy, leadership efficacy, and problem-solving efficacy. Problem solving presents an interesting

A Guide to the Scientific Career: Virtues, Communication, Research, and Academic Writing, First Edition.
Edited by Mohammadali M. Shoja, Anastasia Arynchyna, Marios Loukas, Anthony V. D'Antoni, Sandra M. Buerger, Marion Karl and R. Shane Tubbs.

aspect of success as a state of mind because the entrepreneurial spirit that guides success is rooted in overcoming difficulties or problems in your path. Simply put, it is about the mindset, not intelligence or some measure of analytical skill.

In considering the journey to success, the mindset begins with understanding potential pitfalls and then creating a plan to overcome those difficulties. You are essentially selling yourself on the idea that you can be successful, and then realizing this idea because of a successful mindset. You must be prepared to transform, uncover, clarify, and understand the problems in the path to success.

This relationship with your mindset allows you to overcome limitations you may have placed on yourself. The approach to success with a mindset that allows you to solve the problems along your journey is important not only for seeking your goals, but also for improving yourself.

1.2.2   Success in Not Accidental

The idea of success is nebulous for some people. Perhaps this is because of lack of foresight, but I believe it is because we see fate in success. We see a loaded determinism that robs us of agency. Those who achieve success have developed a long-term plan paired with impressive work habits that push them toward success. They do not trip into it, nor do they come upon it by happenstance.

Taking care in your preparation for success, the roadmap on which you depend to reach your goals, is important because attention to detail is the difference between success and failure. Holding yourself to a level of excellence that befits your goals ensures that success will not merely happen upon your path; you will have driven yourself toward it.

1.2.3   Success Is Simple

Simple, not easy – that is the stuff of success. You need to attend to what you wish to accomplish, and then do it correctly at every step. There is no complexity; it is a straight path when walked accurately.

Yet it remains complicated. Why?

I believe it is focus that determines how simply we perceive success. If you do not understand what you are trying to achieve, it should come as no surprise when the process *feels* complicated. Vagueness and uncertainty have no place in the pursuit of success. Clarity and conviction are imperative for simplifying the path. Understanding and explicitly stating what you need allows you to direct your energy toward your goal. You don't want to follow any old road; you want to take the road that leads to *your* destination.

1.2.4   Success Is an Ever-Changing State

Change is the only true constant. No matter how hard you try, things change. If you are not evolving, you are letting the world pass you by. More importantly, if your goals and plans for success and the very definition of your success are not changing, you put yourself in danger of failing.

Innovation, growth, reinvention: these are the tools necessary for overcoming change in pursuit of personal success. You need only look at other people who continue to pursue their dreams using the same methods as everyone else and watch them become disappointed by the outcome.

Disruption and transformation are not new ideas, and are important in the change that ultimately happens along the path to success. A time comes in the pursuit of success when the need to change your paradigm becomes clear. Learn to transform yourself in the process. Consider the following ideas when change presents itself:

1. Will this help me grow?
2. Have I changed the metrics to reflect the evolution of my success?
3. Am I willing to change how I describe my success?
4. Do I understand the landscape of the changes in my path?
5. Do I have long-term and short-term strategies?

1.2.5 Success Is Measurable

In order to measure success, you must define it. For this purpose, you must create long- and short-terms goals that best reflect the steps that lead to your definition. There is no catchall for how to measure success; its subjectivity lends itself to multiple definitions and metrics.

The important thing is to know what *your* definition is – and then measure it accordingly.

This tricky question needs to be addressed first: how do you define success? From there, you move to understanding and setting the metrics that best reflect how to measure it. Choose a measurement that inspires your pursuit of success, something that matches your personal values or business values. Once you have a scale, you need to track your measurements with a dashboard or chart that keeps the information simple and understandable.

1.2.6 Success Makes You a Better Person

Success, however it is defined, changes your life. And in this change, you come to see the importance of giving back. Giving back enriches not only the lives of others but also the successful individual. As such, societal contribution is a metric by which you should define how success has affected you. Do you give back to your community? To those less fortunate?

Having reached a point in life where you can not only take care of yourself but also provide for others, a successful person is positioned to realize the concept of self-actualization, to become a complete being who is satisfied on every level. This level of awareness makes you more receptive to the plight of others and, hopefully, more giving of your excess time and resources.

Success breeds kindness in those who recognize the struggle.

Being true to yourself is important for success, but also for how you see others. If you have taken control of your life, you begin to see the importance of mentoring others to do the same. You need to be fueled by something greater than yourself to be successful, but it also means asking yourself meaningful questions. Who are you? What do you care about? How can you help?

1.2.7 Happiness and Success Are Mutually Inclusive

Happiness and success are interdependent; it is difficult to achieve real success without happiness. Success in life is subjective, but is no doubt predicated on the pursuit of your dreams – of what makes you happy. If the pursuit or achievement of success does not come with happiness, with a balance of what matters, then can we truly call it success?

If you must sacrifice happiness for success, then what is success?

Quality of life can be measured by various metrics, but chief among them is happiness as an indicator. If you work long hours to acquire wealth, and had previously defined that as success, then does that simple math equate to success? What if you cannot pursue anything other than the acquisition of wealth at the expense of sharing and experiencing the joy in life? How do we reconcile the interdependence of happiness and success?

Identification of success and being able to measure it, as described in a previous section, helps us to understand whether we have been successful if we have traded one for the other.

1.2.8 Success and Fame Are Independent

Fame and success are often conflated because they look so similar. The wealth, prestige and access that comes with fame looks very much like what most people would define as success. However, fame is dependent on success in a particular domain. Were you a famous physicist, then you would be well known in your field and would have achieved a level of fame in direct proportion to your success in that field.

The trickiness of this independence arises when we do not fully understand what is we are pursuing. If we do not define success appropriately, we may see other's success and lose the value of our own success; we may confuse extrinsic value with intrinsic value.

1.2.9 Success and Failure Are Self-Perpetuating in Nature

Success and failure are not unlike other objects in the universe: they are guided by and subject to the whims of momentum, physical and otherwise. The more you engage in a behavior, the more frequently that behavior occurs – and more quickly. This momentum creates a sense of success being generative and influencing future success. The same applies to failure; since it is predicated on a series of behaviors, the more you engage in them, the more frequent they become.

The danger lies in seeing the momentum as fate or determinism. You are in charge of changing the momentum, of creating new behavioral tendencies that propagate toward a synergy. Do not allow past mistakes to create a self-fulfilling prophecy that dominates your path to success.

1.2.10 Values and Success Are Not the Same Things

We are governed by values. These values influence nearly every aspect of our lives. And this influence extends to how we set goals and define success. Values are instilled in us, but we can adopt them as well; they can change.

However, values are not targets.

Success is defined by goals, which *are* targets. We make decisions in our lives on the basis of our values and they color how we understand and process the world around us. This foundational set of rules help you discover your purpose, which is closely tied to ideas of success.

1.2.11 Relativity of Success: Success Is a Self-Defined Phenomenon

When we think of success, we often turn to financial metrics. This illustrates the relativity of success very clearly. The amount of money that one thinks of as enough is wildly subjective; so, too, is the idea of success relative to that. Material success does not necessarily offer value, but it does reveal that each individual's idea of it can be assessed along a spectrum.

No matter the form of success, each person will have a different definition, level, and metric by which they experience it. The important thing to understand is that it is *your* definition of success that matters, not anyone else's.

1.2.12 Pareto Principle: The Major Part of Success Comes from a Small Fraction of Our Decisions and Actions

The Pareto principle is colloquially known as the 80/20 rule. Simply put, it states that roughly 80% of the effects come from 20% of the causes. This is often extended into marketing as 80% of your income comes from 20% of your clients. It provides an interesting base for how we understand and measure success.

1.2.13 The "Luck" Paradox

Thomas Jefferson once attested, "I find that the harder I work, the more luck I seem to have." This statement is the key to understanding luck and its relationship to success. Simply put, there is no bad or good luck in effect. It is determination and working hard toward a set goal that create a context in which one can flourish. If there were bad luck, it would be a law of nature. However, were it a law of nature, it would not suffer under the weight of chance and probability. And were there no chance, how could there be any kind of luck at all? We all have an idea of being in the right place at the right time as a quick way of achieving success. The problem with this kind of thinking is that it ignores the importance of goal setting, planning definition, measurement, and the perseverance to continue forward toward your idea of success.

If you depend on luck as a means of achieving success, you are eliminating the possibility of agency in your pursuit of personal success. How can you influence your success if you are at the whim of mere fickle chance?

1.2.14 Opportunities Are Created

This ties in with the difficulty of letting chance guide your thinking. If you believe that opportunities are fated into your path, then you will not follow through on the plan you have set yourself. Unfortunately, many of us believe that things come to those who wait. Nevertheless, it is the interdependency with others that leads to being in the environment where things happen. It is perseverance that creates the situation in which it appears as if manna falls from heaven. Success is about creating, recognizing, and acting on opportunities, not blindly waiting for them to trip into your path.

1.2.15 Failure Is a Key to Success

Failing forward is a concept traded by those who see failure as a stepping stone toward success, the recognition of the ways forward that do not yield success. Failure is only truly failure if you learn nothing from it and abandon your path because you did not achieve what you set out to do on the first try. Success is littered with failures, as success is ever-changing; you must be willing to understand what has not worked in order to achieve what you have set out to do.

1.2.16 Success Is the Outcome of Struggle for Excellence, Not Struggle for Winning

Winning is short-sighted as it encompasses only a single instance overcome, a single goal achieved. Success is about excellence. The pursuit of success is predicated on having defined what it is, understanding how to measure it, adapting as it changes, and overcoming and learning from failure along the way. Were you to only be interested in winning, you would stop the moment you won; you would not overcome failure or adapt as circumstances demanded.

1.3 Prerequisites Are Internal Factors Required to Become Successful

1. *Sense of purpose.* This is the motivation that drives your pursuit of success. It is shaped by what you believe and your values, which are informed by your personal code of behavior.
2. *Motivation and enthusiasm.* Often talked about rather than acted upon. Enthusiasm mimics motivation at times, but only because it is simpler. Motivation is what drives you, what gets you up in the morning. Motivation is more difficult and requires commitment to a cause or your pursuit of success.
3. *Confidence.* Belief in your ability to succeed can be difficult, as balancing unmitigated cockiness against doing too little can prove challenging. In terms of being successful, you need at least to be confident that you can follow through on the journey toward your idea of success.
4. *Focus.* Keeping your eye on the ball and concentrating on the details of your plan are important. Regardless of the subjectivity of your definition, it is important to have a clear and vivid idea of how you will move toward your goals.
5. *Courage.* Courage is often defined as the willingness to confront difficulty in pursuit of a goal. In regard to success, it is being willing to fail forward and overcome obstacles in your path.
6. *Perseverance, persistence, energy,* and *hard work.* Continuing to do what is necessary, no matter how many times you have to do it, is important in order to become successful.

7. *Communication skills.* The ability to communicate effectively is important to being successful. You need to understand how to communicate with others, and to understand their perspectives, in order to pursue your goals.

8. *Preparation.* This matters because success requires short-term and long-term goals. Without taking precautionary measures, you could find yourself unable to overcome pitfalls to which you might otherwise have been able to adapt.

9. *Creativity.* The relationship between success and creativity is bidirectional. Wilson (2017) has characterized five different levels of creativity (Table 1.1).

10. *Open-mindedness.* The receptiveness to new ideas relates to the way in which people approach other views. The importance of this to success is pretty simple: in order to overcome obstacles, you might have to look at things in new and different ways.

11. *Situational awareness.* Situational awareness refers to understanding how the environment has changed or what resources are lacking or in excess. As conditions change relative to short-term and long-term goals, situational awareness allows you to understand the "lie of the land."

12. *Attitude.* Generally, an attitude can be positive or negative relative to people, events, ideas, etc. Relative to success, a positive attitude is important in order to navigate the environment and situations. Positive attitudes open more doors than negative ones.

13. *Disciplines and habits.* Considering that the road to success is both long and difficult, having habits that lead to success, as well as the discipline to follow them, is extremely important. Self-control takes practice, especially when failure is an option.

14. *Integrity.* Being honest and having strong moral principles is important as you move toward your goals. It is easy to get sidetracked from your goals if your values are not aligned with your definition of success.

15. *Self-reliance.* The capacity to rely on one's own capabilities is sometimes essential along the path to success. You need to have confidence in both your decisions and any actions you take toward your short-term and long-term goals.

16. *Executive and leadership skills.* Leadership can be a nebulous concept, but as a practical skill it is important for success. The ability to lead or guide others becomes increasingly critical, even if you are approaching your idea of success

Table 1.1 Levels of creativity.

Expressive creativity	Spontaneous expression of intuition (without the need for conscious reasoning) results in creation of basic and simple concepts.
Productive creativity	Existing concepts are efficiently and eloquently proliferated through learning skills and technical mastery.
Inventive creativity	Exploring different aspects of existing concepts with materials, methods, and techniques.
Innovative creativity	The existing concepts are improved and modified through conceptualizing skills and generating new materials and methods.
Imaginative creativity (or genius)	An entirely new principle or concept is created.

alone. Along the way, you will encounter viewpoints and perspectives that might force you to adapt either your short-term or long-term goals. Without the ability to lead, you could be stuck working through issues that could otherwise derail your intentions.

17. *Passion, energy, and persistence.* Having strong feelings about your goals is important; without them, you are likely to find that your best efforts stall. The energy and persistence to pursue the goals necessary for success are rooted deeply in passion, in having intense emotions and compelling enthusiasm for your continued growth toward success.

18. *Punctuality.* Punctuality is the characteristic of being able to fulfill an obligation on or before the required deadline or schedule. Being able to complete what is required of you is important because you cannot achieve your goals without understanding the obligation to show up or complete a task when you need to. Punctuality has a cultural aspect as well, as some organizations treat punctuality differently. If we are speaking about personal success and the role of your short-term and long-term goals, punctuality is a cardinal virtue, as holding yourself accountable is paramount to moving forward.

19. *Ability.* When we speak of ability, we are talking about aptitudes. An aptitude speaks to competence in particular domain at a level appropriate to a goal. Ability is the development of these aptitudes toward a goal with an eye on the skills that will assist you in reaching your short-term and long-term goals.

20. *Organization.* Organization ties adherence into goal orientation, which is incredibly important for working toward success. Like any other habit, it is the persistent application of organizational skills that enables you to become more and more organized over time. Undoubtedly, it influences your thought processes and decision-making.

21. *Humility.* Being humble focuses you on adapting to the circumstances along your path to success. Knowing your strengths and weaknesses provides the necessary self-restraint from excessive vanity and myopia.

22. *Sense of urgency.* You will hear this spoken about a lot in corporate culture, and for good reason. In a constantly changing world and business environment, it is often very important to make a sense of urgency a priority.

23. *Loyalty.* Loyalty might seem odd in a list of attributes that contribute to success. However, having a positive orientation binds together those who share your passion and common interests. Goal orientation is made simpler with people along your path who share your passion; like-minded people with common interests. This categorical devotion assists in interpersonal relationships that lead to success.

24. *Self-awareness and emotional intelligence.* Introspection is important for knowing yourself, which is paramount when dealing with others and with changes along your path. Being able to empathize with others will not only make you a better leader, it will also help you to overcome failure and difficulty.

25. *Authenticity.* Being true to yourself alludes to the extent with which you are in congress with your values and personality, the content of your character. Authenticity is the ability to weather the external pressures that might stand in the way of your short-term and long-terms goals along your path to success.

26. *Values.* Values are a tricky category of ideas. They are formative and instructive; however, they also guide your decision-making, goal orientation, and perseverance. Often, values refer to appropriate conduct and a life lived well. As values pertain to success, as with life, they are about being in congress with what matters to you.

1.4 Methods Are Conscious Actions One Should Take to Pave the Road to Success

1. *Being visionary.* Having vision is about seeing the path before you and setting short-term and long-term goals accordingly. Being a visionary is taking risks on the basis of this goal orientation when others might not otherwise do so.

2. *Continuous learning and reflection.* I have to distinguish two forms of learning here: active vs. passive. By passive learning, I mean taking the most out of one's mistakes and failures through reflection and trial-and-error. In contrast, active learning is achieved by constant reading, seeking advice from mentors and the experience of others who have been through a certain path. The point of active learning is to prevent errors and failures from happening and to minimize their negative influence if they strike. Reflect on your past experiences, learn from failure. Continual personal development and growth guides you to success.

3. *Reach your potential and maximize it.* Reaching your potential is more than living up to hype; it is about reaching your goals and then setting new ones. Also, understanding your strengths and maximizing them while hedging against your weaknesses allows you to realize what you are capable of achieving.

4. *Respect your time.* Time management is one of the least utilized, but most important, skills available to anyone pursuing success. Time is an artificial construct, but it cannot be replaced or caught up. You need to live in the moment and challenge yourself to make the most of the "now."

5. *Encourage a sense of urgency.* As previously discussed, a sense of urgency needs to be encouraged because seeing the world as a series of goals is only useful if you are working toward them. Believing that the world will wait for your action misrepresents your agency in your success.

6. *Establish priorities.* Your time is valuable, and you need to learn to prioritize behaviors and goals relative to your definition of success. Prioritizing what matters to you means you are placing it first and making time for it.

7. *Be selective.* The things we don't do are often as important as or even more important than those we do. Be comfortable with saying no to something that is not vital to your goals.

8. *Working hard and smart.* Hard work does not mean working until you collapse. Efficiency matters, and being smart about the kind of work you engage in can help you achieve your goals better.

9. *Do what is right. Don't ask for permission!* There is an oft-used phrase: It is better to ask for forgiveness than permission. The idea is simple. If you do what is right, this will lead you to the best possible action. Inaction is more detrimental that risking upsetting others.

10. *Make good habits.* Habits are the bricks and mortar of success. Habits are what create consistency and reaching long-term (and short-term) goals.

11. *Keep a journal.* Journaling is a long-standing tradition, but for success specifically it is useful. Since we want success to be measurable, keeping a journal allows you to track your goals and amend them on the basis of changes you might see.

12. *Think of new ideas and share them with others.* There is an instinct to keep ideas to yourself because of theft, but the reality is that most people share similar ideas. And being able to share, amend, and synthesize fosters greater ideas that lead to innovation.

13. *Don't hesitate to take action if you have an idea.* Nothing is more satisfying for an intellectual person than learning, and nothing is more rewarding to him/her than taking into practice what has been learned.
14. *Establish a meaningful network.* Networking is more than making friends; it is about creating a web of resources, allies, and thought partners who can help you see the full distance to your definition of success.
15. *Actively seek advice and constructive criticism.* Ask for help. Part of establishing a network is having a group of people whom you can seek out for advice and criticism that can help carry you over slow periods on your path to success.
16. *Get out of your comfort zone.* Being comfortable is the enemy of success. If you are comfortable, you are no longer challenging yourself to overcome obstacles. A comfort zone is not about happiness, but about settling for what you have, not what you are seeking.
17. *Adequately credit others for their contributions.* Consider this a rule for yourself, an academic obligation or a divine responsibility. The importance of this in academic collaboration cannot be overstated.
18. *Identify problems and offer solutions.* Problem solving is an important component of long-term and short-term planning. Being able to see problems *and offer solutions* allows you to overcome pitfalls and adapt to a changing environment.
19. *Demonstrate reasonable flexibility and regularly assess your personal development and growth.* Adaptation is a big part of success, largely because life changes around you whether you want it to or not. Being flexible to changing circumstances allows you to assess where you are on your path to success.
20. *Exercise regularly.* Very successful people will appreciate how important a fit and healthy body is for a great mind. Exercise also allows you to focus and reduce anxiety, which are important when dealing with high-stress situations that can arise in day-to-day business.
21. *Don't give up.* In many ways, perseverance and an attitude of not giving up might be the most important component of all. Innovation comes from going where others might not otherwise have gone. Even when times become hard, reassess and adapt in order to reach your definition of success.

1.5 Enhancers or Catalysts Are External Factors that Enhance One's Chance of Success

The right people foster teamwork. The right people share your passion, your interests, and your motivation to succeed. Build a network that allows you to succeed. This can include critics (who challenge you), mentors and role models (who guide and inspire you), friends (who support you), and family (who support you unconditionally).

1.6 Inhibitors Are Internal or External Factors that Diminish One's Chance of Success

1. *Fear of success.* Pursue whatever you wish without fear, or don't wish it at all. Fear stands in the way of your dreams and definition of success because it blocks out the ability to see exactly what you need to do and how far you have come.

2. *Procrastination.* Wasting time delays the process of reaching your definition of success. Time is of the essence and a sense of urgency eliminates the tendency to put off what can be completed now. Don't let a day go by without working toward your goals.

3. *Heavy multitasking.* Being able to multitask is a wonderful skill; however, taking on too much can be disastrous. It places more in your way than you can reasonably handle, making it more difficult to feel accomplished and satisfied. Be efficient with your time and choose your tasks wisely.

4. *Arrogance.* Confidence in yourself is good. Arrogance, however, is the kind of excessive confidence that obfuscates how well (or poorly) you are doing and engenders lack of cooperation by peers and friends. Nothing is a quicker turnoff to potential investors and partners than acting as though you are the center of the universe.

5. *Clutter.* Disorganization of mind and goals is one of the worst things you can do to yourself relative to success. A cluttered understanding of your goals or actions creates difficulties in measuring your definition of success. Ditch the clutter and get organized.

6. *Lack of focus.* This is particularly common, as focus is driven by motivation. If you don't know what you are working toward, it is difficult to know where to focus. Creating short-term and long-term goals allows you to decide where to focus your attention.

7. *Self-doubt.* Confidence is important, and self-doubt is the antithesis of confidence in your abilities and goals. Cast out doubt and embrace action.

8. *Envy.* Envy is operationally characterized by jealousy combined with action to hurt. Envy is the enemy of success because it ignores *your definition of success.* What you need to focus on is what you want from life, not what others are achieving.

9. *Nonconstructive criticism.* There is nothing wrong with criticism if its aim is to improve something. However, criticism intended only to be critical serves no one except the critic. Beware people who only tell you the problems they see without any constructive advice.

10. *Myths and the wrong mindset.* Magical thinking is a propensity to imagine mystical, mythical, or unknown forces to explain behaviors and outcomes. This kind of mindset sets you up to blind you to what you ought to do and to what you are doing wrong. Exchange magical thinking for discipline and goal orientation, and you will be back on the path of success in no time.

11. *Urge to please others.* Being empathetic and having high emotional intelligence is a good thing. However, aiming to please others instead of reaching goals sets you up for failure. Lead by example, set goals, and persevere. Leave people pleasing to those who lack goals.

Reference

Wilson LO (2017) Levels of creativity. [WWW] The Second Principle, The work of Leslie Owen Wilson. Available from: http://thesecondprinciple.com/creativity/creativetraits/levels-of-creativity (Accessed 14 December 2017).

2

Qualities of Research Scientists: Personality and Leadership Attributes of Research Team Members

Fred Bertino[1] and Mohammadali M. Shoja[2]

[1] *Department of Radiology and Imaging Sciences, Emory University School of Medicine, Atlanta, GA, USA*
[2] *Division of General Surgery, University of Illinois at Chicago Metropolitan Group Hospitals, Chicago, IL, USA*

2.1   Leadership

Several leadership types have been described that may be applied to any field of study. The leadership style of a researcher is a component of the personality that often dictates how interpersonal relations between team members result in an effective and productive work. Trait theory suggests that leadership qualities are inborn and usually result in a leader who is charismatic, friendly, motivational, and intelligent (Sims 2009). By contrast, behavior theory of leadership includes four specific types of leaders: autocratic, bureaucratic, participative, and free rein (Sims 2009; Hanna 1999).

Autocratic leaders tend to be inflexible and unwilling to accept input from others. Though rigid, they tend to be confident, knowledgeable, and have significant control over the group and project workflow. Bureaucratic leaders hold rules and regulations in high esteem and use guidelines as primary support for project development. Similarly rigid, bureaucratic leaders hold strict adherence to formal instructions to complete tasks. Participative leaders are more open to group suggestion when it comes to problem solving and decision-making. Though inclusive, the efficiency of the group may be lessened when compared to authoritative and bureaucratic leaders from a high volume of external input. Participative leaders are willing to make definitive decisions to optimize workflow. Free rein leaders allow group members to exercise individuality and work to their strengths independently in their own styles. However, effective free rein leaders must have an understanding of an individual group member's abilities and how to maximize competency during research projects. Self-motivation on the part of each research team member is crucial for a leadership style of this type (Sims 2009; Hanna 1999).

In addition to behavioral leadership theory, other forms of leadership doctrines exist that emphasize interpersonal relationships. Emphasis of empathy (relationship-oriented leaders), team spirit (coaching leaders), and friendship (affiliative leaders) are several varieties of this interpersonal subset. Alternatively, creativity-based organizational leadership styles are able to encourage visionary and inspirational ideas (transformational) or describe directions based on each person's personality for a system of rewards and punishments for adequate or inadequate work respectively (transactional) (Sims 2009).

A Guide to the Scientific Career: Virtues, Communication, Research, and Academic Writing, First Edition.
Edited by Mohammadali M. Shoja, Anastasia Arynchyna, Marios Loukas, Anthony V. D'Antoni, Sandra M. Buerger, Marion Karl and R. Shane Tubbs.
© 2020 John Wiley & Sons, Inc. Published 2020 by John Wiley & Sons, Inc.

Effective leaders in research adopt elements from each leadership style and use many methods to enhance the group experience. No leadership style is mutually exclusive, and a leader with the ability to combine several aspects of many styles is able to engage a larger audience. The goal of the researcher is to produce high-quality, beneficial work in an efficient manner while maintaining the enthusiasm to encourage others to pursue similar goals. As a leader, targeting one's managerial style to the strengths and weaknesses of the research team is bound to boost morale while maintaining a workplace conducive to quality research. Understanding the specific strengths and weaknesses of the team can be best performed by understanding a team member's personality.

2.2 Personality and Interpersonal Relationships

As mentioned previously, leadership style is often a reflection of the personality traits of the leader. A good understanding of the varieties of personalities of the research team is crucial to developing a keen working atmosphere, but perhaps more imperatively, a thorough understanding of one's own personality can help identify strengths and weaknesses in learning, leadership, and communicative ability. Carl Jung and later, Isabel Briggs Myers and Katherine Briggs contributed significantly to the psychological study of personality typology, their research still applied in the workplace to study interactions among co-workers, and even a medical student's affinity to certain subspecialties (Freeman 2004; Stilwell et al. 2000; Myers 1962; Myers and Davis 1965). Developed in the 1950s, the Myers-Briggs Typology Indicator (MBTI), though under some criticism for its reproducibility and validity, can offer a general assessment of personality for an individual if the assessment is taken honestly, and without too much thought into any one question (Freeman 2004; Pittenger 2005). Honest answers from the unique experiences of problem solving, learning style, and communicative ability of the examinee usually result in a genuine assessment congruent to the examinee's temperament. The tool itself can serve as an introductory method to introduce individual differences between group members, which may be beneficial in the beginnings of a group research project, but not meant to be taken too seriously into account (Freeman 2004; Pittenger 2005). There are a variety of exams available online at no charge, yet official ones may be purchased online as well.

The MBTI is a personality assessment tool that, through a series of simple questions, creates a four-letter personality type for the examinee based on four different personality attributes, originally described by Carl Jung (Freeman 2004). Each personality attribute contains two contrasting subtypes, each represented by a letter. With this breakdown, Myers and Briggs hypothesized that everyone could fall into 1 of 16 different personality types (critics often dismissing the MBTI as too narrowly grouping the personalities of the population) (Pittenger 2005). The four major personality attributes measured by the MTBI are described as follows.

2.2.1 Relationships to Others: Extroversion (E) Versus Introversion (I)

This attribute reflects from whom an individual may derive one's energy—from within, or from those in proximity. Introverts are not described as asocial, but, rather, those who prefer to solve problems by thinking to themselves without an inherent drive to bounce ideas off other people (Atanacio 2010; Freeman 2013). Introverts tend to function better in environments where thought often proceeds speech, and the people surrounding them are also good listeners. Extroverts, by contrast, derive their energy from social situations; they become empowered in group settings, and thoughts and ideas are best conveyed by thinking aloud. Extroverts tend to be more impulsive and expressive than their introverted counterparts (Stilwell et al. 2000; Freeman 2013).

2.2.2 How Information Is Gathered and Metabolized: Sensing (S) Versus Intuition (N)

Sensors rely on the physical world around them to obtain information. They are able to take in the details of their surroundings and rely on the hard, practical facts of nature to arrive at conclusions. Sensors have a tendency to rely on previous experiences to predict the next occurrence in a series of events, and similarly tend to be very literal. Intuitives tend to look beyond the facts and details and rely more on meanings, concepts, and bigger-pictures to understand a specific situation. Intuitives rely on their ability to recognize their gut-feelings to understand a series of events. With a few general understanding of their environment, intuitives are comfortable in further exploration of a new concept without the need for excessive detail (Stilwell et al. 2000; Freeman 2013).

2.2.3 Decision-Making Ability: Thinking (T) Versus Feeling (F)

Thinkers are people who rely on evidence-based objective rationality to come to conclusions. A thinker by personality is more likely to choose the more logical and direct path to an answer. Thinkers find reward in analytics and problem solving. Feelers are dependent on subjective assessments of their surroundings. They tend to be empathetic and compassionate people, often causing them to consider how their actions may affect other people before acting. Feelers find reward in other people's satisfaction with a given situation (Stilwell et al. 2000; Freeman 2013).

2.2.4 Organization: Judgment (J) Versus Perception (P)

In interacting with the outside world, this final attribute assesses in which environment one may operate most optimally. Judgers tend to be very organized, schedule-driven people. Their lives tend to be orderly, planned, and controlled. Structure and organization are paramount and with this, a sense of command over their environment leads to quick and effective decision-making. This comes at a risk of close-mindedness, however. Perceivers tend to be more open-minded, relaxed, and capable of dealing with change. Though seemingly irresponsible from their flexibility and spontaneity, their ability to be aware of ideas and events is far higher than that of Judgers, making them more likely to observe the world before coming to a decision. Decisiveness comes more naturally to judgers, however (Stilwell et al. 2000; Freeman 2013).

From these four personality attributes (and eight variations), comes 16 different combinations of personality according to Myers-Briggs. Effective researchers may fall into any one of the 16 different personality types, but a working knowledge of one's traits is crucial for successful self-assessment and team collaboration.

2.3 Continuous Self-Assessment

An honest assessment of one's personality attributes and leadership styles is an effective way to understand the working environment, the temperament, and relationships with co-researchers, and provides a window of opportunity for self-improvement to better develop the qualities in effective leadership that one may lack. As stated in the beginning of this chapter, a researcher must be an enthusiastic leader, highly curious, and motivated to contribute to his or her field and inspire those in proximity. While there are many variables in personality and leadership style, the best qualities for researchers are the ones that unite ideas and minds in a supportive and effective atmosphere. Quality research and a strong body of researchers will emerge when the attitude of the leadership and contributors is full of harmony and mutual intrigue for the research in question.

2.4 Tips for Developing a Leader-Quality Scientist

2.4.1 Be an Entrepreneur

Scientists should develop entrepreneurial skills through exposure to industry and collaborating with business sector. Although the potential conflicts of interests should always be fully disclosed in any research output and presentation, the entrepreneur-scientists tend to have a greater immediate impact on the community. This is achieved by scientific discoveries and new ideas being transformed into available products as quickly as possible. Not only the financial gain and resulting fame can be substantial, but entrepreneurship will also create an atmosphere of enthusiasm and motivation among the new generation of young scientists.

2.4.2 Work Hard and Work Smart

As Cardinal James Gibbons (1834–1921) righteously indicated, "There are no office hours for leaders." The hard rule of leadership proposed by Molinaro (2016) states that if one avoids the hard work of leadership, he will become a weak leader; but if he embraces the hard work, he will become a strong leader. Leadership requires hard work and also smart working to ensure the available resources are not being wasted but instead are being utilized in a meaningful manner that brings efficiency and sustainability to the management system (McCauley-Bush 2013). The sense of purpose and urgency motivates a leader to do his best and to maintain his determination and commitment, and the resultant hard work creates opportunity by itself (Junarso 2008). Carol Dweck, an American psychologist, has distinguished two leadership mindsets, namely, fixed vs. growth mindset (Dweck 2006). Those with a fixed mindset see talent or genius as an inborn quality (i.e. one either has it or not), while people with a growth mindset see talent as a quality that can be achieved by active learning, perseverance, and purposefulness. The growth mindset is the definite precursor for leadership success as it fosters smart, hard work. This concept is consistent with how Vince Lombardi, an American football player (1913–1970), characterized the constitution of leaders: "Leaders are not born; they are made. And they are made just like anything else, through hard work. And that is the price we will have to pay to achieve that goal, or any goal."

2.4.3 Listen, Observe, and Learn on a Daily Basis

Active listening is one of the most important skills of interpersonal communication that requires lifelong training (West and Turner 2009). Five levels of listening have been identified, namely, passive, marginal, projective, sensitive, and active listening (Bhardwaj 2008; Verma 2015). In passive listening, the listener is rather indifferent to the content of discussion, making a change in his thought trends and ongoing ideas improbable. In marginal listening, the listener only engages in superficial understanding of the discussion without allowing a significant change in his thought trends. In projective listening, the listener absorbs the information in accordance with his own frame of reference. The opposite of projective listening is sensitive (or empathetic) listening, in which the listener understands the viewpoint of the speaker without distorting it with his own perspectives. In active listening, the listener demonstrates a genuine interest in speaker's viewpoint, pays full attention to the speaker's words, speech tone, and body language, and accurately analyzes, interprets, and remembers the provided information (Bhardwaj 2008). The most-effective form of listening is active listening, which, when combined with sensitive listening, can bring to fruition a charismatic leadership. Active listening is a dynamic endeavor that requires attributes that include, but are not limited to, serious concentration, empathy, nonjudgmental attitude, interest in the speaker, patience, and willingness to take responsibility to fully comprehend the information, and to interpret the meanings and provide feedback (Bhardwaj 2008; Hoppe 2011). As for active listening, we should refer to a statement by Peter Drucker, an Austrian American management consultant: "The most important thing in communication is to hear what is not being said." In a cross-cultural leadership setting – commonly seen in academic institutions – active listening is a key to closing the cultural gaps (Whitfield 2014), which can ultimately lead to workplace security and more productivity.

Observation goes the same way as with active listening. Observation is operationally defined as looking with a prepared mind for the meanings, patterns and trends of the data one receives – essentially by means of seeing, but through all other senses as well – from the environment (Welter and Egmon 2006). Seeing by itself brings about a mix of relevant and irrelevant information; it is through observation with an analytical and prepared mind that information directly linked to a particular situation is separated from a wealth of irrelevant information, leading to a phenomenon known as pattern recognition. Pattern recognition is the cognitive process of internalizing raw data or intangible information, and organizing and translating them into a blueprint that explains a situation (McKee et al. 2013). In lay terms, it entails observing what does and what does not work in different situations (Owen 2013). Pattern recognition is the prerequisite for situational awareness and strategic planning (McKee et al. 2013). Equipped with the art of observation and pattern recognition, a leader is therefore capable of identifying new opportunities and anticipating the course of action that is most likely to succeed in a given circumstance (Owen 2013). There is a positive relationship between empathy and transformational behavior that inspires followers to achieve more than expected (Skinner and Spurgeon 2005). It has been shown that empathy and emotional intelligence strengthen a leader's pattern recognition skills as the emotional information coming through empathy empowers the leader to understand the range of issues his team faces and to take the actions necessary to coordinate the team (Wolff et al. 2002).

The 70:20:10 model for competency building and leadership development – suggested by McCall and colleagues at the Center for Creative Leadership (CCL) – states that successful leaders learn about 70% of their lessons from challenging assignments and job-related experiences, 20% from their relationships and feedbacks they receive, and 10% from formal training.

2.4.4 Think, Plan, and Take Action

Leaders are both dreamers and doers. The ability to identify problems and think of solutions is a key performance attribute. Although traditional views tend to distinguish between dreamers who "have their heads in the cloud," and doers who "have their feet on the ground," a visionary leader is dreamer and doer at the same time (Williams and Denney 2015). The message here is clear: A great leader thinks, thinks, thinks, plans, and takes action.

2.4.5 Translate Vision into Reality

Warren Bennis (1925–2014) defined leadership as "the capacity to translate vision into reality." This requires establishing priorities, strategic planning, identifying risks, and developing action plans and contingency plans. Start with low-risk actions and gradually increase the magnitude of risk you are taking. Setting two or three attainable, short-term goals that are congruent with your vision will make a perfect starting point.

2.4.6 Empower Your Followers

A great leader empowers his followers and also turns them into leaders. Jack Welch, a retired American business executive, stated, "Before you are a leader, success is all about growing yourself. When you become a leader, success is all about growing others." It is an inherent characteristic of great leaders to promote self-development among their followers and to motivate them into achieving excellence.

In his 1978 book *Leadership*, James MacGregor Burns proposed the concept of transactional vs. transformational leadership. In transactional leadership, "one person takes the initiative in making contact with the others for the purpose of an exchange of valued things," while in transformational leadership, "one or more persons engage with others in such a way that leaders and followers raise one another to higher levels of motivation and morality" (Burns 1987). The transactional leaders function within a framework of self-interest, while the transformational leaders go beyond such a framework to induce positive changes among the followers (Martin et al. 2006). It has been suggested that transactional and transformational leadership styles are, in fact, complementary, that transformational leaders also exercise transactional behaviors at some levels, and that transactional behaviors are also required for a transformational leader to be efficient (Peters 2010). Transformational leadership generates a remarkable influence over followers and harnesses their commitment, thereby leading to accomplishments above expectations (Bass and Avolio 1990). A direct correlation exists between emotional intelligence and transformational leadership (Hunt and Fitzgerald 2013).

2.4.7 Delegate Tasks Whenever Feasible

The art of delegation is an indispensable quality of every successful leader. Andrew Carnegie (1835–1919) asserted, "No man will make a great leader who wants to do it all himself, or to get all the credit for doing it." Delegation is the process of assigning responsibility and authority to someone at a more junior level for accomplishing objectives (Lussier 2008). Good communication skill is the keystone for a successful delegation. The process of delegation involves three steps (Karmakar and Sarkar Datta 2012):

1. *Assignment of work:* A task is clearly communicated to the delegatee and deadline is set for completion.
2. *Granting of authority:* The delegatee understands that he is fully responsible and in charge of performing the delegated task.
3. *Creation of accountability:* The delegatee recognizes that he is the one to accept credit or blame for the results of the delegated tasks.

It is notable that giving others responsibility without authority is a delegation destined to failure, and it makes the delegatee to feel that he is being taken advantages of or that he cannot be trusted (Bixby 2016).

2.4.8 Establish Priorities

The businessman Carlos Ghosn attested, "The role of leadership is to transform the complex situation into small pieces and prioritize them." As difficult as it seems to prioritize short-term (daily and weekly) and long-term (monthly and yearly) actions or to-do lists, it eventually all comes down to practicing a simple discipline on a regular basis: As you begin your day, oblige yourself to itemize all you desire to accomplish on that day. Repeat this task every day, review the list regularly, and also write down a general to-do list for the upcoming weeks, months, and the year ahead of you. Without having such to-do lists, your ability to prioritize your actions is trivial. Once a to-do list is at hand, the next step is to prioritize actions. The key to prioritizing the action items is that it should be executed by thinking about the most crucial outcome that you are looking for, not by focusing on the nature of tasks itself (Kush 2009). The author and motivational speaker Stephen Covey has asserted, "Effective leadership is putting first things first."

References

Atanacio, S. (2010). *Act from the Inside Out*. San Francisco: Bush Street Press.

Bass, B.M. and Avolio, B.J. (1990). The implications of transactional and transformational leadership for individual, team and organizational development. *Research in Organizational Change and Development* 4: 231–272.

Bhardwaj, K. (2008). *Professional Communication*. New Delhi: I.K. International.

Bixby, D. (2016). *Navigating the Nonsense: Church Conflict and Triangulation*. Eugene, OR: Cascade.

Burns, J.M. (1987). *Leadership*. New York: Harper & Row.

Dweck, C. (2006). *Mindset: The New Psychology of Success*. New York: Ballantine.

Freeman, D. (2004). *Presenting Type in Organizations [CD-ROM]*. Melbourne, Australia: Australian Psychologists Press.

Freeman, B. (2013). *The Ultimate Guide to Choosing a Medical Specialty*. New York: McGraw-Hill.

Hanna, L.A. (1999). Lead the way leader. *Nursing Management* 30 (11): 36–39.

Hoppe, M.H. (2011). *Active Listening: Improve Your Ability to Listen and Lead (Center for Creative Leadership)*. Hoboken, NJ: Wiley.

Hunt, J.B. and Fitzgerald, M. (2013). The relationship between emotional intelligence and transformational leadership: an investigation and review of competing claims in the literature. *American International Journal of Social Science* 2: 30–38.

Junarso, T. (2008). *Leadership Greatness: Best Practices to Become a Great Leader*. New York: iUniverse.

Karmakar, A. and Sarkar Datta, B. (2012). *Principles and Practices of Management and Business Communication*. New Delhi: Dorling Kindersley.

Kush, B.D. (2009). *Auditing Leadership: The Professional and Leadership Skills You Need*. Hoboken, NJ: Wiley.

Lussier, R. (2008). *Management Fundamentals: Concepts, Applications, Skill Development*, 4e. Cengage Learning.

Martin, B., Cashel, C., Wagstaff, M., and Breunig, M. (2006). *Outdoor Leadership: Theory and Practice*, 47. Champaign: Human Kinetics.

McCauley-Bush, P. (2013). *Transforming Your STEM Career Through Leadership and Innovation*. London: Elsevier.

McKee, A., Kemp, T., and Spence, G. (2013). *Management: A Focus on Leaders*. Frenchs Forest: Pearson.

Molinaro, V. (2016). *The Leadership Contract: The Fine Print to Becoming an Accountable Leader*. Hoboken, NJ: Wiley.

Myers, I.B. (1962). *Manual: The Myers-Briggs Type Indicator*. Princeton, NJ: Educational Testing Service.

Myers, I.B. and Davis, J.A. (1965). *Relation of Medical Student's Psychological Type to Their Specialties Twelve Years Later*. Princeton, NJ: Educational Testing Service.

Owen, J. (2013). *How to Manage*. London: Prentice Hall.

Peters, A.L. (2010). Rethinking transformational leadership in schools. In: *New Perspectives in Educational Leadership: Exploring Social, Political, and Community Contexts and Meaning* (ed. S.D. Horsford), 7–28. New York: Peter Lang.

Pittenger, D. (2005). Cautionary comments regarding the Myers-Briggs type indicator. *Consulting Psychology Journal: Practice and Research* 57 (3): 210–221.

Sims, J. (2009). Styles and qualities of effective leaders. *Dimensions of Critical Care Nursing* 28 (6): 272–274.

Skinner, C. and Spurgeon, P. (2005). Valuing empathy and emotional intelligence in health leaderships: a study of empathy, leadership behaviours, and outcome effectiveness. *Health Services Management Research* 10 (1): 1–11.

Stilwell, N., Wallick, M., Thal, S., and Burleson, J. (2000). Myers-Briggs type and medical specialty choice: a new look at an old question. *Teaching and Learning in Medicine* 12 (1): 14–20.

Verma, S. (2015). *Technical Communication for Engineers*. New Delhi: Vikas.

Welter, B. and Egmon, J. (2006). *The Prepared Mind of a Leader: Eight Skills Leaders Use to Innovate, Make Decisions, and Solve Problems*. San Francisco: Jossey-Bass.

West, R. and Turner, L.H. (2009). *Understanding Interpersonal Communication: Making Choices in Changing Times*. Boston: Wadsworth Cengage Learning.

Whitfield, D. (2014). Servant leadership with cultural dimensions in cross-cultural settings. In: *Servant Leadership: Research and Practice: Research and Practice* (ed. R. Selladurai and S. Carraher), 48–70. Hershey: IGI Global.

Williams, P. and Denney, J. (2015). *21 Great Leaders: Learn Their Lessons, Improve Your Influence*. Uhrichsville, OH: Barbour Publishing.

Wolff, S.B., Pescosolido, A.T., and Druskat, V.U. (2002). Emotional intelligence as the basis of leadership emergence in self-managing teams. *The Leadership Quarterly* 13: 505–522.

3

Building a Personal Vision Statement

Genevieve Pinto Zipp

School of Health and Medical Sciences, Department of Graduate Programs in Health Sciences, Seton Hall University, South Orange, NJ, USA

3.1 Personal Vision Statement and Portfolio Overview

As healthcare professionals we are faced with the ongoing responsibility of documenting evidence of our competency, productivity, and the goals to which we aspire. Seldin (2009) described the academic portfolio as a conceptual framework that can be used by faculty to record, share, and reflect upon accomplishments. Zubizarreta (2009) described the learning portfolio as a framework that students can use to foster learning and assess their outcomes via self-refection. But how do we begin this journey? Typically, the first part of any portfolio includes material that reflects upon a personal philosophy, which encompasses the "vision statement," mechanisms to achieve goals, and insightful reflection upon the current outcomes of the goals. In a portfolio, insight into the journey taken can be offered, issues can be addressed that may have impeded the journey, and a discussion of paths to be explored to achieve the set goals can be included. The key feature that drives the portfolio is the development of a strong, vibrant vision statement that describes the desired outcome at a specific point in the future. It is through the development of a personal vision statement that one can reflect on his (or her) strengths and weakness, allowing for an ongoing self-assessment that leads to the development of personal insight. A vision statement can open the mind to possibilities and pitfalls associated with future goals. As William Arthur Ward (1970) said, "If you can imagine it, you can achieve it." I would argue that while it is important to imagine the possibilities, it is only through insightful reflection that we can achieve what we have envisioned. Thus, engaging in the creation of a personal vision statement provides an opportunity for insightful reflection. Creating a personal vision statement is no easy task, and it might take several attempts to artfully and accurately reflect your proposed endeavors.

3.2 Getting Started

So how does one get started in developing their own personal vision statement? First, base your vision statement on the "best" outcome you can envision. Have your vision

A Guide to the Scientific Career: Virtues, Communication, Research, and Academic Writing, First Edition.
Edited by Mohammadali M. Shoja, Anastasia Arynchyna, Marios Loukas, Anthony V. D'Antoni, Sandra B. Buerger, Marion Karl and R. Shane Tubbs.

statement be so rich that it inspires you, energizes you, motivates you, and sparks your creative side. After reviewing your vision statement the writer should be ignited to achieve the noted goals. In addition the reader of the vision statement should see clear benchmarks that the writer expects to be measured upon in order to assess their attainment of their vision. These benchmarks are written as measurable and observable long term goals and short term objectives, which are based on the articulated vision.

3.3   Vision Statement in Action

To help us see the importance of creating a vision statement, let's reflect upon this short story:

> *On the first day of class, a high school teacher asked his computer-science class to create a software package that would revolutionize the software industry and earn someone the Nobel Prize. The students were told that they would have to present their ideas to the entire class at the end of the semester. After all the presentations were completed, the teacher then asked the class if they felt that "any of the ideas presented could earn someone a Nobel Prize." The students felt that most (if not all) of the software ideas were not good enough to win a Nobel Prize. At the end of the class, as the students were packing their backpacks to head out to their next class, the teacher left the students with this statement: "You don't get a Nobel-Prize idea from a run-of-the-mill vision."*

The premise of this short story is to illustrate that the quality of your vision shapes your outcome. An inspiring, engaging, and captivating vision statement broadens expectations and aspirations; and pushes someone out of their so-called "comfort zone." Skeptics of writing powerful vision statements might suggest that failing to achieve the aspired outcome can be detrimental to continued growth and development in an organization, profession, or work community. While this can be a true concern for many based on their organization's beliefs and structure, the development of two vision statements might be the answer. The first vision statement can serve as the "idealized" statement and the second as the "realistic" statement. The realistic vision statement is the one that can be shared within your work community and used to assess work performance. However, the idealistic vision statement is the vision that an individual uses as foundational support for their realistic vision statement. Professionals and organizations must look at vision statements as a learning strategy that inspires them to continue to strive to reach new heights in their personal and professional journey.

Several years ago, I came across a quote by Les Brown (2001): "Shoot for the moon! Even if you miss, you'll still be among the stars." Today, I look at my personal vision statement and the moon and the stars are the objectives along the way that led me closer and closer to the moon.

3.4   Rules to Guide Vision Statement Development

Armed with a clear understanding of what a vision statement is, 10 simple rules will help you begin to develop your own personal vision statement:

1. Vision statements are presented in the present tense. State for the reader what you actually hear, see, feel, and think after your ideal outcome was achieved.
2. Vision statements capture emotion and, therefore, describe how you will feel when the outcome is realized. One might think of these emotions as your "inner vision." Inner vision represents how you want to be viewed by colleagues. One might suggest the inner vision is what you want those in your inner circle to see. Specifically, this points to who you are and what you stand for.
3. Provide a variety of sensory details; one might consider these the "outer vision" as they are colors, shapes, sounds, and smells. Sensory details within your statement provide a mental image of the ideal outcome for the reader. In general, the outer vision represents how you would like noncolleagues to view your work and yourself.
4. Be thoughtful in the development of your vision statement. Identify a time and place where you can focus your attention to engage in insightful reflection and thought in order to effectively address the development of your personal vision statement.
5. Use your personal portfolio (which is meant to house a thorough inventory of your goals, strengths, weaknesses, and accomplishments, along with insightful reflection on each of these areas) as a mechanism to assess your personal development and growth and ignite your vision.
6. Based on this thorough and thoughtful review and reflection of your portfolio documents, create a positive mental image regarding activities you would like to engage in at some point in the future.
7. Draft a concise statement, usually no more than three sentences, of your personal vision that represents what you want to achieve in your life.
8. Share your vision statement with mentors who have inspired you. Engage them in dialogue regarding your vision specific to: its written clarity, depth, appropriateness given the available demands and resources, and, most importantly, its ability to motivate you to continue to seek the best in yourself.
9. Reflect on the information provided from your mentors and revise your vision statement if you judge their feedback as warranted.
10. Think of your vision statement as a tool to jolt your inspiration and support you on your personal journey. Review it quarterly to determine if it continues to be accurate in describing the ideal outcome you want to achieve. Often times, our vision statements require no change; but sometimes, circumstances (whether it be resource availability, life constraints, or just an "Aha!" moment) jolt us such that we need to reflect on our vision statement and modify it, or even start a new one.

3.5 Conclusions

Remember that the purpose of the vision statement is to push you to envision the possible. Use your goals and objectives to measure if you have succeeded or failed in reaching your vision. Once you have crafted your personal vision statement, assess its fit with the organization's vision statement. Organizations use their vision statement as one of the guideposts or strategies to help focus the organization's resources, articulate a shared image, and drive employee motivation as the organization charts its path for success in the future. An organization's vision emerges from a true appreciation of its historical background, insightful reflection on past and current opportunities available, and a comprehensive strategic plan that enables the organization to capitalize on the opportunities. A fit between your personal vision statement and that of the organization ensures that both visions flourish rather than perish. The establishment of a strong motivating vision statement enables both the organization and the individuals within it to envision what can be possible rather than what is impossible.

References

Brown, L. (2001). *Live Your Dreams*. New York, NY: Harper Collins Publishing.
Seldin, P. (2009). *The Teaching Portfolio*. Bolton, MA: Anker Publishing Co.
Ward, W. (1970). *Fountains of Faith*. New York: Droke House.
Zubizarreta, J. (2009). *The Learning Portfolio*. San Francisco, CA: Jossey-Bass.

Further Reading

Foster, R.D. and Akdere, M. (2007). Effective organizational vision: implications for human resource development. *Journal of European Industrial Training* 31 (2): 100–111.
Kirkpatrick, S.A., Wofford, J.C., and Baum, J.R. (2002). Measuring motive imagery contained in the vision statement. *The Leadership Quarterly* 13 (2): 139–150.
Lipton, M. (1996). Demystifying the development of an organizational vision. *MIT Sloan Management Review* 37 (4): 83.
Seldin, P. (2009). *The Teaching Portfolio*. Bolton, MA: Anker Publishing Co.
Zubizarreta, J. (2009). *The Learning Portfolio*. San Francisco, CA: Jossey-Bass.
Zuckerman, A.M. and Coile, R.C. Jr., (2000). Creating a vision for the twenty-first century healthcare organization/practitioner application. *Journal of Healthcare Management* 45 (5): 294.

4

Creativity and Novel Hypotheses

Anthony V. D'Antoni

Division of Anatomy, Department of Radiology, Weill Cornell Medicine, New York, NY, USA

4.1 Creativity and Science

The winter of 2014 was a particularly harsh one in New York City with a significant amount of snowfall, and so in February I decided to do something about it and painted the seascape shown in Figure 4.1 (*La Jolla*, oil on canvas, 2014). This painting took about four hours to complete from start to finish (blank canvas to what is shown in the figure). Some might say that a person has to be creative to paint such a composition. Perhaps, but I believe that creativity is an innate part of every human being.

Human intellectualism requires a great deal of creativity, and science is no exception. Science is a very creative endeavor, and in my experience scientists are often the most creative people I know. They have to be. The research process described in the chapters of this textbook (from conducting a literature review to generating novel hypotheses to designing experiments, analyzing data, and ultimately writing the manuscript) necessitates creativity at every stage. Mumford (2003) analyzed two authoritative handbooks on creativity and defined it as the fabrication of a novel and useful product. This definition fits well for my oil painting but how does it relate to the aforementioned research process? The final outcome, reporting novel findings in a scientific manuscript, clearly fits this definition, although it may not be as tangible as my oil painting.

The purpose of this chapter is not to present a treatise on the scholarship of creativity but, rather, to highlight how novel hypotheses (or other forms of knowledge) can be generated using a specific and creative strategy called mind mapping.

4.2 What Are Mind Maps?

Mind mapping was developed by Tony Buzan (Buzan and Buzan 1993) and the inspiration for this creative strategy arose from the notebooks of Leonardo da Vinci. Mind maps, like da Vinci's notes, are multisensory tools that use visuospatial orientation to integrate information, and consequently, help scientists organize and retain information (D'Antoni et al. 2009, 2010). Mind maps can also be used to generate novel hypotheses whereby existing information from the peer-reviewed literature is incorporated with

A Guide to the Scientific Career: Virtues, Communication, Research, and Academic Writing, First Edition.
Edited by Mohammadali M. Shoja, Anastasia Arynchyna, Marios Loukas, Anthony V. D'Antoni, Sandra M. Buerger, Marion Karl and R. Shane Tubbs.
© 2020 John Wiley & Sons, Inc. Published 2020 by John Wiley & Sons, Inc.

Figure 4.1 *La Jolla*, oil on canvas, 2014, by Anthony V. D'Antoni.

a scientist's expertise of the topic and infused into a visual framework that highlights knowledge gaps. These gaps allow the scientist to generate novel hypotheses and then develop experiments to test these hypotheses.

4.2.1 How to Create Mind Maps

A mind map is a nonlinear, creative strategy that encourages the scientist to think radially using visuospatial relationships (D'Antoni et al. 2010). According to Buzan and Buzan (1993), a mind map should be drawn on blank paper that is larger than standard 8½ by 11 in. paper. The rationale behind using larger paper is to allow the scientist to break away from the boundaries inherent in standard-size paper and thus propagate creativity. The use of lined paper is discouraged because it theoretically restricts thought. Once suitable paper is obtained, a medium for drawing the mind map is necessary – namely, colored pens or pencils. The scientist begins by drawing an image in the center of a horizontally positioned blank paper that reflects the central theme, or topic, of the mind map (D'Antoni and Zipp 2006). For example, a mind map on the rules of mind mapping could have an image of the cerebrum in the center of the page. This central image allows the scientist 360° of freedom to develop the mind map. Next, the scientist would draw the main branches with associated key words extending from the central image and these branches represent the different categories relevant to the content of the mind map. In the previous example, some of the key words are *start*, *connect*, *print*, and *association*. It is important to print the words and ensure that their length is the same as the lines underneath them so that the completed map will be easier to comprehend. From these main branches, relevant sub-branches are

Table 4.1 Steps for creating a mind map.

1. Begin with blank paper (preferably greater than $8\frac{1}{2}$ by 11 in.) in landscape format.
2. Draw an image in the center of the paper that represents the theme of the mind map.
3. From the central image, draw main branches with accompanying key words that represent the major headings of the topic.
4. Further divide the main branches into sub-branches that contain key words.
5. Begin to find associations between different areas of the mind map and draw connections between these associations.
6. Always print one key word per line, allowing the word to be the same length as the line upon which it rests.
7. Use different colors throughout the mind map and include as many pictures as possible.

created. Each of the branches and sub-branches should contain accompanying pictures to aid the scientist in recalling the information. Table 4.1 summarizes the steps used to create a mind map. The result is a nonlinear, pictorial representation of information that highlights interconnections between concepts. As more sub-branches are created, scientists can recognize patterns between key words that should be connected, which may result in the integration of different parts of the mind map. The final map is a creative illustration that can highlight gaps in the literature or novel hypotheses. Mind maps can be created using any type of information or even experimental data.

4.2.2 Mind Map of a Textbook Chapter

Figure 4.2 is a mind map that I created based on a textbook chapter on the assessment and treatment of stroke from a rehabilitative perspective (D'Antoni and Zipp 2006). The central image is an inferior view of the brain emphasizing the cerebral arterial circle (of Willis). I drew this image because stroke is a result of either ischemia or hemorrhage of an artery that supplies blood to the brain and in essence the image reminds the viewer of the central pathogenesis of the disease. Emanating from the central image are five main branches (green lines) that represented the main topics of the chapter: anatomy, rehabilitation, patient impairments, history and physical examination, and epidemiology. Each main branch can be subdivided into sub-branches or a list of examples. Related to the history and physical examination, I have listed the important diagnostic tests for stroke such as urinalysis (UA), complete blood count (CBC), glucose, etc. Note the connection between the diagnostic tests and the main branch called epidemiology. The sensitivity and specificity of these tests are statistical concepts, which are a basis of epidemiology. Therefore, this connection reminds the viewer that epidemiology not only has global applications but also affects decision-making in the clinical setting. Another link was made between a risk factor for stroke (heart disease) and the proposed reasons for the differences in prevalence of stroke between the United States and other countries. This connection highlights the role of lifestyle choices in the genesis of heart disease.

These are examples of how patterns emerge when one begins to construct a mind map, and this is the most powerful aspect of mind mapping because it promotes integration, which is a basis for critical thinking and creativity.

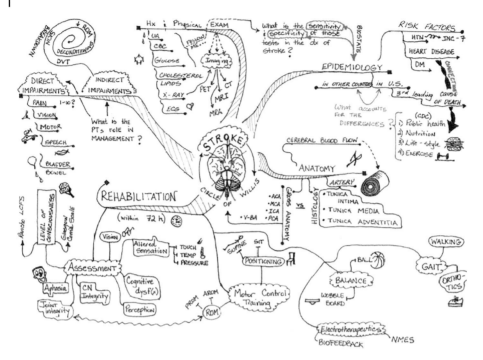

Figure 4.2 Mind map of a textbook chapter on the assessment and treatment of stroke from a rehabilitation perspective. Used with permission by D'Antoni and Zipp (2006).

4.3 Mind Maps and Novel Hypotheses

Mind mapping can help promote creativity (D'Antoni et al. 2009). A scientist can incorporate existing knowledge of a topic with the latest published research in order to create a mind map that identifies gaps in the literature, and these gaps can become the basis of novel hypotheses. Having painted and mind mapped for years, I can say that both become more useful and creative with practice. Therefore, I encourage you to begin mind mapping and remember that your creativity has less to do with how the map looks and more to do with the thinking that was used in order to create it.

References

Buzan, T. and Buzan, B. (1993). *The Mind Map Book*. London: BBC Books.

D'Antoni, A.V. and Zipp, G.P. (2006). Applications of the mind map learning technique in chiropractic education: a pilot study and literature review. *Journal of Chiropractic Humanities* 13: 2–11.

D'Antoni, A.V., Zipp, G.P., and Olson, V.G. (2009). Interrater reliability of the mind map assessment rubric in a cohort of medical students. *BMC Medical Education* 9: 19.

D'Antoni, A.V., Zipp, G.P., Olson, V.G. et al. (2010). Does the mind map learning strategy facilitate information retrieval and critical thinking in medical students? *BMC Medical Education* 10: 61.

Mumford, M.D. (2003). Where have we been, where are we going? Taking stock in creativity research. *Creativity Research Journal* 15 (2–3): 107–120.

5

Confidence and Its Impact on Your Aspiring Career

Toral R. Patel

Department of Cardiology, University of Virginia, Charlottesville, VA, USA

5.1 Introduction

Most students, graduates, and professionals are aware that confidence contributes to their success. Therein lies the question, "What is confidence?" Confidence describes the beliefs you hold about your potential to achieve an outcome. Confidence differs from self-efficacy by referring to a broader belief in the sum of your abilities rather than referring to a specific domain or task. Self-efficacy, on the other hand, is task specific and refers to your belief in your ability to succeed in specific situations (Bandura 1986). Confidence contributes to the decisions made in your life, goals, and career; the approach taken to achieve those goals or career; and how you adjust and adapt to obstacles.

Confidence may not be equally represented in all aspects of one's life. In your personal life, confidence may present as how much you believe in yourself, also known as self-confidence; decisions you make regarding everyday life; the people you surround yourself with; and how you develop your self-image. In academic and medical settings, confidence may refer to your interactions with patients, the relationship between your knowledge and experiences, the quantity or number of papers or books published, the number of grants or presentations, and/or the recognition that you receive from peers. This means that personal confidence may not translate into your career or that your confidence in an academic or medical setting may not make you more confident in your home life. The key to understanding confidence lies in understanding what confidence means to you.

Confidence is expressed by self-awareness, knowing your strengths and weaknesses, prior and current successes, trusting your capabilities, embracing the unknown, taking risks, learning to receive praise or criticism, and practicing tenacity. Amy Lee Tempest once stated about confidence, "I feel it is something that is always there, something you are born with that gets lost along the way, or stolen by others. Sometimes you have to dig deep to find it again" (cited by Deschene 2014). Forming doubts and insecurities and heavily focusing on how others perceive you may cause you to avoid constructive criticism and lower your self-confidence. When this happens, remembering that you were born with confidence, and the ability to use it to your advantage can atten-

A Guide to the Scientific Career: Virtues, Communication, Research, and Academic Writing, First Edition.
Edited by Mohammadali M. Shoja, Anastasia Arynchyna, Marios Loukas, Anthony V. D'Antoni, Sandra M. Buerger, Marion Karl and R. Shane Tubbs.

uate the thoughts that threaten it. As you develop your self-image, you can begin to gain confidence in your strengths and learn from your weaknesses. Making an effort to understand and accept the weaknesses and transform them into strengths can boost confidence, bringing a sense of success and fulfillment.

> Confidence comes from success… But confidence also combines another quality because you can be successful, yet lack confidence. It requires a mental attitude shift to an expectation of success. And this alone, can bring about more success, reinforcing the confidence. It spirals from there.
>
> – Jason Hihn (cited by Rose 2017)

In trusting your capabilities to overcome challenges and obstacles, you can conquer more than you may expect. Confidence stems from the willingness and capacity to learn and not from the security of knowing what you can and cannot do. Constructive criticism becomes an opportunity for you to learn from experiences. Confidence comes from a space of humility, growing when you embrace new and challenging experiences. It is not knowing that you can create the outcome you desire, but understanding your competence and acknowledging that the undesirable outcomes that may happen may have nothing to do with you or what you should have done. This allows you to feel more confident in what you may have to offer. This is very similar to accepting that confidence only matures from taking risks and understanding the associated changes.

> Confidence is a funny thing. You go out and do the thing you are most terrified of, and the confidence comes afterwards.
>
> – Christopher Kaminski (cited by Deschene 2014)

By practicing certain tasks and exercises, you can both improve your tenacity skills and build your confidence in all aspects of life. For example, if you feel inadequate in your professional endeavors but are extremely confident in your personal relationships, by remembering how you feel and how comfortable you are in your personal life you can emit similar confidence into your professional career by drawing from other areas of confidence. The confidence gained from the aforementioned scenario is then cemented by positive recognition and reinforcement. You can trust in yourself, your skills, and your abilities when you believe in yourself and recognize that others believe in you as well.

5.2 Sources of Confidence

Research has shown four major sources of confidence (Bandura 1986, 1997, 2004; Bandura and Locke 2003). These sources can promote or undermine your confidence and its influence on your life and career. The first source of confidence is a history of prior success. This confidence refers to the degree in which you have experienced success in a given domain or task. As students are shaped by changes and challenges, they go through a self-appraisal process that includes comparing themselves to others and to standards that have become more demanding. The second source of confidence is observing others. Role models and peers can affect your confidence. In your personal life, observing parents, siblings, other family members, and friends with their struggles or coping mechanisms, successes and achievements, can affect your confidence. In the medical field, observing fellow medical students, residents, and physicians can have the same influence. This is mirrored in an academic setting through observing fellow students, professors, and researchers. Your peers and mentors can promote thinking skills, provide constructive feedback, and share struggles as well as coping mechanisms to change your sense of self-efficacy and confidence.

> No one can make you feel inferior without your consent.
>
> – Eleanor Roosevelt (1939)

The third source of confidence is persuasion, which entails the efforts to ensure that you can be successful. Verbal encouragement and constructive feedback are most often received from your role model or mentor in your personal life or profession. Unrealistic expectations from peers or role models can affect the level of confidence that you achieve. At this time, it becomes important for you to utilize self-confidence techniques or to seek feedback from others whom you trust, while not allowing unrealistic expectations to jeopardize your ability or level of confidence. The final source of confidence reviewed by Bandura (1997) consists of emotional experiences and physiological responses. Feelings of stress, anxiety, and pressure can undermine your confidence while feelings of enjoyment, interest, and engagement can promote self-confidence.

> Be who you are and say what you feel, because those who mind don't matter, and those who matter don't mind.
>
> – Bernard M. Baruch (cited by Cerf 1948)

5.3 Influence of Confidence on Your Career

Life experiences and other sources of confidence can influence your career choice. Confidence and experiences during the first 12–15 years of education affect the choice of career and the time spent in that career. Finding an interest, developing skills in that interest, managing a mentor-mentee relationship, and receiving recognition on your skill set is a common method in which people build confidence toward an aspiring career. However, maintaining a successful career depends on maintaining and maturing your confidence. By building confidence, you can fully realize and approach your potential (Holgate 2012). An appropriate level of self-confidence allows for you to know what you are expected to deliver. This can allow you to plan carefully and work toward an overall goal within the realm of your potentials. Other characteristics such as determination, persistence, tenacity, modesty, and optimism can allow you to defend your work and move forward regardless of what obstacles may develop.

During the initial period of development in a new endeavor, setting reasonable goals and an achievable vision while surrounding yourself with positive people who assist in expanding skills and experiences can allow you to achieve recognition and rewards. This serves both to create and nurture your confidence in your respected field.

5.4 Confidence Spectrum

Confidence in yourself and in your skills allows you to push your limits, achieve more than you otherwise could, and have the increased drive to continue to pursue your goals. However, inappropriately low or high levels of confidence can negatively affect your perceptions and achievements and create unpleasant results.

> Confidence comes not from always being right but from not fearing to be wrong.
> – Peter T. Mcintyre (cited by Friesen 2014)

5.4.1 Low Confidence and Insecurity

Early on, the development of confidence may be affected by experiences, feelings, and/or mentors and peers. Low levels of confidence precipitate into insecurity, self-doubt, and stress. These feelings can hinder your ability to concentrate, underestimate your abilities, quit activities or projects, and undervalue your achievements. People with a lack of self-confidence also tend to neglect their potentials by believing that their good work was a matter of luck and not their skills or experience. This phenomenon of under-confidence due to successes being attributed to luck is known as impostor syndrome. Clance and Imes (1978) found that women with imposter syndrome, who had notable professional and academic accomplishments, felt that they were not intellectually strong and that they had fooled everyone who believed in their success. People with imposter syndrome may begin to misjudge their strengths and weaknesses by relying on their weaknesses and undermining any of their strengths. Insecure individuals can get passed over for promotions and miss chances to publish their research findings or gain new work experiences. They also miss opportunities for career development and promotions, or participating on committees that could help further their careers.

> Be careful not to mistake insecurity and inadequacy for humility! Humility has nothing to do with the insecure and inadequate! Just like arrogance has nothing to do with greatness!
> – C. JoyBell C. (2014)

5.4.2 Overconfidence

False confidence may also have consequences in your personal life and career. Such feelings can precipitate into a sense of entitlement, arrogance, negligence, prejudice, and presumption. Such feelings hinder your ability to be productive, cause you to overestimate your achievements and skills, and can cause you to participate in activities and projects that you may not qualify for. The overconfident also tend to overestimate their past and present successes, leading to poor decision-making, ignorance or avoidance of constructive criticism, failure to use critical thought, and an overestimation of accuracy and depth of knowledge. People with overconfidence tend to believe that all of their achievements and successes are due to their own entitlement and not due to other factors such as their environment. They begin to misjudge their strengths and weaknesses by believing they have more strengths than weaknesses, negatively impacting any future growth. They tend to oversee opportunities to learn, assuming they know the best methods and practices. An overconfident academician may apply for fellowships, scholarships, grants, and positions without preparation or the necessary skill set or experiences. They may also submit papers or research presentations prematurely, alienate their colleagues, or receive negative feedback or letters of recommendation from their supervisors.

> Confidence turns into pride only when you are in denial of your mistakes.
> – Criss Jami (2015)

5.5 Dunning-Kruger Effect

The Dunning-Kruger effect is the cognitive bias in which people perform poorly on a task but lack the metacognitive capacity to properly evaluate their performance (Figure 5.1). The 1999 study predicted and demonstrated that incompetent individuals would dramatically overestimate their ability and performance relative to the objective criteria (Kruger and Dunning 1999). This suggests that poor problem solvers might not recognize their limitations or situations in which modifications or extra steps are required, or when a problem or project should be referred to a more capable person. This lack of awareness arises because poor performers' lack of skill deprived them not only of the ability to produce correct responses but also of the insight that they are not producing them (Dunning et al. 2003). This genre of studies extends from Confucius's observations centuries ago: achieving and maintaining an adequate measure of the good life means having insight into your limitations. The basis of the studies reflects

Figure 5.1 The Dunning-Kruger effect. (a): The unbiased relationship between the confidence or perceived ability and performance or actual ability is represented by a linear line. Imposter syndrome is a component of Dunning-Kruger effect, in which a highly performing individual presents a low confidence and has inappropriately low perception of his abilities. (b): The biased relationship between confidence and experience is represented by a curve, reflecting the Dunning-Kruger effect. At one end of the spectrum are inexperienced individuals who have inappropriately inflated confidence toward the task at hand. At the other end of this spectrum are experts whose levels of experience are closely related to their perceived ability and confidence. In the middle are a large proportion of professionals with an inappropriately low confidence toward the performed task.

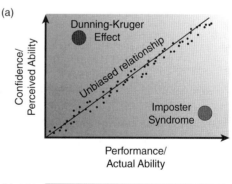

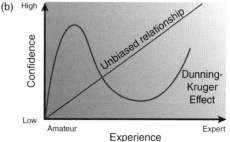

the idea that the skills needed to produce logical arguments are the same skills that are necessary to recognize when a logical argument has been made. If one lacks the skills to produce the answers, they are also unable to know when their answers or anyone's answers are right or wrong. Thus, incompetence means that individuals cannot successfully fulfill the ability to evaluate responses as correct or incorrect. When those who did not produce the correct responses were taught how to solve the problem or received the skills needed to distinguish accurate from inaccurate, they were able to reevaluate their previous tests to provide more realistic self-ratings than they had originally given (Kruger and Dunning 1999).

Kruger and Dunning (2009) noted that there was an undue modesty of top performers and an inflated assessment of lower performers. This study leads to questions of where perceptions of competence come from. It has been reported that the esteem of performances arise from a top-down approach. People start with a preconceived belief about their skills and use those beliefs to estimate how they are doing on a specific task (Kang 2014). This does not correlate with their actual performance. The top-down nature of esteem associated with performance can thus have important behavioral consequences. Starting in adolescence, beliefs about oneself generally rise, causing these beliefs to be integrated into the inner aspects of people's lives. This is true in real-life examples, such as women who rate themselves less scientifically talented than men; with this mindset, they become less enthusiastic in performing scientific or medical activities even though tests show no such difference in their actual performance (Eccles 1987; Ehrlinger and Dunning 2003). It is postulated that perception of performance, and not reality, influences decisions about future activities.

> Ignorance more frequently begets confidence than does knowledge.
> – Charles Darwin (1871)

5.6   Importance of Feedback

The Dunning-Kruger effect describes how people are unaware of their incompetence and fail to take self-improvement measures to rid the incompetence. Feedback plays a very crucial role in reversing this thought process and improving self-awareness. Individuals who are overconfident will perform tasks that they are not adequately equipped to undertake and attempt tasks without evaluating for risk. For those without confidence, they may not be able to work independently or may experience severe levels of anxiety and stress when working alone, even when others consider them to be competent. Competence is defined as the "habitual and judicious use of communication, knowledge, technical skills, critical reasoning, emotions, values, and reflection in daily practice for the benefit of the individual and community being served" (Epstein and Hundert 2002). Competence is a cognitive and integrative function: acquiring and using knowledge to solve problems that require confidence and vice-versa. However, personal confidence and experience have been found not to serve as a direct predictor of competence (Morgan and Cleave-Hogg 2002; Stewart et al. 2000). Confidence represents a judgment, which influences whether an individual is willing or not to undertake an activity; therefore, it is not necessarily based on actual levels of competence.

Confidence, like art, never comes from having all the answers; it comes from being open to all the questions.

– Earl Gray Stevens (cited by Merkel and Al-Falaij, 2003)

The Dunning-Kruger effect has been replicated among undergraduate students completing a classroom exam (Dunning et al. 2003), medical students assessing their interviewing skills, medical students evaluating their performance on clerkships (Edwards et al. 2003), and medical laboratory technicians evaluating their job expertise and problem-solving ability in the workplace (Haun et al. 2000). Consistent with the Kruger and Dunning studies, other studies have shown that overconfident students, yet incompetent, evaluated their performance nearly the same as students that performed well. In medicine, self-evaluation and self-assessment can measure one's confidence in relation to competence (Stewart et al. 2000). Medical students' levels of confidence in the performance of specific skills and patient management correlated to their clinical experience, but neither clinical experience nor level of confidence was able to predict outcomes in standardized performance exams (Morgan and Cleave-Hogg 2002). These findings may be accounted for by the differences in educational experiences; however, the lack of correlation between educational experience and outcomes on standardized assessments requires further explanation and research.

The aforementioned self-assessment and self-evaluation tools are helpful because they enable advisors to focus on the individuals who scored poorly; thus providing feedback to help them improve or at least begin to recognize their own limitations. It is an essential practice to recalibrate your ability to self-assess, as it provides motivation to seek training or assistance and better yourself. Furthermore, the ability to assess your knowledge and learning needs is important in continuing educational activities to address specific gaps in knowledge. Experience may improve the accuracy of self-assessment with more intensive feedback. This can help in identifying and correcting deficiencies. Some ideas to incorporate such feedback include self-assessment of your ability during or prior to training, modeling of self-assessment and self-directed learning by teachers and mentors, introduction of self-assessment skills during early education, and continuous feedback and constructive criticism during performances of any sort.

5.7   Overcoming Confidence Issues

Many methods have been developed to help overcome confidence issues arising from underconfidence or overconfidence. Counseling can assist people to become more self-aware and to analyze the impact that experiences and the environment have had on them. This can allow you to make a plan to overcome feelings of doubt or arrogance and become more confident in your abilities as well as acknowledge the factors that impact your achievements and personal life. Counseling can also be effective with improving your self-image. It is important to acknowledge the difference between being evaluated and being valued. Recognizing this difference can help people to remove themselves from cycles of decreased or inflated confidence and improve their self-image.

Other methods of improving your self-image include acknowledging your strengths and weaknesses, making an effort to improve your weak points, and reminding yourself of your strengths through regular affirmations. Additionally, seeking out trusted mentors for feedback, mapping goals, dealing with criticism constructively, and avoiding overanalysis help in overcoming confidence issues. An example of turning a weakness into a strength is turning worry into a positive driving force. Utilizing worry as a source of motivation imbues an intense focus on the situation at hand. This strategy allows you to both address the current situation productively and improve your confidence.

During your career, or even in your personal life, you will come across circumstances that will affect your confidence. Not only do these situations affect your reputation, but they may also deteriorate your confidence in the long-term. Examples of such situations consist of, but are not limited to, bullying, verbal or physical abuse, gossiping, and public shaming. Additional steps you can take to overcome confidence issues include promoting success by enhancing experiences and taking risks, making use of role models, and seeking encouragement.

> Man often becomes what he believes himself to be. If I keep on saying to myself that I cannot do a certain thing, it is possible that I may end by really becoming incapable of doing it. On the contrary, if I have the belief that I can do it, I shall surely acquire the capacity to do it even if I may not have it at the beginning.
> – Mahatma Gandhi (cited by Deats 2005)

Self-confidence – formed by self-efficacy and self-esteem – plays a very important role in the development of confidence. While confidence is a general understanding of your abilities and may be triggered from any event, self-confidence comes from within you and does not require a trigger. During the process of self-efficacy, one sees themselves mastering and achieving skills in specific areas. The confidence gained through success is the type of confidence that leads people to accept difficult challenges and persist in the face of setbacks. Self-esteem is more of a general sense of coping with whatever goes on in daily life. This internal quality tends to develop from the approval of people in your environment, virtuous behavior, competence, and completion of goals. Affirmations and positive thinking can influence self-esteem and self-confidence. Achieving and setting goals can build competence, which aids in confidence. Some steps that you can take to construct confidence include looking at what you have already achieved, thinking about your strengths, thinking about what is important to you and where you see yourself in the future, rational and positive thinking, and committing yourself to your goals. During this progression, it is very important that you keep a sense of balance between your life and career. Underconfidence can cause you to avoid taking risks or testing your abilities, and overconfidence may cause you to take on more than you can achieve properly or stretch yourself beyond your capabilities. You may also find that you do not try hard enough to focus or complete your tasks to truly succeed. Overcoming false self-images of perfection, blocking out any self-rejection, and changing what you believe and feel can aid in this route.

> Believe in yourself and there will come a day when others will have no choice but to believe with you.
> – Cynthia Kersey (cited by Sharma 2013)

Research has shown that the way people perceive themselves affects how others perceive them (Carney et al. 2010; Cuddy 2009; Cuddy et al. 2011, 2013; Fiske et al. 2006; Sherman et al. 2012). It has also shown that body language affects others' judgment. Amy Cuddy, a social psychologist, has shown how a few simple poses can affect a person's confidence in their personal life, at job interviews, and even in their career (Carney et al. 2010). Such poses include spreading your arms wide to appear more powerful, lifting your chest, holding your head high, and propping your arms either up or on the hips (Carney et al. 2010).

> Our bodies change our minds, and our minds can change our behavior, and our behavior can change our outcomes.
>
> – Amy Cuddy (2012)

This is similar to the concept that regulating and balancing your feelings of confidence requires considerable self-awareness and knowledge. When knowing a subject or stance, speaking with poise and conviction can exude confidence, while uncertainty can emit doubt. Feedback from colleagues, friends, and supervisors on your performance, and identification of strengths and weaknesses can help you utilize and improve on them, respectively. People perceived as warm and competent exhibit uniformly positive emotions and behavior, and those perceived as lacking the warmth portray uniformly negative emotions, such as overconfidence (Fiske et al. 2006). Those with confidence exhibit traits of friendliness, helpfulness, trustworthiness, and morality; competence reflects traits of perceived ability, intelligence, skill, creativity, and efficacy. Practicing these traits and going through the process can mature your confidence.

References

Bandura, A. (1986). *Social Foundations of Thought and Action*. Upper Saddle River, NJ: Prentice Hall.

Bandura, A. (1997). *Self-Efficacy: The Exercise of Control*. New York, NY: W.H. Freeman.

Bandura, A. (2004). Cultivate self-efficacy for personal and organizational effectiveness. In: *Handbook of Principles of Organizational Behavior* (ed. E.A. Locke), 120–136. Malden, MA: Blackwell.

Bandura, A. and Locke, E.A. (2003). Negative self-efficacy and goal effects revisited. *Journal of Applied Psychology* 88 (1): 87–99.

Carney, D.R., Cuddy, A.J.C., and Yap, A.J. (2010). Power posing: brief nonverbal displays affect neuroendocrine levels and risk tolerance. *Psychological Science* 21 (10): 1363–1368.

Cerf, B.A. (1948). *Shake Well before Using: A New Collection of Impressions and Anecdotes, Mostly Humorous*. New York: Garden City Books.

Clance, P.R. and Imes, S.A. (1978). The impostor phenomenon in high achieving women: dynamics and therapeutic intervention. *Psychotherapy: Theory, Research and Practice* 15 (3): 241–247.

Cuddy, Amy. "Just Because I'm Nice, Don't Assume I'm Dumb." Breakthrough Ideas of 2009. Harvard Business Review 87, no. 2 (February 2009).

Cuddy, A. (2012). *Your Body Language Shapes Who you Are*. Edinburgh, Scotland: TEDGlobal.

Cuddy, A.J.C., Glick, P., and Beninger, A. (2011). The dynamics of warmth and competence judgments, and their outcomes in organizations. *Research in Organizational Behavior* 31: 73–98.

Cuddy, A.J.C., M. Kohut, and J. Neffinger. "Connect, Then Lead." Harvard Business Review 91, nos. 7/8 (July–August 2013): 54–61.

Darwin, C. (1871). *The Descent of Man*. New York: D. Appleton and Company.

Deats, R. (2005). *Mahatma Gandhi, Nonviolent Liberator: A Biography*, 108. Hyde Park: New City Press.

Deschene, L. "8 ways to be more confident: live the life of your dreams." Web log post. Tiny Buddha. Web. 16 Jun. 2014.

Dunning, D., Johnson, K., Ehrlinger, J. et al. (2003). *Why People Fail to Recognize their Own Incompetence*, 83–87. American Psychological Society.

Eccles, J.S. (1987). Gender roles and women's achievement-related decisions. *Psychology of Women Quarterly* 11: 135–172.

Edwards, R.K., Kellner, K.R., Sistrom, C.L. et al. (2003). Medical student self-assessment of performance on an obstetrics and gynecology clerkship. *American Journal of Obstetrics Gynecology* 188 (4): 1078–1082.

Ehrlinger, J. and Dunning, D. (2003). How chronic self-views influence and potentially mislead estimates of performance. *Journal of Personality and Social Psychology* 84: 5–17.

Epstein, R.M. and Hundert, E.M. (2002). Defining and assessing professional competence. *JAMA* 287 (2): 226–235.

Fiske, S.T., Cuddy, A.J.C., and Glick, P. (2006). Universal dimensions of social cognition: warmth and competence. *Trends in Cognitive Science* 11 (2): 77–83.

Friesen, T. (2014). *Ride the Waves*, vol. II, 62. Bloomington: Balboa Press.

Haun, D.E., Zeringue, A., Leach, A. et al. (2000). Assessing the competence of specimen-processing personnel. *Laboratory Medicine* 31 (11): 633–637.

Holgate, S.A. Successful careers: a matter of confidence. 2012. Science career magazine (November 2012).

Jami, C. (2015). *Killosophy*, 80. Unknown City.

JoyBell C.C. Quotes. Goodreads. Web. 16 Jun. 2014.

Kang H. (2014) When Ignorance Begets Confidence: The Classic Dunning-Kruger [WWW] http://typezen.com/when-ignorance-begets-confidence-the-classic-dunning-kruger/ 2969. [Accessed 5/9/2017]

Kruger, J. and Dunning, D. (1999). Unskilled and unaware if it: how difficulties in recognizing one's own incompetence lead to inflated self-assessment. *Journal of Personal Social Psychology* 77: 1121–1134.

Kruger, J. and Dunning, D. (2009). Unskilled and unaware of it: how difficulties in recognizing one's own incompetence lead to inflated self-assessments. *Psychology* 1: 30–46.

Merkel, J.H. and Al-Falaij, A.W. (2003). *On the Art of Business*, 101. Coral Springs, FL: Llumina Press.

Morgan, P.J. and Cleave-Hogg, D. (2002). Comparison between medical students' experience, confidence and competence. *Medical Education* 36: 534–539.

Roosevelt, E. (1939). *This Is my Story*. New York: Garden City Publishing Co.

Rose, M. (2017) "Being" Confidence! [WWW] Noomii Coach Directory. http://www .noomii.com/articles/7008-being-confidence. [Accessed 5/9/2017].

Sharma, N.P. (2013). *Let's Live Again*, 78. Gurgaon: Partridge Publishing.

Sherman, G.D., Lee, J.J., Cuddy, A.J.C. et al. (2012). Leadership is associated with lower levels of stress. *Proceedings of the National Academy of Sciences of the United States of America* 109: 17903–17907.

Stewart, J., O'Halloran, C., Barton, J.R. et al. (2000). Clarifying the concepts of confidence and competence to produce appropriate self-evaluation measurement scales. *Medical Education* 34: 903–909.

6

Career Satisfaction and Its Determinants

Nihal Apaydin

Department of Anatomy and Brain Research Center, Faculty of Medicine, Ankara University, Ankara, Turkey

6.1 Introduction

Career satisfaction is the main, and perhaps the only, preventive remedy for professional "burnout" and attrition. The concept of career satisfaction has been defined in many ways by many different researchers and practitioners. Early definitions of career satisfaction have been suggested by Vroom (1964), as well as Hackman and Oldham (1975); however, one of the most widely used definitions was proposed by Locke (1976), who described career satisfaction as "a pleasurable or positive emotional state resulting from the appraisal of one's job or job experiences." Three indispensable elements of career satisfaction are (i) passion and motivation, (ii) proficiency and skill, and (iii) opportunity and benefit (Figure 6.1). A person's attitude toward their job is usually a result of personal experience, which includes emotions and judgments. Within academia, career satisfaction can also be seen in the broader context, which includes determinants that affect an individual's professional experience and quality of working life. There is often a disparity between personal expectations of the job and the possibilities or opportunities presented by the job's environment.

Career success is almost always divided into different domains: extrinsic success versus intrinsic success, objective success versus subjective success, or material elements of success versus psychological elements of success. Markers of extrinsic success can be a financial reward, a job promotion, a position of leadership, grants, and publications. Intrinsic success is measured by a more general satisfaction with career and life (Rubio et al. 2011). This chapter explores the factors influencing career satisfaction, especially among academics and physicians.

6.2 Determinants of Career Satisfaction

There are several determinants for career satisfaction. In general, determinants found to affect satisfaction include age, one's profession, job proficiency, education level, working hours, workplace size, income, gender, marital status, and cultural intelligence, among others (D'Addio et al. 2007; Bender and Sloane 1998). For example, it is reported that

A Guide to the Scientific Career: Virtues, Communication, Research, and Academic Writing, First Edition.
Edited by Mohammadali M. Shoja, Anastasia Arynchyna, Marios Loukas, Anthony V. D'Antoni, Sandra M. Buerger, Marion Karl and R. Shane Tubbs.

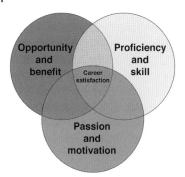

Figure 6.1 Three elements of career satisfaction.

more educated people have a lower job satisfaction. Perhaps because people who are more educated have higher expectations, and therefore are more susceptible to disappointment and dissatisfaction. Married people report greater job satisfaction and those with health problems report lower job satisfaction. The relationship between age and career satisfaction follows a U-shaped pattern in which the youngest and oldest individuals report the greatest job satisfaction. People with higher earnings are generally more satisfied; however, those reporting more hours of work demonstrate the same level of satisfaction as those reporting fewer hours of work (Clark et al. 1996; Clark and Oswald 1996). Academics are happier than nonacademics (Bender and Heywood 2006).

The size of the establishment is also correlated with employee satisfaction. For example, large establishments organize work and production in a less flexible fashion than smaller establishments, and lower levels of job satisfaction have been reported in larger establishments (Idson 1990).

Among the academic disciplines, those in mathematics and engineering are least satisfied with their job, while those in management and the health sciences report relatively high job satisfaction ratings (Table 6.1). It is also worth mentioning that government employees report less job satisfaction than people in the private sector (Bender and Heywood 2006). A study has shown that while earnings contribute to job satisfaction, nonpecuniary benefits – such as relationships with colleagues, the nature of teaching, and publication success – are actually more important determinants of job satisfaction among Scottish academics (Ward and Sloane 2000).

6.2.1 Gender and Career Satisfaction

It has been shown that women in general have a greater job satisfaction than men in the United States and United Kingdom, where it is suggested that women's job expectations are generally lower (Bender et al. 2005; Clark 1997; Sousa-Poza and Sousa-Poza 2000). That said, the gender influence on career satisfaction must vary in the context of the sector and subdiscipline being examined. For example, Bender and Heywood (2006) reported that among economists working in academia, females were less satisfied than their male counterparts. The negative relationship between age and career satisfaction is less for females than males. And except for impaired vision, there is less reported influence of disabilities on job satisfaction among women. Income appears to play no role in the career satisfaction of female academics (Bender and Heywood 2006). The impact of race on career satisfaction among academic is also less strong among females than among males.

Table 6.1 Job satisfaction by discipline and sector.

Discipline	Sector	Average job satisfaction [a]	% Very satisfied
All	Academic	3.43	52.3
	Nonacademic	3.37	49.8
Management	Academic	3.56	60.9
	Nonacademic	3.47	56.8
Health	Academic	3.44	52.3
	Nonacademic	3.46	56.5
Social Science	Academic	3.45	53.2
	Nonacademic	3.44	54.3
Engineering	Academic	3.43	52.6
	Nonacademic	3.31	42.7
Economics	Academic	3.48	52.3
	Nonacademic	3.44	52.7
Computer	Academic	3.39	51.0
	Nonacademic	3.28	44.3
Basic Science	Academic	3.41	51.5
	Nonacademic	3.35	49.2
Computer	Academic	3.39	51.0
	Nonacademic	3.28	44.3
Math	Academic	3.39	50.9
	Nonacademic	3.33	49.1

a) On a Likert scale of 1–5.
Source: Adapted with modifications and from Bender and Heywood 2006.
Reproduced with permission from Wiley.

6.3   Career Satisfaction in Medicine

The career satisfaction of physicians has important implications for the delivery of quality medical care. It has been reported for years that physicians who are satisfied with their careers provide better overall healthcare and have more appeased patients compared to physicians dissatisfied with their careers. Satisfied physicians are more likely to stay in their practice, thus enhancing patient continuity – which is an important component of healthcare quality. In contrast, dissatisfaction among physicians can lead to increased rates of medical errors related to job stress, and dissatisfied physicians are more likely to leave the healthcare profession (Lichtenstein 1984; Landon et al. 2003; Landon et al. 2006). Therefore, satisfaction may be considered an indirect measure of patient outcomes. Although the overall level of satisfaction among physicians in the United States has been consistently reported around 80%, there is a noticeable variation of satisfaction levels among different medical specialties.

6.3.1   Career Satisfaction in Primary Care Physicians

Primary care physicians (including family medicine, internal medicine, pediatrics, and general practice specialists) consistently rank in the bottom tier of satisfaction reports

(Leigh et al. 2002; Leigh et al. 2009; Duffy and Richard 2006; Katerndahl et al. 2009). While primary care is a cornerstone for developing a cost-effective, high-quality health-care system that reduces health disparities, there is a growing shortage of primary care physicians. In parallel to these findings, the interest in primary care among graduating medical students has also declined (Pugno et al. 2006).

The career satisfaction of primary care physicians has thus become the subject of studies that reveal some interesting trends and raise important questions regarding this medical specialty. In a survey conducted in 2006, it was found that more than 50% of primary care physicians considered themselves "second-class citizens" compared with surgical and diagnostic specialists (Physicians Practice 2007). In addition, almost two-thirds of primary care physicians reported that they would choose another field if they could start their careers over. Approximately 39% of the 508 primary care physicians said they would become surgical or diagnostic specialists, and approximately 22% said they would not choose medicine as a career. The only explanation for this situation seems to be career dissatisfaction. In a separate study, Katerndahl et al. (2009) found that doctors who worked more hours per week, more weeks per year, and who maintained solo practices reported significantly lower satisfaction than other physicians – all of these characteristics commonly defined primary caregivers. The results of this study suggested a consistent association between perceived autonomy and career satisfaction.

Deshpande and DeMello (2010) published an analysis to further examine factors that influence career satisfaction of three types of primary care physicians, specifically internal medicine physicians, family/general medicine practitioners, and pediatricians. This study illuminated another issue: malpractice lawsuits. According to the results of this survey, the threat of malpractice lawsuits significantly lowered career satisfaction. On the other hand, quality patient interaction and high incomes significantly increased career satisfaction reported by primary care physicians. With all of these influencing factors considered, pediatricians had the highest overall career satisfaction among the group. The authors suggested that steps like giving adequate time for interaction with patients, reducing potential threats of malpractice lawsuits, and improving income may contribute to increased career satisfaction among primary care physicians (Deshpande and DeMello 2010).

6.3.2   Clinical Specialties and Career Satisfaction

There is a huge range of interests and aptitudes among physicians, which influences their choices from a diverse field of specialties including primary care, surgery, psychiatry, public health, and more. Goldacre et al. (2012) conducted a study on doctors who considered but did not pursue specific clinical specialties as careers. Graduates in 2002, 2005, and 2008 from all UK medical schools were included in the study one year after qualification. According to the results of this study, 2573 of 9155 respondents (28%) had seriously considered, but not pursued, a specialty field. The issue of "work-life balance" was the single-most common factor, particularly for women, in not pursuing a specialty. Competition for positions, difficult examinations, stressful working conditions, and poor training were reported as minor concerns. Unusual work hours – deemed *unsocial hours* – and excessive on-call hours were suggested to be more important determining factors in exclusion of certain specialties by the doctors. However, in some specialties, working unusual or excessive hours are unavoidable. It should be considered by workforce planners, health service administrators, and senior medical professionals that the intensity of working hours will dissuade some doctors from specializing. The study concluded that any approach to policy changes must address the imperatives of service needed, the importance of continuity of patient care, and the quality of training and learning experiences for doctors (Goldacre et al. 2012).

In a different study by Leigh et al. (2002), career satisfaction across 33 specialties among 12 474 US physicians was analyzed. It was suggested that geriatric internal medicine, neonatal-perinatal medicine, dermatology, and pediatrics were all significantly more satisfying than family medicine. The researchers noted a nonlinear relationship between age, income, and satisfaction among surveyed physicians. Increased work hours were found to be associated with dissatisfaction reported by all specialties. A follow-up study published by Leigh et al. (2009) supported the original findings and established a trend: Doctors who practice family medicine reported less job satisfaction than their peers.

6.3.3 Demographic Determinants of Career Satisfaction Among Medical Graduates

Demographic characteristics also seem to be an important determinant for career satisfaction in the healthcare field. For example, international medical school graduates have substantial representation among primary care physicians in the United States and consistently report lower career satisfaction compared with US medical school graduates (Morris et al. 2006). Chen et al. (2012) reported that career satisfaction among primary care physicians was 75.7%; but when US medical graduates were singled out, 82.3% expressed career satisfaction. This statistically significant difference persisted even after adjusting for a broad range of potential explanatory factors, including personal characteristics and variables related to the practice environment. Although international and US graduates differed on several potential explanatory factors – including language, cultural customs, board certification, and experience working in large metropolitan cities – no significant relationship was determined among these factors that would point toward lower career satisfaction reported by international medical school graduates.

Additional domains of experience may contribute in important ways to differences in career satisfaction between international and US graduates. These may include the impact of discrimination in the workplace, stresses of being an "outsider," and lack of appropriate supportive structures in the workplace (Chen et al. 2011). Previous reports indicated that international medical graduates were more likely than US graduates to report experiences of discrimination in the workplace (Chen et al. 2010). Given the central role of international graduates in the US healthcare system, particularly for vulnerable populations, improved understanding of the causes of this differential satisfaction is important to appropriately support the healthcare workforce.

6.4 Research and the Physician-Scientist

A challenge for the medical profession is to preserve the future of academic researchers in the next generation of physicians in a way that is compatible with their personal lives.

There is an increasing number of clinicians who would prefer to work part-time. Hoesli et al. (2013) conducted an exploratory survey to assess predictors for an academic career in a population of physicians working full-time or part-time in the northwestern part of Switzerland. These Swiss physicians were evaluated to determine the individual attitudes, influences, and motivations toward working part-time. The results of this survey suggested that the opportunity to do research on the side was the most significant factor positively influencing the physicians' decisions to work part-time in a clinical setting.

Whether practicing medicine full- or part-time, physicians have always found it challenging to balance the demands of clinical care with the time required to perform research or publish their findings. Many physicians have also found it challenging to obtain the time they need to start and maintain a research program; one hurdle being the long lag in time between beginning a research career and receiving funding. Furthermore, it can be hard to cope with the career uncertainties related to the reliance on other grants for salary support (Nathan 1998). For these reasons, many clinicians have not been attracted to a career that combined clinical work with research.

With today's growing focus on the translation of basic science discoveries into clinical practices, and the increased need for comparative effectiveness research, the demand for physician-scientists is likely to grow – as is the demand for new ideas and programs to increase the supply of physician-scientists. Some universities have established various types of career development programs. However, the success of these programs to substantially increase the cadre of well-trained physician-scientists must address the problems that commonly deter physicians who might be interested in pursuing investigative careers.

6.5 Career Satisfaction and Productivity

The motivation for studying career satisfaction is, in part, the perception that higher satisfaction should be associated with greater productivity. Thus, if the determinants of career satisfaction are understood, researchers may be able to contribute to creating conditions or models that enhance productivity for a multitude of industries including academia. By example, managers concerned with maximizing the impact of their research and development teams are specifically interested in findings revealed by the aforementioned research (Kim and Oh 2002).

The opportunity to perform research remains the most important predictor for a successful academic career. At the University of Pittsburgh's Institute for Clinical Research Education (ICRE), The Research on Careers Workgroup, a development program, created a comprehensive theoretical model to address career success. The Workgroup's model, and consecutive submodels, would allow for the evaluation of personal factors (e.g. demographics, education, and the psychosocial milieu), organizational factors (e.g. financial resources, infrastructure, training, and mentoring), and the interplay between factors that contribute to career success. The authors have suggested that with this model, leaders of training programs could identify potential physician-scientists and provide early opportunities for intervention – thus ensuring career success. The authors concluded that their model may serve as a highly flexible template for concise and testable analytic models used to develop a positive career trajectory for aspiring physician-scientists (Rubio et al. 2011). However, demonstrating how these theoretical associations may apply to important questions related to career success is limited in scope.

6.6 Conclusions

Career satisfaction is an important issue that directly affects a person's productivity and quality of life. There are many determinants, which are all likely play a role in career satisfaction. The role of these determinants have been the subject of many studies; no single determinant seem to outweigh the others, especially since personal expectations regarding one's job cannot be objectively measured. While personal expectations are subject to change over time, continuous research is warranted to evaluate the job and workplace factors that are associated with low satisfaction and burnout, and can then be remedied through new programs and job redesign. Investigating determinants of career satisfaction is an important means for academic institutions not only to achieve better performance outcomes, but to improve the quality of working life.

References

Bender, K.A. and Heywood, J.S. (2006). Job satisfaction of the highly education – the role of gender, academic tenure, and earnings. *Scottish Journal of Political Economy* 53 (2): 253–279.

Bender, K.A. and Sloane, P.J. (1998). Job satisfaction, trade unions and exit-voice revisited. *Industrial and Labor Relations Review* 51: 222–240.

Bender, K.A., Donohue, S.M., and Heywood, J.S. (2005). Job satisfaction and gender segregation. *Oxford Economic Papers* 57: 479–496.

Chen, P., Nunez-Smith, M., Bernheim, S. et al. (2010). Professional experiences of international medical graduates in primary care. *Journal of General Internal Medicine* 25 (9): 947–953.

Chen, P.G., Curry, L.A., Nunez-Smith, M. et al. (2012). Career satisfaction in primary care: a comparison of international and US medical graduates. *Journal of General Internal Medicine* 27 (2): 147–152.

Clark, A.E. (1997). Job satisfaction and gender: why are women so happy at work? *Labour Economics* 4: 341–372.

Clark, A.E. and Oswald, A.J. (1996). Satisfaction and comparison income. *Journal of Public Economics* 61: 359–381.

Clark, A.E., Oswald, A.J., and Andwarr, P. (1996). Is job satisfaction U-shaped in age? *Journal of Occupational Psychology* 69: 57–81.

D'Addio, A.C., Eriksson, T., and Frijters, P. (2007). An analysis of the determinants of job satisfaction when individuals' baseline satisfaction levels may differ. *Applied Economics* 39 (19): 2413–2423.

Deshpande, S.P. and Demello, J. (2010). An empirical investigation of factors influencing career satisfaction of primary care physicians. *Journal of American Board of Family Medicine* 23 (6): 762–769.

Duffy, R.D. and Richard, G.V. (2006). Physician job satisfaction across six major specialties. *Journal of Vocational Behavior* 68: 548–559.

Goldacre, M.J., Goldacre, R., and Lambert, T.W. (2012). Doctors who considered but did not pursue specific clinical specialties as careers: questionnaire surveys. *Journal of the Royal Society of Medicine* 105 (4): 166–176.

Hackman, J.R. and Oldham, G.R. (1975). Development of the job diagnostic survey. *Journal of Applied Psychology* 60: 159–170.

Hoesli, I., Engelhardt, M., Schötzau, A. et al. (2013). Academic career and part-time working in medicine: a cross sectional study. *Swiss Medical Weekly* 143: W13749.

Idson, T.L. (1990). Establishment size, job satisfaction and the structure of work. *Applied Economics* 11: 606–628.

Katerndahl, D., Parchman, M., and Wood, R. (2009). Perceived complexity of care, perceived autonomy, and career satisfaction among primary care physicians. *Journal of American Board of Family Medicine* 22 (1): 24–33.

Kim, B. and Oh, H. (2002). Economic compensation compositions preferred by R&D personnel of different R&D types and intrinsic values. *R&D Management* 32: 47–59.

Landon, B.E., Reschovsky, J.D., and Blumenthal, D. (2003). Changes in career satisfaction among primary care and specialist physicians, 1997–2001. *Journal of the American Medical Association* 289 (4): 442–449.

Landon, B.E., Reschovsky, J.D., Pham, H.H. et al. (2006). Leaving medicine: the consequences of physician dissatisfaction. *Medical Care* 44 (3): 234–242.

Leigh, J.P., Kravitz, R.L., Schembri, M. et al. (2002). Physician career satisfaction across specialities. *Archives Internal Medicine* 162 (14): 1577–1584.

Leigh, J.P., Tancredi, D.J., and Kravitz, R.L. (2009). Physician career satisfaction within specialities. *BMC Health Services Research* 16 (9): 166–177. https://doi.org/10.1186/1472-6963-9-166.

Lichtenstein, R.L. (1984). The job satisfaction and retention of physicians in organized settings: a literature review. *Medical Care Review* 41 (3): 139–179.

Locke, E.A. (1976). The nature and causes of job satisfaction. In: *Handbook of Industrial and Organizational Psychology* (ed. M.D. Dunnette), 1297–1349. Chicago, IL: Rand Mcnally.

Morris, A.L., Phillips, R.L., Fryer, G.E. Jr., et al. (2006). International medical graduates in family medicine in the United States of America: an exploration of professional characteristics and attitudes. *Human Resources for Health* 4: 17.

Nathan, D.G. (1998). Clinical research: perceptions, reality, and proposed solutions. National Institutes of Health Director's Panel on Clinical Research. *JAMA* 280: 1427–1431.

Physicians Practice. (2007) Survey: 60% of primary care physicians would choose another field. Available at: http://www.medicalnewstoday.com/articles/81499.php. Accessed 5 February 2015.

Pugno, P.A., Schmittling, G.T., Mcgaha, A.L. et al. (2006). Entry of us medical school graduates into family medicine residencies: 2005–06 and 3-year summary. *Family Medicine* 38: 626–636.

Rubio, D.M., Primack, B.A., Switzer, G.E. et al. (2011). A comprehensive career-success model for physician-scientists. *Academic Medicine* 86 (12): 1571–1576. https://doi.org/10.1097/ACM.0b013e31823592fd. Review.

Sousa-Poza, A. and Sousa-Poza, A.A. (2000). Taking another look at the gender/job satisfaction paradox. *Kyklos* 53: 135–152.

Vroom, V.H. (1964). *Work and Motivation*. San Francisco, CA: Jossey-Bass.

Ward, M.E. and Sloane, P.J. (2000). Non-pecuniary advantages vs. pecuniary disadvantages: job satisfaction among male and female academics in Scottish universities. *Scottish Journal of Political Economy* 47: 273–303.

7

Spiritual Dimensions of Biomedical Research

Mariana Cuceu[1], Beatrice Gabriela Ioan[2], Marcel Cuceu[3], Kyle E. Karches[4], Cristinel Stefanescu[2] and Aasim I. Padela[5]

[1] *Program on Medicine and Religion, The University of Chicago, Chicago, IL, USA*
[2] *Grigore T. Popa University of Medicine and Pharmacy, Iasi, Romania*
[3] *Romanian Catholic Byzantine Diocese of Saint George, Canton, OH, USA*
[4] *Internal Medicine Residency Program, The University of Chicago Medicine, Chicago, IL, USA*
[5] *MacLean Center for Clinical Medical Ethics, The University of Chicago, USA*

7.1   Introduction

> I have found there is a wonderful harmony in the complementary truths of science and faith. The God of the Bible is also the God of the genome. God can be found in the cathedral or in the laboratory. By investigating God's majestic and awesome creation, science can actually be a means of worship
>
> (Francis Collins 2007).

The aim of this chapter is to comment on the underlying motivations, values, and assumptions in scientific inquiry and biomedical practice as part of a quest for the truth about the world and our place within it. It is possible for the medical scientist to seek the truth and bear witness to the truth as it is reflected in the empirical reality of the world. In so doing, the spiritual scientist acts in harmony with God's created order, and his or her research will bring that order more fully to light.

While scientific endeavors reveal new knowledge and researchers labor to exchange this information, wisdom is necessary to discern between knowledge that brings about benefit and that which brings about harm. Hence, the goals of any research activities are informed in part by the researcher's consideration of what is beneficial and what requires prudence; for example, how medical science contributes to the health of a population. When we forthrightly address the question of what is beneficial, we begin to discern the animating purposes of research. To conduct medical research that better serves patients' needs, researchers must come to a deeper, more accurate understanding of what practices are truly beneficial to the welfare of their patients.

Some scientists view their research practices in the light of spirituality. We propose that, as these scientists seek out harmony in God's created order, their practice takes on a vertical dimension linking their medical practice and research with service to the Divine. For such individuals, their development as scientists coexists with a belief in God that may enrich their experiences and embolden their research endeavors.

A Guide to the Scientific Career: Virtues, Communication, Research, and Academic Writing, First Edition.
Edited by Mohammadali M. Shoja, Anastasia Arynchyna, Marios Loukas, Anthony V. D'Antoni, Sandra M. Buerger, Marion Karl and R. Shane Tubbs.
© 2020 John Wiley & Sons, Inc. Published 2020 by John Wiley & Sons, Inc.

The ways in which religion and spirituality impact health, medical care, and research are intensively studied and vary across the globe (Flannelly et al. 2004). At minimum, medical researchers should see themselves as an integral part of the healthcare system and are accordingly obliged to reflect on how their scholarship can be applied to improve patients' lives, or the well-being of society in general. Such purposes can guide research activities and remind the scientist that serving humanity is an expression of serving one's God.

We must bear in mind that we may never obtain a completely objective grasp on reality, nor can science answer all of our questions. Scientific research remains powerless to provide answers to profound human questions such as *what is the meaning of life* or *what happens after we die*? Modern scientific inquiry focuses on that which is quantifiable, and such questions do not lend themselves to scientific methods of quantification.

To care for the sick is a privilege and a responsibility at the same time The responsibility of those involved in the act of care, doctors and researchers, is to ensure that the medical research together with the practice of medicine benefits the good of their patients. It is not enough merely to advance science and increase knowledge; science should ultimately serve the welfare of individuals and their communities. Medical research and medical practice are inseparable dimensions of caring for patients. Research remains an essential part of the clinical profession and the quality of research is directly reflected in the development of the discipline. Therefore, the education and growth of both researchers and practitioners should aim not only to teach concepts, but also to form character. The virtues that apply to clinical practice also apply to research. As such, we propose that scientists develop their character along two planes.

The vertical plane is concerned with internal moral formation of the virtues that a scientist must acquire to be a good person (e.g. justice, honesty, humility, graciousness), as well as cultivating the virtues expressed in serving others (e.g. kindness and love) – all while nourishing a relationship with God.

The horizontal plane is concerned with societal influence and the role that medical research plays in serving humanity. This plane aims to align research endeavors in ways that benefit society while respecting appropriate limits to the scientific quest for progress and knowledge.

7.2   Virtues in Research

Those responsible for training new researchers should seek to develop not only their students' research skills, but also their characters. Contemporary moral philosopher MacIntyre (1984) argues that new participants in any complex social activity, such as scientific research, must develop virtues in order to learn their craft Students or novices must accept the authority of established standards of excellence in research and recognize the inadequacy of their own work in light of those standards. In time, these students may gain the capacity to criticize or further develop these standards.

The process of initiation into a field requires the practice of particular virtues, which we will discuss. New researchers require *justice* in order to discern what they owe to whom. They must determine, for example, how to show their gratitude to their teachers, how to treat animal subjects of research humanely, and how to compensate human subjects appropriately. They require *courage* to take the risks necessary for good research.

Perhaps the most crucial virtue for researchers is *honesty* or truthfulness. The division of labor among many researchers requires clear and complete presentation of data, even when such data contradicts an attractive hypothesis. The entire enterprise of scientific research would collapse without the commitment to a truthful sharing of results. Consequently, the scientific community levies harsh punishments on scientists who betray this trust. The virtue of *honesty* also promotes a high regard for, and even relentless pursuit of, the truth. The truthful researcher constantly seeks to expand his or her knowledge and avoids sources of information that may be misleading. Such researchers candidly evaluate their own performances and, by practicing the virtue of *humility*, they seek out criticism from those who can correct their errors.

In order to cultivate such virtues in their students, educators cannot merely talk about them in the classroom. Because the virtues are not abstract intellectual concepts, but rather habits of good character and behavior, students will acquire them only through active participation in research under the close guidance of mentors who possess these virtues themselves.

Research from the "positive psychology" movement indicates that under the right circumstances, virtues can be taught, and MacIntyre's work suggests that virtuous behavior will flourish only within communities that hold the virtues at the core of their common life (Bryan and Babelay 2009). Therefore, academic departments must recruit and promote faculty members who exemplify these virtues.

This emphasis on the moral formation of researchers is meant to supplement, not replace, the existing guidelines on ethical research. Indeed, fidelity to fundamental research statutes, such as those regarding informed consent, could be considered a virtue. However, these guidelines may not always provide a clear direction in challenging situations and, from time to time, institutions will promulgate the guidelines as unethical. In such cases, a morally acceptable outcome depends on virtuous persons recognizing what is required of them and acting rightly – thereby displaying virtues such as justice and courage.

7.3   A Christian-Platonic Background of Virtues

We assume that a student who embraces a research career will seek a practical pathway by which the student can develop virtuous habits of character. Lessons from literature that reflect on virtues would further allow researchers to learn from the long history of ethics (e.g. authors such as Homer, Aristotle, Benjamin Franklin, and Jane Austen). Several religions consider the virtues necessary for a good and moral life; thus, researchers can enrich their understanding of the virtues by engaging with these traditions and welcoming religious perspectives in discussions about research ethics. This so-called *open pluralism* can contribute to the professional development of novice researchers, particularly if they belong to one of these religious traditions (Kinghorn et al. 2007).

To illustrate, we will offer a few considerations from the ascetic tradition of the Near East. In the Platonic model of psychology, the human psyche consists of three key elements: the noetic, or intellect; the incensive, describing desires and animations; and the appetitive, the ability to satisfy bodily needs. These elements are further reiterated in the Patristic writings of the Fathers of the Christian Church as the powers of the human soul and can be characterized by the virtues of justice in the search for truth, courage, and temperance, respectively – with wisdom as the overruling principle of the human soul.

A contemporary French philosopher of Christian-Platonic understanding, John Claude Larchet, builds on the philosophies of early Christian Patristic Fathers of the Church such as Saint John of Damascus and Saint Maximus the Confessor. Larchet describes three essential powers of the human soul as vegetative (nutrition and growth), animal (aggressiveness, desire, and affectivity), and reasoning (reason being the principal characteristic that distinguishes humans from all other creatures). The power of reasoning has two important faculties: the spiritual faculty (intellect, or nous, linked to a human's noetic ability and responsible for the moral and psychological senses of consciousness) and the faculty of self-determination (Larchet 2011).

Although it adopts the Platonic virtues, the Christian ascetic tradition, with respect to the incensive element of the soul, focuses on the virtue of love – cultivating practices that exhibit brotherly love and compassion (Vitz 2014). Early Christian-Platonist thinker, Saint John Damascus, described the contrasting "sins" to the incensive powers as including "heartlessness, hatred and lack of compassion" (Palmer et al. 1981). Peter of Damaskos stated that love of neighbor is highest among all virtues (Palmer et al. 1983). And Saint Isaac of Nineveh stated that a person whose incensive powers are in the right order has a heart "burning with love towards the whole creation: towards men, birds, animals, demons and every creature…cannot bear to hear or see the least harm done to or misfortune suffered by creation" (Popovich 1994; Alfeyev 2000).

To summarize, the pursuit of virtue in the Christian ascetic tradition is ultimately the struggle to acquire a compassionate heart and demonstrate mercy toward every human being and creation. Clement of Alexandria (150–215 AD) and John of Damascus (676–749 AD) have described the "passions" as barriers to manifesting love. The passions should not be understood as human emotions but, rather, as vices precluding obedience to the Word of God (see 'The Stromata' in Roberts and Donaldson 1885 and 'An Exact Exposition of the Orthodox Faith' in Schaff and Wace 1899). Examples of these passions are avarice, bitterness, resentfulness, anger, pride, jealousy, lustfulness, or what the ascetic fathers called "self-love." The Church teaches that such "passionate" persons will be preoccupied with satisfying their own desires for pleasure, wealth, and praise – and as such will fail to attend to or cultivate the true virtues of compassion and mercy. Particularly relevant to the practice of research, those who give in to their passions will regard others as rivals or as impediments toward satisfying their self-love (Vitz 2014): a pattern that is likely to erode the spirit of teamwork and undermine a research team's effectiveness and success.

An important reason for bringing to light these traditional thinkers in contemporary practice, and in regard to the ethical growth of researchers and scientists, is that we believe a similar model can be used as an impetus toward understanding the purpose and virtuosity of performing scientific research.

7.4 Skills Versus Wisdom

There is a distinction between the two types of knowledge – wisdom and skill – that applies to every realm of life, including medical research. It may not be surprising if someone who is highly *skilled* in laboratory work or in writing lacks basic wisdom to discern which endeavors his or her research will bring about the most benefit to others.

The distinction between *phronesis*, or practical wisdom, and *techne*, or technical skill, as used in Aristotle's language is important. The practical wisdom is the development or internal formation of the Self as someone cognizant of his or her purpose who acts with wisdom in all areas of life. The technical skill is involved more with the production or formation of an outside object or external good (e.g. a drug, a car, a hospital). *Phronesis* and *techne* are required to aim towards a moral and ethical understanding and virtuous living and to create a virtuous output (Hammalis 2013).

Those who teach medical science or research techniques should exhibit "technical" knowledge − the product of their teachings being the skills that are imparted to their students. Teachers should further develop practical wisdom by applying and practicing virtuous behavior in their own lives so they can become role models for their students.

We constantly make decisions in medical research. What is the quality and character of those decisions? What shapes the researcher's decision-making process? It requires wisdom to exercise discernment and make good decisions in research; in order to aquire wisdom, the researcher needs first to know what wisdom is, and then to diligently work towards fostering it.

7.4.1 Wisdom Is the Space Where Science and Religion Can Meet

Since antiquity, wisdom has been understood as a psychological trait and has been examined in many humanities disciplines. In modern times, wisdom has been variously described as involving social attitudes or behaviors, pragmatic knowledge of life, emotional homeostasis, reflection or self-understanding, value relativism and tolerance, and the acknowledgment of uncertainty with the ability to cope with it effectively. Although broadly defined, wisdom is consistently recognized as including thoughtful decision-making, compassion, altruism, and insight (Meeks and Jeste 2009).

In ancient philosophy, wisdom was considered attainable in the context of practicing *askezis,* which implied exercises of self-control, self-discipline, self-denial, and meditation guided by a mentor.

Many of the world's religions emphasize that adhering to a spiritual path increases emotional maturity. It brings hope of a transcendent wisdom that is accompanied by a heightened consciousness, a sense of inner silence, joy, gratitude, and morality. Both Western and Eastern religions teach that in order to develop wisdom, one needs to practice constraint, disciplined morality, charitable concern, and compassionate actions (Meeks and Jeste 2009).

According to Pavel Florensky (1882−1937), a Russian Orthodox theologian and thinker, wisdom is divine and is a direct representation of God's creative love. "Sophia is the original nature of creation, God's creative love… For everything exists truly only insofar as it communes with the God of Love, the Source of being and truth" (Florensky 1997). As a result of this act of love, wisdom is imprinted in each one of us, "in our hearts by the Holy Spirit who is given unto us" (Romans 5:5, New King James Version).

The art of medical research is to integrate science with wisdom for the benefit of patients. In a clinical setting, the purpose may not always be to cure, but to help the patient throughout the process, while instilling a realistic hope or peace in individual suffering. Very often, medical science cannot provide that hope to the patient. Therefore, medical disciplines serve a greater purpose: to understand the needs and sufferings of patients and address these problems through medical practice and research. We suggest that the training of practitioners and researchers needs to challenge the development of the Self as a prerequisite of developing the sensitivity to others.

7.5   A Crystallizing Example

As we have noted, religion and spirituality may enter the medical research sphere on a personal level, where researchers see their work as part of a Divine calling. The individual may view medical research as part of an obligation to serve humanity and thereby indirectly attain Divine grace. A researcher may decide to work toward ameliorating a particular disease, like diabetes, because it ravages a large swath of humanity – or works to serve an underprivileged community with worse health outcomes in order to bring about health equality. A spiritual researcher may even engage to serve his religious community by investigating health conditions that are more prevalent among a population within his or her own religious community. These sorts of research endeavors encompass a drive to bring about maximal scientific benefit, but also contain an element of Divine service.

A medical researcher who is motivated by religious concerns may also strive to focus his activities upon a more vertical plane, whereby he works to develop within himself the sort of moral character that is pleasing to God and other humans. For example, he may practice prayers that help him stay in connection with God and through that connection instill virtuous behaviors – thereby becoming a better servant of God and as well the community that he serves.

These two dimensions (*horizontal* and *vertical*) involve the formation of the researcher and are often pursued simultaneously; as such, there may be places for overlap. A case example is a researcher who employs his skills to introduce virtues into the medical education curricula, affecting physician behaviors and patient care in the process. Another example is a researcher who examines the ways in which religion impacts patient behaviors and works to develop healthcare systems that serve the religious and spiritual needs of patients, clinicians, and researchers alike.

The content presented in this chapter may enable a future researcher to render a deeper consideration of the idea that a fundamental relationship exists between the rigorous scientific growth and spirituality, and to encourage further reflection and meditation on the universal and particular aspects of this perspective.

Acknowledgment

We thank Dr. Farr Curlin at Duke University for reviewing an earlier version of the manuscript and making many helpful suggestions.

References

Alfeyev, H. (2000). *The Spiritual World of Isaac the Syrian*, 35–43. Collegeville: Liturgical Press.

Bryan, S.C. and Babelay, A. (2009). Building character: a model for reflective practice. *Academic Medicine* 84 (9): 1283–1288.

Roberts, A. and Donaldson, J. (eds.) (1885). *Ante-Nicene Fathers: Translations of the Writings of the Fathers Down to A.D. 325*, vol. 2, 299–568. Buffalo: Christian Literature Publishing.

Collins, F. (2007). Collins: Why this scientist believes in God. *CNN*, [online]. Available at: www.cnn.com/2007/US/04/03/collins.commentary/index.html [Accessed 9/21/18].

Flannelly, K.J., Weaver, A.J., and Costa, K.G. (2004). A systematic review of religion and spirituality in three palliative care journals, 1990–1999. *Journal of Palliative Care* 20 (1): 50–56.

Florensky, P. (1997). *The Pillar and Ground of the Truth*, 233–237. Princeton: Princeton University Press.

Hammalis, T.P. (2013) *Greek philosophy as a way of life: historical shifts and contemporary implications*. Hellenic Link-Midwest Meeting, 4–10 December, Chicago, Illinois.

Schaff, P. and Wace, H. (eds.) (1899). *Nicene and Post-Nicene Fathers*, Series II, vol. 9. Buffalo: Christian Literature Publishing.

Kinghorn, A.W., McEvoy, D.M., Michel, A. et al. (2007). Professionalism in modern medicine: does the emperor have any clothes? *Academic Medicine* 82 (1): 40–45.

Larchet, J.C. (2011). *Mental Disorders and Spiritual Healing*, 27–33. Hillsdale: Angelico Press.

MacIntyre, A. (1984). *After Virtue: A Study in Moral Theory*, 187–196. Notre Dame: University of Notre Dame Press.

Meeks, W.T. and Jeste, J.D. (2009). Neurobiology of wisdom. A literature overview. *Archives of General Psychiatry* 66 (4): 355–359.

Palmer, G., Sherrard, P., and Ware, K. (1981). *The Philokalia*, vol. 2, 337–338. New York: Faber and Faber.

Palmer, G., Sherrard, P., and Ware, K. (1983). *The Philokalia*, vol. 3, 163–169. New York: Faber and Faber.

Popovich, J. (1994). *Orthodox Faith and Life in Christ*, 161–162. Belmont: Institute for Modern Greek Studies.

Vitz, R. (2014) *Situationism, skepticism, and asceticism*. Annual Meeting of the Society of Orthodox Philosophy in America, 11–16 February, Houston, Texas.

8

Publishing in Academic Medicine: Does It Translate into a Successful Career?

Bradley K. Weiner[1], Paige Vargo[2] and Joseph Fernandez[3]

[1] *The Methodist Hospital Research Institute, Weill Cornell Medical College, Houston Methodist Hospital, Houston, TX, USA*
[2] *Department of Internal Medicine, Northeast Ohio Medical University, Rootstown, OH, USA*
[3] *Houston Methodist Hospital, Department of Surgery, Houston, TX, USA*

8.1 Introduction

The "outcomes" movement in healthcare and the subsequent emergence of evidence-based medicine over the past 30 years have been predicated upon a single question: *Why are we doing what we're doing?* It started as a general inquiry and evolved to include a more specific line of questioning. Why are we ordering this test? Does it provide the information we need to guide our therapies? Is it sensitive? Specific? Accurate? Have a good positive and negative predictive value? Beyond diagnostics, we began to further question our therapeutic approaches. Why are we treating this disease this way? What are the patients' real outcomes? Are there randomized trials to support what we are doing?

Thus, a shift has occurred in our approach to practicing medicine from doing what we have been taught by mentors (directly or via their textbooks) or what personal experience has taught us, to doing what we assess may provide the most desirable outcome. Surprisingly, a similar perspectival shift has *not* occurred inside the world of academic medicine, and specifically with regard to the publication of scientific articles. While there is plenty of literature available to support academic physicians on the nuts and bolts of getting their work published, very little addresses the philosophical question of why we do what we do. What are the motivations for our publishing in the medical scientific literature? And what are the potential outcomes?

In this chapter, we will explore why publishing our observations and findings is important and how the act of publishing can impact both our audience or community and our careers.

8.2 Dissemination of Knowledge and Impacting Patient Care

A key component of evidence-based medicine is that the evidence itself is to be found primarily, if not exclusively, in medical literature. What physicians are taught by their mentors and what they experience during their practices, while important, are

A Guide to the Scientific Career: Virtues, Communication, Research, and Academic Writing, First Edition.
Edited by Mohammadali M. Shoja, Anastasia Arynchyna, Marios Loukas, Anthony V. D'Antoni, Sandra M. Buerger, Marion Karl and R. Shane Tubbs.

Table 8.1 Karl Popper's pluralist concept of three interacting worlds or sub-universes.

	Composition	Example
World 1	Physical objects and entities	Nonliving physical and living, biological objects
World 2	Mental or psychological entities	Thoughts, decisions, perceptions, and observations
World 3	Objective knowledge (the products of the human mind)	Myths and scientific conjectures or theories

considered lesser forms of evidence informing the medical decision-making process. The trump card of evidence-based research is the published results of quality cohort studies, randomized trials, and systematic reviews.

One of the major reasons why we do what we do – why we write with the goal of publication – is to impact patient care. If the results of what one is publishing are to form the foundation for decision-making by other physicians, whether locally, nationally, or around the world, then one is able to impact individual patients whom they have never met through their published writings. Therefore, publishing offers the tremendous opportunity – and at the same time the responsibility – to do good beyond the confines of personal location, time, and energy.

The philosopher Karl Popper (1978) described "world 3" as manmade ontological entities that take on a life of their own with the ability to impact the world for better or worse (Table 8.1); the characteristic of world 3 objects or entities is that they *can be improved by criticism* and they may *stimulate people to think*. Popper maintained that "My thesis was that world 3 objects such as theories play a tremendous role in changing our world 1 environment and that, because of their indirect causal influence upon material world 1 objects, we should regard world 3 objects as real." A published, scientific article is one such entity. An article that is of high-quality and high-impact can change the world, even if it is a simple case series. For example, consider the impact of an article identifying the antibiotics for bacterial meningitis, or the first report of reverse transcriptase blockers for HIV. In fact, the ability to impact a population requires dissemination of knowledge in the form of publication. If one feels they can help people beyond their own limits, they should write. Even in the publishing world, it is a good reminder that altruism and obligation are inextricably tied.

8.3 Becoming a Recognized Expert

While the vast majority of scientific publications may be thought of as research projects put onto paper – one-and-done documents of clinical observations or therapeutic recommendations – many truly motivated academicians think of a particular article they have authored as merely a chapter in an ongoing *program*. It is a mantra of our research institute and laboratories; long before the first experiment is conducted, before clinical research is gathered, the academic physician thinks of the *program*, as opposed to the *project*. A *program*, in this sense, is a *pathway* of investigation that may take many years to develop, with the end result being a direct and substantial impact on patient care.

Publication serves as a public document of the pathway undertaken; a story that can be read by others that establishes the authors as experts. The audience may include a very important constituency: those who control the flow of money to support promising medical initiatives, those who review and assign grants (federal, foundations, and the like), and those who will invest entrepreneurially (venture capitalists, pharmaceuticals, and device companies). As such, publication of medical and scientific literature has the secondary effect of providing the financial support for further research along the pathway, an absolute requirement if novel ideas or observations are going to be properly explored and elucidated to impact clinical practices and hence patients.

8.4 Academic Promotion

The act of publishing medical or scientific literature, to a point, serves as passage to academic promotion. There are only a few such opportunities for advancement available to those in academic medicine. Teaching is one means of promotion for the academic physician, but it can be hard to measure and even harder to observe, as teaching moments often happen within the clinic or operating room. Thus, while there may be superstar educators who might achieve promotion through teaching or service alone, the majority of academicians will ascend via publication. Grants are, of course, a valuable means to promotion, but as we outlined above, important grants are almost always predicated on a solid foundation of a publication pathway.

Notice that we did not say *research* but *publication*. It may be said that in the eyes of the ambitious academic world, research that does not lead to publication or patents did not happen. Presentations and abstracts are often not considered as reliable and effective as a published, peer-reviewed article. Thus, the most achievable road to academic advancement is paved primarily by scholarly publication.

8.5 Professional Standing

Respect within a professional or academic community, nationally and internationally, for a physician is often defined by his or her publication record. Publishing quality articles in scientific journals most often requires sound methodology (academic or clinical) coupled with excellent writing skills. These two commonly underrated factors will allow the physician-author to stand out amongst his or her peers. Consistently well-written, well-constructed content will garner the author respect at the local, national, and international level. Much comes with that, including professional advancement and recognition – being deemed an expert within the greater medical or research community.

8.6 Personal Satisfaction

There are, of course, some simpler pleasures associated with academic publication. The "byline" phenomenon means something on the local level, a source of pride for the academic department, institution, and so on. But it also means something closer to home.

Speaking from the most personal level, we delight when our children or our friends – on their own – have searched us on Google Scholar or PubMed and shared their admiration of our accomplishments. In a romantic sense, publication is our legacy, both professional and personal.

8.7   Editorial Benefits

Just as lifting weights is a form of training that directly enhances physical athletic performance and reading great novels helps the novelist improve his own works, the act of publishing medical and scientific content allows one to significantly improve their reading and critical thinking skills. As a published author, the practice of being self-critical when it comes to writing better prepares one to critically assess other scientific literature and to distinguish the "wheat from the chaff." It also allows one to develop improved editorial skills, for example, as a reviewer or editor for a scientific journal, advising other authors on how to improve their writings. Without a developed critical eye and practiced editorial skills, many great scientific papers of the past might have failed to make the impact necessary to directly improve patient care and our community today. How it is written does matter, and those who write well can help others get there, too.

8.8   Professional Contacts

The publication of an article can often open the door to unexpected opportunities. Another researcher with similar interests contacts the author and collaboration is born. It may also lead to consultant positions in government or industry, or in a company or venture capitalist interest. A well-written paper can be career-, even life-changing for the author as well as the audience. The professional networking potential presented by publication is not uncommon, and the impact of scholarly publication extends far beyond the confines of academics to all facets of society.

8.9   Summary: Does Publishing Scholarly Articles Translate into a Successful Scientific Career?

For many residents, young physicians, surgeons, and scientists, the answer to "why are we doing what we're doing?" as it pertains to publication is simply because it is what we do. It is just part of the game, in a sense. However, when we carefully reflect on the consequences of publication, we realize that it is *not* a game but rather, it is serious business. First, the author can directly impact patient care far beyond his or her local confines by enhancing the evidence-base, which serves as a guide for the proper care of patients and development of new technologies. Second, a well written, well-received paper can blossom from a one-and-done project to a flourishing program. Third, publications can serve as a foundation for academic promotion, enhanced professional standing, and personal satisfaction. Fourth, by writing one becomes a better reader and critical thinker. And fifth, important publications can create a network of professional contacts leading the author down new, often unanticipated paths, with unexpected and extra-academic implications. Therefore, it is without doubt that academic writing and publishing can directly translate into a successful career. Young physicians, surgeons, and scientists wishing to advance in their fields should be advised to ardently hone their writing skills in the greater pursuit of publication. Do it and do it well.

Reference

Popper K. Three Worlds–The Tanner Lecture on Human Values, Delivered at The University of Michigan April 7, 1978. Available at: http://tannerlectures.utah.edu.

9

Assessing a Researcher's Scientific Productivity and Scholarly Impact

John Panaretos[1] *and Chrisovalantis Malesios*[2]

[1] Department of Statistics, Athens University of Economics and Business, Athens, Greece
[2] Department of Environment, University of the Aegean, Mytilene, Greece

9.1 Introduction

Evaluating the scientific performance of researchers has always been a beneficial yet difficult task. Over the last 20 years, a steep increase in the number of scientific journals and publications has necessitated useful metrics to accurately capture the scientific productivity of the researchers. These metrics are used to quantify both the individual levels of research activity as well as the researcher's overall impact on the scientific community. A simple approach to measuring the specific scientific impact is to evaluate the number of articles published by a researcher or an institution and the consequent number of citations. However, these numbers alone fail to capture the manifold aspects of a researcher's scientific record and impact. Unfortunately, because of their simplicity, these unidimensional indices are used constantly (and sometimes misused) by administrators to make critical decisions.

More and more we see attempts to provide rankings of researchers, universities, academic departments and programs, and institutions in general. Policy makers all over the world make frequent references to the Academic Ranking of World Universities (published by the Shanghai Jiao Tong University, China),[1] the THES-QS World University Rankings (published by the Times Higher Education Supplement and Quacquarelli Symonds),[2] the Webometrics Ranking of World Universities (produced by the Cybermetrics Lab [CINDOC], a unit of the National Research Council [CSIC]),[3] and Professional Ranking of World Universities (established by the École Nationale Supérieure des Mines de Paris in 2007),[4] among others. Recently, the European Union (EU) established its own rankings of research institutions[5] and universities.[6] The need to assess research performance and its impact is rapidly expanding. In the United Kingdom, a new initiative was introduced to develop metrics that evaluate the success of research

1 Available from http://ed.sjtu.edu.cn/ranking.htm.
2 Available from http://www.paked.net/higher_education/rankings/rankings.htm.
3 Available from www.webometrics.info.
4 Available from http://www.mines-paristech.eu/About-us/Rankings/professional-ranking.
5 Available from http://www.researchranking.org/index.php.
6 Available from www.umultirank.org.

A Guide to the Scientific Career: Virtues, Communication, Research, and Academic Writing, First Edition.
Edited by Mohammadali M. Shoja, Anastasia Arynchyna, Marios Loukas, Anthony V. D'Antoni, Sandra M. Buerger, Marion Karl and R. Shane Tubbs.
© 2020 John Wiley & Sons, Inc. Published 2020 by John Wiley & Sons, Inc.

organizations for accountability purposes (Department for Business, Innovation and Skills 2014).

Rankings of institutions provide important information for interested students, funding agencies, and even university administrators (e.g. in attracting potential faculty). These rankings, however, have also generated concern. Criticism is mostly due to the lack of a common, universal authority, and a consistent methodology used to establish the rankings (Van Parijs 2009). The static nature of the rankings (because of an institution's relatively steady staffing profile) is also concerning (Panaretos and Malesios 2012). The aforementioned rankings, which are conducted annually and have global reach, are not concentrated solely on the research quality of the institutions. A broad number of indicators not directly associated with the research are additionally considered, including the student/faculty ratio and the percentage of employed graduates.

In 2005, Hirsch proposed a metric based on number of articles published by a researcher and the citations received by them. This metric is now called the *h*-index, and today is the choice single metric for assessing and validating publication/citation output of researchers. The *h*-index can also be applied to any publication set, which includes the collective publications of institutions, departments, journals, and more (Schubert 2007). Following the introduction of the *h*-index in bibliometrics (the statistical analysis of written publications), numerous articles and reports have appeared either proposing modifications of the *h*-index or examining its properties and theoretical background.

The enormous impact of the *h*-index on scientometric analysis (the study of measuring and analyzing science) is illustrated by Prathap (2010), who argues that the history of bibliometrics can be divided into a pre-Hirsch and a post-Hirsch period. Between 2005 and 2010, there were approximately 200 papers published on the subject (Norris and Oppenheim 2010). Since then, applications of the *h*-index go well beyond the bibliometric field, ranging from assessing the relative impact of various human diseases and pathogens (McIntyre et al. 2011) to evaluating top content creators on YouTube (Hovden 2013). There is even a website available where you can obtain a prediction of your own personal *h*-index between 1 and 10 years in the future based on regression modeling (see Acuna et al. 2012).[7] However, such predictive models have been the subject of criticism (Penner et al. 2013).[8]

9.2   The *h*-Index

The *h*-index (a.k.a., the Hirsch index or the Hirsch number) was originally proposed by Hirsch as a tool for determining theoretical physicists' relative academic productivity. Since its inception, this index has attracted the attention of the scientific community for assessing the scientific performance of a researcher based on bibliometric data. Prior to widespread use of the *h*-index, the individual scientific performance was assessed using unidimensional metrics, such as the number of articles published, the number of citations received by the published articles, or the average number of citations per article. The *h*-index has a bidimensional nature, simultaneously taking into account both

---

7 Available from http://klab.smpp.northwestern.edu/h-index.html.
8 For an extensive and critical review of the *h*-index and other similar indices see Panaretos and Malesios (2009).

the quality and quantity of scientific output, because it is based on an aggregate set of the researcher's most cited papers along with the associated citations received by those publications. The *h*-index can also be applied to quantify the productivity and impact of a group of researchers belonging to a department, university, or country.

Among the advantages of the *h*-index is its simplicity and ease of calculation. It aims at reflecting high-quality works, as it combines both citation impact (citations received by the papers) with publication activity (number of papers published). The *h*-index is not influenced by a single, successful paper that has received many citations. Nor is the *h*-index sensitive to less frequently cited publications. Furthermore, increasing the number of publications will not necessarily affect the *h*-index. By definition of Hirsch (2005), "A scientist has index *h* if *h* of his *N* papers have at least *h* citations each, and the other $(N - h)$ papers have at most *h* citations each."

9.3 Criticisms of the *h*-Index

Besides the popularity of the *h*-index, some criticism has been drawn and an enormous number of modifications and extensions of the *h*-index have since appeared (Costas and Franssen 2018; Meho 2007; Schreiber 2007; Vinkler 2007; Adler et al. 2008; Schreiber et al. 2012; Waltman and van Eck 2012). The *h*-index is not as objective as the research community would like it to be. By definition, it is biased in favor of mature researchers over younger researchers. A mature researcher with moderate research impact is expected to have a higher *h*-index than a young researcher at the beginning of his or her career, even if the latter eventually develops into a researcher with a higher impact factor.

There are a number of other situations mentioned in the literature where the *h*-index may provide misleading information about a researcher's impact and productivity. For example, the lack of sensitivity of the *h*-index to the excess citations of the *h*-core papers (the set of papers whose citations contribute toward *h-index*) is a frequently noted disadvantage (Egghe 2006a,b,c; Kosmulski 2007). The *h*-index does not take into account important factors that differentiate the ways research activity develops and is transferred, such as the distinction between research fields and specialties (van Leeuwen 2008). For example, an *h*-index of 20 for an applied physicist would be a fair score, whereas the same figure would be wishful thinking for a theoretical mathematician.

Abramo et al. (2013) reveal yet another example of the problematic use of the *h*-index for measuring research performance of institutions. The most profound argument against using the *h*-index for ranking larger bodies (such as institutions, departments, etc.) is the influence of faculty size in calculating the *h*-index value. Because the organizations are comprised of greatly varying numbers of faculty and research staff, the *h*-index value is significantly affected. Thus, various modifications and extensions of the *h*-index have appeared in literature starting almost immediately after its introduction.

9.4 Modifications and Extensions of the *h*-Index

In an effort to address the shortcomings of the *h*-index, several modifications and extensions of this index have been proposed. At least 37 *h*-index variants (a.k.a. *h*-type indicators or *h*-index related indices) are found in literature (see Panaretos and Malesios 2009;

Bornmann et al. 2011; Schreiber et al. 2012; Zhang 2013). Of course, various authorities favor certain indicators over others. Each of these indicators are intended to address one or more of the limitations presented by the original *h*-index. For example, some of the *h*-index variants are no longer robust to the number of excess citations of the highly cited articles in the *h*-core (Jin 2006), meaning that excess citations can actually have an affect on the ranking. Other *h*-related indices weigh the paper's contribution toward the index based on the number of authors (Schreiber 2008a). In a multilevel meta-analysis, Bornmann et al. (2011) noted that "some *h*-index variants have a relatively low correlation with the *h*-index" and "can make a non-redundant contribution to the *h*-index." These variants are primarily included in the modified impact index (Sypsa and Hatzakis 2009) and the *m*-index (Bornmann et al. 2008).[9]

Table 9.1 Describing some of the popular *h*-type indicators. Among the vast literature on *h*-index variants, we single out the *g*-index, the *A*-index, the *R*-index, the h_w-index, and the h_m-index. Alonso et al. (2010) proposed the use of the geometric mean of the *h*- and *g*-indices, which they called the *hg*-index, as a remedy to the high sensitivity of the *g*-index to single highly cited papers. Despite the fact that *h*-index variants fix some of the problems of the *h*-index (e.g. the problem of the *h*-index being robust to the number of citations of the *h*-core's highly cited articles), there are situations where the features of the new index itself constitute a drawback. For example, some *h*-index variants are influenced by the presence of one highly cited paper (Alonso et al. 2010; Costas and Bordons 2008). The *g*-index specifically is limited by its extreme sensitivity to highly cited papers in a scientist's portfolio (Costas and Bordons 2008).

Attempts have been made to classify all of the proposed indices. Schreiber et al. (2012) suggested a classification system based on two dimensions of scientific performance: quality and quantity. This classification system considers that some indices have a stronger tendency than the rest to measure the quantity of research output, while other indices (*g*-, *A*-, and *R*-indices) have a stronger tendency to characterize the quality of research output. It is suggested that two complementary indices, for example, *h*- and *A*-indices, should be used in the scientometric analysis of an individual researcher's productivity and impact (Schreiber et al. 2011).

9.5 A General Criticism on the Use of Metrics

Although the use of single metrics (based on bibliometric measurements) for the comparison of researchers has steadily gained popularity in recent years, there is an ongoing debate regarding the appropriateness of this practice. The question is whether single measures of research performance are sufficient to quantify such complex activities. A report by the Joint Committee on Quantitative Assessment of Research argues strongly against the use of citation metrics alone as a tool in the field of mathematics. Rather, the committee encourages the use of more complex methods for judging the impact of researchers; for example, using evaluation criteria that combines citation metrics with other relevant determinants, including membership on editorial boards, awards, invitations, or peer-review activities (Adler et al. 2008). Along these lines, Egghe (2007) stated

---

9 For an up-to-date description on the *h*-index and some of its most important variants, see Schreiber et al. (2012).

Table 9.1 List of h-type indices and their descriptions.

Indicators	Definition/significance	References
w-index	The highest number *w* of articles that each received 10*w* or more citations.	Wu (2010)
h2-index	The highest number $h(2)$ of articles that received at least $[h(2)]^2$ citations.	Kosmulski (2006)
h-index	The highest number *h* of articles that each received *h* or more citations.	Hirsch (2005)
f-index	The highest number of articles that received *f* or more citations on average, where the average is calculated as the harmonic mean.	Tol (2009)
t-index	The highest number of articles that received *t* or more citations on average, where the average is calculated as the geometric mean.	Tol (2009)
ħ-index	The square root of half of the total number of citations.	Miller (2006)
s-index	Measures the deviation from a uniform citation record.	Silagadze (2010)
h_T-index	The sum of weights $w(i, r) = \begin{cases} (2i-1)^{-1}, & r \le i \\ (2r-1)^{-1}, & r \ge i \end{cases}$ of the *i*th citation to the *r*th paper.	Anderson et al. (2008)
x-index	Maximum of the product of rank and citation frequency.	Kosmulski (2007)
A-index	Average number of citations received by the articles in the *h*-core.	Jin (2006)
g-index	The highest number *g* of articles that together received g^2 or more citations.	Egghe (2006c)
m-index	The median number of citations received by the articles in the *h*-core.	Bornmann et al. (2008)
h_w-index	The square root of the total number S_w of citations received by the highest number of articles that each received S_w/h or more citations.	Egghe and Rousseau (2008)
R-index	The square root of the total number of citations received by the articles in the *h*-core.	Jin et al. (2007)
π-index	The one-hundredth of the total number of citations received by top square root of the total number of papers ("elite set of papers").	Vinkler (2009)
e-index	Reflects excess citations of the *h*-core that are ignored by the *h*-index.	Zhang (2009)
hg-index	The geometric mean of the *h*- and *g*-indices. $hg\text{-index} = \sqrt{h \cdot g}$	Alonso et al. (2010)

earlier, "The reality is that as time passes, it's not going to be possible to measure an author's performance using just one tool. A range of indices is needed that together will produce a highly accurate evaluation of an author's impact." Shortly thereafter, Bollen et al. (2009) offered empirical verification of Egghe's intuitive hypothesis. Based on the results of a principal component analysis (PCA) on a total number of 39 existing indicators of scientific impact, an argument was made that ideal scientific impact should be multidimensional and cannot be effectively measured by a single numeric indicator.

9.6   Citation Data Sources

There is an ongoing debate on the issue of multiple citation data sources (Jacso 2008). Citation data sources, or databases, are web-based data sources that can be accessed freely or through a subscription cost. These databases provide the meta-data of scientific publications and their citation information. The debate mainly concerns the fact that the various available data sources are likely to produce different citation data for the same publication. In fact, a comparison has been made to test the robustness of citation outputs from Thomson Reuters Web of Science (WoS) – formerly known as Thomson Corporation's Institute for Scientific Information (ISI) Web of Knowledge – and Google Scholar. The latter uses Publish or Perish, a software program that retrieves citations from Google Scholar and analyses them to present various metrics, a common application in bibliometric literature.[10] The WoS results tend to underestimate the citations because WoS covers solely journals included in the ISI list. Google Scholar, on the other hand, tends to overestimate the citations because in addition to covering more journals, it also retrieves citations to working papers, books, and more (Falagas et al. 2007; Meho and Yang 2007; Jacso 2008; Franceschet 2010). Although several groups have supported the idea of using Google Scholar to implement citation-based statistics, many scientific authorities claim that the Google Scholar data are often inaccurate (Adler et al. 2008). Meho and Rogers (2008) further examined the differences between Scopus and WoS, and reported that no significant differences exist between the two databases if only journal citations are compared. Nevertheless, we have to stress that the specific database used for collecting bibliometric citation data can, in fact, influence scientometric analysis. Many authors have cautioned against the use of citation data without further evaluating the database for validity and verification (Dodson 2009).

9.7   Discussion

An important consideration in evaluating academic performance and impact of a researcher is the manifest aspects of scientific work. As many authors have argued, the use of indices to assess only one component of a researcher's work, like citations, is unfair (Kelly and Jennions 2006; Adler et al. 2008; Sanderson 2008). Measuring the scientific performance of a researcher by using only bibliometric data is already more or less restrictive by default, as is measuring the citation performance with only a single one of the metrics previously described. The idea that using only one or two

---

10 Publish or Perish User's Manual (2008) is available at http://www.harzing.com/resources.htm#/pop.htm.

indicators may be adequate for assessing research performance has been increasingly criticized. The h-index and the h-type indicators still have strong intrinsic problems and limitations that make them unsuitable as "unique" indicators for this purpose (see Costas and Bordons 2008). Other problems such as an age bias, dependence on the research field (van Leeuwen 2008), or the influence of self-citations on these indices make their use questionable in principle (Schreiber 2008b). The h-index has also been criticized on a theoretical basis as well (Waltman and van Eck 2012).

Where do we currently stand in terms of single metrics? Is their standard application effective for evaluation purposes? The common use of bibliometric indicators is currently illustrated by the incorporation of the h-index (and in some instances related indices) in bibliometric databases such as the WoS and Scopus (van Eck and Waltman 2008). In addition, use of search engines by university administrators to determine the h-index of their faculty is increasingly popular. For example, the h-index is frequently used for departmental reports or advancement to tenure (Dodson 2009), signifying that at least this indicator is here to stay. However, caution should be exercised, mainly because of discrepancies between the various search engines.

Unfortunately, even faculty members can fall into this trap. In Greece, where the research performances of the faculty were recently evaluated using their h-indices, the discrepancy between h-indices retrieved from various citation search engines was huge, raising serious doubt as to the validity and usefulness of the evaluation method (Hellenic Quality Assurance and Accreditation Agency 2012).

Of course, it is convenient to measure the research activity and impact in terms of a single metric, and perhaps this is why the h-index has achieved almost universal acceptance within the scientific community and by academic administrators. Most of us want to see where we stand in relation to the rest of scientific community in our research field; however, it is widely acknowledged that this is an oversimplification. The advent of the h-index was accompanied by primarily positive but also some negative reception. Among the positive outcomes was the revival of research assessment practices and the development of valid tools to support them. On the negative side, *the $h*-index strongly influenced promotions and grants, which has led many researchers to be preoccupied with the h-index beyond its true value. This explains why extensive efforts are underway to come up with a better index.

If we require indices or metrics to reliably assess research performance, we should look at more than one indicator. Efforts toward defining additional measures should continue, and these measures should provide a more general picture of the researcher's activity by examining different aspects of academic contribution. For example, a measure should be found that reflects and recognizes the overall lifetime achievements of a researcher, and favors those with a broader citation output compared to others who have published one or two papers with a large number of citations. A new index favoring researchers who constantly publish interesting papers would also be beneficial, in comparison to scientists that have only published a few highly cited articles. In other words, emphasis on consistency over a single outbreak is preferable. Another useful development would be constructing a measure that can be utilized for interdisciplinary comparisons (Malesios and Psarakis 2014). To date, there is no universal quantitative measure for evaluating a researcher's academic stance and impact. The initial proposal for incorporating the h-index for this purpose – although it was adopted with great enthusiasm – has proven in practice to suffer from several shortcomings.

References

Abramo, G., D'angelo, C.A., and Viel, F. (2013). The suitability of *h* and *g* indexes for measuring the research performance of institutions. *Scientometrics* 97 (3): 555–570.

Acuna, D.E., Allesina, S., and Kording, K.P. (2012). Future impact: predicting scientific success. *Nature* 489: 201–202.

Adler, R., Ewing, J. and Taylor, P. (2008). Citation statistics. Joint IMU/ICIAM/IMS-Committee on Quantitative Assessment of Research. [WWW] Available at: http://www.mathunion.org/fileadmin/IMU/Report/CitationStatistics.pdf. [12/14/2017]

Alonso, S., Cabrerizo, F.J., Herrera-Viedma, E. et al. (2010). *hg*-index: a new index to characterize the scientific output of researchers based on the *h*- and *g*-indices. *Scientometrics* 82: 391–400.

Anderson, T.R., Hankin, R.K.S., and Killworth, P.D. (2008). Beyond the Durfee square: enhancing the *h*-index to score total publication output. *Scientometrics* 76: 577–588.

Bollen, J., Van de Sompel, H., Hagberg, A. et al. (2009). A principal component analysis of 39 scientific impact measures. *PLoS One* 4 (6): e6022.

Bornmann, L., Mutz, R., and Daniel, H.-D. (2008). Are there better indices for evaluation purposes than the *h*-index? A comparison of nine different variants of the *h*-index using data from biomedicine. *Journal of the American Society for Information Science and Technology* 59 (5): 830–837.

Bornmann, L., Mutz, R., Hug, S.E. et al. (2011). A multilevel meta-analysis of studies reporting correlations between the *h*-index and 37 different *h*-index variants. *Journal of Informetrics* 5 (3): 346–359.

Costas, R. and Bordons, M. (2008). Is *g*-index better than *h*-index? An exploratory study at the individual level. *Scientometrics* 77 (2): 267–288.

Costas, R. and Franssen, T. (2018). Reflections around 'the cautionary use' of the h-index: response to Teixeira da Silva and Dobránszki. *Scientometrics* 115 (2): 1125–1130.

Department for Business, Innovation and Skills (BIS) (2014) *Triennial review of the research councils (BIS/14/746)* [WWW] UK Government. Available from: https://www.gov.uk/government/publications/triennial-review-of-the-research-councils [Accessed 4/19/2017]

Dodson, M.V. (2009). Citation analysis: maintenance of the *h*-index and use of *e*-index. *Biochemical and Biophysical Research Communications* 387 (4): 625–626.

Egghe, L. (2006a). How to improve the *h*-index. *The Scientist* 20 (3): 14.

Egghe, L. (2006b). An improvement of the *h*-index: the *g*-index. *ISSI Newsletter* 2 (1): 8–9.

Egghe, L. (2006c). Theory and practice of the *g*-index. *Scientometrics* 69 (1): 131–152.

Egghe, L. (2007). From *h* to *g*: the evolution of citation indices. *Research Trends* 1 (1): https://www.researchtrends.com/issue1-september-2007/from-h-to-g/.

Egghe, L. and Rousseau, R. (2008). An *h*-index weighted by citation impact. *Information Processing & Management* 44: 770–780.

Falagas, M.E., Pitsouni, E.I., Malietzis, G.A. et al. (2007). Comparison of PubMed, Scopus, Web of Science, and Google Scholar: strengths and weaknesses. *The FASEB Journal* 22 (2): 338–342.

Franceschet, M. (2010). A comparison of bibliometric indicators for computer science scholars and journals on Web of Science and Google Scholar. *Scientometrics* 83 (1): 243–258.

Hellenic Quality Assurance And Accreditation Agency (HQAA) (2012). Guidelines for the members of the external evaluation committee (Available at: http://www.hqaa.gr/data1/ Guidelines%20for%20External%20Evaluation.pdf)

Hirsch, J.E. (2005). An index to quantify an individual's scientific research output. *Proceedings of the National Academy of Sciences, USA* 102 (46): 16569–16572.

Hovden, R. (2013). Bibliometrics for internet media: applying the *h*-index to YouTube. *Journal of the American Society for Information Science and Technology* 64 (11): 2326.

Jacsó, P. (2008). The plausibility of computing the *h*-index of scholarly productivity and impact using reference-enhanced databases. *Online Information Review* 32 (2): 266–283.

Jin, B.-H. (2006). *H*-index: an evaluation indicator proposed by scientist. *Science Focus* 1 (1): 8–9.

Jin, B.-H., Liang, L., Rousseau, R. et al. (2007). The *R*- and *AR*- indices: complementing the *h*-index. *Chinese Science Bulletin* 52: 855–863.

Kelly, C.D. and Jennions, M.D. (2006). The *h*-index and career assessment by numbers. *Trends in Ecology and Evolution* 21 (4): 167–170.

Kosmulski, M. (2006). A new Hirsch-type index saves time and works equally well as the original *h*-index. *ISSI Newsletter* 2 (3): 4–6.

Kosmulski, M. (2007). MAXPROD: a new index for assessment of the scientific output of an individual, and a comparison with the *h*-index. *International Journal of Scientometrics, Informetrics and Bibliometrics* 11 (1): 5.

Malesios, C.C. and Psarakis, S. (2014). Comparison of the *h*-index for different fields of research using bootstrap methodology. *Quality & Quantity* 48 (1): 521–545.

McIntyre, K.M., Hawkes, I., Waret-Szkuta, A. et al. (2011). The *h*-index as a quantitative indicator of the relative impact of human diseases. *PLoS One* 6 (5): e19558.

Meho, L.I. (2007). The rise and rise of citation analysis. *Physics World* 29 (1): 32–36.

Meho, L. and Rogers, Y. (2008). Citation counting, citation ranking, and *h*-index of human–computer interaction researchers: a comparison of Scopus and Web of Science. *Journal of the American Society for Information Science and Technology* 59 (11): 1711–1726.

Meho, L.I. and Yang, K. (2007). Impact of data sources on citation counts and rankings of LIS faculty: Web of Science vs. Scopus and Google Scholar. *Journal of the American Society for Information Science and Technology* 58 (13): 2105–2125.

Miller, C.W. (2006). Superiority of the h-index over the Impact Factor for Physics, *arXiv: physics / 0608183*.

Norris, M. and Oppenheim, C. (2010). The *h*-index: a broad review of a new bibliometric indicator. *Journal of Documentation* 66 (5): 681–705.

Panaretos, J. and Malesios, C.C. (2009). Assessing scientific research performance and impact with single indices. *Scientometrics* 81 (3): 635–670.

Panaretos, J. and Malesios, C.C. (2012). Influential mathematicians: birth, education and affiliation. *Notices of the American Mathematical Society* 59 (2): 274–286.

Penner, O., Pan, R.K., Petersen, A.M. et al. (2013). On the predictability of future impact in science. *Scientific Reports* 3: 3052.

Prathap, G. (2010). Is there a place for a mock *h*-index? *Scientometrics* 84: 153–165.

Sanderson, M. (2008). Revisiting *h* measured on UK LIS academics. *Journal of the American Society for Information Science and Technology* 59 (7): 1184–1190.

Schreiber, M. (2007). Self-citation corrections for the Hirsch index. *EPL* 78: 30002, 1–6.

Schreiber, M. (2008a). A modification of the *h*-index: the h_m-index accounts for multi-authored manuscripts. *Journal of Informetrics* 2 (3): 211–216.

Schreiber, M. (2008b). An empirical investigation of the *g*-index for 26 physicists in comparison with the *h*-index, the *A*-index, and the *R*-index. *Journal of the American Society for Information Science and Technology* 59 (9): 1513–1522.

Schreiber, M., Malesios, C.C., and Psarakis, S. (2011). Categorizing Hirsch index variants. *Research Evaluation* 20 (5): 397–409.

Schreiber, M., Malesios, C.C., and Psarakis, S. (2012). Exploratory factor analysis for the Hirsch index, 17 *h*-type variants and some traditional bibliometric indicators. *Journal of Informetrics* 6: 347–358.

Schubert, A. (2007). Successive *h*-indices. *Scientometrics* 70 (1): 201–205.

Silagadze, Z.K. (2010). Citation entropy and research impact estimation. *Acta Physica Polonica B* 41: 2325–2333.

Sypsa, V. and Hatzakis, A. (2009). Assessing the impact of biomedical research in academic institutions of disparate sizes. *BMC Medical Research Methodology* 9: 33.

Tol, R.S.J. (2009). The *h*-index and its alternatives: an application to the 100 most prolific economists. *Scientometrics* 80 (2): 317–324.

Van Eck, N.L. and Waltman, L. (2008). Generalizing the *h*- and *g*-indices. *Journal of Informetrics* 2 (4): 263–271.

Van Leeuwen, T. (2008). Testing the validity of the Hirsch-index for research assessment purposes. *Research Evaluation* 17 (2): 157–160.

Van Parijs, P. (2009). European higher education under the spell of university rankings. *Ethical Perspectives* 16 (2): 189–206.

Vinkler, P. (2007). Eminence of scientists in the light of the *h*-index and other scientometric indicators. *Journal of Information Science* 33: 481–491.

Vinkler, P. (2009). The π-index: a new indicator for assessing scientific impact. *Journal of Information Science* 35: 602–612.

Waltman, L. and van Eck, N.J. (2012). The inconsistency of the *h*-index. *Journal of the American Society for Information and Technology* 63 (2): 406–415.

Wu, Q. (2010). The *w*-index: a measure to assess scientific impact by focusing on widely cited papers. *Journal of the American Society for Information Science and Technology* 61: 609–614.

Zhang, C.T. (2009). The *e*-index, complementing the *h*-index for excess citations. *PLoS One* 4: e5429.

Zhang, C.T. (2013). The *h*-index, effectively improving the *h*-index based on the citation distribution. *PLos One* 8 (4): e59912.

Further Reading

Bastian, S., Ippolito, J.A., Lopez, S.A. et al. (2017). The use of the *h*-index in academic orthopaedic surgery. *Journal of Bone & Joint Surgery* (American Volume) 99 (4): e14.

Bornmann, L., Mutz, R., Daniel, H.-D. et al. (2009). Are there really two types of *h*-index variants? A validation study by using molecular life sciences data. *Research Evaluation* 18 (3): 185–190.

Dunnick, N.R. (2017). The *h*-index in perspective. *Academic Radiology* 24 (2): 117–118.

Masic, I. (2016). Scientometric analysis: a technical need for medical science researchers either as authors or as peer reviewers. *Journal of Research in Pharmacy Practice* 5 (1): 1–6.

Saraykar, S., Saleh, A., and Selek, S. (2017). The association between NIMH funding and *h*-index in psychiatry. *Academic Psychiatry* https://doi.org/10.1007/s40596-016-0654-4.

Ruscio, J. (2016). Taking advantage of citation measures of scholarly impact: hip hip *h*-index! *Perspectives on Psychological Science* 11 (6): 905–908.

Section II

Communication

10

Manners in Academics
R. Shane Tubbs

Seattle Science Foundation, Seattle, WA, USA

10.1   General Aspects

Unfortunately, there is not a manual that one can consult on the topic of manners in an academic career. Although not totally inclusive, I have used the mantra of "Accommodate, Collaborate, Facilitate, and Communicate" in my career as an academician. On the surface, this may sound overly simplistic and flowery. However, to date, following this mantra has resulted in my continued enjoyment of (and employment in) academia and the recruitment of others into the field. This chapter will discuss manners in academics with a focus on academic writing and publishing.

10.1.1   Accommodate

Follow the Golden Rule.
   A career in academics should not be an isolated endeavor. Too many academics see themselves as living in an ivory tower and shun not only their colleagues but also others in the field. Treat others as you would want them to treat you. This should go beyond your peers and should also include anyone that you deal with on a day-to-day basis such as secretarial support, librarians, students, and research technologists.

10.1.2   Collaborate

Strengthening one's own academic career can be achieved through a comprehensive chain of collaborators. Such collaborations take time to establish but are well worth the investment. Collaborators not only enhance your own academic pursuits but also allow your work to be disseminated to those who may have otherwise not have appreciated it fully. Collaborators should be viewed as part of the team, and although not at your own institution, they can have a profound effect on the discipline in which you study.

10.1.3   Facilitate

By facilitating, I mean fully engaging and assisting academic colleagues. For example, if a peer requests a pdf of one of your publications, don't wait to reply but find the paper

A Guide to the Scientific Career: Virtues, Communication, Research, and Academic Writing, First Edition.
Edited by Mohammadali M. Shoja, Anastasia Arynchyna, Marios Loukas, Anthony V. D'Antoni, Sandra M. Buerger, Marion Karl and R. Shane Tubbs.
© 2020 John Wiley & Sons, Inc. Published 2020 by John Wiley & Sons, Inc.

and send it the same day, if possible. This not only assists the colleague but also shows that you are willing to go the extra mile for colleagues. In other words, don't let the sun set on requests from colleagues if at all possible. If the Golden Rule stands up, then when you need the same favor, it is more likely to be returned to you. Another example is reviewing papers or chapters for colleagues. Especially if you are a co-author, never hold up the colleague by procrastinating and not reviewing the document for them in a timely manner. These examples all boil down to greasing the academic wheel. This "wheel" not only moves you around in academic circles but also supports your colleagues who will, in turn, support you when the time comes.

10.1.4 Communicate

Communicating with your peers can be one of the largest hurdles for some in regards to manners in academics. Many researchers are often introverts, and this can lead to miscommunication and lack of communication between peers. Without constant and timely communication with your peers, such things as collaboration can come to a grinding halt. Therefore, I advise students and colleagues who are interested in optimizing their academics and, specific for this chapter, manners in academics to prioritize and maximize communications. This can be as simple as updating collaborators as to the timeline of a publication or keeping them abreast of abstracts when they are published.

Lastly, if a colleague calls or emails you, respond to them in as reasonable a time as possible. A rule that has yet to be well defined is how quickly an academic colleague should respond to missed phone calls or emails from other colleagues, students, or research assistants. My rule of thumb is to do so within one day of the initial contact. This is not always possible, such as when one is out of the country or at a meeting, but for me, it has resulted in a fruitful co-existence with colleagues. Collaborators will not only appreciate your timely response but will equate these to your enthusiasm for their contact and as a potential future collaborator. In other words, don't burn bridges by simply being inattentive to your peers and their phone calls/emails.

10.2 Manners in Academic Writing and Publishing

What are academic manners in writing and publishing? Do they exist in scholarly writing? Are manners in academic writing necessary? Many writers in academia may have never considered that there are unsaid "rules" of academic writing that not only establish a researcher/writer in their scientific community but also can help in propelling the writer in their field. Some aspects of being "polite" in academic writing may seem trivial but others are so important, that not following them may result in a writer being labeled as difficult. Such a label will not only potentially taint the writer's work but also the department/institution that they hail from. Herein, I will discuss several aspects of this topic that I believe, if followed, can help the academic writer become well known in their field and have their work more widely appreciated and potentially, published. The focus of this chapter will be on the academician and their writing, submitting and publishing in the peer review process. As a writer, author, reviewer, and editor, I believe that I have a unique perspective from which to address this topic.

10.2.1 Do Academic Manners in Writing Exist?

The simple answer to this question is yes. However, there is no standardized source that one is able to consult for rules or regulations. As there are no guidelines on this topic, many authors are simply clueless on how they should interact with publishers or editors. Most of the time, simple common sense for how to best interact with a journal or publisher will be sufficient in making the process of academic writing and publishing straightforward.

10.2.2 What Are Manners in Academic Writing?

Manners in academic writing exist as an unsaid code of conduct that if followed, will behoove the scholarly writer. Putting one's finger on exactly what these manners are can be difficult. I tell junior authors that they should treat the academic process of writing just as they would treat someone in person. Extending simple kindness and consideration to a journal's staff and editor is not meant to "grease the wheel" or ensure an acceptance but is rather a simple courteously. Don't be misled that such courtesies will guarantee acceptance of your paper or an expedited review process. Do not think that your submission is the only one that the editor is considering at that moment. Most editors will be involved with many other paper submissions at the same time as your paper is being evaluated. Therefore, never ask for or expect special attention.

10.2.3 Are Manners in Academic Writing Necessary?

Necessary is probably too strong for this answer, but academic manners in the writing process cannot hurt and if anything, especially for a prolific writer, will help. A prolific writer will become known by both the quality of their work and the attitude/style by which they submit their work and respond to the review process or other academic publishing questions. On the flip side, an unmannered writer who ignores common courtesies or comes across as rude and inflammatory will become known for such traits and, eventually, may become labeled a difficult author. This doesn't mean that this type of author may never be able to publish, but most academic specialties and their associated journals are small, and such author characteristics will soon become well known. Officially, this will not be a roadblock to publishing for such an author, but being human, unofficially, this will have ramifications for the author – and whether anyone wants to admit it or not, negatively perceived academic authors will have a tougher time publishing, at least with those who know of them from past experiences or through others.

10.2.4 Thanking the Editor and Reviewers

I begin every cover letter with a short thank-you to the editor for considering my paper. This should not be a long or overly sugar-coated. This is not intended to bribe or convince the editor to take a special look at your paper. It is, however, a simple academic manner that acknowledges that the editor is taking the time and resources to have your paper reviewed. Also, when a paper is returned to you after peer-review, do not hesitate to briefly thank the reviewers for their time in reading and critiquing your paper.

Even if the reviewer never sees this acknowledgment, the editor does and may appreciate that you are taking the time (or a least thinking about) the process that has been undertaken to evaluate your submission. As an editor, I can say that I always appreciate seeing such comments from the authors. As a writer, I make this a reflexive part of every submission, not as a token but as a sincere thanks to those who have taken time out of their day for the academic peer-review process. Most editors and certainly most reviewers do not receive any compensation for their time in evaluating academic submissions. Therefore, often the only acknowledgment of their contribution is a simple thanks from the authors!

10.2.5   Do Not Take Reviewer Comments Personally

I have all too often seen authors who, after reading suggestions/comments from reviewers about their paper, take the reviews to heart, i.e. personally. Some will develop a "conspiracy theory" response, which is very unhealthy and will potentially taint future submissions. Such a response can lead to nothing good. Reviewers are human and are often your peers. Sure, some reviewers can be quite forward and abrasive in their reviews, and some will be down right prickly. However, it is the job of the editor to take all reviews into consideration and mesh all reviewer comments into a single overall "review." If there are outliers in the process, the editor must appreciate these and use them sparingly in their final decision. With this said, the author should take nothing the reviewers say personally, even if it feels very personal. Most of the time, this is not the case and especially, when the peer review process is blinded. On rare occasions, and especially in smaller journals with a niche-type readership, the reviewer may "figure out" who the author is (if it is a blinded review), and in such cases, personal biases may come into play. However, this is usually not the case, and pointed or overly critical comments may just be the way a particular reviewer will critique your paper.

In the end, a negative response from the author may result in a reviewer taking it personally and being overly critical with their recommendation. Moreover, as an editor, author responses that come across as arrogant or hostile are interpreted as being unprofessional, even if they are founded on truth.

10.2.6   Try to Accommodate the Reviewer's Suggestions

I have often seen authors respond to a reviewer's comments in a negative way. This not only comes across as disrespectful but rude when read by the editor who, remember, is usually the one who makes the final decision on your paper. The reviewer may or may not see your response to their comments, depending or whether your revision is sent back out for re-review. I always try to respond to each comment from the reviewer as positively as possible. Now, comments like "This is the worst paper I have ever seen…" will be difficult to respond to in a positive way. In these situations, it is best to say less versus more. In these cases, let the editor be the judge of the reviewer's comments. Comments to the authors that are negative should be responded to briefly and honestly. If changing something in your paper based on the reviewer's comment will not take away from the essence of the study, then do it. If you cannot accommodate a reviewer's suggestion, then respond reasonably and without any negativity or sense that you are taking the response personally. A short and well-thought-out response is the best way to respond to a reviewer's comment that you might not agree with. Remember, if you can't say it nicely, then don't say it at all.

10.2.7 Respect to Editorial Staff

The editorial staff for a journal are the "movers and shakers" for moving papers through the review process. They not only verify basic things such as if your paper is formatted correctly but also check that all forms and other documents are in the correct order. You should regard the editorial staff of a journal as you would the administrative assistant to a CEO of a company that you would like to work for. Being polite and considerate of these staff goes a long way for the writer of an academic paper. Responding to their emails promptly and courteously is good form. The point of being kind to editorial staff is not necessarily that they will remember your future submissions but that they will NOT remember you in the future as a inconsiderate author!

One of the worst things an academic writer can do is to harass the editorial staff or worse, the editor who is handling their paper. Calling or emailing to check on the status of your submission on a frequent basis will not help you and could potentially hurt the process. You can imagine that an editorial staff member who is pestered by you could easily put your paper "at the end of the list." Most journals will give you a rough idea of the review process time and it is appropriate to check on the status of your paper if this time comes and goes. However, checking on the status prematurely is considered rude by most academic journals.

10.2.8 Respect Your Co-Authors

Courteousness to your co-authors not only ensures that they will want to work with you in the future but that they will also respect you more as an academic writer. The primary author of an academic paper should respect their co-authors and make sure that they are included in all decisions and correspondence for a paper. Allow your co-authors to see all responses to a reviewer response letter.

10.2.9 Respect to the Publisher

It is important to respond to publisher queries in a timely manner. This not only facilitates the speed at which your paper will be handled and eventually published, but it also keeps your name off of any "black list" that the publisher may have, mentally or physically. Unless your publication is seriously delayed, it is not a good idea to bother the publisher with "when will my work be published" questions. Not being listed as a tardy or difficult author may help those who wish to publish with the publisher in the future with, for example, a book.

10.2.10 Respect Authors Who Contribute

If you are editing an issue of a journal or book or simply writing a paper where authors are invited to contribute, it is important to maintain good communication. A simple "thank you" goes a long way and should not be used too sparingly. Remember, most invited authors will receive nothing in return for their contributions and give of their time for no other purpose than to support academia.

10.2.11 Academic Manners as a Reviewer

Whether you are reviewing for a journal or book proposal, there are some loose guidelines that, although not "rules," can make your efforts more useful to the authors and publisher that you are working for.

As a reviewer, you hold a unique position that carries with it a lot of weight in that your comments can make or break a submission or proposal. Although there are usually multiple reviewers for say a journal article, a single reviewer can do much damage to a submission and the authors' feelings. Most authors believe that their academic submission is the best the world has ever seen. Therefore, there are often statements in the paper that may come across as grandiose. I think that a good reviewer will ignore such statements and simply and courteously mention that the authors might want to tone down their paper in this regard. Alternatively, the reviewer can just inform the editor that some of the comments are too bold and take themselves out of the "bad guy" role. Earlier in this chapter, I mentioned that author's should never take what a reviewer says about their paper personally. The reverse is also true. The reviewer should not take anything that the authors say to heart. There are often cultural or other idiosyncrasies that authors come to the table with. As a reviewer, one should try to overlook these and focus on the essence of the submission. For example, many papers will come to reviewers in English that is very poor. Instead of holding this against the authors, I always consider how my paper would read if I tried to write an academic paper in for example, Romanian. Another perspective is to view the submission by a non-native speaker as courageous. This technique, I think, helps look beyond the construction of the paper and focus on the content. This being said, the reviewer's comments can often come across as too professorial or harsh. A good review will provide the authors with a constructive criticism and should always strive to guide the authors to a better submission. Even when a paper or book proposal is so bad that it will be rejected no matter what the authors do, e.g. flawed methodology, a good reviewer will try to point out the major problems. Making inflammatory comments or attacking an author's ability to write in your language never leads to anything positive. Always remember when you first began writing and publishing and put yourself in the place of the authors.

10.3 Conclusions

Manners in academic writing are real and can have a real influence on your work as an author. Although there are no go-to guidelines for authors to follow, common sense and manners that you would follow in person should be used in academic writing. In the end, an academically well-mannered author will be much more likely to have their work "facilitated" through the peer review process than an author who lacks basic good manners.

Manners should permeate your academics and interactions with others. However, most would agree that manners in academics are simply a bridge of manner practiced in one's daily life. In other words, such considerations do not exist merely in a professional world and not in one's private life. Therefore, I think that one of the first steps of practicing good manners in academics is to take a up close and personal look at your daily life and appreciate whether you would be considered a well-mannered person by your friends and peers. This step is critical in the process of establishing good manners in academics!

11

Emotional Intelligence: Its Place in Your Professional and Academic Careers

Sandra Buerger and Gernot Buerger

Department of Natural Sciences and Mathematics, Boston University, College of General Studies, Boston, MA, USA

11.1   Background

There has been considerable interest in measuring intelligence for a long time. However, there has been debate over what exactly constitutes intelligence and how this might contribute to overall success. The Intelligence Quotient (IQ) test, developed in the early twentieth century sought to measure and quantify these mental abilities with respect to age. Initially, IQ tests that measured analytical ability were embraced as predictors of overall intelligence and potential future success (Salovey and Mayer 1990; Davis 2004). These models, however, have some major limitations. While they may (or may not) provide an accurate analysis of a certain type of intelligence, they ignore many so-called "soft skills," such as interpersonal relationships, maturity, empathy, and ability to deal with setbacks and challenges (Salovey and Mayer 1990; Mayer and Salovey 1995; Goleman 1995).

These limitations were recognized early on, but a comprehensive and cohesive definition of these other aspects of intelligence did not emerge until the late twentieth century. Before that, literature addressing these issues remained scattered and largely ignored by the popular press. This was due in large part to the myriad of challenges that are present in accurately measuring and analyzing such skills (Sayer and Mayer 1990). The first definition of emotional intelligence (EQ) was provided in a seminal 1990 paper written by Salovey and Mayer. In this paper, they gathered together ideas from the literature, provided distinct categories, analyzed contemporary methods of measuring aspects of EQ, and, importantly, provided a definition for the term they coined EQ. This original definition of EQ was stated as the "ability to monitor one's own and others' feelings and emotions to discriminate among them and to use this information to guide one's thinking and actions." One important aspect of this quote is the definition of EQ as not only an awareness of these emotions but an ability to apply these skills to solve problems and regulate behavior both in oneself and in others (Salovey and Mayer 1990; Goleman 1995).

Since the original paper by Salovey and Mayer, two distinct models of EQ – the ability model and the trait model – have been developed. Stated simply, the ability model proposes a viewpoint that the skills that comprise EQ can be learned and developed, while the trait model states that EQ is largely personality based and unchangeable

A Guide to the Scientific Career: Virtues, Communication, Research, and Academic Writing, First Edition.
Edited by Mohammadali M. Shoja, Anastasia Arynchyna, Marios Loukas, Anthony V. D'Antoni, Sandra M. Buerger, Marion Karl and R. Shane Tubbs.

(Fox 2013; Monroe and English 2013). In this chapter, we will focus on the ability model, based on the original paper by Salovey and Mayer and popularized by the work of Daniel Goleman.

11.2 The Importance of EQ in Academia and on the Job

The idea of EQ grew substantially in popularity with the publication of Daniel Goleman's 1995 book *Emotional Intelligence: Why It Can Matter More than IQ*. Goleman expanded upon the ideas presented by Salovey and Mayer and brought them to a wider audience. His ideas are represented by broader definitions and notably by the applicability of the "soft skills" in the classroom and the workplace. The idea that the ability-based model of EQ is applicable and scientifically valid has been widely accepted in the popular press, academic psychological field, and in the field of organizational behavior (Daus and Ashkanasy 2005).

Various publications have shown that EQ constitutes a significant, and at times even a majority, predictor of success in the workplace (Goleman 1995; Davis 2004). While this appears to be true at all levels of careers, EQ seems to play an even greater role at higher levels, e.g. in leadership positions.

Interestingly, some studies have indicated that EQ may be less important in the more technical fields (e.g. in the sciences). Some have even suggested that EQ may be a liability (Grant 2014). However, while technical positions may be less focused on these skills initially, movement to higher levels in the workplace, including management, requires development of these skills. Further, patient-physician interaction in medicine and communication of results in research (all vital parts of the career of a medical student, clinician, or scientist) rely heavily on the skills discussed in this chapter.

11.3 Major Aspects of Emotional Intelligence

Salovey and Mayer (1990) first identified four major components of EQ. They included: recognizing emotions, understanding emotions, regulating emotions, and employing emotions. Later, through the popularization of the idea of EQ, Goleman broke EQ into more general domains, some major categories of which are listed in this section. We will provide here general definitions and descriptions for each category.

11.3.1 Self-Awareness

Awareness of the self is vital to understanding, identifying, and changing nonoptimal behaviors. Understanding how you are feeling about a situation is important. It is a good idea to pause during stressful situations or before making a big decision and access your own state of mind.

Many successful people report being guided by gut-feelings when making a difficult decision. Identifying your emotional response (conscious and subconscious) should be an integral part of decision-making. In addition to using your own emotional response to guide your career path and decision-making, it is important to understand how your emotional response affects other people.

This self-awareness further extends into identifying your own weaknesses, accepting them and dealing with them. The ability to change means that you need to be able to accept criticism and adapt your path accordingly.

11.3.2 Managing Emotions

Recognizing your own emotional response in itself is not enough to display EQ. You need to be able to manage your emotions. This includes not only how you display your emotions (e.g. losing your temper, coming across as out of control, angry, etc.), but also your own internal management of the emotions. People respond differently when they are confronted with a difficult or stressful situation. People with a high competency in this area (high EQ) will remain in control of their emotions. In contrast, people who struggle in this aspect will allow these emotions to spin out of control, essentially highjacking their response. In this scenario, a person's thoughts spin out of control and he or she becomes increasingly agitated and frustrated with the situation, preventing the person from thinking clearly about solutions and strategies.

Regulating emotion is not merely a function of your personality. There are a number of coping strategies. Seeking out activities that put you into a positive frame of mind is a strategy that has been suggested by numerous experts (Salovey and Mayer 1990). This may include engaging in a sport, social activity, or more solitary activity, such as meditation.

Additionally, your surroundings can affect how you are able to regulate your emotion. It is important to regulate your choice of associates. You should surround yourself with people who are supportive and give you positive (but realistic) feedback. Of course, you must also present yourself in a way that encourages people to have a positive view. This includes acting in a professional manner, being on time, and being respectful of your colleague's time. Another interesting way of regulating your emotion is to engage in altruistic acts. This idea has been gaining more attention in recent years.

Finally, the ability to recognize that mood is temporary and under your control is vital to obtaining a mental state of emotional control. You should not seek to change your mood or suppress negative feelings. Rather, use your current mood to your advantage. In a positive state of mind, you might imagine achieving more lofty goals, while in a negative state of mind, you might be more likely to think of an alternative plan in the case these goals don't work out. Use these different moods as an opportunity for creative thinking and formulation of alternative plans. A negative feeling about a particular task may also serve to focus your attention on an area that needs attention. This can help you prioritize.

11.3.3 Motivation

Motivation, or persistence, in the face of challenging tasks is a major indicator of success. This is the result of a belief in the self (self-confidence). People who displays high competence in this area have a strong sense of self, including a strong definition of personal values, but avoid overestimating or being unrealistic about their own capabilities. Both an overblown sense of self and an underestimation of your own skills can result in low levels of motivation and persistence. People with overblown senses of their own capabilities may overreach, for example, applying for a position to which they are not qualified. This overblown sense of self can result in discouragement (when they are not accepted for the position) and an inability to accept criticism. The ability to accept criticism and learn from these mistakes and criticisms can result in inability to improve and grow in one's career. Conversely, having a lack of belief in one's own skills can result in trepidation. This can hold an individual back from contributing to workplace discussions and trying new procedures. This fear of seeming inept can have a negative impact on the career path.

It is important to be aware of your own strengths and weakness, and then to believe in your underlying abilities even in the face of failures, setbacks, and challenges. Various studies have shown that the ability to continue with a challenging task is instrumental in success (Babineaux and Krumboltz 2013).

Achievement of this motivation can be fostered by choices in your career path. Various studies have indicated that intrinsic motivation is more effective than extrinsic. A desire to help people (e.g. patients or the group your research focuses on) or a sense of interest in your field will help you maintain motivation over the long term. Recognition, social status, and money – all extrinsic motivators – will have less of an impact on your motivation.

All of this ties into the awareness of your own emotions. An analysis of how you feel about your career path, positions you are considering, and which subfield to enter should all be influenced by the level of intrinsic motivation you can muster.

11.3.4   Empathy/Social Skills

Those who perceive the emotions of other around them will be more likely to have smoother interpersonal relationships, including workplace relationships. This is important in having the people you work with operate as a team. One poignant area today is the ability to give the person your full attention (the principle of presence). This means actually listening to the person, rather than just planning what you are going to say next, or avoiding looking at a laptop screen or phone during a conversation (Ioannidou and Konstantikaki 2008; Goleman 2007, 1995).

In a workplace setting, such presence should be implemented by listening to the viewpoints of your colleagues and considering their opinion genuinely and with interest. Hearing, examining, and commenting on other people's experiences can help you develop this skill. If you are teaching, then listening to and understanding the views of the students can demonstrate active listening and has been shown to increase the EQ of the students in the classroom (Ioannidou and Konstantikaki 2008).

11.4   Developing EQ

EQ is an important, but also a learnable quality. The ability to develop this skill hinges on a willingness to accept and deal with areas in which you may be lacking. Ironically, those with a lower EQ tend to be less willing to work on these skills (Sheldon et al. 2014). Those with higher EQ expressed a greater willingness to develop these skills and improve. Thus, the first step in developing a high EQ is to be willing to identify areas in which you may be lacking and a willingness to improve upon these areas.

While professional programs do exist and are often part of medical school training, there are a few things you can do to improve your EQ on your own. Some are suggested above and include regulating your emotion by engaging in activities that you find pleasurable, surrounding yourself with colleagues that are supportive, making career choices that focus on your core values, and actively engaging and listening to people you interact with. All of these skills take time to develop. The typical time period is considered to be between three and six months.

EQ is an important quality to develop regardless of your planned career path. In the scientific and medical field, the development of EQ-related skills can allow individuals to effectively communicate with colleagues, supervisors, patients, and the general public. Importantly, one should remember, whatever weaknesses or strengths one possess in this area, they can be improved upon and learned with significantly positive results.

References

Babineaux, R. and Krumboltz, J. (2013). *Fail Fast, Fail Often: How Losing can Help you Win.* New York, NY: Tarcher.

Daus, C.S. and Ashkanasy, N.M. (2005). The case for the ability based model of emotional intelligence in organizational behavior. *Journal of Organizational Behavior* 26 (4): 453–466.

Davis, M. (2004). *Test your EQ.* London: Piatkus.

Fox, M. (2013). Putting emotional intelligence to work. *Journal of The Academy of Nutrition and Dietetics* 113 (9): 1138–1143.

Goleman, D. (1995). *Emotional Intelligence: Why it Can Matter More than IQ.* New York, NY: Bantam Books.

Goleman D (2007) Daniel Goleman: Why aren't we more compassionate? [Video File] Retrieved from http://www.ted.com/talks/daniel_goleman_on_compassion.

Grant A (2014, Jan 2) The Dark Side of Emotional Intelligence. *The Atlantic* Retrieved from http://www.theatlantic.com/health/archive/2014/01/the-dark-side-of-emotional-intelligence/282720.

Ioannidou, F. and Konstantikaki, V. (2008). Empathy and emotional intelligence: what is it really about? *International Journal of Caring Sciences* 1 (3): 118–123.

Monroe, A. and English, A. (2013). Medical education: fostering emotional intelligence in medical training: the SELECT program. *American Medical Association Journal of Ethics* 15 (6): 509–513.

Salovey, P. and Mayer, J. (1990). Emotional intelligence. *Imagination, Cognition and Personality* 9 (3): 185–211.

Mayer, J.D. and Salovey, P. (1995). Emotional intelligence and the construction and regulation of feelings. *Applied and Preventive Psychology* 4: 197–208.

Sheldon, O., Duning, D., and Amers, D. (2014). Emotionally unskilled, unaware, and uninterested in learning more: reactions to feedback about deficits in emotional intelligence. *Journal of Applied Psychology.* 99 (1): 125–137.

12

Communication Skills

Sandra Buerger

Department of Natural Sciences and Mathematics, Boston University, College of General Studies, Boston, MA, USA

12.1 Introduction

Communication extends to both your personal life and the workplace. The aim of this chapter is to examine workplace and academic interactions; however, many of these topics will overlap into both realms of life. The main aspects of effective communication to be discussed briefly here include:

1. Emotional intelligence
2. Open-mindedness/tolerance
3. Objectivity/awareness
4. Consistency/honesty
5. Confidence/assertiveness
6. Flexibility
7. Clarity

First, we must define communication. Communication is described by Pfeiffer (1998) as the ability to make thoughts, feelings, and needs known to another and the reception of that information from the other person. However simple that may sound, there are many factors that influence how those messages are sent and received. It is the aim of this section to give an overview of some of these factors and help you become aware of how they affect communication. The sections below will take the seven factors listed above into account when discussing what constitutes effective communication.

12.2 Effective Communication

Emotional intelligence (EI) is defined as the ability to understand the motivations of the self and of others that you interact with. Emotional intelligence also includes the ability to express empathy and understanding to those around you (Salovey and Mayer 1990). The cultivation of emotional intelligence and the use of emotional intelligence in your interactions with peers and colleagues is a theme that runs through the other tenets of communication listed above. You can also refer back to Chapter 11 where the topic of emotional intelligence was covered in more detail.

A Guide to the Scientific Career: Virtues, Communication, Research, and Academic Writing, First Edition.
Edited by Mohammadali M. Shoja, Anastasia Arynchyna, Marios Loukas, Anthony V. D'Antoni, Sandra M. Buerger, Marion Karl and R. Shane Tubbs.

When communicating with others, it is important to keep an open mind and listen to what others have to say. This brings us back to the idea of emotional intelligence. Listening to others speak has been shown to relate to a higher level of emotional intelligence and more effective communication. Studies have shown that deliberate attention to listening and modeling of conscious listening have been effective tools in increasing effective communication (Ioannidou and Konstantikaki 2008).

In fact, there is much interest in the idea of active listening and tolerance of the viewpoint of the other person and how that influences effective communication. When people act aggressively or insult each other, the predictable outcome is a hostile situation in which the two speakers perceive each other as adversaries. This results in a shutdown of communication. One possible way around this situation is to avoid insulting people when expressing your viewpoint, look for common ground, and resist the urge to define a difference of opinion as an insurmountable obstacle to effective communication (Waugh 2014).

One particularly good demonstration of effective communication that involves tolerance and open-mindedness is the relationship between the teacher and the student. Thomas Gordon (2003) describes effective communication between teachers and students. Although his work focuses on the teacher–student relationship, we can apply many of the lessons to other situations. Gordon discusses the need for awareness of the other person's needs in the conversation and how to ensure you are sending messages that will result in effective communication with the other person. He describes a number of phrases that can shut down communication by sending messages of lack of acceptance, inadequacies, and faults. Some of these include ordering or commanding, moralizing, judging, stereotyping, and labeling. These types of messages, along with others, shut down communication and don't result in an effective exchange (Gordon 2003).

The solution is to engage in active listening. Verbalizing what the other person is saying can be helpful in promoting an understanding between people. This involves mirroring or restating what the other person has said. This helps in two major ways. One, the person hears back what they are trying to communicate and can revise their statement if what they hear is not what they were trying to say. Two, it sends the message that you are truly listening to what the other person is saying and shows concern and interest (Gordon 2003).

In addition to listening, you must also put aside your personal biases to interpret the message. Here, again, we see the importance of tolerance of a different viewpoint. Once you feel you have understood what the other person is saying and looked at it from that person's viewpoint, you can respond. Your response should be careful and thought out and avoid insults. The idea is to avoid hostile negativity, such as defining an idea as "stupid," but to still express your own thoughts (Gordon 2003; Waugh 2014).

Active listening and tolerance of others' viewpoints is a skill that should be practiced and demonstrated to others (Ioannidou and Konstantikaki 2008). Active listening, along with thoughtful responses that avoid hostility, will promote objectivity in the response. All of these strategies will help reduce conversations that are charged with emotion or characterize the speaker based on past experiences or stereotypes without hearing the message that is being expressed.

In addition to the interpretation part of communication, you must also be able to effectively express your own ideas. A number of factors can help your ideas come across clearly and advance communication. These verbal skills are vital to effective communication (Pfeiffer 1998; Mikoluk 2014).

Consistency and honestly in the expression of your thoughts is vital. Consistency refers to the ability to stay on point. Avoid going off on tangents that will cause the other person in the exchange to lose interest or become confused. This is also directly related to the idea of clarity. Use words and terms that the listener will be familiar with and avoid the use of jargon (Mikoluk 2014). Be aware of your audience and their background. You do not want to communicate complex technical ideas to first-year undergraduates and, at the same time, you do not want to explain basic concepts to experts in the field. In order to keep the listener's attention, you must tailor your communication to your audience.

When speaking, you should communicate with confidence in your ideas. You should be assertive, but not aggressive. This will reduce hostility in the response from your audience. Exhibit flexibility in your ideas and your method of communication.

Our final consideration in this overview of effective communication is environmental. People often underestimate the affect that the surrounding environment has on communication. Feeling uncomfortable due to temperature, location, and surrounding noise can all have a negative influence on effective communication. If you can, control the environment in which the communication takes place. If you cannot, take into account these factors (Pfeiffer 1998).

Effective communication will reduce frustration in the workplace and help you advance your career. Effective communication skills can and should be learned and practiced. We have seen here some general points. We will discuss some more specific examples in the section below.

12.3 Communication in the Scientific and Medical Community

In the previous section, we discussed general principles of effective communication. We focused on communication between two or more people engaged in a conversation. However, these lessons can apply beyond simple conversations to written communications or various sorts of group communication. In this final section, we will examine some situations that are especially relevant to those in the scientific and/or medical community.

12.3.1 Written Communications

One of the major forms of communication in the scientific community is through publications. You will likely communicate via formal written communication – papers and posters, for example – throughout your career. In order to communicate effectively, you must apply the tenets of effective communication above to your writing. This includes clarity, honesty, and brevity in preparing scientific papers. You must ensure that your papers effectively communicate the message without overstating your findings. A number of chapters deal with the specifics of how to achieve these goals in the various parts of the formal scientific paper.

Other important written communications are less formal. Among the most common are email and text message communications. You will likely communicate daily with colleagues through email. Again, you can refer to the general description of effective communication above. There are, however, a few points that relate specifically to email communication. When writing an email or text, you must consider the context and the status of the person you are writing to. For example, while informal greetings and words may be appropriate when communicating with peers at the same level, communication with a supervisor or with the head of a department (or any time you do not know the person well) should be more formal. For those communications, it is important to make sure you have used correct grammar and spelling and an appropriate title (e.g. Dr. Smith versus Joe). You should also avoid greetings such as "Hey" or "Hi" unless the person is a close peer. Finally, you should avoid abbreviations of words such as "u" for "you" or "2" for "to" in these more formal email communications. Once again, in these types of communication you must consider the audience and your relationship to them.

Email communication is instant and thus allows us more interaction with people than ever before. General guidelines for response times include 24–48 hours during the business week. On weekends or during holidays, you should be understanding if a response takes longer.

12.3.2 Informal Meetings

During the course of your career, you will often attend informal meetings. These may be lab meetings, for example, that gather members of the wider research team to present data and discuss issues in their research. The goal of these meetings is to practice effective communication of ideas and to suggest alternative strategies to achieve goals. However, these meetings can often be the scenes of hostile interactions. It is important to practice active listening and tolerance of others' viewpoints in these meetings. Respect for the other person's hard work will prevent feelings of defensiveness. On the other hand, if you are the target of criticism, you should attempt to listen to the critique of your work without becoming overly emotional and defensive. Consider the person's thoughts and ideas. If you still disagree with their analysis, express this idea assertively, but without the use of insults or speech that can escalate the situation.

Setting a positive tone is important to prevent the development of an idea that a meeting that takes place on a weekly basis is not useful. If that happens, people will come to the meeting with an idea that this is a waste of their time, already setting the stage for ineffective communication. If you are in the position as head of a lab, keeping the meeting to a scheduled time, making sure everyone has the opportunity to talk and express their ideas, and intervening in situations where conversations become too heated will ensure that a negative feeling is not associated with these meetings.

Finally, we also saw that physical conditions can influence how effective communication is. If you are able, try to schedule meetings in a comfortable place and monitor the overall temperature and other controllable physical conditions in the room where the meeting is taking place.

References

Gordon, T. (2003). *Teacher Effectiveness Training*. First Revised Edition. New York: Three Rivers Press.

Ioannidou, F. and Konstantikaki, V. (2008). Empathy and emotional intelligence: what is it really about? *International Journal of Caring Sciences* 1 (3): 118–123.

Mikoluk K (2014) Principles of Communication: 7 Pillars of Business Communication. Udemy https://blog.udemy.com/principles-of-communication/.

Pfeiffer JW (1998) Conditions that Hinder Effective Communication.

Salovey, P. and Mayer, J. (1990). Emotional intelligence. *Imagination, Cognition and Personality* 9 (3): 185–211.

Waugh, J. (2014). *Real Dialogue Require Tolerance: Hostility Is the Antithesis of Effective Communication*. The Daily of the University of Washington.

13

Learning Charisma
R. Shane Tubbs

Seattle Science Foundation, Seattle, WA, USA

13.1   Introduction

The often enigmatic term *charisma* is important in influencing others and in that regard, useful in such endeavors as business, relationships, leadership, parenting, and academics. One might consider this characteristic as an external projection of one's internal security and confidence. As this is an important quality in a productive and elevated career in academia, the following chapter will define and give methods for developing and improving charisma.

13.2   What Is Charisma?

Charisma comes from the Greek word χάρισμα meaning gift or divine favor, implying it is handed out by the gods. Wiseman has said, "Charisma is hard to pin down; we all have a sense of someone having it, but it is difficult to explain why" (Highfield 2005). In this regard, there are multiple and conflicting viewpoints on how to specifically define this term. Some have considered charisma as a behavior and others as a trait (Owen 2014). Antonakis et al. (2012) have defined charisma as "the ability to communicate a clear, visionary, and inspirational message that captivates and motivates an audience." Kendall et al. (2000) defined charismatic authority as the "power legitimized on the basis of a leader's exceptional personal qualities or the demonstration of extraordinary insight and accomplishment, which inspire loyalty and obedience from followers." Weber defined charisma as "a certain quality of an individual's personality by virtue of which he/she is considered extraordinary and treated as endowed with supernatural, superhuman, or at least specifically exceptional powers or qualities" (Owen 2014). Wiseman believes charisma has deep evolutionary roots and that a person's impact on someone else's emotions can result in the affected person attributing charisma to the person (Highfield 2005).

Werrell (2013) stated that charisma is the rare quality that makes people like you, even when they don't know much about you and an intangible quality that makes people want

A Guide to the Scientific Career: Virtues, Communication, Research, and Academic Writing, First Edition.
Edited by Mohammadali M. Shoja, Anastasia Arynchyna, Marios Loukas, Anthony V. D'Antoni, Sandra M. Buerger, Marion Karl and R. Shane Tubbs.
© 2020 John Wiley & Sons, Inc. Published 2020 by John Wiley & Sons, Inc.

to follow you, be around you, and to be influenced by the things that you say i.e. a personal magnetism. Cabane, author of *The Charisma Myth- How Anyone Can Master the Art and Science of Personal Magnetism*, states that the three main traits of charisma can be broken down into three categories: presence, power, and warmth, and when combined, these three components produce strong personal magnetism. Owen (2014) stated, "Charismatic people possess a potent blend of attractiveness and presence that commands attention with an irresistible magnetic force." Therefore, many have equated charisma with magnetism.

13.3   Learning How to Be Charismatic

Antonakis et al. (2012) asked the question, "So how do you learn charisma?" following this with the statement that many people believe that it's impossible to learn how to be charismatic and that charismatic people are born that way. The authors conclude that charisma "is not all innate; it's a learnable skill or, rather, a set of skills that have been practiced since antiquity." However, other others, such as Wiseman, have estimated that charisma is half trained and half innate. Werrell (2013) believes that charisma is inborn in everyone but in different amounts. Cabane (2012) thinks that charisma is simply a learned and, specifically, nonverbal behavior. Antonakis et al. (2012) have trained professionals in "charismatic leadership tactics" (CLTs). They found that following such training, these individuals became "influential, trustworthy, and more leader-like and that to persuade others, you must use powerful and reasoned rhetoric, establish personal and moral credibility, and then rouse followers' emotions and passions." They concluded that if "a leader can do those three things well, he or she can then tap into the hopes and ideals of followers, give them a sense of purpose, and inspire them to achieve great things."

McKay and McKay (2013) state that presence is a key component of charisma and to develop this, one must do the following:

- Be in the here and now.
- Be physically comfortable.
- Keep eye-to-eye contact.
- Nod to signal that you are listening.
- Ask clarifying questions.
- Avoid fidgeting.
- Wait two seconds before responding in a conversation.
- Don't think about how to respond while person is talking.

Antonakis et al. (2012) have identified key CLTs. Nine of these are verbal: metaphors, similes, and analogies; stories and anecdotes; contrasts; rhetorical questions; three-part lists; expressions of moral conviction; reflections of the group's sentiments; the setting of high goals; and conveying confidence that they can be achieved. Three tactics are nonverbal: animated voice, facial expressions, and gestures. The authors mention that there are certainly other CLTs that leaders can use, such as creating a sense of urgency, invoking history, using repetition, talking about sacrifice, and using humor, but the 12 they describe can have the greatest effect and can work in almost any context. They also found that people who use them appropriately can unite their followers around their vision. They found that in 8 of the past 10 US presidential races, the candidate who used such verbal CLTs most often won the race. Also, these authors found:

> when we measured good presentation skills, such as speech structure, clear pronunciation, use of easy-to-understand language, tempo of speech, and speaker comfort, and compared their impact against that of the CLTs, CLTs played a much bigger role in determining who was perceived to be more leader-like, competent, and trustworthy (Antonakis et al. 2012).

Gustin (1973) has proposed that charisma is an explanation for scientific motivation. Wiseman stated that charisma is key to communicating science. Wiseman stated that charismatic people have the following three main attributes:

1. They feel emotions themselves quite strongly.
2. They induce them in others.
3. They are impervious to the influences of other charismatic people.

Charisma was investigated by an experiment performed by Richard Wiseman, a psychologist at the University of Hertfordshire, who first examined the theory of "emotional contagion" and charisma in a public arena (Highfield 2005). Via a national study, Wiseman had 200 participants fill out a questionnaire in order to measure their charisma (Highfield 2005). For his study, Wiseman asked participants to complete a 13-item questionnaire designed by Friedman at the University of California, Riverside. Questions ranged from whether you like to touch people when you talk to them, enjoy being the center of attention, or can keep still when you hear good dance music. A clear relationship emerged between charisma ratings and questionnaire scores, suggesting, per the author, that individuals with charisma can project their own emotions to those around them (Geoghegan 2005). "Interestingly, those who scored highly for charisma and in the questionnaire on emotional contagion were also those that also did well in the competition," said Wiseman (Highfield 2005). As a group, the FameLab applicants had an average score of 86, which is high, given that the mean of the US population is 71. "The finalists are especially high – as a group, they have a mean of 90," said Wiseman, adding that the number one Toyota salesman in America scored only nine points higher (Highfield 2005). There were no significant differences between men and women or in area of the United Kingdom that they were from.

13.4 Improving Your Charisma

In order to be more charismatic, Reynolds (2013) suggests the following seven characteristics/techniques:

1. Be prepared.
2. Ask questions.
3. Use your hands while speaking.
4. Be genuinely interested in the people you speak with.
5. Develop passions.
6. Be energetic.
7. Be optimistic.

Wiseman has suggested adopting the following in order to improve your charisma:

- Use an open body posture that will help attract other people to you.
- Move around to produce the feeling of energy and enthusiasm.
- Speak in a clear, fluent, forceful, and articulate way that evokes imagery, energy, and action.
- Constantly alter the intonation and pacing of your delivery to maintain interest.

In general, use an upbeat tempo, only occasionally switching to a slow delivery to create tension and to emphasis key points (Highfield 2005). Wiseman has also proposed the following specifics for becoming more charismatic:

- *General.* Open body posture, hands away from face when talking, stand up straight, relax, hands apart with palms forwards or upwards.
- *To an individual.* Let people know they matter and you enjoy being around them, develop a genuine smile, nod when they talk, briefly touch them on the upper arm, and maintain eye contact.
- *To a group.* Be comfortable as leader, move around to appear enthusiastic, lean slightly forward, and look at all parts of the group.
- *Message.* Move beyond status quo and make a difference, be controversial, new, simple to understand, counterintuitive.
- *Speech.* Be clear, fluent, forceful, and articulate, evoke imagery, use an upbeat tempo, occasionally slow for tension or emphasis.

Owen (2014) has suggested the following five characteristics are found in a charismatic individual:

- *High self-esteem.* Confidence, inner calm, self-reliance, and independence
- *A driving force.* Purpose, personal values, and principles
- *Sensory awareness.* Empathy and emotional intelligence
- *A vision.* Positive attitude toward aim, belief, mental picture
- *High energy.* Passion, enthusiasm, commitment, determination

Cabane (2012) suggest the following three tips to increase your charisma in a conversation:

- Lower the intonation of your voice at the end of each sentence.
- Reduce how quickly and how often you nod your head during the conversation.
- Pause for two seconds before you speak.

13.5 Conclusions

Charisma still lacks a precise definition, although several authors have equated this term to magnetism. Regardless, charisma is considered a human characteristic that is often desirable. Multiple strategies and techniques have been suggested that might develop or improve one's charisma. However, there is some debate as to where charisma can be learned or if it is simply a variable trait.

References

Antonakis J, Fenley M, Liechti S (2012) Learning Charisma. http://hbr.org/2012/06/learning-charisma/ar/1

Cabane OF (2012). *The Charisma Myth: How Anyone Can Master the Art and Science of Personal Magnetism*. London: Penguin Books.

Geoghegan T (2005) A step-by-step guide to charisma. *BBC News Magazine* http://news.bbc.co.uk/2/hi/uk_news/magazine/4579681.stm

Gustin, B.H. (1973). Charisma, recognition, and the motivation of scientists. *American Journal of Sociology* 78: 1119–1123.

Highfield R (2005) FameLab and the secrets of an infectious personality. *The Telegraph* www.telegraph.co.uk/technology/3341135/FameLab-and-the-secrets-of-an-infectious-personality.html

Kendall, D., Murray, J., and Linden, R. (2000). *Sociology in Our Time*. Scarborough, Ont: Wadsworth Publishing.

McKay B, Mckay K (2013) http://www.artofmanliness.com/2013/11/06/the-3-elements-of-charisma-presence

Owen N (2014) Releasing Your Hidden Charisma. http://www.businessballs.com/freespecialresources/Charisma_Report.pdf

Reynolds S (2013) http://www.forbes.com/sites/siimonreynolds/2013/09/15/how-to-be-more-charismatic

Werrell, L. (2013). *How to Develop Your Natural Charisma the Easy Way and Learn to Charm the Birds from the Trees*. CEI Publishers.

14

Essence of Collaborative Research: Leadership, Engaging Others, and Cooperativeness

Rebecca Andall, Petru Matusz, Jerzy Gielecki and Marios Loukas

Department of Anatomical Sciences, St. George's University School of Medicine, Grenada, West Indies

14.1   Introduction

Collaborative research is a fairly modern concept in which the pursuit of new knowledge is married to the desire to solve societal problems through the interaction of a multidisciplinary expert team. Scientists and researchers are now encouraged to work with the various stakeholders in the community to which their work applies in order to break down the divide between researchers, healthcare professionals, scholars, decision makers, and community members (O'Sullivan et al. 2010). This approach requires active engagement of these various professionals throughout all stages of the research process to ensure that the research problem is addressed in its societal context (Barkin et al. 2013; Krebbekx et al. 2012).

Collaborations are made between professionals in the same field or across scientific disciplines and institutions in order to take advantage of the unique perspectives and expertise that these individuals can offer (O'Sullivan et al. 2010; Hofmeyer et al. 2012). These collaborations are necessary in this new era of ultra-specialization in medicine and with the increased prevalence of chronic multisystem diseases. It also ensures that clinicians are accountable to their colleagues and to the public to deliver quality, evidence-based care while efficiently using resources and avoiding waste (Miles et al. 2013).

There are numerous examples of multicenter partnerships around the world. The Tuberculosis Network European Trials group (TBNET) is a collaboration, launched in 2006, made up of a network of 500 professionals around the world to address shared research agendas related to tuberculosis in Europe (Giehl et al. 2012). The Early Nutrition Programming Project (EARNEST) is another organization comprising a multidisciplinary international team from 40 institutions around Europe who address issues surrounding nutrition in early childhood (Koletzko et al. 2011). Similarly, the World Health Organization (WHO) has taken the concept of collaboration to a global level with an importance placed on bringing professionals together to help improve health and healthcare policy around the world.

The purpose of this chapter is to provide some background to this approach, guidance on how to make collaborations more effective and to caution on some of the challenges that may be encountered as different teams interact to achieve a common goal.

A Guide to the Scientific Career: Virtues, Communication, Research, and Academic Writing, First Edition.
Edited by Mohammadali M. Shoja, Anastasia Arynchyna, Marios Loukas, Anthony V. D'Antoni, Sandra M. Buerger, Marion Karl and R. Shane Tubbs.
© 2020 John Wiley & Sons, Inc. Published 2020 by John Wiley & Sons, Inc.

14.2   Why Collaborate?

Collaborative research methodologies have been developed to minimize efforts from a single group. It was also implemented to reduce the delays between scientific advances, improvements in healthcare policy and the use of the new knowledge by clinicians (Barkin et al. 2013; Mikesell et al. 2013; Miles et al. 2013).

Minority communities should be included in decisions about healthcare policy because they tend to be compliant with research and benefit from improved health outcomes (Amendola 2011). Studies that engage minority populations in all aspects of research experience higher enrollment rates and fewer dropout rates within these ethnic groups (De las Nueces et al. 2012). Patients benefit from partnerships because they have access to a team of experts that offer "best practice" by continually incorporating up-to-date innovations into their patient care (Miles et al. 2013). Researchers benefit from collaborations because they have access to multidisciplinary perspectives that allow them to tackle a research problem in a well-rounded but relevant way. They can also take advantage of pooled resources, technologies, and funding.

14.3   Challenges to Collaborative Research

Researchers may find collaborative research a challenge because they were not taught how to effectively interact within interdisciplinary teams during their training. It is also difficult to create partnerships amongst academic institutions that subscribe to conflicting theoretical frameworks (O'Sullivan et al. 2010). More often than not, it is the cultural differences or the managerial politics within institutions that create barriers to collaborations (Cummings and Kiesler 2005; Raza 2005).

There can also be conflicts between the different professions within a team. Researchers may focus on hypothesis-driven goals while clinicians may reject these theories for ones that coincide with what they experience in their workplace (Krebbekx et al. 2012). Team members may have different expectations and goals for a study that can impede group synergy (Mikesell et al. 2013). Still, there are special challenges that can be experienced when researchers work with policy makers as there can be frequent changes in professional roles, availability and political interest (Hofmeyer et al. 2012). Also, working with untrained community members may make maintaining study protocols and standards more difficult (Barkin et al. 2013).

The right to authorship and ownership of research can also become a complicated issue as the number of collaborators on a project, with their varying levels of contribution, increase. In practice, a person's right to authorship should not be challenged if that person contributed to the study design, analysis of data or its interpretation, or were contacted to review the manuscript before it is published. These guidelines, however, do not address the rights to authorship for contributions made by other stakeholders in research such as biobanks (Colledge et al. 2013).

14.4   Ethical Considerations

Most scholars agree that collaborative research calls for its own set of ethical guidelines. Community autonomy versus individual autonomy is one topic that is frequently debated in the literature. Some scholars are of the view that the traditional Belmont Report definition of patient autonomy is not comprehensive enough to provide appropriate guidance for community medicine and collaboration. Researchers are asked to recognize a community's right to self-determination about the involvement of its individual members in research. Researchers are also asked to safeguard members of the community with impaired autonomy. A big aspect of respecting community autonomy involves the handling of research findings. However, some researchers argue against disclosing negative findings to community members (Mikesell et al. 2013). Researchers need to find ways of handling research data that does not jeopardize the confidentiality of the individual members in the community.

Collaborative research also imposes unique ethical dilemmas for researchers in terms of justice. For instance, problems can arise where there is a need to fairly distribute resources throughout the community despite limited funding. Minimizing harm to the community as well as managing informed consent can also prove more challenging for collaborative teams (Mikesell et al. 2013).

14.5 How to Make Collaborations Work

In order to make collaborations work, there should be ongoing communications between team members throughout the research process. These meetings should aim to clarify roles and expectations and to foster trust between members. It is important that problems are discussed openly as they arise and that the research goals address the needs of all parties in a relevant way (Happell 2010). It is also important that funding, capacity building (Mikesell et al. 2013), and post-study sustainability (Blevins et al. 2010) in a community are addressed before the study is underway.

14.6 Conclusions

Regional and global health issues have become increasingly more important in society. As a result, academics are asked to make meaningful partnerships with key players within and outside of their field in order to have a more comprehensive understanding on health and healthcare policies. While most scholars would agree that collaborations are a good thing, there are still unique challenges that will need to be overcome to make these interactions beneficial to all stakeholders in the community.

References

Amendola, M. (2011). Empowerment: healthcare professionals' and community members' contributions. *J. Cult. Divers.* 18: 82–89.

Barkin, S., Schlundt, D., and Smith, P. (2013). Community-engaged research perspectives: then and now. *Acad. Pediatr.* 13: 93–97.

Blevins, D., Farmer, M.S., Edlund, C. et al. (2010). Collaborative research between clinicians and researchers: a multiple case study of implementation. *Implement. Sci.* 5:: 76–84.

Colledge, F., Elger, B., and Shaw, D. (2013). "Conferring authorship": biobank stakeholders' experiences with publication credit in collaborative research. *PLoS One* 8:: e76686.

Cummings, J.N. and Kiesler, S. (2005). Collaborative research across disciplinary and organisational boundaries. *Soc. Stud. Sci.* 35: 703–722.

De las Nueces, D., Hacker, K., DiGirolamo, A. et al. (2012). A systematic review of community-based participatory research to enhance clinical trials in racial and ethnic minority groups. *Health Serv. Res.* 47: 363–1386.

Giehl, C., Lange, C., Duarte, R. et al. (2012). TBNET - collaborative research on tuberculosis in Europe. *Eur. J. Microbiol. Immunol.* 2: 264–274.

Happell, B. (2010). Protecting the rights of individuals in collaborative research. *Nurse Res.* 17: 34–43.

Hofmeyer, A., Scott, C., and Lagendyk, L. (2012). Researcher-decision-maker partnerships in health services research: practical challenges, guiding principles. *BMC Health Serv. Res.* 12: 280.

Koletzko, B., Brands, B., and Demmelmair, H. (2011). The early nutrition programming project (EARNEST): 5 y of successful multidisciplinary collaborative research. *Am. J. Clin. Nutr.* 94: 1749S–1753S.

Krebbekx, W., Harting, J., and Stronks, K. (2012). Does collaborative research enhance the integration of research, policy and practice? The case of the Dutch health broker partnership. *J. Health Serv. Res. Policy* 17: 219–226. Accessed June 3, 2014.

Mikesell, L., Bromley, E., and Khodyakov, D. (2013). Ethical community-engaged research: a literature review. *Am. J. Public Health* 103: e7–e14.

Miles, P.V., Conway, P.H., and Pawlson, L.G. (2013). Physician professionalism and accountability: the role of collaborative improvement networks. *Pediatrics* 131: S2049.

O'Sullivan, P., Stoddard, H., and Kalishman, S. (2010). Collaborative research in medical education: a discussion of theory and practice. *Med. Educ.* 44: 1175–1184.

Raza, M. (2005). Collaborative healthcare research: some ethical considerations. *Sci. Eng. Ethics* 1: 177–186.

15

Personal Branding for Physicians and Researchers

Tracy E. Bunting-Early

Bunting Medical Communications, Newark, DE, USA

15.1 Introduction

What is a brand, and how can it relate to a person? Branding originated from cattle branding, which involved marking cattle with a distinctive symbol. These days a brand has come to mean how an entity is characterized in an audience's mind. It can relate to a person, a product, a service, a company, or any entity. In many cases, a personal brand does not have a symbolic logo, but the person's name or image represents who they are in peoples' minds. It may be synonymous with reputation; however, because reputation often has a connotation of something that is defended, rather than something that is actively built, reputation is not exactly synonymous. Personal branding is nothing more than an exercise in how you communicate who you are to the outside world, and is best achieved when your desired brand is authentic and honest.

For a researcher, clinician, or physician, personal branding is important to consider if one wants to advance either a career or a clinical practice. Physicians and researchers can benefit greatly by proactively working toward their own brand identity. Depending on a physician's specialty and practice, it may be about building a greater patient base or it may be about securing career promotions or tenure. It may also simply open doors to collaborations. Once a personal brand is built, a physician/researcher then has an audience who respects him/her and also one that is likely to notice new pursuits or interests to align with and support. Being unknown or uncharacterized might make building these alliances more difficult.

The concept of branding might seem promotional on the surface. But if you are authentic and simply want to ensure that you are clearly communicating who you are, there is nothing dubious about it.

15.2 Personal Branding and Authenticity

Your brand identity is closely tied to what you stand for and what is unique or offers value to an intended target audience. It can be anything that you want to be known for,

A Guide to the Scientific Career: Virtues, Communication, Research, and Academic Writing, First Edition.
Edited by Mohammadali M. Shoja, Anastasia Arynchyna, Marios Loukas, Anthony V. D'Antoni, Sandra M. Buerger, Marion Karl and R. Shane Tubbs.

such as a particular set of skills, a vision, or some other value. Determining a brand strategy is rarely about one's basic skills or strengths alone, but, rather, how those play out within a particular environment and among other individuals in the environment, such as with your peers, colleagues, or upper management.

Authenticity and passion are imperative. No matter how much you may want to be associated with a particular value, if it is not well supported by your competencies, your audience will not accept it. It is important to remain honest with yourself. This does not mean that you cannot focus on or work to develop skills that support your desired brand. Simply start with a solid basis for your true value. Passion is related to your reflection on the world around you and any opinions you may have. The greatest brands are associated with missions to solve an important problem, and work best when the problem is being solved by a true desire and ability.

Just having particular assets and skills are not enough; you need to be *known* for your value. You do this by building your brand. Through building a clear plan and direction, you are able to stay consistent in how you communicate about yourself and ensure that you reach the people that matter to you. Consistency over time and media are key to building one's brand, so having a plan or direction is important.

The purpose of this chapter is to describe how to create a plan for your brand. It will uniquely describe the process specific to physicians and researchers and offer solutions to common questions that make building a personal brand so challenging for many individuals.

15.3 Your Brand Plan: Defining Your Positioning

The process of creating a brand plan and strategy involves organizing your ideas and making decisions about your career directions that you can consistently implement over time. Your strategy and brand plan will evolve, especially as you gain a better understanding of your audience and your niche, or position, among similar services or peers. It may be helpful to write down your plan and revisit it every year to see how you are doing and whether any changes need to be made. In general, your brand plan will involve determining your positioning with respect to your target audience(s), creating brand elements, and strategizing how you will support this plan through activities and communication, also called *tactics* (Figure 15.1).

Positioning is how you will use your identity and attributes to stand out and build your brand. It refers to the niche you fill and it is specific to an audience or set of audiences. Many online or print resources describe exercises for writing a brand positioning, but you do not need to follow any formal or standardized formats. In fact, trying to *write* a positioning statement can detract from the more important process of strategizing. A positioning statement merely needs to describe your goals with respect to a target audience and how you uniquely qualify to fulfill them (Figure 15.2).

For many individuals, this is the most difficult part of the branding process to work through. Most personal branding resources begin with self-awareness exercises, such as identifying key attributes that best reflect who you are from a list of characteristics. The lists may often seem too generic and lack nuance to individuals who are further along in their professional careers. This may be a result of career services professionals

Figure 15.1 Personal branding process.

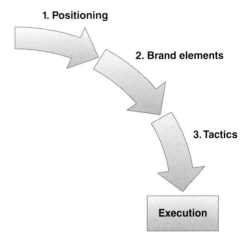

1. Positioning

2. Brand elements

3. Tactics

Execution

Figure 15.2 Elements of a positioning.

Positioning

GOAL

Audience

Audience Problems

Valued Attributes

Competition

authoring most of these resources and because it is traditional to use these tools with someone who is very young and seeking a career. The traditional branding process at an advertisement agency also begins with an assessment of the product attributes. It makes sense to start with learning about a new product if you are a creative director given a new assignment. As a physician/researcher who is already in a career track, you likely already have a good idea about your strengths, interests, and abilities. Because the underlying purpose of the exercise is really about choosing which attributes to focus on, it is easier to start with thinking about one of your goals, the important audience, and what problems that audience needs to solve. This will point you to the attributes that the audience will value. So, begin by formulating your goals and describing the target audience that is important to each goal. In our first case, a community pediatrician identifies a need to brand himself and his group practice.

15.3.1 Positioning Examples

15.3.1.1 Case 1: A Community Pediatrician

A community pediatrician has a unique service approach that ensures patients have short waiting room times, can easily reach a staff member by phone, and can quickly get answers to questions. He has four business partners, who are also pediatrics-focused healthcare professionals. They attend national meetings and meet with each other monthly to stay current with guidelines and new treatments. He is interested in branding himself along with his business. He knows that his audience includes his local community of potential patients as well as his peers, who may make referrals. He has naturally adapted his practice to provide what his key target audiences value most and is ready to begin the positioning and branding exercises.

The most important aspect of personal branding is your positioning, or where you fit in your target audience's mind. This is a process of matching authentic attributes that you possess with a need that is valued by your audience. The ideal place to begin is to think from *the other side of the desk*, or from the perspective of your target audience. Start by creating goal scenarios that you may already have in mind based on inspiration from mentors or that you develop based on decisions you have already made. Process each scenario by imagining that you are a member of your target audience who is tasked with identifying someone of value for a specific need or problem. In other words, take yourself out of the equation and work through each scenario to identify the key characteristics of value. Force yourself to maintain an anonymous perspective, keep you and your attributes out of the picture at this point.

After completing this exercise, for each scenario, assess yourself for fit within the audience's needs. Be realistic; you want to be able to fulfill the audience's perception of value. It may be that you are capable of the tasks but that you need to work on demonstrating the capability to the audience.

The next step in building your positioning is to consider the "competition." This can be peers, colleagues, or others who are attempting to fill the same needs with the same audience. How can you uniquely stand out in a way that your target audience truly finds valuable?

After building all of your scenarios, assess each for how it best fits you, or which feels the most authentic or important to you. You may see how you already fill those needs or how you may strive to add skills or attributes that meet the target audience needs. It may be tempting to decide on what your audience *should* want or need, but it is often easier to address a need that is already appreciated than to try to reeducate your audience on new concepts. Most importantly, the ideal positioning (and your brand) will be based on what is authentic and consistent with who you are and what your interests and skills are. The following case highlights a community surgeon who works hard to set himself apart from potential competitors. By studying what matters most to his audience, he is able to more easily home in on the way to communicate his value in terms they will relate to.

15.3.1.2 Case 2: A Community Surgeon

A surgeon who has been known to his mentors as an adept, meticulous, and medically knowledgeable physician wants to establish a successful community practice. He specializes in neck surgery and continues to follow advancements in underlying clinical conditions. He sees his audience as local specialists and perhaps primary care physicians. Through networking with potential target audience members, he learns that many of them fear unnecessary surgery or inappropriate surgery with outcomes worse than the initial concerns. Many local primary care physicians describe how they field patient questions about geographically distant surgeons who have a strong online presence and marketing. They are often unsure who to recommend in the local area.

Our surgeon puts a lot of energy into staying up to date with the latest techniques and warranting surgery through accurate diagnosis. He does not wish to build a practice around handling direct patient calls at this time. He builds his positioning around the specialist and primary care physicians and his ability to offer discerning and appropriate surgery, expertly performed. In this way, he sets himself apart because he determines a need or a problem in the community, thereby accepting the values of his target audience. He has appropriately positioned himself in relation not to what he himself wants, but strategically in line with the values he has assessed in his target audience.

When preparing your positioning, consider three key components: your target audience and the problems that they value solutions for, your attributes that solve their problems, and some acknowledgment of the competition (how you *uniquely* qualify). This is expressed as a "promise" to your audience and provides the basis for your brand. Your brand promise is how you capture this in writing. It is ideal for you to have only one or two brand positionings. Consider the case study of an academic physician who wants to advance her career and learns that her value to clinical research organizations may also be valued among her institutional administrators.

15.3.1.3 Case 3: A Clinical Oncologist/Researcher at an Academic Hospital

A clinical oncologist/researcher is inspired to become an expert in designing study protocols after participating on a drug manufacturer's advisory board. Through that experience, she learned that the manufacturer's medical directors need insight into current treatment paradigms and patient characteristics. They need oncologists who understand the drug discovery, clinical study process, and how to anticipate potential problems that can arise with study protocols. They simply cannot readily find individuals who specialize in this. Our oncologist has experience in designing local clinical trials and collaborating with colleagues at her institution. She has collected a wealth of literature regarding clinical study design and in particular within oncology.

That same clinical oncologist is an active researcher who is interested in moving up a career ladder within her academic institution. She begins researching how promotion decisions are made, who decides, and what attributes are important to those decision makers. She uses her network and mentors to gain a better perspective on how she measures up and what attributes she has or can build to make her a standout candidate.

She creates her positionings separately based on her unique attributes that solve the problems already identified by her audiences.

A SWOT (strengths, weaknesses, opportunities, and threats) analysis is another exercise that can evaluate your positioning with respect to time and movement toward a goal. To conduct a SWOT analysis, simply label a sheet of paper with four quadrants representing each word of the acronym (Figure 15.3). Then list qualities in each quadrant as they relate to your positioning(s). Weaknesses can reflect how you need to build skills or credentials to support your positioning. Opportunities and threats can capture

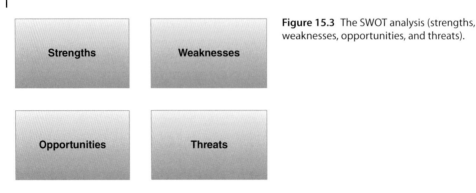

Figure 15.3 The SWOT analysis (strengths, weaknesses, opportunities, and threats).

how you may want/need to evolve your brand in the future. For instance, threats can represent how a negative incident can affect your brand image. By performing a SWOT analysis, you can acknowledge threats while focusing on the big picture and the value of your brand.

15.4 Creating Your Brand Elements

Once you have determined your one or two key positionings, you should think about your brand elements for each. When branding a company or product, brand elements include a name, logo, tagline, colors, etc. For personal branding, visual elements may not be necessary. Here is where you can begin to articulate your brand promise, though. True personal branding will often include a résumé or curriculum vitae (CV), headshot photo, and various length biographies. All of these materials need to clearly support your positioning. You may want to include other information, but the theme should be consistent. You may also want to consider writing a tagline or catchphrase, even if it is not communicated verbatim externally.

If you have more than one positioning, you may need to build two sets of personal branding elements to clearly communicate with different audiences. Additionally, if you are a physician working in the community, you may want to consider your personal brand as more of a small business brand and include a company name, logo, etc. This is particularly useful if you expect your business to grow and intend to hire other healthcare providers.

15.4.1 Brand Elements Examples

15.4.1.1 Case 1: A Community Pediatrician

For years the community pediatrician's practice referenced above has not had a particularly unique or memorable logo. They have called themselves Pediatric Associates. Half of the partners have professional photos of themselves on hand. Working with a brand strategist, he creates a new, more memorable company name and unique, appealing logo. All of the partners work on their personal branding to ensure that they have resumes, bios, and photos at the ready that support their company positioning. Their new company name becomes Parents' Choice Pediatrics Associates with the tagline, "providing prompt, leading-edge care." By creating a unique name that doesn't just describe their general service, they take advantage of the opportunity to describe what sets them apart — their understanding of what matters to parents. And it implies far more than just excellent care.

15.4.1.2 Case 2: A Community Surgeon

The neck surgeon ensures that his CV is in order as well. He obtains testimonials from his mentors (other healthcare professionals) and patients, especially those that mention his unique and valued skills. He arranges professional photos (headshots, during surgery, and with happy, health-looking patients taken at follow-up visits). He collates statistics about his performance that are accurate and truly reflect his skill. He prepares short videos answering commonly asked questions. And he ensures his practice and staff are ready to educate and facilitate the surgical process.

15.4.1.3 Case 3: A Clinical Oncologist/Researcher at an Academic Hospital

The clinical oncologist/researcher mentioned earlier knows that her CV will be an important part of her brand elements. For her external audience, those individuals and companies who organize early stage advisory boards, she will highlight her work in designing clinical trials and her department's infrastructure that allow her to recruit, enroll, and monitor trial subjects. To her internal audience, she ensures that her CV is kept current with all necessary information to support tenure and promotion. Through networking with her institution's communications office, she hires a writer to write a biography to keep on hand. She also arranges for a few professional photos to be taken, including some that show her working with patients, giving presentations, and working with colleagues. She collaborates to prepare articles (both peer-reviewed and white papers) that can help build her reputation as an expert in designing clinical studies.

15.5   Strategizing Your Tactics

Once you have determined how you want to be positioned and have some clear ideas for how to communicate this to your target audience(s), you are ready to strategize your tactics. Simply, this means how you will reach your target audience and let them know about who you are and what you do. This exercise also works best when you think *from the other side of the desk*. You may want to focus on a segment of your audience. Imagine clearly where your audience is likely to learn about you or your competition. The possibilities are endless and vary by cost, effort to maintain, and return on investment. Perhaps your audience would see you in a Super Bowl commercial, but that is prohibitively expensive and you will not likely recoup the costs. Table 15.1 lists a number of tactics that are available, but it is by no means comprehensive. As you work through potential tactics, prioritize them based on those that are absolutely necessary and that are easiest to implement. You may also need to consider deadlines, such as those for specific meetings and conferences.

15.5.1   Tactical Examples

15.5.1.1   Case 1: A Community Pediatrician

Our community pediatric practice melds a combination of personal branding with traditional business branding. With less emphasis on clinical research, the physicians in the

Table 15.1 Tactics for personal branding for physicians.

LinkedIn or Other Audience Relevant Social Media Sites
Institution, Company, or Personal Website
Personal Networking
Research Collaboration
National Conference Presentations
Peer-Reviewed Journal Articles
Book Chapters
Local Speaking Engagements
Speaking at Industry-Focused Conferences/Meetings
Blogs, White Papers, Newsletters

practice all maintain profiles on the clinic website, which has their unique promise and branding clearly evident. They also have a dedicated staff member to monitor patient grading sites to ensure that they are represented. The same staff member posts to and monitors the practice's active Facebook page and engages patients during visits to obtain feedback and act on that feedback. The practice makes sure that its unique promise of caring for children and the convenience for caregivers is reinforced when networking with obstetricians and other family practices, knowing that most of their patients come via referrals.

15.5.1.2 Case 2: A Community Surgeon

Our neck surgeon ensures that he has a solid presence through many of the same websites that our oncologist uses. He has his own company/clinic site as well. As a new key feature of his strategy, he arranges and delivers local presentations for community primary care providers on a quarterly or semi-annual basis to show advancements in diagnosis and treatments for relevant conditions (both surgical and medical). Between these events he sends out a newsletter updating his audience of his recent presentations, upcoming presentations, national engagements or publications, and reminding these physicians of his services and competence.

15.5.1.3 Case 3: A Clinical Oncologist/Researcher at an Academic Hospital

Our clinical oncologist/researcher applies a strategy that focuses on her positioning toward industry clinical trial design. She feels that the efforts to build this positioning will translate to unique value for internal promotion. Her goal is to maintain a strong presence in her institution through networking and collaborating. She already has an institution website that is kept up to date. Her growing number of peer-reviewed publications demonstrate her competence as a clinical study design expert and an active researcher and thought leader in oncology. She sets up accounts on LinkedIn.com, WebMD.com, ReachMD.com, and several other similar sites based on information from her CV. She also monitors her profiles on Vitals.com, HealthGrades.com, and other physician rating sites. She speaks regularly at national oncology conferences and uses that opportunity to network with industry medical attendees. As a new key feature of her strategy, she begins seeking out speaking opportunities at smaller conferences geared toward clinical research trial design that bring in an industry audience.

15.6   Executing Your Brand Plan

Once you have your positioning, brand elements, and tactics planned, you are ready to execute your plan. Set milestones and goals. You may already be given deadlines, such as those for tenure review. Work backward from those dates to ensure that you are able to complete your tactics in time and adjust and streamline your plan to accommodate. Know that the plan, including tactics, will develop as you see what works best or as new ideas come to mind.

Building your brand requires patience and your brand will likely evolve as you learn more about your profession and are exposed to new concepts and ideas. It is recommended to review your brand plan annually and re-SWOT to see how you are doing over time and to keep tabs on any new threats. It is important to remember to remain true to yourself, be authentic, and continue to be a valuable contributor to your profession.

15.7   Conclusion

As a clinician/researcher, building a personal brand can be important to your goals. When building your brand, the ideal place to start is determining your positioning or niche by finding your value in your audience's mind. You can do this by visualizing moving around to "the other side of desk" and imagining what problems you might have if you were an actual member of your target audience. Build materials to communicate your value and plan tactics that help you reach your audience.

Further Reading

Chritton, S. (2012). *Personal Branding for Dummies*. Hoboken, NJ: Wiley.

Kang, K. (2013). *Branding Pays: The Five-Step System to Reinvent your Personal Brand*. Palo Alto, CA: BrandingPays Media.

Kapferer, J.-N. (2012). *The New Strategic Brand Management: Advanced Insights and Strategic Thinking*. London, UK: Kogan.

Ries, A. and Trout, J. (2001). *Positioning: The Battle for Your Mind*. New York, NY: McGraw-Hill.

16

Dealing with Manipulative People

George K. Simon

Character Development, Forensic, and Clinical Psychology, Little Rock, AR, USA

For several years I was both a consulting and supervising psychologist to a private company managing a women's correctional facility in a relatively remote part of the state. One day, the nursing supervisor asked me to check out a situation in which an inmate had secured two prescriptions from a consulting psychiatrist for medicines that could easily be abused. I arrived at the inmate's housing unit just as correctional officers were following up on a tip that the inmate in question might have recently sold Valium and Trazodone to another inmate, who, when she did not pay as agreed for the drugs, was viciously assaulted. It appeared the inmate who was earlier prescribed the medications "cheeked" them deftly before leaving the clinic and attempted to make the sale very shortly after returning to her housing pod. She also was overheard bragging about how easy it was to "score" the drugs she obtained and was advising other inmates on strategies they might use to do the same. Because I had substantial experience working with disturbed and disordered characters, and because the consulting psychiatrist in question was a private practitioner with little experience with such a population, the nursing supervisor asked if I might facilitate a discussion with him about the incident and emphasize the need for greater caution when prescribing certain medications. The facility was desperate for medical personnel, and the physician (whom I'll call "Dr. James") was one of only a small handful both available and willing to meet the need. So, after carefully reviewing the inmate's chart and gathering as much information as I could about her background, I agreed to visit with Dr. James.

I remember ever so vividly the sincere yet surprising statement Dr. James made to me when I asked him if he'd considered the possibility that the person he interviewed earlier might have misrepresented her symptomatology, harbored hidden agendas, or perhaps even successfully feigned pathology for untoward purposes. He looked me straight in the eye and without flinching asked: "Why would she lie?" I was really taken aback. This woman's character pathology was such that you could write an entire book on all the reasons she might lie at any given moment, including the possibility of lying purely for "sport" and even when the truth would do just as well. But the nature of Dr. James's practice and experience didn't really prepare him for dealing with someone with this level of character disturbance, so for several days thereafter, over coffee and donuts, we had a collegial discussion on personality and character disturbances, and especially, the art of manipulation.

A Guide to the Scientific Career: Virtues, Communication, Research, and Academic Writing, First Edition.
Edited by Mohammadali M. Shoja, Anastasia Arynchyna, Marios Loukas, Anthony V. D'Antoni, Sandra M. Buerger, Marion Karl and R. Shane Tubbs.
© 2020 John Wiley & Sons, Inc. Published 2020 by John Wiley & Sons, Inc.

It was only five years prior to this incident that my first book *In Sheep's Clothing: Understanding and Dealing with Manipulative People* had been published. I wrote the book for two reasons: First, character pathology was rapidly replacing "neurosis" as the dominant psychological problem coming to the attention of mental health professionals; and second, I wanted to provide both professionals and lay persons a relatively comprehensive yet straightforward and easily digestible framework for understanding and dealing with some of the more troubling personality types among us. To avoid being taken advantage of, you have to be able to rudimentarily assess the character of a person and to know how to spot and appropriately respond to the most common tactics of manipulation and responsibility-resistance that impaired characters use to gain advantage over others. So, in the book, I gave folks some practical ways to determine what kind of person they might be dealing with and outlined the most common tactics of manipulation, responsibility-avoidance, and impression management, as well as the problematic thinking patterns and attitudes that frequently predispose and accompany these behaviors.

These days, it's pretty safe to assume that everyone has at least some degree of character pathology (and the use of term *character* here is not meant to be synonymous with *personality* – character being that aspect of personality that reflects a person's moral fiber). As I emphasize time and again in my book *Character Disturbance: The Phenomenon of Our Age*, character pathology exists along a spectrum or continuum, with the malignantly narcissistic, highly manipulative, and empathy-devoid predators we sometimes call psychopaths or sociopaths at an extreme end. Fortunately, such folks are relatively uncommon. But there are many other character-impaired folks out there, and some are quite skilled in the art of manipulation and responsibility-evasion.

In your professional career, you won't just find manipulators or other disturbed characters among your patients. Character disturbance of some type or degree is simply too prevalent to avoid encountering it no matter what setting you work in. You may find yourself dealing with manipulative or otherwise character impaired supervisors, facility administrators, pharmaceutical company representatives, and colleagues. And even if you're planning a strictly academic career, you're likely to find the politics of the "publish or perish" and turf-sensitive environment dominated by unspoken "angles" and hidden agendas of one type or another. You're also likely to encounter subtle undermining, backstabbing, and various other forms of covertly aggressive behavior as your associates jockey for power, positions of influence, and territory.

One of my dearest friends and mentors had an experience I'll never forget while we were working at the same hospital. My friend was of somewhat advanced age, but he brought with him not only years of seasoned experience but also a most stellar reputation. His presence at the hospital as the primary attending psychiatrist on one of the behavioral health units was probably the biggest single reasons its beds stayed full and it managed to recruit and keep a stellar support staff. Both the administrator of the hospital and a group of other physicians wanted to "phase out" my friend's influence and presence. Theirs was primarily a teaching hospital, and not only that, one increasingly invested in building a name for itself in two areas of psychiatric research. Although my friend had done a bit of research as a younger man, running subjects and publishing was not his main interest. He was at heart a clinician and not a researcher, and was fast becoming a poor fit for an organization seeking to recruit and groom scientist-practitioners. But the powers that be didn't want to get rid of my friend too soon, because they knew what it would cost them in the way of goodwill and bed count. And they didn't want to state their intentions openly to him because they feared he might quit and possibly even sign on with a competing entity. So, they had a series of meetings with my friend in which they laid out plans for him to "free up" some of his precious time to do much more of the hands-on work that he loved and less training (which for me was a very big loss because of how much more his years of experience and wisdom brought to the table as opposed to solely reviewing dry research findings). In the end, they "used" him well, and when they no longer had need for him, they unceremoniously let him go. And as is so often the case, the manipulation involved was in all the little things weren't said or weren't done. The real agenda was clear from the start, but no one wanted my friend to know the truth – until, that is, they were ready for it to be known.

16.1   Tips for Avoiding the Traps of a Manipulator

Whether you are dealing with patients, colleagues, or supervisors, your best defense against being misled, exploited, conned, or otherwise hoodwinked is to know well the telltale signs of character disturbance and the tactics manipulative characters are most fond of using to get the better of others. Spotting these tactics the moment they're displayed and having the skills to deal effectively with them is your best insurance against victimization. And while it goes without saying that whole books can and have been written on the subject, being mindful of a few general rules can really help you avoid falling prey to a manipulator's ploys.

16.1.1   Know the Kind of Persons You're Dealing With

Naturally, you can't do a comprehensive personality assessment on everyone you meet. But in these days of widespread character disturbance, it's incumbent upon you make at least a rudimentary assessment of a person's basic makeup. Know the most common personality types and their key features. Be especially on the lookout for those aspects of personality that tip you off to someone's basic character. Know the difference between someone's basic personality or "style" of relating (i.e. their distinctive manner of perceiving and interacting with the world) and a style that is either so extreme in its manifestation, deviant from the norm, or inflexible that it causes unnecessary tension and dysfunction in a person's interpersonal relations. And if something about a person's "style" of relating doesn't strike you just right, scrutinize a little further. Knowing the kind of person you're dealing is your single best defense against being exploited. This is especially true when it comes to the most severely disturbed characters. Many character-impaired individuals have keen predatory instincts and an uncanny ability to size up other people. Knowing you better than you know yourself and deceiving you about what they're really like is a major way they can get the better of you.

16.1.2   Educate Yourself about Manipulative Tactics

Learn about the most common power "tactics" certain character types use to influence the behavior others and to simultaneously manage the impressions others have of them. This is the heart of manipulation. And because the tactics are generally so effective, the folks who use them get a lot of reinforcement for doing so, and their use of the tactics easily becomes habitual – even in situations when using them is not all that necessary. So, you can expect manipulative people to have a hard time not displaying certain tip-off behaviors. If you know what to look for, you'll have a better shot at not being duped. Now, there are an awful lot of tactics people can use to manipulate others. And skilled manipulators can use just about any behavior you can think of to con and deceive.

In the clinical setting, you're likely to encounter certain tactics more often than others (although you'll certainly find the same tactics used in other settings as well). Here are six primary ones to watch out for:

1. *Evasion and diversion.* Manipulators will give half-answers that only seem like whole answers unless you think about them for a minute. Be careful not to take things at face value too readily. If it seems that someone is sidestepping or dodging an issue in some way, avoiding a topic, or balking at full disclosure, don't assume that they're simply anxious or being justifiably apprehensive or guarded. They could be being deliberately evasive for manipulative purposes. Press gently, while remaining respectful of boundaries. Deliberate and potentially malicious sidestepping will most likely then reveal itself. Manipulators are also good at redirecting focus to something irrelevant when they suspect you're zeroing in on something. Sometimes you don't even realize they've gotten you off topic until you're well into dealing with issues completely different from those you started to address. It's up to you to keep the focus where it needs to be, no matter what other diversionary paths they might tend to lead you down.

2. *Distortions, inconsistencies, and calculated omissions.* There are many ways to lie, and the most effective of these are subtle and covert. Sometimes the most effective way a person can lie (and therefore succeed at manipulating you) is to recite a litany of completely true, verifiable things while deliberately leaving out a key bit of information that would shed an entirely different light on things. Be especially on the lookout for *vagueness*, especially when it seems to occur at particular times. I once asked a person I suspected had a drinking problem if he'd ever been diagnosed or treated for substance abuse. The response: "They sent me to one of those places where they do the steps and all that and they told me I didn't have a problem and let me go," was classic in its evasive quality. First, there's the use of the nebulous "they." And when this vagueness is accompanied by no further clarifying information (a calculated "omission") it can be a successful manipulation tactic (i.e. "I know what you're going for here, Doc, but really there's nothing there so let's move on to something else, okay?"). But through polite, gentle, yet persistent follow-up, I eventually learned that the person had indeed been referred for inpatient treatment several times and to a very reputable clinic at that but was discharged for noncompliance each time.

3. *Rationalizations and excuses.* Some people have what seems to be an answer for everything. That is, until you reflect on things for a bit. Then, all of a sudden, their "explanations" don't make as much sense. Be mindful when someone is giving their explanations of events or reasons for doing certain things and don't take anything purely at face value. Trust your instincts when things just don't seem to be adding up. Don't be afraid to ask follow-up questions. Even the most seasoned con artists can dig themselves a pretty deep hole when you follow-up on answers that on second thought don't really explain anything. Most of us were taught to view a rationalization

as something a person unconsciously does to avoid the anxiety and pain associated with feelings of guilt. But artful excuse making can also be a fully conscious and deliberate way to make you think a person didn't do anything wrong or had just cause to do something that might appear wrong. Be leery when someone has too many answers for behaviors you find troubling in some way.

4. *Minimization and magnifying.* Earlier I gave the example of the inmate who "scored" drugs highly prone to abuse from an unsuspecting physician. The physician would later tell me how distressed the woman he treated appeared during the clinical interview. Her hands were shaking, her voice quivered, and she appeared to be holding back tears much of the time. And she reported all the things you'd expect for someone who was dealing with a fair degree of anxiety and depression: She couldn't get to sleep, couldn't stay asleep, wasn't eating right, was withdrawn, felt edgy and confused much of the time, and all the joy had gone out of her life. But these symptoms were all highly magnified. The records and reliable collateral information would have indicated that she routinely slept and ate well, was regarded as one of the more jovial, socially gregarious, and active individuals among her peers. And while she magnified any level of distress she might be experiencing for manipulative purposes, she also minimized the nature of other things, especially with regard to her history of substance misuse and other criminal activity. While it was true that she had never been incarcerated before and came in on a "technical" violation of the terms of her probation (she provided this information during her clinical interview), it was not true that she was a relatively minor offender who just happened to make a careless mistake. In fact, this individual had a lengthy history of all sorts of criminal activity for which she had been craftily able to avoid successful prosecution. And authorities were more than pleased that at least her violation of the terms of her probation for the one major charge that "stuck" allowed them to take her off the streets for a bit. With sufficient access to and careful examination of the official record and other reliable collateral data, the physician would have had a much better idea the kind of person he was dealing with. In today's world, with character dysfunction so much more prevalent than in years past, having sufficient historical and reliable collateral information is essential. Armed with such information, you're better able to spot minimization, magnifying, and other tactics when they occur.

5. *Seduction.* Manipulators know how to turn on the charm. Sometimes, it's so clearly superficial that an astute person can see through the charade. But the more crafty a manipulator is and the more intent they are on taking advantage of you or putting one over on you, the more careful they are about currying your favor. So sometimes they can appear quite genuine. They hope to endear your trust so that your guard will be down when they start advancing their real agenda. They will promise you things they have no intent on delivering, appeal to or stroke your "ego" to make you think they really like and value you, etc. when all the while their real intention is to eventually take advantage of you in some way. And, if you've allowed yourself to be charmed enough, you won't realize what they're really all about in character until long after you've been exploited.

6. *Shaming and guilting.* By and large, physicians are a fairly conscientious lot. It takes more than a fair amount of conscientiousness to excel in one's studies, to endure rigorous training, and to work long hours for rewards that are often quite delayed in coming. Character-impaired people know this intuitively. And they also know that

the best way to manipulate someone is to prey on their level of conscientiousness. Conscientious people want to do the right thing. So, if someone "invites" them to feel guilty about either doing something or failing to do something, or to feel ashamed about something they've done, they're likely to feel particularly motivated to make things right. Shaming and guilting work as manipulation tactics when the intended target of manipulation has a well-developed conscience rooted in empathy for others. Just try using the same tactics with a character-impaired individual and see how far you get. Manipulators prey on your good nature, pushing your guilt and shame buttons. Most of the time this is done very subtly. You won't hear a manipulator say something stark and overt like, "I think you're a horrible person," or call you a horrible doctor. Rather, they'll imply such indirectly. Still, the effect is the same. And the bigger conscience you have, the more susceptible you are to this kind of manipulation. It's important to develop what some clinicians refer to as your "third eye" and "third ear" in your interactions with folks who might be manipulative or otherwise character disturbed. At workshops, I like to advise attendees to take a step back and pay less attention to what the person says and more attention to the kinds of things they're saying, keeping that third ear open for the kinds of tactics described above. The same thing is true for the interpersonal processes taking place in an encounter with them. Sometimes, if you take a step back and look at the process critically with that third eye, you can more readily spot the jockeying for power and advantage that might otherwise slip your notice.

16.1.3 Divest Yourself of Harmful Misconceptions

Misconceptions about human nature that stem from traditional perspectives that for the most part have outlived their usefulness. Times have changed since Sigmund Freud and his theories about what makes most of us tick. He formulated his theories of personality based on the symptoms his patients reported and the observations he made of certain behaviors they displayed. For the most part, his patients were suffering varying degrees of what he called "neurosis." At its heart, neurosis is the result of a person's unsuccessful mitigation of the anxiety they experience in their unconscious struggle to curtail their primal urges in accordance with social norms. And in his time, one would expect there to be a whole lot of neurosis. The Victorian era was one of the most sexually and socially repressive times in modern history. If there were a motto or slogan that would best typify the "zeitgeist" or sociocultural milieu of the time, it would be: Don't even *think* about it! And the patients Freud worked with suffered from all sorts of bizarre maladies that were fueled not by physiological disease but by the anxiety (and inadequate "defenses" against the anxiety) associated with the frustration and stifling of their baser instincts. These folks were literally overly conscientious and deeply repressed "nervous wrecks." And neurosis, in one degree or another, was the defining psychological phenomenon of the era. Unfortunately Freud generalized his findings to an absurd extreme, thinking he'd discovered universal laws for human psychological functioning.

16.2 Neurosis versus Pathology: A Continuum

Whether you choose a specialty area like psychiatry where understanding and appropriately intervening with character disturbance is essential or go into family practice where you simply must take into account psychological health issues in order to provide sound overall medical intervention, you must keep in the forefront of your mind that we live in a vastly different age. The dominant sociocultural milieu is not one of massive repression but one of permissiveness and narcissistic indulgence and entitlement. If there were a motto best describing the times, it would be much like one heard in an old commercial: "Just *do* it!" For this reason, our populace is not as much teeming with "nervous wrecks" as it is liberally dotted with folks of impaired character. There are still neurotics among us, to be sure. But most neurosis these days is at a level most of us would consider *functional*. That is, the neurotics among us generally experience just enough compunction with respect to their primal urges that they make civilized life possible. By and large, the neurotics among us are functioning reasonably well (i.e. they're not suffering from the extreme and debilitating symptoms Freud's patients were) themselves and holding up society at the same time. You'll encounter neurosis, no doubt, but you'll more often encounter folks whose problems stem not so much from their neurotic conflicts but from their character deficiencies.

In *Character Disturbance*, I conceptualize a continuum of psychological adjustment with pure "neurosis" lying at one end of the spectrum and "character pathology" lying at the other end. Most folks fall somewhere along this continuum. And those whose problems arise mostly out of their character deficiencies are vastly different from their more "neurotic" counterparts on just about every dimension of interpersonal functioning you can imagine. This is extremely important to remember in your professional work.

In my years of private practice, I provided services to both folks who would rightfully be considered the "victims" in abusive, manipulative relationships as well as those character-deficient and responsibility-challenged individuals who were largely responsible for all the problems. And I quickly came to appreciate how destructive our traditional yet still dominant models of understanding human behavior really are. Victims, you see, spend a lot of time and energy desperately trying to figure out why the disordered person in their lives behaves in a manner they view as so irrational. Sometimes, they start to question their own sanity, because what their gut tells them about their manipulator goes against the popular psychology with which they are familiar. If, for example, they viewed their serial philandering spouse as "an insecure person underneath, struggling with low self-esteem, needy of affirmation, and fearful of commitment," they might be more likely to endure the situation, blame themselves in part for it, and strive to better understand as opposed to merely setting firm limits. But if they allowed themselves to believe what they'd always suspected in their gut – that their partner was a self-indulgent, entitled, ravenous sensation-seeker devoid of both the emotional maturity and the level of conscience to really love someone and not only knew it and didn't care to boot – it would be a whole different story. So, when a professional like myself finally gave them permission to trust their instincts, a whole new world opened up for them. And for that alone most were eternally grateful.

Working with neurotics is relatively easy, you see. Show trustworthiness, build positive expectations, and give them a safe place to work through issues, and all is good. Disturbed and disordered characters are an entirely different matter. They're harder to work with by nature and traditional approaches to understanding and intervening with them are virtually useless. As difficult as they are, you can work with them, especially if you're willing to adopt a radically different approach. I mentioned before that neurotics and disturbed characters are different on just about every dimension you can think of. Some of the dimensions on which the two groups differ most strikingly include the following:

- *Anxiety*. Anxiety is like fear except it's not attached to any identifiable source. When it is, we call it a phobia. And when someone is riddled with it, almost anyone with a decent degree of sensitivity can detect it. The symptoms neurotics report and the signs they display with regard to their psychological problems are fueled by anxiety. By and large, neurotics are an overly anxious bunch; their anxiety is inadequately relieved by the defense mechanisms they employ to manage it. It's their anxiety that makes them sick. In contrast, disturbed and disordered characters are not only generally lacking in anxiety but also some even lack the kind and degree of anxiety that would be adaptive. That is, they don't experience enough internal disquiet when contemplating the actions that get them into trouble. If they had even a little of the neurotic's typical apprehension, their lives would not be such a shipwreck.
- *Conscience*. Neurotics are generally conscientious types who have well-developed consciences, and perhaps even overly active, oppressive consciences. That's in large measure what fuels their anxiety. Disturbed characters, on the other hand, are lacking in conscience development and maturity. They're not conscientious enough to function responsibly. And in the case of extreme character pathology (e.g. psychopathy), any semblance of what most of us would consider a conscience can be absent altogether.
- *Shame and guilt sensitivity*. Having intact consciences, neurotics generally avoid doing things they'll feel badly about or suffer pangs of guilt or feel a sense of shame when they fall short of their internal moral standard. By contrast, disturbed characters don't have much sensitivity to guilt or shame. And they're well aware that their neurotic counterparts are very different from them, which is why "guilting" and "shaming" work so well as manipulation tactics (they only work, however, on neurotics).
- *Level of awareness*. Freud was right when it comes to neurotics. Most of the time, they're completely unaware of the conflicts raging inside of them because their unconsciously employed "defense mechanisms" keep the emotional realities behind those conflicts out of their conscious awareness. Disturbed characters have disturbing behavior patterns to be sure, but they generally know what they're doing and why. As I like to say in workshops: "It's not that they don't *see* (that what they're doing is socially disdained), it's that they *disagree*" (with what they know are society's generally accepted expectations). Many a clinician has wasted hours of precious time and energy laboring under the delusion that if they could just get a person to "see" the error of his ways, he'd change for the better. When it comes to a disturbed character's irresponsible behavior, it's not a matter of awareness but a matter of acceptance of and acquiescence to society's legitimate expectations.
- *Role of feelings*. The internal conflicts with which neurotics struggle are emotional in nature and most often involve feelings that have been repressed. Sorting out and working through these feelings is a principal focus of good psychotherapy. But when it comes to disturbed characters, while their feelings may be of some concern, the much bigger issue (and the reason for all the problems they have in their interpersonal relations) is the way they *think* about things. It's the attitudes they hold, their core beliefs about the world and how to get along in it, etc. and the problematic behaviors

in which they habitually engage that are predisposed by those dysfunctional attitudes, beliefs, and ways of thinking that need to be focused on when interventions are tried.

- *Nature and level of discomfort.* Neurotics by and large are distressed by the symptoms of their neurosis. They aren't comfortable with themselves in their unhappiness and not only seek help on their own but also tend to appreciate it greatly when they get it. Impaired characters are generally quite comfortable with all the signs of their character disturbance. They generally come to the attention of professionals under pressure from someone else. They're happy with who they are and how they operate. The symptoms of their dysfunction primarily cause distress to those around them. There's an adage in clinical circles: If they're making themselves miserable, they're probably neurotic. If they're making everybody else miserable, they probably have a character disorder. There's a lot of truth to that adage.

- *Response to adverse consequence.* Neurotics question themselves and strive to make changes when things go wrong. They tend to become unnerved easily and to blame themselves all too readily. Disturbed characters, by contrast, tend to be relatively unfazed when their actions invite negative consequences, and even if they are perturbed to some degree, it's generally a short-lived reaction. Some of them more severely impaired characters even become more emboldened and determined in their manner of coping as consequences mount. They blame others, not themselves when things go wrong anyway, and as a result, they don't appear to profit from experience.

- *Nature of self-presentation.* The traditional conceptualization of personality as a "mask" a person wears to hide the true (often unconscious) self is still a reasonably valid perspective when it comes to understanding and dealing with neurotics. And for far too long, all personality styles have commonly been considered, as David Shapiro aptly titled his landmark book, "neurotic styles." But certain personality manifestations tend to lie more toward the character-disturbed end of the neurosis – character disturbance spectrum, so it's not only inaccurate but sometimes even dangerous to view such individuals within the neurosis framework. For example, the ego-inflation associated with narcissism has been commonly viewed as an outward and unconscious "compensation" for underlying feelings of inadequacy and deficient self-esteem. We used to think similar things about bullies as well. But research is proving these assumptions mistaken most of the time. There are folks who are just exactly as they appear. When such folks act like they're all that, it's because they really sincerely believe they are. It's neither an act nor pretense. Sometimes, there's some objective basis for their beliefs about their greatness, and sometimes there's not. But even when there isn't, that fact isn't necessarily proof that the person must be "compensating" for anything. Some people simply tend to overestimate their value regardless of the objective realities. Even those narcissists with substantial resumes will still tend to believe they know more than they actually know or can do more than they actually can do. When you're dealing with character disturbance, most of the time what you see is what you get. If they're presenting a façade, it's usually done consciously and deliberately as part of the game of impression management or some other confidence scheme.

- *Needs in therapy.* The emotional conflicts causing neurotics distress are mostly unconscious to them. In insight-oriented therapy, we establish a trusting relationship that facilitates mindful reflection to help bring these conflicts into greater conscious awareness. Neurotics both need and generally benefit from insight-oriented interventions. Character-disturbed individuals need something entirely different in any therapeutic encounter. Already having insight but lacking in motivation to change course, they need "confrontation" and therapist-facilitated "correction" of their maladaptive ways of thinking, attitudes, and behavior patterns. And if change is ever to occur, it will happen in the here-and-now moment of interpersonal interaction when the therapist benignly but firmly shines a light on a dysfunctional attitude, way of thinking, manipulation tactic, or other responsibility-avoidant behavior and both proposes and reinforces appropriate alternatives.

Now, there are many, many more dimensions on which neurotics and disturbed characters differ, and in my book *Character Disturbance*, I outline several more than I've mentioned here. But there are even more than the additional ones I mention in the book, and it would be fair to say that just about everything you've ever learned about what goes on inside a person and why folks do the kinds of things they do probably doesn't apply when it comes to dealing with an impaired character. Narcissists, antisocial personalities, covert-aggressors, and other manipulators are simply a different breed for the most part, and the rules for dealing effectively with them are very different, too. It's when folks (professional and lay persons alike) refuse to accept that fact that they put themselves at a distinct disadvantage in any encounter with such individuals.

16.3   Aggressive Personalities

At the heart of most manipulation is covert-aggression (i.e. aggression "under cover"). Now, you'll notice I did not say "passive-aggression," and there's a very big reason for that. The kind of aggression that underlies most manipulation is anything but passive. Human aggression can be expressed in a number of different ways (i.e. directly vs. indirectly, overtly vs. covertly, reactively vs. predatorily, etc.), principle among them actively or passively. And by "aggression," I don't mean violence. Rather, I mean the forceful energy we all expend in pursuit of a goal. Aggression is different from *assertiveness* because when someone is aggressive they don't take deliberate care to place limits on their aggressive quests out of respect for the legitimate rights, needs, and boundaries of others.

Not only is covert-aggression the major vehicle for manipulating others but also most folks who manipulate as a lifestyle can be conceptualized as covertly aggressive personalities. They're part of a group that I and others (c.f., for example, Millon, *Personality Disorders in Modern Life*) prefer to label the "aggressive personalities" (all of whom can be viewed as aggressive variants of the narcissistic personality). And in *Character Disturbance*, I outline five major subtypes of this group: the unbridled aggressive (those who habitually break the major rules, are often in trouble with the law, and whom we've traditionally labeled antisocial personalities), the channeled-aggressive (ruthless individuals who for purely practical reasons keep their behavior within the confines of the law unless they're confident they can get away with breaking it), the sadistic aggressive (whose primary objective is not merely to "win," control, or dominate, as is the case with the other aggressive personalities, but rather, to inflict pain and relish both in that pain and the power they exerted over the victim), the covert-aggressive (manipulative) personality, and the predatory aggressive (the severely empathy-devoid, without conscience, malignantly narcissistic individuals who view all "lessor" creatures as their rightful prey), whom we have variously labeled as psychopathic or sociopathic in the past, and many of whom also show features common to the other subtypes (especially the sadistic and manipulative features).

As mentioned earlier, there are many ways to aggress. One can aggress by not doing (i.e. passively) or by doing something *actively* to injure, exploit, hoodwink, or control another. And one particularly effective way to aggress is to do it so subtly, or to disguise your behavior in such a manner that your intended target can't readily detect the victimization you have planned for them. That, in a nutshell is what covert-aggression is all about, and it's at the heart of most manipulative behavior. Hopefully, you will not join the already swelled ranks of folks (professionals being among the worst offenders) who mistakenly use the term *passive-aggressive* to describe covertly aggressive behavior. Covert-aggression is decidedly *active*, albeit *veiled*, aggression.

Covert-aggressors rely on certain behaviors I call power tactics to manipulate and control. Some of the common ones were mentioned earlier and some others will be mentioned a bit later. These tactics are effective because they put you on the defensive while simultaneously cloaking the aggressive nature of and intent behind the behaviors. At the very moment when a person is employing these behaviors, they're attempting to get the better of you, look good while doing it, and stiffening their resolve not to do as they know society wants them to do all at the same time. This is really important to remember, because many of the behaviors I'm talking about have been traditionally conceptualized as *defense mechanisms*. But when you have a strong gut feeling that someone is doing you wrong, and then when you confront them they offer an excuse for it, rather than assume that they're "rationalizing" (unconsciously as a way to assuage feelings of guilt), you might want to consider that they simply don't want you to keep your guard up, to recognize what kind of person they really are and what they're really trying to do, and, ultimately, making a statement that they believe they have a perfect right to keep doing the very thing they've done even though they know that you and many others would regard it as wrong. And if you buy the excuse and back off or back down, it's game, set, and match. That's how manipulation works.

Some of you are embarking on careers that don't necessarily involve working directly with patients. Perhaps you're planning on working in academia or some other non-clinical setting. But you're still likely to encounter all the different types of aggression I outlined earlier on at least some occasions. Our professional working environments are by nature highly competitive and relatively insecure, so they frequently condone, enable, and even promote aggressive behavior of all types. By far the most common form of aggression you're likely to encounter is covert-aggression (i.e. manipulation).

16.4 Tactics Used by the Covert-Aggressor

There are several kinds of covertly aggressive (manipulative) behaviors, and you're likely to encounter some more often than others in nonclinical professional settings. These are discussed next.

16.4.1 Covert Intimidation

Of course, every workplace has its bully. But overt intimidation is not generally the manipulator's game. Covert-aggressors find it more effective to make carefully veiled threats to get people to cave in to their demands. There are many ways to stealthily intimidate others. Principal among them are:

1. *Selective treatment.* Singling the target out for "special treatment."
2. *Systematic exclusion or ostracizing.* Leaving the target to feel abandoned and alone if he or she doesn't "play ball" in the manner the manipulator wants.
3. *Covert rewards and punishers.* Subtly providing for pleasant rewards to come a person's way for compliance while stealthily imposing negative consequences for non-compliance.

When covert means of intimidation fail, manipulators will often turn to more overt forms, still careful, however, not to be so blatantly obvious or easily discoverable in their behavior that they'll tarnish their image or invite sanction. Some of the more overt tactics include emotional, personal, occupational, and career "blackmail."

16.4.2 Lying

At first glance, this appears to be an obvious thing to include, and it should be easily detected. But in actuality, skilled manipulators are expert in lying in many ways that are not obvious enough to give you forewarning. Sometimes, the most effective way to lie is to state a virtual litany of verifiably true things while deliberately omitting a key piece of information that would shed an entirely different light on things. Manipulators are calculated and selective in their omissions, which allows them to come across as sincere and truthful while fully intending to deceive and take advantage of you. Manipulators will not just lie *to* you, they'll also lie *about* you. They'll also subtly encourage others to spread lies about you as well to diminish your standing or to gain some other advantage over you. Lying is, perhaps, one of the most telltale signs of character disturbance. And the severity of someone's character impairment is often most evident in *how* and *why* they lie. The most severely disordered characters (e.g. psychopaths, sociopaths) are particularly "pathological" in their lying in that they lie liberally and without compunction, and without any apparent necessity or rationality (i.e. when the truth would easily suffice or would seem of more practical benefit). Rampant lying, for the sole purpose of remaining a step ahead of someone else, should always raise a red flag. You'll want to keep your distance from such folks.

16.4.3 Denial

Now the denial I'm talking about here is the "tactic" of manipulation, impression-management, and control. Many (professionals are again among the worst offenders, here) mistake it for the ego-defense mechanism of denial. I give examples of the two radically different types of denial in *Character Disturbance*.

The defense mechanism of denial is best illustrated in the example of a woman whose husband was working gleefully with her outdoors in their garden when he complained of the heat, seemed to get unsteady, could barely talk, and ended up being taken to the hospital. Not long after he arrives there, the woman is told he's suffered a severe stroke and, while the life support makes it appear he is alive, he is virtually gone. His wife holds his hand and talks to him even though the doctors and nurses tell her he can't listen. And she comes every day and waits for him to awaken, even though she's been told he will never do so. This woman is in a unique psychological state. Her unconscious mind has put her into a state of denial because the reality of circumstances is simply too emotionally painful to bear at the moment. But in time, and with gradual acceptance, her denial mechanisms will break down. And when that happens, the anguish and grief that her mind was trying to keep her from experiencing will gush forth.

The manipulation and responsibility-evasion tactic of denial is of a very different kind. I illustrate it with the example of Joe, the class bully, who derives satisfaction out of roaming the school halls during class exchanges, and pushing the books out from the arms of unsuspecting students. One day, a hall monitor catches the event and calls him out. "What?" he exclaims, throwing his arms wide open and putting an innocent look on his face. "What are you looking at me for?", his charade continues. Is this person so riddled with emotional pain that his unconscious won't let him accept what he's done? Does he really think he didn't do anything? Probably not. He probably thinks that if he acts innocent enough he might avoid detention hall. Maybe the hall monitor didn't quite see enough. Maybe if he acts self-righteous enough, the conscientious monitor will entertain some doubts about what she saw. It's a tactic – a lie – pure and simple. There is likely to come a time when you'll have to confront someone you think has done you wrong. And when you do, he's likely to deny it. You might assume he feels badly about it and is trying to save face. But this is a potentially dangerous assumption. Denial can be tactical as opposed to protective. And some of the most ardent aggressors can be the staunchest deniers when caught. Sometimes just the level of apparent conviction when someone is denying can lead you to doubt yourself. And denial is often even more effective as a tactic when coupled with the following tactics.

16.4.4 Feigning Confusion/Ignorance and Innocence

When you have to confront character-impaired people with something they've done to injure you, they might pretend they have no idea what you're talking about, act confused, or otherwise imply that their hands are clean. And if they're convincing enough about this, given your own level of conscientiousness, you might end up feeling like the bad guy for even suspecting them. That's why this tactic is so effective. It's always helpful to have your facts straight and your supporting documentation at your fingertips. Still, a staunch manipulator might hold onto a story. You have to be prepared to discount the tactic and set the boundaries and limits you need to protect yourself. Putting your energy into getting them to "fess-up" is both unnecessary and pointless most of the time.

16.4.5 Playing the Role of Victim

The one thing that disturbed characters know about sensitive, conscientious, "neurotic" people is that they can't stand to see anyone in pain or a position of disadvantage. And the very best way to play on someone's sympathy, and thus manipulate them into doing something they might not otherwise do, is to successfully cast yourself as a victim of some unfortunate circumstance or someone else's behavior. This is an especially good trump card for manipulators to play when someone's on to their games and strongly suspects them as the victimizer. Claiming victim status is a way to turn the tables and take the heat off. This tactic is often coupled with the tactic of feigning innocence, and it's effective because most of us hate to think we're not only falsely accusing but also adding unnecessarily to an already wounded party's pain.

In my books *In Sheep's Clothing* and *Character Disturbance*, I outline many of the more common tactics that disturbed characters use to manipulate and control others. It's a good idea to become familiar with as many of these tactics as you can so you can spot them quickly. But it's also important to heighten your awareness about what makes all the disturbed and disordered characters tick because, frankly, there's virtually no limit to the kinds of behaviors they can use as tactics. So your greatest security is in knowing what such people are really like. Fortunately, most of these folks are relatively easy to understand.

In the real estate business. there's an adage that only three things really matter: location, location, and location. For the kinds of characters we've been talking about, no matter what the situation, only three things really matter to them: position, position, and position! As unfathomable to most folks as this seems, it's the most crucial thing for you to keep it in mind. Whether you're dealing with a manipulative or otherwise character-impaired client, co-worker, supervisor, or colleague, just remember that their life script is all about gaining advantage, so you always need to be proactive in keeping the field of play (i.e. the nature of the interpersonal interaction) as level as possible. You can expect them to seek the upper hand. But you can't trust them with any position of advantage and you certainly can't allow them to set the rules of engagement. You have to do that, and both proactively and *early on* in any encounter. That's your best protection.

Scott Peck (1983) called many of these folks "people of the lie" and claimed that because they fear being exposed more than anything else, calling them on their game is enough to arrest matters. But I have found this to be both untrue and really bad advice. Remember, folks who always have to be in the dominant position can be particularly vindictive when they think you've called them on their game, gained the upper hand, or especially if they feel you've defeated them. And when "outed," covert-fighters can become more openly, fiercely, and unscrupulously aggressive. That's why it's so important for you to set the terms of engagement and to do so early on in your encounters with character-impaired people.

16.5   Tips for Setting the Terms of Engagement

To set the terms of engagement effectively, you must observe these general rules:

- Do your best to avoid direct contests, and avoid trying to expose. Covert characters will eventually expose themselves, and contests with them inevitably escalate to everyone's disadvantage. Do your best to look for the potential "win-win" in your relations with them. And when fashioning a win-win scenario appears impossible and it looks like they might be facing the possibility of defeat, you need to be prepared for consequences and have an action plan ready if recriminations ensue.
- When you assert yourself on principle, stand your ground and don't let any of the tactics a manipulator might throw your way sway you. The manipulator's tactics will "invite" you to mistrust your gut and to buy into their distortions and misrepresentations. Stick to your principles and trust your instincts. They will guide you.

- Be open and direct in your dealings and communications. Expect the same from those with whom you engage. When you ask a simple direct question, don't settle for anything less than a simple, direct answer. And when you lay your cards openly on the table, don't accept an evasive or noncommittal response.
- Make sure you have a strong support system and have proactively fashioned for yourself some viable alternatives if the situation involving the character-impaired person you've been dealing with deteriorates or becomes impossible.

It would be simply impossible to list all the steps you might take to empower yourself in your relationships and lessen your chances of being manipulated (in *In Sheep's Clothing*, I outline 12 of the more important "tools of empowerment"). But your intuition will guide you to a host of other empowering possibilities once you keep firmly in your mind that we live in a very different age – the age of character disturbance – and you stop assuming that everyone you meet is a frightened, insecure, unconsciously well-defended, anxious neurotic like Freud believed, accepting instead that there are more self-centered, grandiose, entitled, unscrupulous, calculating, and sometimes surreptitious fighters out there these days just waiting to get the better of you. Once you've made peace with that reality, heightened your awareness about such folks, know how to spot them, and understand how they tend to think and act, you've already empowered yourself substantially. And once you accept the notion that it's incumbent upon you to proactively define the terms of engagement with them and you understand the basic principles of doing so, it will become clearer to you with each encounter how to respond to their manipulations and maximally empower yourself.

Reference

Peck, M.S. (1983). *People of the Lie*. New York: Simon and Schuster.

Section III

Research Ethics

17

Honesty and Truth in Academic Research and Writing

Thomas F. Lüscher

Heart Division, Royal Brompton and Harefield Hospitals and Imperial College, London, UK
Center for Molecular Cardiology, University Zürich, Zürich, Switzerland
European Heart Journal Editorial Office, Zurich Heart House, Zürich, Switzerland

17.1   Introduction

Academic writing is an essential activity for researchers and has many standards that need to be fulfilled, i.e. formal requirements, quality standards, as well novelty of the content. In this chapter, I will dwell on honesty and truth as integral component of any scientific process. The prime goal of a scientific endeavor is search for truth, for which the researcher's honesty is a prerequisite. But what is honesty? If we were to explain it to our children, we would probably say: Honesty is when you speak the truth and act truthfully. This leads just to another question: What is truth and truthful acting? Thus, let's start with defining the truth.

17.2   Truth

As simple as it seems, there are many definitions that seem obvious. Ever since humans began to think for themselves, defining truth has been a challenge for any philosopher and theologian. Initially, the truth came from prophets, ancestors, and priests. The world was ruled by gods, spirits, and angels. Thales of Miletus (624–547 BC) is considered to be the first philosophers who set the basis of what we call science today (Hawking and Mlodinow 2010). Nature was no longer the product of the unpredictable will and temper of gods, but something that was governed by laws that can be discerned by observation and logical thinking. Thales' vision was that nature could be understood, and hence would be eventually predictable, once its laws are discovered. Truth in this sense is nothing else than the successful discovery of such laws.

For the advancement of what we call the scientific process today, another important component comes into play: *Doubt*. If we believe everything that we are told, we are believers, not scientists. Discovery often begins with an important question: Is this really true? Rene Descartes (1596–1650), the philosopher of doubt, is another founder of the modern science. His philosophical writings are classics (see Descartes 1984). His approach was to put in doubt any belief that falls prey to even the slightest uncertainty

A Guide to the Scientific Career: Virtues, Communication, Research, and Academic Writing, First Edition.
Edited by Mohammadali M. Shoja, Anastasia Arynchyna, Marios Loukas, Anthony V. D'Antoni, Sandra M. Buerger, Marion Karl and R. Shane Tubbs.

in order to allow for an unprejudiced search for the truth. He then searched for something that lies beyond all doubt and ended up with his famous statement: "je pense, donc je suis": *I think, therefore I am.* Once this certain ground was reached, Descartes then proceeded to rebuild his world on the basis of this absolute certainty. The facts that he reestablished on this absolute certainty included the existence of a world of bodies external to the mind, the dualistic distinction of the immaterial mind from the body. This distinction between mind and body made man an observer of nature, who then, with the principles of logic, mathematics, and geometry, set out to discover the world. The scientific search for truth was a major break with the scholastic Aristotelian tradition, as it was now based on the concept of observation and analysis using a mechanistic approach applying the principles of physics, geometry, and mathematical quantification.

The next step in understanding truth was provided by Sir Karl Raimund Popper (1902–1994), one of the most influential philosophers of science. In his seminal work, *The Logic of Science*, Popper introduced the principle of falsification (Popper 2002). According to this concept, science advances with conjectures (hypotheses) and refutations (falsifications). Truth, therefore, is what has not been refuted based on novel observations and findings. Thus, the statement "All swans are white" was no longer true as soon as black swans were discovered in Australia. To summarize, truth is what has survived the test of time. This puts observations at the center of the scientific process. Indeed, natural science and medicine are based on this very principle. However, science is not simply observation of facts. Facts are what we observe, but observation is not fully devoid of theoretical assumptions. Indeed, observation is based more and more on highly sophisticated tools, and what we observe in science and medicine, in particular, is much more than what we can see with the naked eye. Eventually, however, all findings, whether obtained by the naked eye or tools of any sort, have to pass the test of time.

17.3 Honesty

Let's consider honesty now: why is it so important in the context of science? The goal of science is continuous production and extension of certified knowledge. Certified knowledge is based on empirical observation, precise and consistent expression of facts (Wittgenstein 2005), and their interaction supported by modern statistical analysis (Lüscher 2012); it is continually up for confirmation or, rather, falsification, as Popper has taught us (Popper 1989).

The intention of scientists is to pursue truth and nothing but the truth. Without such a mindset, the aim of science cannot be reached. As outlined by Immanuel Kant (1724–1804), it is the intention that counts, not necessarily the consequences of one's doing. In his seminal work, *Critique of Practical Reason*, published in Königsberg in 1788, Kant declared that nothing is as undisputedly good as a good will (Kant 1996). We all may err (and we are all prone to it), but is should not be based on intention, on the will to deceive. As the Scottish writer Samuel Smiles (1812–1904) put it "He who never made a mistake, never made a discovery." In fact, we err more often than we are right. Mistakes are inevitable experiences on the road to discoveries. Honesty and truth must remain as the foundations of the scientific process; without that, it would fail.

Let's define honesty more precisely: What does it mean in the scientific process? Certainly we have to report what we have found and how we have found it (and document it in the study report or manuscript) as precisely as we can. The data should be appropriately analyzed, quantified, and statistically tested. But, which data should be reported? Only the important ones? All data? Are we allowed to delete what does not fit the expectations or data that is not in line with the appealing concepts? Certainly not – unless there is a good reason to do so. The experiments may fail for obvious reasons, the cells may have been infected, the mice might get sick or belong to another than the expected strain, blood samples may have been mixed up, among other reasons. These should not alter a scientist's clear judgment. Researchers don't like outliers, but they are part of biology and should not be deleted–otherwise we twist the reality. In medicine, there is another reason for absolute honesty: fraudulent data not only endangers the mission of science and the reputation of journals and authors, they may harm patients and cause unnecessary costs for useless studies trying to reproduce fabricated data.

17.4 Dishonesty

Obviously, contrary to the codex of science, things can go wrong and undermine the scientific integrity. How common is scientific misconduct? While data on the incidence of fraudulent research are probably not reliable, available surveys suggest that the behavior might be much more common than expected. Fanelli (2009) reported that 2% of scientists admitted fabricating data, and up to a third admitted other misconduct such as dropping data points that did not fit their expectations, changing study designs retrospectively, using inappropriate methodologies, or altering results in response to pressures from competitors or funding sources.

Steen (2011) found in his study that rate of article retractions increased markedly from 2000 to 2010. Disturbingly, more than a quarter of retractions were for scientific fraud and many retracted papers involved high impact journals. In a subsequent analysis by Fang and Casadevall (2011), the rates of retractions by prestigious journals correlated well with their impact factors, i.e. higher-impact journals had higher retraction rates (Figure 17.1). This trend has increased recently (Nallamothu and Lüscher 2012). *The New York Times* even proclaimed an epidemic of retraction (Marcus and Oransky 2012).

Some very recent fraud scandals in the Netherlands (Wise 2011; Anon 2012a) and Japan (Shimokawa 2010; Anon 2012b; Yui 2012) sparked intense discussion about the issue. Why has fraud become more common in today's scientific community? If this is indeed the case, several explanations are possible: First, the prestige associated with publishing in higher impact journals and the importance of such publications for the future careers of the authors may jeopardize scientific integrity. Humans thrive for recognition, visibility, and glory, and some cross the red line in this context. Second, the increasing competitiveness for grants from funding agencies as well as bonuses that certain universities provide to their faculty members who publish in high impact journals may encourage scientific misconduct. On the other hand, the high visibility of papers published in high-impact journals facilitates detection of fraudulent data.

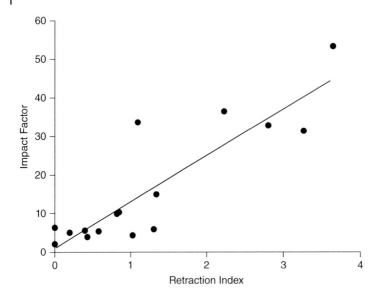

Figure 17.1 A direct relationship exists between journal impact factor and retraction index (a measure of the frequency of retracted articles). Source: Adopted from Fang and Casadevall (2011).

17.5   Spectrum of Fraud

Fraud is a harsh word – it can damage the reputation of entire institution, halt research programs, and destroy promising careers. When using the term, we should therefore do so with great care. At first, we have to remind ourselves of what the word means (Lüscher 2013a). The spectrum of its connotation is broad: It ranges from duplicate publication, self-plagiarism, sloppiness and data suppression to true plagiarism, data theft, hoaxing, forging, trimming, and cooking (Figure 17.2). It is obvious that not all misconducts are of similar severity. Some such as data suppression are not uncommon practices. For instance, there are good reasons to exclude certain pieces of data of a study, e.g. if there

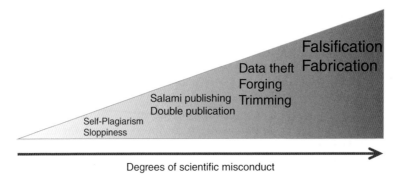

Figure 17.2 The spectrum of scientific misconduct. Source: Lüscher (2013b). Reproduced with permission.

have been well-documented technical problems with the assay, if the cell culture happened to be contaminated, or if patients have been mislabeled. However, such measures must be reported in the laboratory book, the project report and if appropriate in the methods section of the paper. Sloppiness such as different numbers in the results and tables, inappropriate description of methods and other aspects of a study are certainly misconduct, but it is not fraud in the proper sense, rather, mistakes that we all can make unintentionally. Certainly, this is not good scientific practice, but from an ethical standpoint not a severe misconduct. Finally, in clinical science, proper informed consent may not have been obtained from all patients and other ethical standards may have been neglected – again inappropriate, but not fraud in the proper sense.

Thus, deviations from accepted good scientific practice in obtaining, analyzing, and reporting results may vary in the degree of severity with falsification, fabrication, and invention at the top (Figure 17.2). If integrity and fraud are not distinct entities, but, rather, two extremes on a continuum of academic behavior, how can we distinguish the two? Again Kant shows us the way: It is the intention to deceive that is essential; it is that very mindset that separates mistakes from trimming and sloppiness from fraud. Similarly, if the work of others is used or even incorporated, but not cited, it matters whether it is done by intention to deceive or just negligence. Certainly, the proper way is to cite appropriately, as it acknowledges the work and contributions of others. Thus, from a moral standpoint, scientific misconduct may be minimal or severe, and we have to consider this aspect when discussing the issue.

17.6 Learning from the Past

Deviations from good academic practice are not new. They have occurred throughout the history of science involving even the very best discoverers. For example, in 1857 the Augustinian monk Gregor Mendel (1822–1884) began to carry out experiments with peas in his monastery garden. He presented his findings at the Brünn *Natural-History Society* in 1865. He published his intriguing results in the journal of the Brünn *Natural-History Society* in his native language German, as was common in the Austrian-Hungarian Empire. What was later to become the basis of modern genetics was largely ignored by the scientists of his time. In 1900, more than a decade after his death, his seminal work was acknowledged by the plant biologists Hugo de Vries, Carl Correns, and Erich Tschermak. However, in 1936, Aylmer Fisher, a population geneticist, charged Mendel of fudging. While he acknowledged Mendel's role as a true pioneer, he claimed that Mendel's data were too good to be true, as the distribution of dominant-to-recessive traits in his papers was amazingly close to the ideal values predicted by his hypothesis.

Was Mendel fraudulent? Although Fisher's accusations were not unfounded, we have no real proof of any misconduct. Whether Mendel was selecting data intentionally or whether he felt that he was entitled to omit experiments that he considered out of range for some reason remains uncertain. In any case, he was right. Certainly, he was unaware of the probability and statistics, posttranscriptional regulation and penetrance of genes, microRNAs, methylation, and other factors that contribute to a greater-than-expected variability of genetic experiments.

Also, as in many current cases, scientific illustrations have been manipulated and improved by many giants of the past, among them Charles Darwin (1809–1882). In 1872, more than a decade after the publication of *On the Origin of Species* (Darwin 1859), Darwin published *The Expression of the Emotions in Man and Animals* (Darwin 1872). The book was illustrated by figures of human faces expressing emotions such as grief, joy, or anger. Today it is obvious that some of the pictures had been intentionally modified or retouched (Judson 2004). If we take into consideration that cameras of that time were slow and it was difficult to catch the right expressions at the very right moment in time, are the pictures fraudulent? Or are they just educational prototypes? Again, Darwin was mostly right, as were other icons such as Sir Isaac Newton (1643–1727), who tended to adjust his calculations to fit best with his models (Westphal 1980).

But there are cases of true fraud as well. In 1971, William T. Summerlin, a transplant immunologist, claimed that rejection of skin grafts was preventable if the transplanted tissue was held in tissue culture before transplantation. This was considered breaking news at the time. Based on his findings, he was offered a position at the Sloan-Kettering Institute in New York as chief of transplantation immunology. Unfortunately, he could not reproduce his findings in his new lab – obviously this happens as any researcher knows. In desperation he colored the grafts of his white mice black. The misconduct was soon discovered and during the inquiry, doubts fell also on his previous work. The investigative committee found that he had manipulated data in his past studies as well. He was fired few days later (Judson 2004).

The famous case of John Darsee had a similar pattern (Braunwald 1987; Stewart and Feder 1987). By the age of 33, the intellectually brilliant and charming Darsee had published more than 100 articles in prestigious journals such as the *New England Journal of Medicine, Circulation,* and others. He continued on his path until his colleagues in the Cardiac Research Laboratory at Harvard grew suspicions. They informed Robert Kloner, the head of the Animal Research Laboratory at the Brigham, that they were convinced that Darsee was inventing experiments. After an investigation, Darsee's abstracts and papers had to be withdrawn. Eventually he was fired and lost his medical license.

Elias Alsabti used another at-the-time sophisticated technique. He copied articles from less prominent journals, changed titles and wording, made himself the first author, and submitted his plagiarism to other journals. This was a form of blunt plagiarism that would no longer be feasible in the era of the internet and specialized software.

Furthermore, insider knowledge of the review process may still be misused, as was the case for Helena Wachslicht-Rodbard: On November 9, 1978, she had submitted a paper to the *New England Journal of Medicine* about insulin receptors in anorexia nervosa (Judson 2004). Arnold Relman, the then editor-in-chief, sent the paper out for review, as is still common practice today. One of the invited reviewers, Philip Feig, vice-chairman of the Department of Medicine at Yale, asked his post-doc Vijay Soman to have a look at the manuscript. The two of three reviewers recommended rejection, and Relman, with divergent reviews at hand, asked for a major revision. Meanwhile, Vijay Soman and Philip Feig, who worked on the same subject, submitted their own paper to the *American Journal of Medicine* where Feig was associate editor. The editor of that journal sent the paper to Helena Wachslicht-Rodbard for review (Broad 1980a,b). This unfortunate event led to the uncovering of a major fraud. Indeed, Soma – inspired by the work of Wachslicht-Rodbard that he had seen confidentially as a reviewer – had fabricated most of the results and he had copied parts of Wachslicht-Rodbard's article into his own paper without citing it.

Finally, a more recent case shook the cardiology community: On November 17, 2011, Erasmus University fired its internationally known expert in perioperative care, Don Poldermans. As outlined in a report the investigative committee made available in 2014, Poldermans was charged with misconduct on several accounts (Hazes et al. 2014). The first studies that have been questioned were the Dutch Echocardiographic Cardiac Risk Evaluation Applying Stress Echocardiography (DECREASE) studies, i.e. DECREASE VI, IV, III, and II (Feringa et al. 2006a,b; Poldermans et al. 2006, 2008; Dunkelgrun et al. 2009). Specifically, the committee found that Poldermans did not adhere to the proper procedures of informed consent. Although none of the patients were harmed and the data of these papers appeared to be solid, Poldermans' actions were considered a breach of scientific conduct. Furthermore, the committee concluded that he had not collected data according to scientific protocol. In two cases examined, i.e. DECREASE II and DECREASE VI, there were serious errors. Finally, in several other instances, Poldermans apparently fabricated data. As a consequence, the *Journal of the American College of Cardiology* announced concerns about the scientific validity of three published papers. The *European Heart Journal,* who had published 11 of Poldermans' papers, published an expression of concern related to papers where Poldermans was the responsible author (Lüscher 2013b). Furthermore, the editors placed a similar statement on their website related to the DECREASE V study's two-year follow-up, which was initially published in the *New England Journal of Medicine.* Indeed, while other studies such as POISE found an increased mortality and stroke rate in patients treated with beta blockers prior to high-risk surgery (Devereaux et al. 2008), the DECREASE trial reported a striking benefit of bisoprolol under these conditions. A meta-analysis published in HEART in July 2013 casted further doubts on the validity of DECREASE V (Bouri et al. 2014). Consequently, the *European Society of Cardiology* decided to revise their guidelines on perioperative care, which had been written under the leadership of Don Poldermans and his colleagues in 2009 (Poldermans et al. 2009). In September 2014, the revised version was eventually published in the *European Heart Journal* (Kristensen and Knuuti 2014). Lastly, stem cell research had its case. In 2018, Harvard University investigated the scientific conduct of one of its well-funded stem cell researchers, Piero Anversa, and determined that many of his published data had been falsified and/or fabricated. As a consequence, more than 30 papers of his were retracted and a large sum of funds from federal grants had to be returned (Lüscher 2019).

17.7 The Pattern of Fraud

Although they usually can hide themselves well, fraudulent scientists often demonstrate a typical pattern. Often, they are highly ambitious, talented, charming, and often charismatic with a brilliant mind and an impressive publication record. Based on their remarkable achievements, they move up the academic ladder quickly and become part of highly productive research centers at prestigious institutions typically led by a famous mentor. The atmosphere of such centers is highly competitive where positive results, if not breakthroughs are expected. To fulfill such expectations, the fraudulent scientists immediately exhibit an impressive productivity and publish a large number of papers in high impact journals at a young age. One of their typical strategies is to invite notable established scientists from other institutions to join their papers in order to increase the credibility to their findings. Numerous collaborations with various institutions make it difficult for invited authors to assess the quality, if not the sheer existence of the reported data. Duplicate publications with similar or identical figures and tables are also not uncommon – a practice that can now easily be detected with today's search engines. Indeed, many journals check novelty and the degree of overlap with previous publications of each submission considered for publication using a medline search or plagiarism detection tools.

Fortunately, sooner or later fraud comes to an end: colleagues, students, post-doctoral fellows or technicians eventually become aware of the misconduct. Whistle-blowers and the scientific process itself play a crucial role in exploring varying forms of scientific misconduct. Most importantly, as other scientists are unable to reproduce the data, doubts about the validity of such data arise and the fraudulent findings are eventually uncovered. That is the good news: *Lies also do not live long in science*. Indeed, they do not pass the test of time.

Can the peer-review process detect scientific misconduct? The peer-review is not designed to do so, as it assumes honesty as a prerequisite of scientific works and aims at improving submitted data through expert advice (Lüscher 2012). Peer-review is not designed to be an inquiry; suspicion is not its mindset, nor it is a legal entity.

When selecting manuscripts based on their novelty, importance, and interest, editors must be aware of scientific misconduct in one form or another and must assess the precision and consistency of the submitted work: Have the data been accurately reported? Precision and consistency refers to the principle that all data have been obtained as described in the methods section, are consistent in the text, figures, and tables, and have been analyzed with proper statistical methods. The latter must be assured by a mandatory statistical review, a practice followed by the majority of high-ranking journals. This is particularly important in clinical research where power calculations, superiority and noninferiority designs, as well as various types of analyses of variance, including propensity analysis, must be used (Lüscher et al. 2009; Heinze and Jüni 2011; Head et al. 2012).

Are editors truly able to detect "cooking," "trimming," or blatant forgery? This is obviously challenging. What they can do is to use computer-assisted programs not only to assess novelty but to exclude plagiarism, self-plagiarism, and *salami* publishing, where data from a single research project is reported in multiple end publications. Further, inconsistency of the numbers reported in the methods, results, tables, and figures is a sign of potential misconduct; this is often overlooked by co-authors, reviewers, and editors. Finally, plausibility should not be forgotten. A embarrassing example is one of John Darsee's fabrications, in which he described a rare form of familial taurin-related cardiomyopathy. In the paper published in 1981, in the *New England Journal of Medicine* and later retracted, any careful reader (or again dedicated co-authors, reviewers, or editors) should have noticed that a 17-year-old male could not possibly have had four children ages 4, 5, 7, and 8 (see Darsee and Heymsfield 1981; Heymsfield and Glenn 1983). *This is a lesson to us all, to carefully read papers we are involved in.*

17.8   Conflicts of Interest

Another deviation from the truth has recently come into focus: biases due to financial incentives. Are we free enough to follow the path of good scientific practice, if we are supported or even paid by those producing the product to be tested? Obviously, most large studies are not feasible without major investments provided by pharmaceutical and device industries. How are we going to handle this? The current mindset is transparency. But does transparency really help? Yes, it makes readers aware of the possibility of bias (regardless whether they exist or not). A recent study suggested that we thereby create a bias of a new kind: in this study involving over 500 internists to whom trials of different rigor were presented and in a randomized fashion different sponsorships were mentioned, showed that sponsorship by industry, even for trials with the highest rigor, induces doubts and disbelief among readers beyond the objective quality (Kesselheim et al. 2012)! Thus, at the end of the day, transparency alone only helps partially. It is again the test of time that separates truth from random findings or even fraud. Furthermore, conflicts are not only financial, but may be intellectual (mainly in basic and pathophysiological research) and professional (mainly in device- and equipment-based research) in nature. Thus, as scientists, we must be aware of these problems to remain on the path of truth and honesty.

17.9 Lessons for the Future

In response to the unfortunate trend of recent scientific misconducts, C. Glenn Begley (2013) put together the red flags to assess scientific findings (Table 17.1). As experienced scientists know, bench experiments are rarely performed in a blind fashion. Furthermore, suppression of data – as discussed earlier – is not uncommon (Kesselheim et al. 2012). As outlined above, there may be good reasons to do so, but then it must be stated in the methods. That reagents have to be validated has recently been stressed by a study showing that dimethyl sulfoxide (DMSO), a commonly used solvent, has profound biological effects (Camici et al. 2006). Statistics are obviously essential, and most peer-reviewed journals have statistical editors that scrutinize all papers considered for publication. Finally, in "hot" scientific fields, researchers may be pushed to publish too quickly and too enthusiastically, as exemplified by fraud scandals in stem cell research (Francis et al. 2013). Importantly, and not mentioned by C. Glenn Begley, is the lab book: *All experiments need to be documented and accessible for the entire team and the supervisor in particular.*

In clinical research, the red flags are slightly different (see Table 17.1): here, the study design is of utmost importance and it should be defined beforehand, preferentially published in a design paper ("Study Protocol") and registered as required by the *International Collaboration of Medical Journal Editors.* Indeed, non-randomized and small randomized studies are more likely to be refuted (Head et al. 2012). For instance, an initial case-control study involving 1334 individuals claimed that an angiotensin-converting enzyme (ACE) polymorphism was associated with an increased rate of myocardial infarction (Cambien et al. 1992), a finding that became less and less convincing and eventually no longer notable in subsequent studies with over 10 000 individuals (Keavney et al. 2000). Hence earlier reports may reflect the enthusiasm of the investigators than reality. Furthermore, the deletion of just a few patients' data, justified or not, distort small data sets. Registry data, even when analyzed using modern statistics such a propensity analysis are also less reliable and prone to biases and overreporting, although they do reflect clinical practice more closely. For instance, the Nurses' Health Study suggested that hormone replacement therapy was protective in

Table 17.1 Red flags in basic and clinical research.

Basic research	Clinical research
Has ethical approval of the protocols been obtained from the responsible committee?	Has ethical approval of the protocols been obtained from the responsible committee?
Have the experiments been performed in a blinded fashion?	Has the study design been registered (in case of a comparative clinical trial)?
Were experiments appropriately repeated?	Has the design of a randomized trial or registry been previously published?
Were all results presented? And if not, was this indicated in the methods?	Have the data been obtained prospectively or retrospectively?
Were there positive and negative controls?	Was there an independent events adjudication and/or data and safety committee?
Were reagents validated?	Were the steering committee, the investigators, and/or patients blinded?
Were statistical tests appropriate?	Was there a predefined statistical analysis plan with power calculation and defined subgroup analysis?
Were all experiments documented in a lab book?	Was there appropriate and independent monitoring of the center(s)?
Did all the co-authors have access to the original data?	Were any data excluded from analysis?

postmenopausal women (Stampfer et al. 1991), a finding refuted by large randomized trials (Hulley et al. 1998). In retrospect, hormone use reflected health consciousness of participants rather than having a casual effect. Furthermore, postmarketing registries of novel compounds are prone to overreporting providing a distorted estimate compared to established treatments (Southworth et al. 2013).

A major drawback of clinical trials – when eventually translated into clinical practice – is the fact that only a minority of qualifying patients are enrolled and that those who are enrolled differ from non-participants. Recruiting physicians may further distort the characteristics of enrolled patients compared to the real world by preferences, prejudices, or frank intentions. As a result, study patients often have different baseline characteristics and a lower mortality and event rate than non-participants (Hordijk-Trion et al. 2006; Vist et al. 2008; de Boer et al. 2011, 2013).

What can we do make science honest and reliable? Colin Norman outlined measures to sustain scientific integrity in 1984: (i) trainees need to be supervised by an experienced mentor; (ii) the results obtained by researchers should be discussed regularly at research meetings with presentation of the raw data; (iii) all authors should carefully read and approve papers they are involved in and have access to the original data; (iv) ethical and animal research approval should be available and checked by a responsible person in the department at the beginning of any study; (v) certified courses on good clinical practice and animal experimentation for clinical and basic researchers should be mandatory; and last but not least (vi) we must stress that science is a commitment to honesty and the pursuit of truth – nothing else (see Norman 1984).

Acknowledgment

This chapter is based on and updates an article published in the European Heart Journal by the author (T.F. Lüscher: The codex of science – honesty, precision, and truth – and its violations. European Heart Journal. 2013; 34: 1018–1023).

References

Anon. (2012a). Notice of concern. *Journal of the American College of Cardiology* 60 (25): 2696–2697.

Anon. (2012b). Expression of concern. *Circulation Research* 110: e47.

Begley, C. (2013). Six red flags for suspect work. *Nature* 497 (7450): 433–434.

de Boer, S.P.M., Lenzen, M.J., Oemrawsingh, R.M. et al. (2011). Evaluating the "all-comers" design: a comparison of participants in two "all-comers" PCI trials with non-participants. *European Heart Journal* 32 (17): 2161–2167.

de Boer, S.P.M., van Leeuwen, M.A.H., Cheng, J.M. et al. (2013). Trial participation as a determinant of clinical outcome: differences between trial-participants and every day clinical care patients in the field of interventional cardiology. *International Journal of Cardiology* 169 (4): 305–310.

Bouri, S., Shun-Shin, M.J., Cole, G.D. et al. (2014). Meta-analysis of secure randomised controlled trials of β-blockade to prevent perioperative death in non-cardiac surgery. *Heart (British Cardiac Society)* 100 (6): 456–464.

Braunwald, E. (1987). On analyzing scientific fraud. *Nature* 325 (6101): 215–216.

Broad, W.J. (1980a). Imbroglio at Yale (I): emergence of a fraud. *Science (New York, N.Y.)* 210 (4465): 38–41.

Broad, W.J. (1980b). Imbroglio at Yale (II): a top job lost. *Science (New York, N.Y.)* 210 (4466): 171–173.

Cambien, F., Poirier, O., Lecerf, L. et al. (1992). Deletion polymorphism in the gene for angiotensin-converting enzyme is a potent risk factor for myocardial infarction. *Nature* 359 (6396): 641–644.

Camici, G.G., Steffel, J., Akhmedov, A. et al. (2006). Dimethyl sulfoxide inhibits tissue factor expression, thrombus formation, and vascular smooth muscle cell activation: a potential treatment strategy for drug-eluting stents. *Circulation* 114 (14): 1512–1521.

Darsee, J. and Heymsfield, S. (1981). Decreased myocardial taurine levels and hypertaurinuria in a kindred with mitral-valve prolapse and congestive cardiomyopathy. *The New England Journal of Medicine* 304 (3): 129–135.

Darwin, C. (1859). *On the Origin of Species*. London: John Murray, Albemarle Street. ... C. Darwin, Charles, J. F. Duthie, and William Hopkins.

Darwin, C. (1872). *The Expression of the Emotions in Man and Animals*. London: John Murray.

Descartes, R. (1984). *The Philosophical Writings of Descartes*, Translated by John Cottingham, Robert Stoothoff, Dugald Murdoch and Anthony Kenny, vol. 3. Cambridge: Cambridge University Press.

Devereaux, P.J., Yang, H., Yusuf, S. et al. (2008). Effects of extended-release metoprolol succinate in patients undergoing non-cardiac surgery (POISE trial): a randomised controlled trial. *Lancet (London, England)* 371 (9627): 1839–1847.

Dunkelgrun, M., Boersma, E., Schouten, O. et al. (2009). Bisoprolol and fluvastatin for the reduction of perioperative cardiac mortality and myocardial infarction in intermediate-risk patients undergoing noncardiovascular surgery: a randomized controlled trial (DECREASE-IV). *Annals of Surgery* 249 (6): 921–926.

Fanelli, D. (2009). How many scientists fabricate and falsify research? A systematic review and meta-analysis of survey data. *PloS One* 4 (5): e5738.

Fang, F.C. and Casadevall, A. (2011). Retracted science and the retraction index. *Infection and Immunity* 79 (10): 3855–3859.

Feringa, H.H.H., Bax, J.J., Elhendy, A. et al. (2006a). Association of plasma N-terminal pro-B-type natriuretic peptide with postoperative cardiac events in patients undergoing surgery for abdominal aortic aneurysm or leg bypass. *The American Journal of Cardiology* 98 (1): 111–115.

Feringa, H.H.H., Elhendy, A., Bax, J.J. et al. (2006b). Baseline plasma N-terminal pro-B-type natriuretic peptide is associated with the extent of stress-induced myocardial ischemia during dobutamine stress echocardiography (DECREASE VI). *Coronary Artery Disease* 17 (3): 255–259.

Francis, D.P., Mielewczik, M., Zargaran, D. et al. (2013). Autologous bone marrow-derived stem cell therapy in heart disease: discrepancies and contradictions. *International Journal of Cardiology* 168 (4): 3381–3403.

Hawking, S. and Mlodinow, L. (2010). *The Grand Design*, 18. New York: Bantam Books.

Hazes, J., van der Maas, P., Peters, R., et al. (2014) Report Follow-up of the Investigative Committee on Academic Integrity 2013. Rotterdam.

Head, S.J., Kaul, S., Bogers, A.J.J.C. et al. (2012). Non-inferiority study design: lessons to be learned from cardiovascular trials. *European Heart Journal* 33 (11): 1318–1324.

Heinze, G. and Jüni, P. (2011). An overview of the objectives of and the approaches to propensity score analyses. *European Heart Journal* 32 (14): 1704–1708.

Heymsfield, S. and Glenn, J. (1983). Retraction. Darsee JR, Heymsfield SB. Decreased myocardial taurine levels and hypertaurinuria in a kindred with mitral-valve prolapse and congestive cardiomyopathy. N Engl J Med 1981;304:129–135. *The New England Journal of Medicine* 308 (23): 1400.

Hordijk-Trion, M., Lenzen, M., Wijns, W. et al. (2006). Patients enrolled in coronary intervention trials are not representative of patients in clinical practice: results from the euro heart survey on coronary revascularization. *European Heart Journal* 27 (6): 671–678.

Hulley, S., Grady, D., Bush, T. et al. (1998). Randomized trial of estrogen plus progestin for secondary prevention of coronary heart disease in postmenopausal women. Heart and Estrogen/progestin Replacement Study (HERS) research group. *JAMA* 280 (7): 605–613.

Judson, H. (2004). *The Great Betrayal: Fraud in Science*, 63–64. Orlando: Harcourt Inc., 104–108, 110–112.

Kant, I. (1996). Critique of practical reason. In: *Practical Philosophy* (ed. M.J. Gregor). Cambridge: Cambridge Universiety Press.

Keavney, B., McKenzie, C., Parish, S. et al. (2000). Large-scale test of hypothesised associations between the angiotensin-converting-enzyme insertion/deletion polymorphism and myocardial infarction in about 5000 cases and 6000 controls. International Studies of Infarct Survival (ISIS) collaborators. *Lancet (London, England)* 355 (9202): 434–442.

Kesselheim, A.S., Robertson, C.T., Myers, J.A. et al. (2012). A randomized study of how physicians interpret research funding disclosures. *The New England Journal of Medicine* 367 (12): 1119–1127.

Kristensen, S.D. and Knuuti, J. (2014). New ESC/ESA Guidelines on non-cardiac surgery: cardiovascular assessment and management. *European Heart Journal* 35 (35): 2344–2345.

Lüscher, T.F. (2012). Good publishing practice. *European Heart Journal* 33 (5): 557–561.

Lüscher, T.F. (2013a). In search of the right word: a statement of the HEART Group on scientific language. *European Heart Journal* 34 (1): 7–9.

Lüscher, T.F. (2013b). The codex of science: honesty, precision, and truth--and its violations. *European Heart Journal* 34 (14): 1018–1023.

Lüscher, T.F., Gersh, B., Brugada, J. et al. (2009). The European heart journal goes global: the road ahead of the editorial team 2009–2011. *European Heart Journal* 30 (1): 1–5.

Lüscher, T.F. (2019). Back to square one: The future of stem cell therapy and regenerative medicine after the recent events. *European Heart Journal* 40 (13): 1031–1033.

Marcus, A. and Oransky, I. (2012) A retraction epidemic, New York Times, 1st April, p. 7.

Nallamothu, B.K. and Lüscher, T.F. (2012). Moving from impact to influence: measurement and the changing role of medical journals. *European Heart Journal* 33 (23): 2892–2896. https://doi.org/10.1093/eurheartj/ehs308.

Norman, C. (1984). Reduce fraud in seven easy steps. *Science (New York, N.Y.)* 224 (4649): 581.

Poldermans, D., Bax, J.J., Schouten, O. et al. (2006). Should major vascular surgery be delayed because of preoperative cardiac testing in intermediate-risk patients receiving beta-blocker therapy with tight heart rate control? (DECREASE II). *Journal of the American College of Cardiology* 48 (5): 964–969.

Poldermans, D., Schouten, O., Benner, R. et al. (2008). Fluvastatin XL use is associated with improved cardiac outcome after major vascular surgery. Results from a randomized placebo controlled trial: DECREASE III. *Circulation* 118: S792.

Poldermans, D., Bax, J.J., Boersma, E. et al. (2009). Guidelines for pre-operative cardiac risk assessment and perioperative cardiac management in non-cardiac surgery. *European Heart Journal* 30 (22): 2769–2812.

Popper, K.R. (1989). *Conjectures and Refutations: The Growth of Scientific Knowledge*, 5e. London: Routledge and Kegan Paul.

Popper, K. (2002). *The Logic of Scientific Discovery*. London/New York: Routledge Classics.

Shimokawa, H. (2010). Urgent announcement from the editor-in-chief regarding duplicate publication. *Circulation Journal: Official Journal of the Japanese Circulation Society* 74 (9): 2026.

Southworth, M.R., Reichman, M.E., and Unger, E.F. (2013). Dabigatran and postmarketing reports of bleeding. *The New England Journal of Medicine* 368 (14): 1272–1274.

Stampfer, M.J., Colditz, G.A., Willett, W.C. et al. (1991). Postmenopausal estrogen therapy and cardiovascular disease. Ten-year follow-up from the nurses' health study. *The New England Journal of Medicine* 325 (11): 756–762.

Steen, R.G. (2011). Retractions in the scientific literature: is the incidence of research fraud increasing? *Journal of Medical Ethics* 37 (4): 249–253.

Stewart, W.W. and Feder, N. (1987). The integrity of the scientific literature. *Nature* 325 (6101): 207–214.

Vist, G.E., Bryant, D., Somerville, L. et al. (2008). Outcomes of patients who participate in randomized controlled trials compared to similar patients receiving similar interventions who do not participate. *The Cochrane Database of Systematic Reviews* (3): MR000009.

Westphal, R. (1980). *Never at Rest: A Biography of Isaac Newton*. Cambridge: Cambridge University Press.

Wise, J. (2011). Extent of Dutch psychologist's research fraud was "unprecedented". *BMJ (Clinical Researched.)* 343: d7201.

Wittgenstein, L. (2005). *Tractatus Logico-Philosophicus*, 5. New York: Routledge.

Yui, Y. (2012). Concerns about the Jikei heart study. *Lancet (London, England)* 379 (9824): e48.

18

Writing and Scientific Misconduct: Ethical and Legal Aspects

Marcoen J.T.F. Cabbolet

Department of Philosophy, Free University of Brussels, Brussels, Belgium

18.1   Introduction

When writing a contribution of any type to the enterprise called "science," one must keep in mind that this contribution has to meet criteria of scientific quality. Although Hemlin and Montgomery (1990) reported that there will perhaps never be consensus on what scientific quality means up to the last detail, there are some quality standards of reporting that are widely agreed upon in all branches of science: gross violations thereof are forms of *scientific misconduct*. This does not necessarily have to be *intentional*: such is merely a severe degree of scientific misconduct. The following two questions then arise:

1. How can we avoid scientific misconduct?
2. What are the consequences if we nevertheless do commit scientific misconduct?

These questions touch on the ethical and legal aspects of the relation between scientific writing and scientific misconduct. This chapter deals with these two aspects.

18.2   Ethical Aspects

A written contribution is guaranteed to meet the required basic quality standards of reporting by adhering to widely accepted ethical norms. These are the *basic principles of good scientific practice*, which Van der Heijden et al. have nicely formulated (2012); see Table 18.1. We can thus *avoid* scientific misconduct in writing by making a decent effort to adhere to these principles. The treatment of the basic principles of good scientific practice below will be split into two parts: *avoiding type-one scientific misconduct* and *avoiding type-two scientific misconduct* – type one leads to falsely positive conclusions about one's own work, while type two leads to falsely negative conclusions about someone else's work (Cabbolet 2014).

A Guide to the Scientific Career: Virtues, Communication, Research, and Academic Writing, First Edition.
Edited by Mohammadali M. Shoja, Anastasia Arynchyna, Marios Loukas, Anthony V. D'Antoni, Sandra M. Buerger, Marion Karl and R. Shane Tubbs.

Table 18.1 The Netherlands Code of Conduct for Scientific Practice: principles of good scientific practice and their descriptions.

Principle	Description
Scrupulousness (Carefulness)	Scientific activities are performed scrupulously, unaffected by mounting pressure to achieve results.
Reliability	A scientific practitioner is reliable in the performance of his research and in the reporting, and equally in the transfer of knowledge through teaching and publication.
Verifiability	Whenever research results are publicized, it must be made clear what the data and the conclusions are based on, where they were derived from and how they can be verified.
Impartiality	In a scientific endeavor, the scientist heeds no other interest than the scientific interest.
Independence	Scientists do their work in academic freedom and independence. It must be made clear when limits to that freedom are unavoidable.

Source: Adapted from Van der Heijden et al. (2012).

18.2.1 Avoiding Type-One Scientific Misconduct

In the first place, we have the *principle of carefulness* (or *scrupulousness*):

> Scientific activities are performed scrupulously, unaffected by mounting pressure to achieve
>
> (Van der Heijden et al. 2012).

For scientific writing, this means, first and foremost, that sources of the material should be properly identified, to avoid someone else's work being passed off as one's own. This form of misconduct ranges "from gross intentional plagiarism to 'inaccurate referencing' and from deliberately stealing other people's ideas to the careless 'use' of other people's thoughts" (Drenth 1999). Of course, one can use ideas and results of others, but the golden rule as formulated by Drenth (1999) is this:

> When the final result of scientific endeavor is presented … there should be a clear distinction between that which is a product of personal reflection, analysis, data gathering and interpretation, and that which should be attributed to others. And the latter should be clearly indicated by means of proper references.

This principle further means that authorship has to be properly acknowledged, to avoid on the one hand someone who has significantly contributed to the work being excluded as an author, and on the other hand someone who hasn't contributed being mentioned as an author. In the first case, one is passing off someone else's work as one's own, and in the second case we speak of *gifted authorship*: both cases are explicitly mentioned as examples of scientific misconduct by KNAW,[1] NWO,[2] and VSNU[3] (2001).

---

1 Koninklijke Nederlandse Academie der Wetenschappen (Royal Dutch Academy of Sciences).
2 Nederlandse Organisatie voor Wetenschappelijk Onderzoek (Dutch Organization for Scientific Research).
3 Vereniging van Samenwerkende Nederlandse Universiteiten (Association of Cooperating Dutch Universities).

However, given a proper list of co-authors, there is still the question of the *order* of authorship. The rule here is as follows: if all authors have contributed equally, then the order is alphabetically; if not, the order is by magnitude of the contribution, with the group leader last (Heilbron 2005). So if there are multiple co-leaders in the latter case, these should be mentioned last in alphabetical order.

Last but not least, this principle means that material from one's own previous publications also has to be identified, to avoid that one's own work is passed off more than once as original research (self-plagiarism). However, while multiple publication of one and the same result is an obvious example, self-plagiarism is not as clear-cut as plagiarism. For example, in a large project that results in several publications, it is unavoidable that some text will have to be reused (Nijkamp 2014). The general rule is to "adhere to the spirit of ethical writing and avoid reusing … previously published text, unless it is done in a manner consistent with standard scholarly conventions" (Roig 2006, p. 24).

In the second place we have the *principle of reliability*:

> … A scientific practitioner is reliable in the performance of his research and in the reporting, and equally in the transfer of knowledge through teaching and publication
>
> (Van der Heijden et al. 2012).

For scientific writing, this means that the presented empirical data must have been experimentally obtained, and not made up or tampered with: A violation of this principle is the most severe form of misconduct in scientific writing. There are no nuances involved here, and there is no need to waste many words on a matter that is so abundantly clear.

Furthermore, this principle means that *all* experimental data must be presented. One might be tempted to leave out negative results and report only those data that support the desired outcome, but the selective omission of experimental results is another serious form of misconduct. Of course, it can be interesting to show what the outcome would have been if certain results are left out of the analysis. But the rule is then that all results that were obtained according to the initial plan should be reported, and that any manipulations of these results afterwards should be clearly described and motivated (Roig 2006, p. 35).

Finally, we have the *principle of verifiability*:

> … Whenever research results are publicized, it is made clear what the data and the conclusions are based on, where they were derived from and how they can be verified
>
> (Van der Heijden et al. 2012).

This means that the experimental section of a scientific publication must contain enough detail for an independent research group to reproduce the results, and the rationale for the conclusions must be sufficiently detailed to be understandable for a third party.

The principle of verifiability helps researchers avoid the pitfall of enthusiasm illustrated by the cold-fusion case. On March 23, 1989, Martin Fleischman and Stanley Pons announced in a press release of the University of Utah that they had "successfully created a sustained nuclear fusion at room temperature" in a "surprisingly simple experiment that is equivalent to one in a freshman-level, college chemistry course" (University of Utah 1989). This announcement was shocking because at the time nuclear fusion was thought to require a temperature of millions of degrees on the Kelvin scale. However, hundreds of other researchers could not reproduce the results reported by Fleischman and Pons, which soon casted doubt on their main claim. Eventually, the idea that one can produce cold nuclear fusion with the experimental set up of Fleischman and Pons was rejected. Fleischman and Pons were so eager to proclaim their findings that they didn't want to await publication in a peer-reviewed journal, and so their enthusiasm *unintentionally* led to a violation of the principle of verifiability. For further details, see Taubes (1993).

18.2.2   Avoiding Type-Two Scientific Misconduct

Of particular importance in this context is the *principle of impartiality*:

> In his scientific activities, the scientific practitioner heeds no other interest than the scientific interest. …
>
> (Van der Heijden et al. 2012).

Among other things, that means that critical comments on someone else's work have to be written with the scientific interest in mind, leaving room for a different intellectual stance: one has to avoid publishing venomous, false allegations that arise as an emotional reaction to a new development that dissents from one's own view. This is about understanding the difference between *skepticism* and *pseudoskepticism*, a term made popular by Truzzi (1987). There is nothing wrong with a healthy skepticism toward a new idea. It can be in the interest of science to rigorously point out why a new theory is not convincing or might even be wrong. Pseudoskepticism, on the other hand, has nothing to do with a scientific discussion: it is gravely discrediting somebody else's work without even trying to prove the allegations, e.g. bluntly alleging "all his formulas are syntactically ill-formed" without that actually being the case. Thus, when writing a critical comment about someone else's work, one should avoid being on the wrong side of the border between skepticism and pseudoskepticism. Falsely accusing someone else of bad research belongs to the greatest impudencies a scientist can commit. Official measures against pseudoskepticism are currently still in their infancy, but making up negative conclusions about someone else's work ought to be treated the same way as fabricating data in one's own work (Cabbolet 2014).

This pertains in the first place to scientific writings that are not subject to peer-review, such as:

- Monographs
- Preprints
- Peer-review reports themselves, which too fall under the realm of scientific writing

The point is that not only arguments must be given when one puts forward negative conclusions about someone else's work, but also these arguments must be *of scientific substance*: the rule is here that "clear reasons with appropriate references [must] be provided to justify any claims that impugn either the methods, data or conclusions of the work under consideration" (Cabbolet 2014). Gross violations of this rule are a form of type-two scientific misconduct (Cabbolet 2014).

In the second place, this pertains to opinion pieces in the mass media: one has to realize that writing an opinion piece on someone else's work is also a scientific activity, as one does it in one's capacity as a scientist. There is nothing wrong with explaining a scientific controversy in understandable language for the general public: this, too, can be in the interest of science. But it is something else when one falsely discredits a dissenting view in the mass media: *even if* the allegations are false, the general public will accept such a smear as true when the article is written by a university scientist. This is a form of type-two scientific misconduct, and those that engage in it ought to be eliminated from academic circles (Cabbolet 2014).

To avoid pseudoskepticism, it is important to understand that it arises from an automatic, negative emotional reaction on a piece that implies that one's own beliefs are false (Cabbolet 2014); this simply has to be seen as a part of human nature. To not let one's actions be guided by this emotional reaction, the following two imperatives might be helpful. The first is based on Thomas More's *Utopia*: Don't give a written reaction to a piece on the same day that one has read it for the first time. The second applies Fuller's exercise in self-reflection (1981): Before submitting a comment, get some distance from it intellectually and emotionally, look at it as if one is an outsider, and reflect on the question, "Isn't this pseudoskepticism?" This way, at least, *unintentional* cases of pseudoskepticism could be avoided.

As a future measure against reviewer misconduct (pseudoskepticism) in peer review, it has been suggested that the journal editor reveals the identity of the otherwise anonymous referee to the author(s) (Cabbolet 2014).

18.3   Legal Aspects

Nowadays, employment contracts for scientists at accredited universities usually contain a clause that the undersigned has to comply with an ethical code of conduct, a kind of Hippocratic oath. That means that even *unintentional* scientific misconduct already implies breach of contract. Furthermore, like any other person, scientists can be prosecuted for criminal acts: if the scientific misconduct is so severe that it falls under criminal law, the legal consequences can be severe as well. In the following overview, known consequences of scientific misconduct are divided in three groups: *breach of contract*, *criminal law violations*, and *additional consequences*.

18.3.1   Breach of Contract

A scientist can receive a reprimand, which can take the form of a written remark in the personnel file. This is nothing but a metaphorical slap on the wrists meant to get the scientist to reflect on his own behavior, in the hope that the scientist will have a change of heart and never repeat that behavior in the future. From published cases, e.g. ANP[4] (2008) and DFG[5] (2012), it is known that both type-one and type-two scientific misconduct have led to reprimands for scientists.

---

4 Algemeen Nederlands Persbureau (General Dutch Press Agency).
5 Deutsche Forschungsgemeinschaft (German Research Union).

PhD students – often employees of the university – can be blocked from graduation when the dissertation shows evidence of writing misconduct. See Van Kolfschooten 2006, 2014 for a few examples.

A scientist can be dismissed. There are many highly publicized cases, although thus far no cases are known involving type-two scientific misconduct. For some examples, see Bartlett 2008 and Reich 2011.

When submitting a grant proposal, one often has to agree explicitly with the terms and conditions of the funding agency. Consequently, scientists who have been caught faking data in grant proposals can be banned from submitting further proposals by the funding organization. For some examples, see Interlandi 2006; DFG 2012 and Swiss National Science Foundation (SNSF) 2013.

Likewise, when submitting a paper to a journal, one also has to agree to the terms and conditions of the journal. Consequently, when an automated test for plagiarism turns out a positive result, authors can get banned from submitting further papers to that journal.

18.3.2 Criminal Law Violations

Fabrication of data has given rise to several criminal prosecutions resulting in a conviction of the scientist (see, e.g. Reich 2011; Parrish and Mercurio 2011; Stroebe et al. 2012). Scientists who falsely discredits another scientist in the mass media can be sued for libel. Although nowadays courts are *likely* to rule in favor of vested interests (Martin 1998), this probability should not be interpreted as a *guarantee* that this form of type-two scientific misconduct is free of legal consequences.

Grants may have to be repaid. Cases are known where it came out that grants had been obtained with a proposal that contained faked data: the institutions that had received the grants subsequently had to pay them back. For a few examples, see Parrish and Mercurio 2011.

18.3.3 Additional Consequences

In some high-profile cases, the PhD degree of the scientist in question has been withdrawn. For a few examples, see Löwenstein and Müller (2011) and Stroebe et al. (2012). Especially in the medical sciences, a scientist can be banned permanently or temporarily from the profession after being found guilty of serious misconduct. For examples involving the fabrication of data, see Meikle and Bosely (2010) and Stroebe et al. (2012). Students can also be expelled from the university for plagiarism in essays (Ross 2012).

Last but not least, any scientist who commits scientific misconduct, regardless of whether it concerns type one or type two, has to reckon with a *loss of reputation* once the misconduct surfaces. Although not exactly a legal consequence, it nevertheless is *justice after all*.

References

ANP. 2008. Wetenschappers onder vuur. De Telegraaf, Edition September 20, TA010 (in Dutch)

Bartlett, T. 2008. Columbia U. Fires Teachers College Professor Accused of Rampant Plagiarism. The Chronicle of Higher Education, Edition June 23.

Cabbolet, M.J.T.F. (2014). Scientific misconduct: three forms that directly harm others as the modus operandi of Mill's tyranny of the prevailing opinion. *Science and Engineer Ethics* 20 (1): 41–54.

DFG, 2012, *DFG Imposes Sanctions for Scientific Misconduct: Exclusion from Submission Process and Written Reprimand*, Press Release #3, February 9.

Drenth, P.J.D. (1999). Scientists at fault: causes and consequences of misconduct in science. In: *European Science and Scientists between Freedom and Responsibility* (ed. P.J.D. Drenth, J.E. Fenstad and J.D. Schiereck), 41–52. Luxembourg: Office for Official Publications of the European Community.

Fuller, L.L. (1981). Philosophy for the practicing lawyer. In: *The Principles of Social Order: Selected Essays of Lon L. Fuller* (ed. K.I. Winston), 287–290. Durham: Duke University Press.

van der Heijden, P.F., Fokkema, J., Lamberts, S.W.J., et al. (eds.) (2012). *The Netherlands Code of Conduct for Scientific Practice*. Amsterdam: VSNU.

Heilbron, J. (2005). *Wetenschappelijk onderzoek: dilemma's en verleidingen*. Amsterdam: KNAW (in Dutch).

Hemlin, S. and Montgomery, H. (1990). Scientists' conceptions of scientific quality. An interview study. *Science Studies* 3 (1): 73–81.

Interlandi, J. 2006. An Unwelcome Discovery. *The New York Times*, Edition October 22

KNAW, NWO and VSNU (2001). *Notitie Wetenschappelijke Integriteit*. Amsterdam, Den Haag, and Utrecht: KNAW, NWO and VSNU (in Dutch).

van Kolfschooten, F. 2006. Fraude door anesthesist 'aannemelijk'. *NRC Handelsblad*, Edition May 18 (in Dutch)

van Kolfschooten, F. 2014. Onderzoek NRC: topeconoom VU pleegde meerdere keren zelfplagiaat. *NRC Handelsblad*, Edition Januari 7 (in Dutch)

Löwenstein, S. , Müller, R. 2011. Wir sind einem Betrüger aufgesessen. *Frankfurter Allgemeine Zeitung*, Edition February 25 (in German)

Martin, B. (1998). Strategies for dissenting scientists. *Journal of Scientific Exploration* 12 (4): 605–616.

Meikle, J. , Bosely, S. 2010. MMR row doctor Andrew Wakefield struck off register. *The Guardian*, Edition May 24

Nijkamp, P. (2014). Het is van de zotte dit zelfplagiaat te noemen. *Ad Valvas* 61 (10): 23–25. (in Dutch).

Parrish, D.M., Mercurio, T. 2011. Research misconduct: potential monetary recoveries and exposures. Paper read at the annual conference of the NACUA, June 26–29, at the Marriott Marquis hotel in San Fransico, California.

Reich, E.S. (2011). Biologist spared jail for grant fraud. *Nature* 474 (7353): 552.

Roig, M. (2006). *Avoiding Plagiarism, Self-Plagiarism, and Other Questionable Writing Practices: A Guide to Ethical Writing*. New York: St. John's University Press.

Ross, J. 2012. Nine expelled for plagiarism at Deakin. *The Australian*, Edition October 6.

SNSF (2013). *Plagiarism and Incorrect Citation in Applications Submitted to the Swiss National Science Foundation*, 4. Bern: SNSF.

Stroebe, W., Postmes, T., and Spears, R. (2012). Scientific misconduct and the myth of self-correction in science. *Perspectives on Psychological Science* 7 (6): 670–688.

Taubes, G. (1993). *Bad Science: The Short Life and Weird Times of Cold Fusion*, 503. New York: Random House.

Truzzi, M. (1987). On pseudo-skepticism. *Zetetic Scholar* 12–13: 3–4.

University of Utah. 1989. 'Simple experiment' results in sustained n-fusion at room temperature for first time. *Press release*, March 23.

19

Plagiarism and How to Avoid It
Izet Masic

Faculty of Medicine, University of Sarajevo, Sarajevo, Bosnia and Herzegovina

19.1   Introduction

The end product of scientific research is publication, and the key to a scientist's promotion and success. In order to contribute to scientific knowledge, an article should be based on optimal research design and credible data reporting practices (Masic 2012a,b). Authors of academic works are responsible for what they publish, as they can influence the future of publication, science, and education. Currently, many recognized associations strongly advocate publishing standards that are specific to different scientific fields and promote ethical reporting and writing practices aimed at the preservation of scientific integrity. In recent years, scientific misconduct and academic dishonesty have become crucial problems in the field of medical research; this includes the fabrication and falsification of data, plagiarism, and deception, among others (Masic 2012a,b). This chapter provides a glimpse into the ongoing discussion surrounding plagiarism, the burden of plagiarism in academia, and methods of detecting and preventing plagiarism.

19.2   Definition of the Plagiarism Problem

Plagiarism is a relatively common form of inappropriate academic behavior, and comes from the Latin word *plagiare* meaning "to kidnap." Plagiarism is defined as the "unauthorized misappropriation of another's work, ideas, methods, results, or words without giving the original source" (Ghajarzadeh et al. 2013). Plagiarism in research is a serious form of misconduct and a violation of the scientific norms. Certain forms of plagiarism, especially when it involves another's idea, are more complicated than the above definition may suggest (Anderson and Stenack 2011). To counter plagiarism, it is not enough that the offending authors simply change a few words of the source material and transform the borrowed text or concept into their own. Plagiarism is the use of materials from another author without proper attribution of merit. In fact, this violation of scientific norms can be regarded as a crime or theft. Absence of intent to plagiarize is not an acceptable excuse, and even trivial errors in citation or turbid expression can be interpreted as plagiarism (Cameron and McHugh 2012). Taking into account the gravity of

A Guide to the Scientific Career: Virtues, Communication, Research, and Academic Writing, First Edition.
Edited by Mohammadali M. Shoja, Anastasia Arynchyna, Marios Loukas, Anthony V. D'Antoni, Sandra M. Buerger, Marion Karl and R. Shane Tubbs.
© 2020 John Wiley & Sons, Inc. Published 2020 by John Wiley & Sons, Inc.

the plagiarism problem – and its relationship with culture and language – it might be hard to imagine how the issue of plagiarism can be adequately addressed.

Plagiarism dates back to the founding of scientific communication as a discipline. According to the World Association of Medical Editors (WAME), one operational definition of plagiarism is when six consecutive words are copied from the original text (Masic 2012a,b). Various factors have been identified as contributing to the act of plagiarism; for example, lack of writing skills and feeble grasp of the English language in countries where other languages are spoken, a poor state of social and academic well-being, and ignorance of respecting intellectual property (Masic 2012a,b). There has been a disturbingly large number of plagiarism cases in recent years. Plagiarism can occur in several forms, all of which violate professional ethical standards, agreements on copyright, and ethical conduct in the publication of research. There must be a significant component of novelty in the research to be publishable as an original work. Copying large segments of text from previously published papers with only minimal cosmetic changes is not acceptable and can – and should – lead to rejection of a submission by the scientific community.

Plagiarism occurs not only in the form of taking another author's words, but also in plagiarizing one's own work, otherwise called self-plagiarism. This involves the publication of substantially similar content of one's own work in the same or in different journals. An example of self-plagiarism is duplicate publications, which is a breach of publisher-author agreement and a waste of time, effort, and resources by peer-reviewers, editors, and publishers. Plagiarizing someone else's work is the most offensive form of plagiarism. Any text, equations, ideas, or figures taken from another work must be appropriately acknowledged. Figures, tables, or images reproduced from other sources require permission of the original author (The Optical Society of America Board of Editors 2005).

19.3   Academic Integrity and Plagiarism

Academic integrity demands honesty, trust, fairness, and responsibility. These are the ideals that should be upheld by all educational institutions. "Academic integrity involves ensuring that in research, and in teaching and learning, both staff and students act in an honest way. They need to acknowledge the intellectual contributions of others, be open and accountable for their actions, and exhibit fairness and transparency in all aspects of scholarly endeavor" (Bretag 2013). Violations of academic integrity encompass a range of unfair and unethical practices, including plagiarism, cheating on exams and assignments, the theft of another student's work, paying third parties to perform tasks, falsification of data, incorrect presentation of information, and other dishonest activities (Bretag 2013).

Plagiarism is one of the most serious violations of academic integrity because it undermines the assumption that academic work will lead to a genuine and honest contribution to existing knowledge and science. Although plagiarism occurs on all levels of academic research and writing, recent focus is on the outbreak of student plagiarism. Walter Bagehot, a physicist and politician, has said, "The tendency of a human to imitate what is in front of him is one of the strongest parts of his nature."

A facetious definition of plagiarism attributed to Wilson Mizner (1876–1933), a playwright and entrepreneur, is still, unfortunately, popular among students today: "To copy from one book is plagiarism, from two an essay, from three a compilation, and from four a dissertation." Unfortunately, this witty quip may be contributing to the illusion that plagiarism – particularly from multiple contents published on the internet – is a normal part of research and not a serious offense (Chowhan et al. 2013).

Students are learning quickly that finding and manipulating data available on the internet is a convenient practice. With huge amounts of information freely accessible online, finding resources that support an original idea or offer a valid interpretation may seem like a daunting task. Rather than looking for it in published works, students should learn skilled analysis and how to effectively process information. The best advice for students on avoiding plagiarism is to follow the rule, "When in doubt, quote sources" (Chowhan et al. 2013).

19.3.1 Plagiarism by Students

Plagiarism rates for secondary school students vary from 20% to 80%. This disparity can be attributed to differences in language and culture. Students for whom English is an Additional Language (EAL) are much more likely to be involved in a serious form of plagiarism compared to non-EAL students (Marshall and Garry 2006). On the other hand, college students who have already spent 15 years in an English-speaking educational system are often familiar with the requirements of academic integrity. These students are more apt to consciously avoid plagiarism; it is a safe to assume that, as a result of high school or college education, they are being taught to appropriately credit the work of others (Bretag 2013).

19.3.2 The Complexity of Plagiarism among Students

While defining what constitutes plagiarism is complex, identifying plagiarized content may prove even more difficult. This problem is particularly pronounced in rookie academics: in two separate studies, Roig (2006) asked students to identify plagiarized text and found that 40–50% of students were not able to successfully recognize plagiarism. Other authors agree that many students have difficulty identifying plagiarism and do not adequately understand how to paraphrase text with appropriate citations in order to avoid plagiarism (Bretag 2013). Students often want to know exactly when "awkward referencing" becomes "serious plagiarism." The criteria by which academics can determine whether the student's plagiarism is "serious" – requiring a punitive response – or whether it can be best corrected by educational intervention are twofold (James et al. 2002 as cited by Bretag 2013):

1. Whether the student's intent is to defraud (intentionally present the work of others as their own).
2. The extent or scope of the plagiarism – an essay that is paraphrased and submitted as the student's own as opposed to a few sentences – is an example of serious plagiarism.

Recent proposals by the Exemplary Academic Integrity Project suggest that even severe punishments, such as suspension or expulsion from the study, are appropriate for certain types of violations of academic integrity (Bretag 2013).

19.3.3 Plagiarism in Medical Schools and Colleges

Plagiarism has become quite common in universities and secondary schools in the medical field. Undoubtedly, the accessibility of large quantities of electronic medical documents is one explanation for the increasing incidence of plagiarism among medical students. Although many schools and universities have regulations on plagiarism, the methods available for its detection may sometimes be insufficient. Furthermore, seminars that educate students and teachers about responsible research practices may not be in place or upheld. Lack of time is another reason for plagiarism, as it affects the academic practice of the medical students (Annane and Annane 2012).

19.4 Intellectual Dishonesty and Plagiarism in Science

Trust and reliability are at the core of scientific research. In order to get reliable results and avoid inappropriate research practices such as plagiarism, researchers should use optimal study design and observe ethical standards (Masic 2012a,b). Specific guidelines exist to help students and scientists design their study using proper methods of research conduct. Scientific misconduct may be unintentional or intentional. The first is usually a result of improper research methodology, while the latter is pure dishonesty and a violation of ethical standards, otherwise known as intellectual dishonesty (Masic 2012a,b).

Academic misconduct can result in legal recourse. Inappropriate claims of authorship and manipulation of research are viewed as serious ethical and legal violations, and are subject to penalties by international institutions and organizations. Regardless of their background or affiliation, these organizations are aimed at improving and maintaining the integrity of the scientific work (Masic 2012a,b). Among these international organizations are two highly influential groups, the Committee on Publication Ethics (COPE) and the European Association of Science Editors (EASE). By stringently following the guidelines and standards put forth by such organizations, every person engaged in scientific work would greatly improve the value of their research and contribution and avoid the disruption of academic integrity.

19.4.1 Committee on Publication Ethics (COPE)

COPE was established in 1997 by a small group of medical journals in England. It now has over 7000 members worldwide from all academic fields. Membership is open to editors of academic journals and others who are interested in the ethics of publication. This organization provides advice to editors and publishers on all aspects of publication ethics, in particular, how to deal with cases of scientific and research misconduct including plagiarism.[1] In collaboration with the Open Access Scholarly Publishers Association (OASPA), Directory of Open Access Journals (DOAJ), and the WAME, COPE has compiled a minimum set of criteria against which journals will be evaluated in order to become members of one of these organizations (Table 19.1). This set of criteria is based on the principles of transparency and should be clearly exhibited by members.

---

1 The URL for COPE's website is http://publicationethics.org

Table 19.1 COPE's minimum set of standards for journals.

1	A standard review peer-process should be in place extending to the entire content of the journal and carried out by experts who are not employees of the journal.
2	A governing body, such as an editorial board, should include members who are recognized experts in the specific fields with which the journal is associated.
3	Contact information of the journal editorial team with full names and affiliations must be provided.
4	The ownership and management personnel of the journal should be clearly stated on the journal website.
5	Author fees or costs necessary for the journal to process or publish papers should be clearly stated.
6	The author guideline or journal website should include information about authors' duties and responsibilities as contributors.
7	Authorship, copyright, and licensing information should be clearly specified by the journal, for example, on the journal's website.
8	The journal should identify a clear policy for resolving potential conflicts of interest of editors, authors, and third parties.
9	Journal archiving, electronic support, and accessibility to journal content in the event that the journal is no longer issued should be detailed.
10	The name of the journal should be unique and should not be interchangeable with other journals already in publication.
11	Schedule of publication or journal periodicity should be clearly noted.
12	Sources of revenue should be clearly outlined (e.g., royalties, subscriptions, advertising, institutional and organizational support).
13	Access to journal content and individual articles should be indicated, along with subscription details and payment terms (e.g., pay per view).
14	The journal should declare its policy of advertising if relevant, including the consideration and acceptance of individual advertisers and if the advertisements are associated with the specific content or randomly displayed.
15	Any text contained on the journal website should exhibit high ethical and professional standards.
16	A process identifying and addressing allegations of research misconduct should be established, for which journals take reasonable steps to identify and prevent the publication of papers that demonstrate misconduct, including plagiarism, citing manipulation, and falsification/fabrication of data. In no event shall journals and their editors support such misconduct or knowingly permit the same. If the journal and its editors recognize the allegations of research misconduct associated with an article published in their journal, the publisher or the editor will follow the COPE guidelines (or equivalent) in solving these allegations.

As gathered from these principles, which address plagiarism among other points, strict guidelines are in place demanding a professional and ethical commitment by the journals, publishers, authors, and others associated with the publication of scientific articles. If these standard practices are violated, COPE organizations will first try to solve the issue together with the journal. In case of an inability or unwillingness of the journal and editor to cooperatively solve a breach or a nonfulfillment of standards, their membership in the organization will be suspended or permanently terminated.

Among the most interesting and useful resources and guidelines offered by COPE are the numerous flowcharts and diagrams designed to help editors and authors follow COPE's code of conduct and help implement standards of practice. They are available in several languages and can be downloaded individually or as a set. The flowcharts relevant to this chapter and available on the COPE website are titled "What to do if you suspect plagiarism: Suspected plagiarism in a submitted manuscript" (Figure 19.1) and "What to do if you suspect plagiarism: Suspected plagiarism in a published manuscript" (Figure 19.2).

19.4.2 The European Association of Science Editors (EASE)

EASE was founded in Pau, France, in May 1982 by two respected groups of editors, namely, the European Life Science Editors' Association (ELSE) and the European Association of Earth Science Editors (Editerra). It is an international community of individuals with different backgrounds, linguistic traditions, and professional experience, who share an interest in scientific communication and editorial practices. Membership is accepted from all parts of the world. EASE allows members to stay up to date with trends in the rapidly changing landscape of scientific publishing, traditional or electronic.[2] As a member of EASE, a person can sharpen their writing skills and editorial thinking, broaden horizons by meeting people with different backgrounds and experiences, and deepen their understanding of important international publishing issues. EASE promotes cooperation and mutual respect among members, while holding them to the highest ethical standards in publishing. EASE, among other things, provides guidance to authors and translators publishing scientific articles in the English language. These guidelines are accessible to a worldwide audience, available in PDF format, and can be downloaded for free from the official EASE website.

Besides best writing practices, a special focus of EASE is prevention of plagiarism. It contains practical advice for junior researchers on how to avoid plagiarism and what constitutes authorship. Each article submitted for publication in any biomedical or scientific journal must meet specific qualifications as defined by EASE. Authors submitting an article to EASE-affiliated organizations should take into account the recommendations in Table 19.2. By following these guidelines, authors can emulate ethical standards in their scientific work and writing, and conscientiously avoid and prevent unethical behavior, especially plagiarism.

19.5 Detection of Plagiarism: Electronic Tools

The plagiarism detection tools (Table 19.3) allow systematic detection of plagiarism. One of the biggest responsibilities of journal editors is to reveal plagiarism if the need arises regarding any submission. While strict anti-plagiarism policies exist, detecting plagiarism in the first place has come to lean on specific anti-plagiarism tools. There are several online tools available for detecting plagiarism; however, there is concern about the reliability and accuracy of each of these tools. The current tools cannot easily detect

---

2 The URL for EASE's website is www.ease.org.uk. See, e.g. EASE *Guidelines for Authors and Translators of Scientific Articles to Be Published in English*, November 2016.

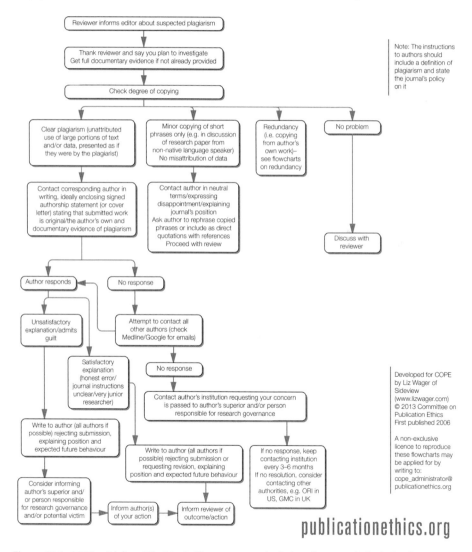

Figure 19.1 COPE guideline: What to do if you suspect plagiarism: Suspected plagiarism in a submitted manuscript. Source: Liz Wager. (2013) https://publicationethics.org. Licensed under CC BY-NC-ND 3.0 and reproduced with permission of COPE in this textbook.

What to do if you suspect plagiarism
(b) Suspected plagiarism in a published manuscript

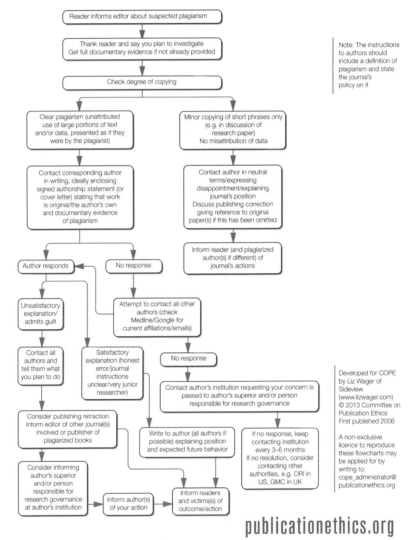

Figure 19.2 COPE guideline: What to do if you suspect plagiarism: Suspected plagiarism in a published manuscript. Source: Liz Wager. (2013) https://publicationethics.org. Licensed under CC BY-NC-ND 3.0 and reproduced with permission of COPE in this textbook.

Table 19.2 EASE's recommendations for authors.

1	The acquisition of financing, collection of data, and the general supervision of a research group alone do not qualify one for authorship.
2	All listed authors should definitively qualify for authorship.
3	Sometimes it is necessary to specify a description of the contribution of each author individually.
4	In some cases, one or more authors is required to take responsibility for the integrity of the research and work as a whole.
5	All persons who contribute to the work but do not qualify as authors should be acknowledged (e.g., persons providing technical assistance, financial support, etc.).
6	Authors should disclose their contribution to the study and the article.
7	Authors should state the existence of a known conflict of interest.
8	Authors should know that it is often necessary to extend authorship to the publisher for professional and legal purposes.
9	Authors should ensure that no permissible material is copied without the proper recognition of said material.
10	Publishing of others' work as your own is not allowed. This applies to the entire text as well as to *ideas* taken from other authors, websites, and databases.
11	Duplication of an article for different publications should be avoided.
12	Authors looking to publish should understand how unethical conduct is dealt with and study the sanctions created to avoid such behavior, including duplicate publication, plagiarism of others, and self-plagiarism.
13	Authors should consult recommendations and guidelines established by recognized organizations (International Committee of Medical Journal Editors, Committee on Publication Ethics, World Association of Medical Editors, European Association of Science Editors, etc.).
14	Published authors should present their work and the result of their research at scientific meetings so that it becomes available to the general scientific community and can be reviewed by appropriate persons.
15	It is advisable to include relevant articles in the references section even when no content is derived from these specific works.

some types of plagiarism, such as that of numbers either with or without modification. It should also be noted that if the plagiarized work is not available in electronic version, the electronic tools are useless in detecting plagiarism (Bamidis et al. 2010).

Anti-plagiarism software compares the article in question against a large database of publications. These software platforms use algorithms that compare relevant keywords, statistically unusual phrases, sentences or sorted words to compute a measure of similarity, integrating the results for the user. A journal publisher or editor then examines the article manually against the text found by the software in order to declare the suspected plagiarism valid or invalid. In short, the less similarity with the database content, the more credible the work (Garner 2011). Plagiarism detection software, like other software products, are either open source (free of charge) or proprietary (users must pay to get software). Students or novice authors, who only make use of detection software on occasion, may be fully satisfied with the effectiveness of open-source tools.

In order for the journal to achieve a high standard of integrity and quality, it should carry out a systematic survey of submitted manuscripts. One of the major limitations

Table 19.3 Examples of plagiarism detection tools.

Tool	Website
ArticleChecker	www.articlechecker.com
Copyleaks	https:copyleaks.com
Copyscape	https://copyscape.com
CrossCheck/iThenticate	www.ithenticate.com
DupliChecker	https://www.duplichecker.com
Dustball	http://www.dustball.com
Eve	http://www.rankfisher.com
Grammarly	https://www.grammarly.com
HelioBLAST	https://helioblast.heliotext.com
Noplag	https://noplag.com
PaperRater	http://www.paperrater.com
Plagiarisma	http://plagiarisma.net
PlagiarismCheck	https://plagiarismcheck.org
PlagiarismChecker	http://www.plagiarismchecker.com
PlagiarismDetect	http://plagiarismdetect.org
Plagium	http://www.plagium.com
Plagly	https://plagly.com
PlagScan	https://www.plagscan.com
PlagTracker	www.plagtracker.com
Quetext	http://www.quetext.com
SafeAssign	https://help.blackboard.com
SmallSEOTools	http://smallseotools.com/plagiarism-checker
TurnItIn	http://turnitin.com
Viper	https://www.scanmyessay.com

of electronic database tools is that original texts do not always exist in full in electronic databases. In fact, the relatively large number of publications available in full-text electronic versions represents only a fraction of scientific and biomedical publications existing to date (Garner 2011). While detection software has its deficiencies, journals may benefit from directly educating their contributing authors on appropriate conduct.

19.5.1 Choosing a Tool to Detect Plagiarism

A few things to consider before selecting an anti-plagiarism tool are its compatibility with your computer system, the integrity of the database that is used to reference comparisons, security, and of course, price. Although many plagiarism detection tools claim to have an edge over the competition, there have been no comparative reviews officiated by an independent entity that determine the relative performance of each tool.

Table 19.3 lists several proprietary and open-source plagiarism detection tools. Each tool has its own advantages and disadvantages; however, the overarching question is how to account for the scientific papers that are not available in electronic form. Even more important is what to do about plagiarized articles that possibly already exist in the databases that these platforms use as a basis for comparison (Garner 2011). When we take into account the bulkiness of the plagiarism dilemma, although useful, these software tools do not address the whole problem. Other questions, such as who the responsible party is, on what criteria is the accusation of plagiarism based, and whether the offender unintentionally or deliberately committed plagiarism need to be further investigated. Strengthening the systems for plagiarism detection alone will not have a sustained impact on preventing plagiarism in the long term. Educating researchers and writers on the consequences of plagiarism and how to avoid it is the best approach to solving the problem at the root.

19.6 Prevention of Plagiarism: A Better Remedy

Like with most things, prevention still has an advantage over the curative measures. Taking into account the limitations and reliability of detection software, plagiarism is best solved by concentrating on the instructional system of prevention. Many institutions have begun to implement courses that teach responsible conduct in research and openly discuss a number of problems in the current review of research integrity. Although this is a step in the right direction, these educational programs are currently provided to a fraction of researchers, while many other researchers, both experienced and inexperienced, operate without such instruction. Integrity in research must be a value incorporated in continuing education and should target all researchers (Roig 2006). Conversations about responsible conduct in research must go beyond the "crime" of fabrication, falsification, and plagiarism to talk about how behaviors of misconduct can undermine research efforts and disestablish the very qualities of validity and accuracy that are so vital to the field of science and medicine. Instruction on preventing plagiarism should begin with the principles of ethical writing (Roig 2006).

19.6.1 Teaching Ethical Writing

The dissemination of ideas is extremely important in the current culture of sharing information, especially in advancing the scientific knowledge. However, the responsibility required in sharing these ideas must distinguish between what is general knowledge and what is a refined idea or theory expressly proposed by another. While professional and educational institutions need to exercise a low tolerance for plagiarized materials, it is also important to help students and seasoned writers alike maintain good research ethics. Chowhan et al. (2013) recommends the following tips to avoid plagiarism:

1. The original source of ideas, text, or illustrations must always be acknowledged.
2. If the text is taken verbatim, it must be enclosed within quotation marks.
3. An idea that has been reinterpreted by the author or a text that has been paraphrased should still recognize the original source.
4. If the author is not sure whether the idea or fact is common knowledge, it is better to cite the source.
5. If a piece of text is taken from one's own previous work, it must be clearly indicated.

6. Written permission is required to use previously published drafts, data, images, etc.

7. Any time after submitting the manuscript or after it is published, if the author finds that he has unintentionally used somebody else's ideas or text without appropriate referencing, he should immediately inform the journal editor and ask for advice.

19.6.2   Screening for Plagiarism

Authors themselves can perform plagiarism screening before submitting their work. First, they should thoroughly review the article, seeking random errors in references or omitted references. Next, they should use free online anti-plagiarism tools to check their work against content found in published databases; this will help them quickly discover a potential problem in their writing (Wiwanitkit 2011). Following a simple screening process will greatly benefit authors in their practice of ethical writing. Given that it is almost impossible to eliminate all acts of plagiarism or other intentional unethical behaviors, an effective system to prosecute and punish offenders should be in place. Education on the ethical conduct of writing and research, together with plagiarism detection tools and defined sanctions against offenders, will certainly lead to a reduction in the volume of plagiarism incidence.

19.6.3   Reducing Student Plagiarism

Many studies on plagiarism and other violations of academic integrity focus on the impact of educating researchers on the ethics of writing, especially at the student level. These studies encourage the targeted introduction of appropriate conduct, availability of support systems, and facilitated instruction for all students. Strategies for reducing and preventing plagiarism include promoting the importance of ethical behavior and modifying poor academic behaviors. These strategies are recommended together with vigilant detection practices and defined policies on punishment and imposed sanctions. Used by educational institutions, electronic detection tools that combat plagiarism can discourage copy-and-paste plagiarism among students, improving their ethical writing skills (Bretag 2013).

19.7   Penalties for Plagiarism

Many institutions and organizations implement specific policies to resolve issues of plagiarism (Masic 2016). An example is the policy adopted by the Optical Society of America (OSA) publications. The OSA Board of Editors requires that upon discovery of plagiarism, the editor of the journal (or equivalent personnel) will make a preliminary investigation into the allegations, in which an explanation will be requested of the author accused of plagiarism. If further action is justified, the editor will recommend the case to the OSA Editorial Ethics Review Panel, which consults standardized guidelines set by reputable institutions such as COPE and EASE. Appropriate punishment will be allocated from a simple warning to barring the author from submitting other manuscripts to the journal, disclosure of plagiarist activity, or being blacklisted from further publication. Penalties range in severity, depending on the nature of violation, and are generally determined by the scope of plagiarism — a few copied sentences, plagiarism of an entire work, or, more seriously, a stolen idea. For example, self-plagiarism is considered less serious than plagiarism of another's content. And while taking another's words verbatim is serious, plagiarized ideas and results are even more dangerous and yield the most severe consequences (OSA Board of Editors 2013).

19.7.1   The Dilemma of Who, What, When, and How

Detecting plagiarism raises additional questions: Who is the responsible party? (This can be difficult to ascertain when multiple authors are involved.) On what basis (criteria, standards, rules) is the work considered plagiarized? When and how should a person be judiciously declared a plagiarist? Only after plagiarism is proven can an institution or academic entity take corrective action based on the recommended sanctions (Masic 2013).

There is a significant movement among academics and publishers for establishing a uniform and transparent approach to reveal plagiarism and share information on the incident with the scientific community. Creating a blacklist of plagiarists on an international level can lead not only to the exclusion of offending authors from a specific journal but from the great majority of scientific venues, effectively eliminating these individuals from the publishing community all together. Based on "the principle of fear of possible sanctions," the consequences of transparency could have an enormous impact on the prevention of plagiarism (Masic 2013). Consequently, public disclosure of plagiarism can tarnish the reputation of the offender. The promise of ostracism not only meets the intrinsic need for social justice, but can also serve as an intimidation tactic and warning for novice writers and publishers to follow the rules of conduct. Unfortunately, many scientific journals, professional organizations, and academic institutions do not have the necessary resources, or in some cases even the will to adequately investigate and sanction misconduct allegations (Roig 2006).

19.7.2   Preventing and Monitoring Student Plagiarism

In summary, many possible solutions exist to prevent and monitor plagiarism among students:

- Promote, advertise, and mentor students on the principles of ethical writing and standards of scientific work within the university.
- Establish new coursework that exclusively addresses the ethics of scientific work, including the current methods of scientific research and writing, and provide continuing education for students from their first year through post-graduate years on the proper approach to scientific writing.
- Train faculty and teaching assistants on prevention and detection of plagiarism.
- Establish an academic board that represents all faculty, departments, and courses in an institution, and task them with reviewing and verifying student work. This academic board should use up-to-date plagiarism detection tools and subscribe to internationally recognized policies on dealing with breaches in academic conduct. Such an authority should be guided by recognized experts in ethical behavior, academic writing, and research methodology, ensuring a high standard of academic integrity.

- Impose sanctions on students for plagiarism and other forms of unethical conduct; this may include additional assignments, repeating exams, suspension, and in severe cases, expulsion.

19.8 Conclusions

Plagiarism appears in many forms, like self-plagiarism, copying someone else's work (including the popular copy-and-paste technique used by students), duplicating publications, or taking another's idea as one's own. A responsible author and researcher should always adhere to best practices in ethical writing, consulting the guidelines and policies defined by organizations such as COPE and EASE. While there are many tools available for detecting plagiarism, including anti-plagiarism software and web-based platforms, their application without preventive measures (such as continuing education of researchers) cannot have a long-term impact on reducing plagiarism. Starting in primary education, a student's curriculum should cultivate the tradition of academic integrity and ethical writing practices so that one day the student may become an active contributor to the scientific community. Furthermore, these future researchers will continue to propagate an example of responsible conduct in research and writing. While students are not expected to reinvent the wheel, their work must honor and recognize that of others through proper referencing and citation. Only by learning proper research conduct and best-writing practices can students learn to preserve academic integrity. There is an evident problem of plagiarism in developing countries where current education reform allows for gaps in academic integrity, including the large momentum of copy-and-paste plagiarism. Through the combined influence of education, detection, and prevention of plagiarism, along with imposing sanctions on the plagiarist, we can expect a significant reduction of plagiarism incidences and bring light to the movement of scientific integrity.

References

Anderson, M.S. and Steneck, N.H. (2011). The problem of plagiarism. *Urol. Oncol.* 29 (1): 90–94.

Annane, D. and Annane, F. (2012). Plagiarism in medical schools, and its prevention. *Presse Med.* 41 (9 Pt 1): 821–826.

Bamidis, P.D., Lithari, C., and Konstantinidis, S.T. (2010). Revisiting information technology tools serving authorship and editorship: a case-guided tutorial to statistical analysis and plagiarism detection. *Hippokratia* 14 (Suppl. 1): 38–48.

Bretag, T. (2013). Challenges in addressing plagiarism in education. *PLoS Med.* 10 (12): e1001574.

Cameron, C. and McHugh, M.K. (2012). Publication ethics and the emerging scientific workforce: understanding plagiarism in a global context. *Acad. Med.* (1): 87. https://doi.org/10.1097/ACM.0b013e31823aadc7

Chowhan, A.K., Nandyala, R., Patnayak, R. et al. (2013). Plagiarism: trespassing the Grey zone between searching and researching. *Ann. Med. Health Sci. Res.* 3 (Suppl. 1): S56–S58.

Garner, H.R. (2011). Combating unethical publications with plagiarism detection services. *Urol. Oncol.* 29 (1): 95–99.

Ghajarzadeh, M., Mohammadifar, M., and Safari, S. (2013). Introducing plagiarism and its aspects to medical researchers is essential. *Anesth. Pain* 2 (4): 186–187.

Marshall, S. and Garry, M. (2006). NESB and ESB students' attitudes and perceptions of plagiarism. *Int. J. Educ. Integ.* 2: 26–37.

Masic, I. (2012a). Plagiarism in scientific publishing. *Acta Inform. Med.* 20 (4): 208–213.

Masic, I. (2012b). Ethical aspects and dilemmas of preparing, writing and publishing of the scientific papers in the biomedical journals. *Acta Inform. Med.* 20 (3): 141–148.

Masic, I. (2013). The importance of proper citation of references in biomedical articles. *Acta Inform. Med.* 21 (3): 148–155. https://doi.org/10.5455/aim.2013.21.148-155.

Roig, M. (2006). Ethical writing should be taught. *BMJ* 333 (7568): 596–597.

The Optical Society of America Board of Editors (2005). From the board of editors: on plagiarism. *Opt. Express* 13 (5): 1351–1352.

Wiwanitkit, W. (2011). Plagiarism: pre-submission screening. *Perspect. Clin. Res.* 2 (4): 149–150.

Further Reading

Al Lamki, L. (2013). Ethics in scientific publication: plagiarism and other scientific misconduct. *Acta Inform. Med.* 28 (6): 379–381.

Ferguson, P.R., Mike, M., Griffin, S.M. et al. (2012). Perspective on plagiarism. *J. Community Hosp. Intern. Med. Perspect.* 2 (1): https://doi.org/10.3402/jchimp.v2i1.18048.

Kim, S.Y. (2013). Plagiarism detection. *Korean J. Fam. Med.* 34 (6): 371.

Panda, A. and Kekre, N.S. (2013). Plagiarism: is it time to rethink our approach? *Indian J. Urol.*

20

Conflicts of Interest: A Simple Explanation

Bradley K. Weiner[1], Leslie A. Duncan[2] and Paige E. Vargo[3]

[1] Weill Cornell Medical College, Houston Methodist Hospital, The Methodist Hospital Research Institute, Houston, TX, USA
[2] Dartmouth College, Hanover, NH, USA
[3] Northeast Ohio College of Medicine, Rootstown, OH, USA

20.1   Introduction

It is now part of the everyday life of academic physicians and scientists to avoid or disclose, explain, and manage conflicts of interest. Forms are filled out detailing them. Research talks include conflicts-of-interest statements on one of the first two slides. Journal submissions require disclosure of conflicts of interest before the peer-review process even begins. It is part of the water in which we swim, yet like fish swimming through that water, the conflict of interest is simply pushed out of our way because of obligation on our journey through academia. The importance has been lost in the ubiquity. In this chapter, we aim to explain conflict of interest and to touch on their importance.

20.2   What Is a Conflict of Interest?

The simplest and clearest explanation of a conflict of interest is the one put forth by the Institute of Medicine (2009):

> A conflict of interest is a set of circumstances that creates a risk that professional judgment or actions regarding a primary interest will be unduly influenced by a secondary interest.

By using this definition, we can break down and analyze its components to see things more clearly.

20.2.1   What Are Our Primary Interests?

Within academic medicine, we have five primary interests. They are (i) patients and their care, (ii) research and its integrity, (iii) students, residents, post-docs, etc. and their education, (iv) the institutions (hospitals, universities), and (v) our communities in which we live and work.

A Guide to the Scientific Career: Virtues, Communication, Research, and Academic Writing, First Edition.
Edited by Mohammadali M. Shoja, Anastasia Arynchyna, Marios Loukas, Anthony V. D'Antoni, Sandra M. Buerger, Marion Karl and R. Shane Tubbs.
© 2020 John Wiley & Sons, Inc. Published 2020 by John Wiley & Sons, Inc.

20.2.2   What Are the Secondary Interests?

Needless to say, the main secondary interest is financial gain. That said, other interests such as academic promotion, local and national recognition, favors to friends and family, obligations to our institution, etc. must also be considered.

20.2.3   What Is Meant by Unduly Influenced?

Implicit in the use of *unduly* is the fact that the secondary interests are inextricably tied to the primary interests. We expect to get paid to appropriately take care of patients –regardless of the healthcare system in which we practice. We hope to profit appropriately if we make a significant research contribution that becomes a product that positively impacts patients. We expect to advance academically if we have appropriately excelled at clinical, educational, and research work. Thus, the concept of "appropriateness" is central.

Most would agree that doing appropriate good work that benefits others should be rewarded – and the aforementioned secondary interests are often the rewards themselves. Indeed, they may well be the main drivers of bringing our best and brightest into medicine, and, thereby, of ensuring that patients receive the best available and ever-improving treatments. Altruism is an important element for all of us who care about medicine, but each of us needs to pay the light bills, to keep the hospitals open, to be able to afford to develop expensive new effective drugs for challenging problems, etc.

It becomes "inappropriate," however, when the secondary interests trump the primary interests. The scenarios are well-known. The surgeon offers surgery because it pays him more than nonoperative care (despite the fact that nonoperative care would work equally well). He sends the patient for unnecessary or excessive preoperative labs and an imaging studies at a facility he owns and profits from. He uses the medical devices from companies from which he receives considerable income as a "consultant" and chooses the most expensive one despite no impact on outcomes. He sends the patient for postoperative rehabilitation at a center in which he is a financial partner. Similarly, the researcher skews results to fall in line with the company whose product he is testing, or to get published for academic promotion, etc.

20.3   Why Does Avoidance or Full Disclosure of Conflicts-of-Interest Matter?

It is well known that something as small as a free, inexpensive logo pen given to a physician as a gift can result in a change in physician behavior such as writing prescriptions for a particular medication. It is equally known that, quite often, the source of these changes in behavior are opaque to the physician and researcher, i.e. it is not a conscious decision. As such, simple conflicts that might shift physician behavior away from evidence-based care or appropriate treatment based on value analysis should simply be avoided. Instead, literature about the medicine or device should be carefully analyzed. The role of pharmaceutical and device company representatives should be limited to the introduction of new products to physicians so that they, independently, can assess their worth. Avoidance, as it pertains to gifts that may unconsciously affects one's decision-making, is best way to go.

As noted above, however, some conflicts are necessary, and are simply part of what we do. Getting paid to treat patients, getting paid to consult for companies so that we might work with them to improve care, getting paid to do research for a company since we are the only ones to get the job done, etc. Under these circumstances, full disclosure of the sources of conflict-of-interest is important so that those impacted recognize that the suggested treatment plan or research result *might* be unduly influenced. The disclosure should include the value of the secondary interest. If a surgeon is using a particular device for surgery, he must disclose to the hospital his work as a consultant and the money made so that the hospital can determine if the evidence-based care and costs supports the use of that product. If he is guiding a patient toward a particular outpatient center in which he has a financial interest, he must let the patient know so that they (hopefully) can assess the quality and safety there. If a researcher is receiving money from any pharmaceutical or medical device company, the amount and scope of the relationship should be readily available to the institutes and readers so that they can assume any potential biases of the investigators – whether conscious or unconscious – and dig more deeply.

In the end, it is vital to remember that the end-result of inappropriate conflicts of interest is *harm*. Patients are harmed – or, at least, exposed to undue treatments or risks – by inappropriate nonevidence-based treatments. All of us are harmed by excess costs that do not impact outcomes. Evidence-based care suffers and patients are harmed when our research is biased in any way. This is of utmost importance, as our first obligation to all primary interests is to *do no harm*. With respect to the conflicts-of-interest, one should foster a sense of responsibility to recognize, and avoid or fully disclose them.

Reference

Institute of Medicine (US) (2009). Committee on Conflict of Interest in Medical Research, Education, and Practice. In: *Conflict of Interest in Medical Research, Education, and Practice* (ed. B. Lo and M.J. Field). Washington, DC: National Academies Press (US).

21

Gender Differences in Medical Research Productivity

Peter F. Svider[1], Priyanka Shah[1], Adam J. Folbe[1] and Jean Anderson Eloy[2, 3, 4]

[1]*Department of Otolaryngology–Head and Neck Surgery, Wayne State University School of Medicine, Detroit, MI, USA*
[2]*Department of Otolaryngology–Head and Neck Surgery, Rutgers New Jersey Medical School, Newark, NJ, USA*
[3]*Department of Neurological Surgery, Rutgers New Jersey Medical School, Newark, NJ, USA*
[4]*Center for Skull Base and Pituitary Surgery, Rutgers New Jersey Medical School, Newark, NJ, USA*

21.1   Introduction

A diverse array of factors impact prospects for appointment and advancement in academic medicine. Clinical productivity and the ability to effectively guide medical students, house staff, and even younger faculty members are among the most significant factors playing a role. One drawback in using these factors, however, relates to the issue that measuring success in these domains may be difficult and subjective. For several reasons, a third pillar of academic success often considered by promotions committee is scholarly productivity, not least important of which involves the relatively objective measures available to assess faculty in this regard (for additional information on assessing a researcher's scientific productivity and scholarly impact, see Chapter 9). Consequently, previously identified asymmetric promotion practices may be a consequence of gender differences in scholarly productivity.

The proportion of female physicians has increased dramatically over the past four decades in the United States. Evolving societal norms and attitudes regarding gender in the workplace are two of many factors responsible for this trend, and comprehensive discussion of these demographic trends can be extensive and is beyond the scope of this chapter. Approximately half of US medical students are now women. Despite these strides, women remain underrepresented in several areas of medicine. While they comprise approximately one-third of faculty at teaching institutions, fewer than 19% of full professors are women (Association of American Medical Colleges 2011, 2012a,b; Rotbart et al. 2012). Furthermore, women are considerably underrepresented in surgical specialties compared to nonsurgical specialties, although increased recruitment efforts have diminished this discrepancy in recent years (Borman 2007; Benzil et al. 2008).

This chapter is not meant as an all-encompassing review of gender dynamics in the contemporary healthcare environment, as many volumes can be written about the subject. Rather, the authors aim to briefly survey recent literature covering how gender differences in medical research productivity have been characterized, as well as explore factors that have been suggested to be responsible for measured disparities.

A Guide to the Scientific Career: Virtues, Communication, Research, and Academic Writing, First Edition.
Edited by Mohammadali M. Shoja, Anastasia Arynchyna, Marios Loukas, Anthony V. D'Antoni, Sandra M. Buerger, Marion Karl and R. Shane Tubbs.

21.2   Gender Differences in Scholarly Productivity

Gender disparities in research output among academicians have been documented in numerous analyses, with several studies noting that the higher aggregate scholarly productivity of male faculty may be due to the fact that women have lower *early* career productivity (Reed and Buddeberg-Fischer 2001; Reed et al. 2011; Eloy et al. 2013a,b; Pashkova et al. 2013). In fact, several studies have noted that women have mid- and later-career productivity equivalent to and even exceeding that of their male colleagues (Eloy et al., 2013a; Lopez et al. 2014; Tomei et al. 2014).

Career trajectory is often strongly influenced by early-career contributions, and early-career scholarly deficits may be responsible for women constituting a smaller proportion of academic physicians with successive academic rank. Importantly, leadership appointments typically occur at first or second decades of service.

A retrospective longitudinal analysis of female physicians serving as faculty at the Mayo Clinic noted that women published less frequently than their male counterparts when considering the entirety of their careers (Reed et al. 2011). Upon focused examination of annual publication rates among faculty in the third and fourth decades of their career, however, women had a significantly greater annual publication rate compared to their male colleagues. Despite this finding, men in this analysis were significantly more likely to hold the rank of professor and contribute in leadership roles.

Another recent analysis examining academic faculty practicing in the field of Otolaryngology – Head and Neck Surgery also noted differing productivity curves (Eloy et al. 2013a). The authors used the *h-index* as a proxy to measure scholarly impact, as this is an objective metric that not only measures the quantity of research one produces, but also the extent to which it is impactful in the peer-reviewed literature (Hirsch 2005). Briefly, an individual's *h-index* is 5 if they have had five papers in the peer-reviewed literature cited at least five times each, regardless of the presence or other papers that have been cited less than their *h* threshold. An *h-index* of 10 means an author has had 10 papers cited 10 times each, and so on. Consequently, this has been a widely used measure to assess the consistency with which an author has had impact on scholarly discourse within his or her field (for more information on *h-index*, see Chapter 9). Analysis among otolaryngology faculty noted that men had higher overall research productivity as measured by the *h-index*. However, in later stages of their careers, women had impact values equivalent to and even exceeding that of their male colleagues (Eloy et al. 2013a). Similar trends have been noted among a sample of academic anesthesiologists (Pashkova et al. 2013). In another analysis of nearly 10 000 academic physicians practicing 34 medical specialties, patterns among career-stage specific productivity rates were less clear (Eloy et al. 2013b). Nonetheless, men had higher aggregate scholarly impact in all specialties examined, while women were significantly underrepresented among promoted faculty and in positions of departmental leadership.

21.3   Gender Differences in Research Funding

In addition to producing publications, procurement of external funding plays an important role in assessing one's scholarly impact. Research grants facilitate scholarly productivity, and may reduce institutional financial pressures to support investigators with internal funding mechanisms. Additionally, an institution or department's success in procuring funding may attract prospective trainees and faculty members interested in contributing to scholarly discourse. Differences in funding attainment, specifically in the forms of grants from the National Institutes of Health (NIH), have been previously noted in several analyses.

One evaluation focusing on grant support mechanisms among Harvard Medical School faculty noted that not only did female faculty apply for fewer grants, but also they had lower funding awards and less success in their applications (Waisbren et al. 2008). Controlling for academic rank diminished disparities in awards and success, leading the authors to suggest that the relative underrepresentation of women among promoted faculty was responsible for the differences identified. It should be noted, however, that even upon controlling for academic rank, women were found to have pursued fewer grants.

Another analysis looking solely at NIH-funded investigators among faculty in otolaryngology – head and neck surgery – departments also found differences in funding. Men were found to have higher individual awards ($362,946) than women in these departments ($287,188) (Eloy et al. 2013c). Upon controlling for academic rank, men had higher funding levels among junior faculty and those with 10–20 years of experience. Furthermore, a larger proportion of awards to men were of the prestigious R-series variety. Similar findings were found in an analysis of NIH funding differences among academic ophthalmologists, in which male investigators had greater awards, with differences most marked among early career investigators (Svider et al. 2014).

21.4   Issues Potentially Facilitating Gender Differences in Research

While women reportedly demonstrate a different productivity curve than their male colleagues, the issues responsible for this discrepancy are less clear. Several potential reasons have been proposed for these varying productivity curves.

21.4.1   Discrimination in Academic Medicine

Although definitive data may be hard to come by due to the contentious nature of the issue, some have suggested that women may be excluded from networking by the traditional organizational cultures that dominate in certain aspects in academic medicine (Longo and Straehley 2008; Zhuge et al. 2011). Although potentially decreasing, some have suggested a culture of sexism in medicine, particularly in surgical fields (Baldwin et al. 1991; Komaromy et al. 1993; Zhuge et al. 2011). A database organized by the Women Physicians' Health Study noted that nearly half of female physicians reported experiencing harassment and sexism, more commonly during medical school and residency training than while in practice (Frank et al. 1998). Furthermore, female practitioners who felt lack of control over their work environments because of harassment experienced significantly lower career satisfaction as a result.

21.4.2 Family Responsibilities

There is a wide body of literature describing how there are perceptions that earlier in academic careers, family responsibilities take priority over academic aspirations, with a disproportionate effect on women. A survey of medical school faculty found that for both men and women, greater than 90% of time spent on "family responsibilities" were directly related to childcare. Upon further examination, however, divergent gender perceptions were noted. Women reported receiving less institutional support, including research funding, secretarial support, and administrative assistance; these perceptions led to subsequently lower career satisfaction (Carr et al. 1998). This analysis concluded that female faculty with children perceived a greater number of career obstacles compared to their colleagues without children, as well as relative to male colleagues with children. Other sources in the literature have noted that women consider availability of flexible family leave accommodations and call arrangements that facilitate family involvement when evaluating career opportunities (Levinson et al. 1991; Eskenazi and Weston 1995; Carr et al. 1998; Sanfey et al. 2006; Achkar 2008; Singh et al. 2008; Kuehn 2012).

Another analysis partially supports the impact of parental status in explaining gender differences in career trajectories. Jagsi et al. (2011) surveyed several hundred NIH Career Development Award (K-Grant) recipients, asking information regarding career path and quality of life. A significantly smaller proportion of women receiving these grants had children compared to their male colleagues. Interestingly, multivariate analysis revealed that gender differences in grant success that were present were *unrelated* to parental status, suggesting other factors may have been responsible for varying career trajectories.

While family considerations have been a recurring theme in literature exploring gender disparities in academic medicine, the authors would be remiss not to mention several analyses that provide skepticism to these claims. Zhuge et al. (2011) noted that the level of motivation present in those pursuing leadership opportunities in academic surgery was equivalent between genders. Furthermore, multiple analyses have noted that the impact of family responsibilities may be diminishing with changes in societal gender norms, and that too much emphasis is placed on this paradigm (Kaplan et al. 1996; Zhuge et al. 2011). Nonetheless, the persistent relative underrepresentation of women in academic medicine as well as within surgical specialties with rigorous training requirements at least facilitate the impression that family considerations may disproportionately impact women (Carr et al. 1998).

21.4.3 Pregnancy During Residency Training and Early in One's Academic Career

The stigma associated with pregnancy among practitioners in academic medicine has been anecdotal, particularly within surgical specialties and during residency, and has been thought to impact career prospects in a negative fashion (Kuehn 2012). Furthermore, in addition to potential professional repercussions, numerous analyses have suggested adverse health effects from pregnancy both during residency training and early in an individual's academic career (Grunebaum et al. 1987; Young-Shumate et al. 1993; Eskenazi and Weston 1995; Warnock 2011; Hamilton et al. 2012; Kuehn 2012; Turner et al. 2012). A survey conducted by the Women Plastic Surgeons' Caucus Committee of the American Society of Plastic and Reconstructive Surgeons noted pregnancy-related complication rates exceeding 50% among female plastic surgeons, as well as increases in elective abortion and infertility rates relative to overall population averages (Eskenazi and Weston 1995). Similar increases in pregnancy-related complication rates have also been noted among female orthopedic surgeons (Hamilton et al. 2012). Questionnaires sent to female obstetricians noted that children of practitioners delivered during or after residency had lower mean birthweights than those delivered *prior* to residency, with significant implications affecting growth; there were no differences incidence of other pregnancy complications, however, related to residency training in this particular cohort (Grunebaum et al. 1987).

21.4.4 Role Models and Mentoring

Comprising only 10% of physicians in 1970, the numbers of women entering medical school increased substantially in the following decades. Hence, the numbers of women in the latter portions of their careers is still disproportionately lower relative to their early and mid-career counterparts. This *pipeline effect* is most pronounced in surgical specialties, where women are still substantially underrepresented (Sexton et al. 2012). Numerous analyses have noted the importance of appropriate mentorship as a necessary factor in facilitating interest in a specialty (Caniano et al. 2004; Barry and Fallat 2011; Healy et al. 2012; Dageforde et al. 2013; Franzblau et al. 2013). Consequently, a paucity of mentors, particularly in surgical specialties, has been suggested as a factor for the under-representation of women at senior and leadership levels (Levinson et al. 1991; Matorin et al. 1997; Sambunjak et al. 2006; Wyrzykowski et al. 2006; Serrano 2007; Zhuge et al. 2011; Moed 2012).

21.4.5 Regional Differences

Another issue potentially impacting gender disparities in scholarly productivity, and subsequently, academic promotion relates to geography. This has been an understudied aspect that may contribute to some of the differences noted. A recent analysis examined whether regional differences exist in scholarly productivity and academic rank among academic otolaryngologists (Eloy et al. 2014). Examining 98 academic departments with residency training programs, an overall disparity in promotion practices was noted, similar to the aforementioned findings in this chapter. Organizing promotion patterns by region, however, the authors noted that these differences were most marked in the Northeast and the South, the two regions where gender disparities in scholarly impact were statistically significant. In contrast, there were no statistical gender differences in academic promotion among academic otolaryngologists in the Midwest and the West (Eloy et al. 2014). The causes and implications of these findings are unclear, although they suggest that further analysis into regional hiring and advancement trends in other medical specialties represent a future area of study.

21.4.6 Other Considerations

Other factors have been reported to potentially contribute to gender disparities in research productivity. Zhuge et al. (2011) noted a body of literature noting deficits both in salary and academic resources for female faculty, likely placing increased pressure for clinical productivity. Furthermore, some have suggested that women are more likely to be given certain administrative tasks and educational responsibilities, taking away from protected time dedicated to scholarly pursuits (Tesch et al. 1995; Wright et al. 2003; Numann 2011; Zhuge et al. 2011).

21.5 Conclusion

Gender differences in scholarly productivity have been demonstrated in numerous medical specialties among academic faculty. The preponderance of studies evaluating this topic have not been able to attribute causation, but have simply demonstrated an association; nonetheless, a common emerging theme is that mid- and later-career productivity among female faculty increases significantly and is equivalent to that of male colleagues, suggesting early career considerations play an important role. Importantly, as the number of women in academic medicine continues to rise, these trends may evolve considerably in the coming years, and efforts may need to be made to encourage mentorship opportunities for younger female faculty. Furthermore, additional inquiry into the mechanisms behind these trends may be important in addressing these disparities.

References

Achkar, E. (2008). Will women ever break the glass ceiling in medicine. *The American Journal of Gastroenterology* 103 (7): 1587–1588.

Association of American Medical Colleges (2011) *Women in U.S. Academic Medicine: Statistics and Benchmarking Report 2009–2010*. Available at: www.aamc.org.

Association of American Medical Colleges (2012a) *2012 Physician Specialty Data Book*. Available at: www.aamc.org.

Association of American Medical Colleges (2012b) *U.S. Medical School Applicants and Students 1982–1983 to 2011–2012*. Available at: https://wwwaamcorg/download/153708/data/charts1982to2012pdf.

Baldwin, D.C., Daugherty, S.R., and Eckenfels, E.J. (1991). Student perceptions of mistreatment and harassment during medical school. A survey of ten United States schools. *The Western Journal of Medicine* 155 (2): 140–145.

Barry, P.N. and Fallat, M.E. (2011). Medical student mentorship in a university setting as a strategy for a career in surgery. *The American Surgeon* 77 (11): 1432–1434.

Benzil, D.L., Abosch, A., Germano, I. et al. (2008). The future of neurosurgery: a white paper on the recruitment and retention of women in neurosurgery. *Journal of Neurosurgery* 109 (3): 378–386.

Borman, K.R. (2007). Gender issues in surgical training: from minority to mainstream. *The American Surgeon* 73 (2): 161–165.

Caniano, D.A., Sonnino, R.E., and Paolo, A.M. (2004). Keys to career satisfaction: insights from a survey of women pediatric surgeons. *Journal of Pediatric Surgery* 39 (6): 984–990.

Carr, P.L., Ash, A.S., Friedman, R.H. et al. (1998). Relation of family responsibilities and gender to the productivity and career satisfaction of medical faculty. *Annals of Internal Medicine* 129 (7): 532–538.

Dageforde, L.A., Kibbe, M., and Jackson, G.P. (2013). Recruiting women to vascular surgery and other surgical specialties. *Journal of Vascular Surgery* 57 (1): 262–267.

Eloy, J.A., Svider, P., Chandrasekhar, S.S. et al. (2013a). Gender disparities in scholarly productivity within academic otolaryngology departments. *Otolaryngology–Head and Neck Surgery* 148 (2): 215–222.

Eloy, J.A., Svider, P.F., Cherla, D.V. et al. (2013b). Gender disparities in research productivity among 9952 academic physicians. *The Laryngoscope* 123 (8): 1865–1875.

Eloy, J.A., Svider, P.F., Kovalerchik, O. et al. (2013c). Gender differences in successful NIH grant funding in otolaryngology. *Otolaryngology–Head And Neck Surgery* 149 (1): 77–83.

Eloy, J.A., Mady, L.J., Svider, P.F. et al. (2014). Regional differences in gender promotion and scholarly productivity in otolaryngology. *Otolaryngology–Head and Neck Surgery* 150 (3): 371–377. https://doi.org/10.1177/0194599813515183.

Eskenazi, L. and Weston, J. (1995). The pregnant plastic surgical resident: results of a survey of women plastic surgeons and plastic surgery residency directors. *Plastic and Reconstructive Surgery* 95 (2): 330–335.

Frank, E., Brogan, D., and Schiffman, M. (1998). Prevalence and correlates of harassment among US women physicians. *Archives of Internal Medicine* 158 (4): 352–358.

Franzblau, L.E., Kotsis, S.V., and Chung, K.C. (2013). Mentorship: concepts and application to plastic surgery training programs. *Plastic and Reconstructive Surgery* 131 (5): 837e–843e.

Grunebaum, A., Minkoff, H., and Blake, D. (1987). Pregnancy among obstetricians: a comparison of births before, during, and after residency. *American Journal of Obstetrics and Gynecology* 157 (1): 79–83.

Hamilton, A.R., Tyson, M.D., Braga, J.A. et al. (2012). Childbearing and pregnancy characteristics of female orthopaedic surgeons. *The Journal of Bone and Joint Surgery* 94 (11): e77.

Healy, N.A., Cantillon, P., Malone, C. et al. (2012). Role models and mentors in surgery. *The American Journal of Surgery* 204 (2): 256–261.

Hirsch, J.E. (2005). An index to quantify an individual's scientific research output. *Proceedings of the National Academy of Sciences of the United States of America* 102 (46): 16569–16572.

Jagsi, R., DeCastro, R., Griffith, K.A. et al. (2011). Similarities and differences in the career trajectories of male and female career development award recipients. *Academic Medicine* 86 (11): 1415–1421.

Kaplan, S.H., Sullivan, L.M., Dukes, K.A. et al. (1996). Sex differences in academic advancement. Results of a national study of pediatricians. *The New England Journal of Medicine* 335 (17): 1282–1289.

Komaromy, M., Bindman, A.B., Haber, R.J. et al. (1993). Sexual harassment in medical training. *The New England Journal of Medicine* 328 (5): 322–326.

Kuehn, B.M. (2012). More women choose careers in surgery: bias, work-life issues remain challenges. *JAMA* 307 (18): 1899–1901.

Levinson, W., Kaufman, K., Clark, B. et al. (1991). Mentors and role models for women in academic medicine. *The Western Journal of Medicine* 154 (4): 423–426.

Longo, P. and Straehley, C.J. (2008). Whack! I've hit the glass ceiling! Women's efforts to gain status in surgery. *Gender Medicine* 5 (1): 88–100.

Lopez, S.A., Svider, P.F., Misra, P. et al. (2014). Gender differences in promotion and scholarly impact: an analysis of 1460 academic ophthalmologists. *Journal of Surgical Education* 71 (6): 851–859.

Matorin, A.A., Collins, D.M., Abdulla, A. et al. (1997). Women's advancement in medicine and academia: barriers and future perspectives. *Texas Medicine* 93 (11): 60–64.

Moed, B.R. (2012). Mentoring: the role of a mentor and finding one. *Journal of Orthopaedic Trauma* 26 (Suppl 1): S23–S24.

Numann, P.J. (2011). Perspectives on career advancement for women. *The American Surgeon* 77 (11): 1435–1436.

Pashkova, A.A., Svider, P.F., Chang, C.Y. et al. (2013). Gender disparity among US anaesthesiologists: are women underrepresented in academic ranks and scholarly productivity? *Acta Anaesthesiologica Scandinavica* 57 (8): 1058–1064.

Reed, V. and Buddeberg-Fischer, B. (2001). Career obstacles for women in medicine: an overview. *Medical Education* 35 (2): 139–147.

Reed, D.A., Enders, F., Lindor, R. et al. (2011). Gender differences in academic productivity and leadership appointments of physicians throughout academic careers. *Academic Medicine* 86 (1): 43–47.

Rotbart, H.A., McMillen, D., Taussig, H. et al. (2012). Assessing gender equity in a large academic department of pediatrics. *Academic Medicine* 87 (1): 98–104.

Sambunjak, D., Straus, S.E., and Marušić, A. (2006). Mentoring in academic medicine. *JAMA* 296 (9): 1103.

Sanfey, H.A., Saalwachter-Schulman, A.R., Nyhof-Young, J.M. et al. (2006). Influences on medical student career choice. *Archives of Surgery* 141 (11): 1086.

Serrano, K. (2007). Women residents, women physicians and medicine's future. *WMJ: Official Publication of the State Medical Society of Wisconsin* 106 (5): 260–265.

Sexton, K.W., Hocking, K.M., Wise, E. et al. (2012). Women in academic surgery: the pipeline is busted. *Journal of Surgical Education* 69 (1): 84–90.

Singh, A., Burke, C.A., Larive, B. et al. (2008). Do gender disparities persist in gastroenterology after 10 years of practice. *The American Journal of Gastroenterology* 103 (7): 1589–1595.

Svider, P.F., D'Aguillo, C.M., White, P.E. et al. (2014). Gender differences in successful National Institutes of Health funding in ophthalmology. *Journal of Surgical Education* 71 (5): 680–688.

Tesch, B.J., Wood, H.M., Helwig, A.L. et al. (1995). Promotion of women physicians in academic medicine. Glass ceiling or sticky floor? *JAMA* 273 (13): 1022–1025.

Tomei, K.L., Nahass, M.M., Husain, Q. et al. (2014). A gender-based comparison of academic rank and scholarly productivity in academic neurological surgery. *Journal of Clinical Neuroscience: Official Journal of the Neurosurgical Society of Australasia* 21 (7): 1102–1105.

Turner, P.L., Lumpkins, K., Gabre, J. et al. (2012). Pregnancy among women surgeons. *Archives of Surgery* 147 (5).

Waisbren, S.E., Bowles, H., Hasan, T. et al. (2008). Gender differences in research grant applications and funding outcomes for medical school faculty. *Journal of Women's Health* 17 (2): 207–214.

Warnock, G. (2011). Pregnancy during postgraduate surgical training. *Canadian Journal of Surgery* 54 (6): 365–365.

Wright, A.L., Schwindt, L.A., Bassford, T.L. et al. (2003). Gender differences in academic advancement: patterns, causes, and potential solutions in one US College of medicine. *Academic Medicine: Journal of the Association of American Medical Colleges* 78 (5): 500–508.

Wyrzykowski, A.D., Han, E., Pettitt, B.J. et al. (2006). A profile of female academic surgeons: training, credentials, and academic success. *The American Surgeon* 72 (12): 1153–1179.

Young-Shumate, L., Kramer, T., and Beresin, E. (1993). Pregnancy during graduate medical training. *Academic Medicine: Journal of the Association of American Medical Colleges* 68 (10): 792–799.

Zhuge, Y., Kaufman, J., Simeone, D.M. et al. (2011). Is there still a glass ceiling for women in academic surgery? *Annals of Surgery* 253 (4): 637–643.

Further Reading

Agarwal, N., Clark, S., Svider, P.F. et al. (2013). Impact of fellowship training on research productivity in academic neurological surgery. *World Neurosurgery* 80 (6): 738–744.

Colaco, M., Svider, P.F., Mauro, K.M. et al. (2013). Is there a relationship between National Institutes of Health funding and research impact on academic urology? *The Journal of Urology* 190 (3): 999–1003.

Eloy, J.A., Svider, P.F., Folbe, A.J. et al. (2014). Comparison of plaintiff and defendant expert witness qualification in malpractice litigation in neurological surgery. *Journal of Neurosurgery* 120 (1): 185–190.

Eloy, J.A., Svider, P.F., Folbe, A.J. et al. (2014). AAO-HNSF CORE grant acquisition is associated with greater scholarly impact. *Otolaryngology–Head and Neck Surgery* 150 (1): 53–60.

Eloy, J.A., Svider, P.F., Mauro, K.M. et al. (2012). Impact of fellowship training on research productivity in academic otolaryngology. *The Laryngoscope* 122 (12): 2690–2694.

Eloy, J.A., Svider, P.F., Patel, D. et al. (2013). Comparison of plaintiff and defendant expert witness qualification in malpractice litigation in otolaryngology. *Otolaryngology–Head and Neck Surgery* 148 (5): 764–769.

Eloy, J.A., Svider, P.F., Setzen, M. et al. (2014). Does receiving an American Academy of Otolaryngology–Head and Neck Surgery Foundation Centralized Otolaryngology Research Efforts grant influence career path and scholarly impact among fellowship-trained rhinologists? *International Forum of Allergy and Rhinology* 4 (1): 85–90.

Kasabwala, K., Morton, C.M., Svider, P.F. et al. (2014). Factors influencing scholarly impact: does urology fellowship training affect research output? *Journal of Surgical Education* 71 (3): 345–352.

Paik, A.M., Mady, L.J., Villanueva, N.L. et al. (2014). Research productivity and gender disparities: a look at academic plastic surgery. *Journal of Surgical Education* 71 (4): 593–600.

Svider, P.F., Choudhry, Z.A., Choudhry, O.J. et al. (2013). The use of the h-index in academic otolaryngology. *The Laryngoscope* 123 (1): 103–106.

Svider, P.F., Husain, Q., Folbe, A.J. et al. (2014). Assessing National Institutes of Health funding and scholarly impact in neurological surgery. *Journal of Neurosurgery* 120 (1): 191–196.

Svider, P.F., Husain, Q., Mauro, K.M. et al. (2014). Impact of mentoring medical students on scholarly productivity. *International Forum of Allergy and Rhinology* 4 (2): 138–142.

Svider, P.F., Lopez, S.A., Husain, Q. et al. (2014). The association between scholarly impact and National Institutes of Health funding in ophthalmology. *Ophthalmology* 121 (1): 423–428.

Svider, P.F., Mady, L.J., Husain, Q. et al. (2013). Geographic differences in academic promotion practices, fellowship training, and scholarly impact. *American Journal of Otolaryngology* 34 (5): 464–470.

Svider, P.F., Mauro, K.M., Sanghvi, S. et al. (2013). Is NIH funding predictive of greater research productivity and impact among academic otolaryngologists? *The Laryngoscope* 123 (1): 118–122.

Section IV

Research Regulations

22

Institutional Review Boards: General Regulations, Institutional Obligations, and Personal Responsibility

Anastasia A. Arynchyna[1], Sarah B. Putney[2] and R. Peter Iafrate[3]

[1] Department of Neurosurgery, Children's of Alabama, University of Alabama at Birmingham, Birmingham, AL USA
[2] Administrative Compliance, Internal Audit Division, Emory University, Atlanta, GA USA
[3] Health Center IRB, University of Florida, Gainesville, FL USA

22.1   IRB Background

The background for institutional review board (IRB) regulations lies in the Nuremberg Code of 1947, the world's first codification of ethical principles for research with human subjects. Emerging from the Nazi experiments on concentration camp prisoners during World War II, the Nuremberg Code emphasizes individual informed consent as essential. In 1964, the World Medical Association published the Declaration of Helsinki, an ethical statement that focuses on separating the meaning between treatment of patients and research.

In spite of these statements, many research studies with human participants were conducted in ethically questionable ways in the preregulated era. Henry Beecher (1813–1887) called attention to the lack of informed consent in American research in an important article published in the *New England Journal of Medicine* in 1966, "Ethics of Clinical Investigation." Beecher believed that the "conscientious investigator" would always obtain informed consent. Probably the most infamous study to contribute to the movement toward regulation was the US Public Health Service study of Untreated Syphilis in the Negro Male, commonly known as the Tuskegee Syphilis Study, which took place in Alabama from 1932 to 1972 and involved questionable consent processes and the withholding of penicillin as treatment for syphilis.

These types of studies prompted action from Congress to establish the National Commission for the Protection of Human Subjects of Biomedical and Behavioral Research to study the ethics of human subjects research (HSR). In 1979 the Commission published The Belmont Report: Ethical Principles and Guidelines for the Protection of Human Subjects of Research. This report established three main ethical principles: (i) respect for persons, (ii) beneficence, and (iii) justice. These principles undergird the federal regulations in regard to protecting human subjects that were assembled under the Department of Health and Human Services (DHHS) in the Title 45 Part 46 of the Code of Federal Regulations (CFR).

A Guide to the Scientific Career: Virtues, Communication, Research, and Academic Writing, First Edition.
Edited by Mohammadali M. Shoja, Anastasia Arynchyna, Marios Loukas, Anthony V. D'Antoni, Sandra M. Buerger, Marion Karl and R. Shane Tubbs.
© 2020 John Wiley & Sons, Inc. Published 2020 by John Wiley & Sons, Inc.

22.2 Regulatory Framework

Research with human subjects has become a highly regulated activity, with especially significant responsibilities assigned to principal investigators (PIs) and their institutions. Even if you are not the principal investigator, it is important to know which legal rules govern your work so that you can be sure to comply with them. You should also know which regulatory agencies have enforcement powers and the potential consequences for noncompliance. If you are a principal investigator, you should realize that regulatory agencies (and most IRBs) will hold *you* accountable for any noncompliance with its regulations by other members of the research team! A wise principal investigator will make extra efforts to provide diligent oversight of every aspect of the research conduct and be a role model for taking the regulations seriously. It's a tall order.

The question of which regulations apply is essentially one of jurisdiction. The answer may be that more than one set of rules applies through different mechanisms. For each research project, the answer depends mainly on these four factors: (i) the source of funding or other support and the terms of the associated grant or contract; (ii) where the research will take place; (iii) whether the human subjects belong to certain vulnerable populations; and (iv) whether the research involves regulated products. Let's take each factor in turn, assuming for our purposes that the investigator will be working in the United States or its territories. (See the following chapter for more information about conducting research outside the United States.)

22.2.1 First: The Source of Funding or Other Support

In the United States, federal agencies such as the National Institutes of Health support and regulate a great deal of HSR. Hence, the key issue is whether the source of support is part of the federal government. Most federal agencies have adopted the Federal Policy and the substance of DHHS regulations at 45 CFR 46.[1] Subpart A of this set of rules, known as "the Common Rule" because 15 federal agencies have adopted it, sets forth important requirements for investigators, institutions, and IRBs about informed consent, IRB composition, and IRB review. Some agencies, such as the Veterans Administration and the Department of Defense, apply additional rules. You should familiarize yourself with your funding agency's rules so that you understand the basis for the IRB's requirements. The Office for Human Research Protections (OHRP), which enforces 45 CFR 46, serves as an educational resource about its rules and the ethical principles underlying them. Explore the OHRP website.[2]

There is another important aspect of federal oversight of HSR that deserves attention: the FederalWide Assurance (FWA) system. The FWA works like a driver's license for institutions receiving federal support to conduct HSR. The FWA is granted upon application to OHRP based on the institution's assuring that it will adhere to certain terms. Like a driver's license, the FWA is subject to sanctions such as restriction and revocation if OHRP finds that an institution is in breach. Without a valid FWA, an institution and its investigators cannot conduct HSR activities.

---

1 http://www.hhs.gov/ohrp/humansubjects/guidance/45cfr46.html.
2 http://www.hhs.gov/ohrp.

State and local governments also provide funding and support for HSR, in which cases state and local rules would apply. The IRB or institutional counsel should provide help with these.

If the source of support is not federal, such as a nonprofit organization or pharmaceutical company, the federal rules would not apply unless the institution to which the award has been made has elected to apply them to all of its human subject research regardless of funding source. To find out whether this is the case at your institution, ask the IRB if the institution has "checked the box" or "unchecked the box" on its FWA. If it has "checked the box," it has elected to apply the federal rules; if it has "unchecked the box," it has not; thus you will need to check with your IRB regarding local regulations.

If the source of support is nongovernmental, the grant or contract terms may bind the investigator and the awardee institution to the jurisdiction of a particular state. Furthermore, the grant or contract may contain terms specifying the investigator's and institution's obligations, for example for reporting unanticipated problems and adverse events to the sponsor, the sponsor's plans for using subjects' data for commercial purposes, and whether the sponsor will pay for expenses of medical care for subjects injured as a result of study participation. These terms should be reflected accurately in the informed consent documents.

22.2.2   Second: The Site of the Research

Depending on the site of the research, state and local laws or rules affecting the age of consent, the availability of funds in case of injury, and other practical issues may apply; the principal investigator should find out before writing the protocol and drafting consent forms. If the research will take place on an American Indian reservation, a tribal advisory board and/or IRB can impose their own requirements. If the study will be conducted in schools, the investigator should check with the principal to obtain permission and find out if any special rules apply. Research conducted in prisons or certain detention centers with prisoners or detainees and subject to federal rules must comply with the special requirements of Subpart C, 45 CFR 46.

The Health Insurance Portability and Accountability Act (HIPAA) Privacy Rule is a federal rule that creates its own geography affecting your research (45 C.F.R. 160 and 164). The Privacy Rule applies to research conducted by or at "covered entities" using subjects' individually identifiable, protected health information (PHI) and billing electronically (Table 22.1). Most healthcare providers and settings are covered entities, but investigators should check with the IRB to be sure they know if their institution is a covered entity, noncovered entity, or hybrid entity including both covered and noncovered components. In most institutions, the IRB acts as the HIPAA Privacy Board and can provide guidance to investigators about how the Privacy Rule applies to a given study. Every covered entity also has a privacy officer who should be available to answer questions.

We could write a book just about HIPAA, but here we will highlight the key points and strongly advise you to read the HIPAA policies of the covered entity from which you will obtain the PHI and your own covered entity if these are not the same. Be sure the members of your research team who will work with PHI complete HIPAA Privacy Rule training and understand their obligations under it, including the duty to report any breaches. And don't hesitate to ask questions of your IRB or privacy officer about how HIPAA works in a given scenario! They are there to help you comply with the rather

Table 22.1 Eighteen protected health information identifiers under HIPAA law.

1	Names
2	Geographical subdivisions smaller than a state, including street address, city, county, precinct, zip code, and equivalent geocodes, except for the first three digits of a zip code
3	All elements of dates (except year) for dates directly related to an individual, including birth date, admission date, discharge date, date of death; and all ages over 89, and all elements of dates (including year) indicative of age, except that such ages and elements may be aggregated into a single category of age 90 or older
4	Phone numbers
5	Fax numbers
6	Email addresses
7	Social Security numbers
8	Medical record numbers
9	Health plan beneficiary numbers
10	Account numbers
11	Certificate/license numbers
12	Vehicle identification and serial numbers, including license plate numbers
13	Device identifiers and serial numbers
14	Web Universal Resource Locators (URLs)
15	Internet Protocol (IP) address numbers
16	Biometric identifiers, including fingerprints and voice prints
17	Full-face photographic images and any comparable images
18	Any other unique identifying number, characteristic, or code (this does not mean the unique code an investigator may apply to data)

complicated rules. In addition to these rules, the Privacy Rule brings with it substantial financial penalties if the rules are broken. Those financial penalties can be levied both on you as the investigator and to the institution you with which you are employed.

If HIPAA applies to a research study, there are a handful of defined pathways an investigator may take to obtain and work with PHI. The most common scenarios involve obtaining a waiver (or partial waiver) of written HIPAA authorization from the IRB (Privacy Board), for example, for a chart review; and obtaining written HIPAA authorization from subjects for the use and disclosure of their PHI in addition to informed consent. The IRB should have a template authorization that contains all the elements required by the law, including the specific research purpose and an expiration date for the authorization or a statement that it does not expire.

Three other pathways to obtaining PHI for research purposes are: de-identified data (which contain none of the 18 HIPAA identifiers), the limited data set (which may contain certain of the 18 identifiers), and decedent's information. Follow the covered entity's HIPAA policies and procedures to work within these pathways.

The Office of Civil Rights enforces the HIPAA Privacy Rule. Violations of HIPAA involving PHI in a research study should be reported promptly to the IRB or privacy officer so that a timely breach analysis can be conducted, the breach mitigated, and other appropriate steps taken. In some cases of breach, subjects may have to be notified.

If the investigator's institution is accredited by the Association for the Accreditation of Human Research Protection Programs, Inc. or Alion Science and Technology Corporation, it means that the institution has implemented standards even higher than those set by federal rules. These would be incorporated into the operations of the IRB, office of sponsored programs, and other research administration offices.

22.2.3   Third: Human Subjects in Certain Vulnerable Populations

Under the Federal Policy, children, pregnant women/fetuses, and prisoners are identified as populations vulnerable to giving consent that is not fully informed and voluntary because of their reduced autonomy. The Belmont principles call for additional protections in the IRB review and consent processes to make up for their relative lack of power. See Section 22.5 of this chapter for more information.

22.2.4   Fourth: Research or Clinical Investigations with Regulated Products

Clinical investigations using investigational products (drugs, devices, biologics) are *highly* regulated by the Food and Drug Administration (FDA) (21 CFR 50, 56, 312, and 812). Investigator responsibilities for obtaining the necessary permissions from the FDA, documenting informed consent, strictly adhering to the approved protocol, and properly reporting adverse events to sponsors and the IRB are substantial. Even more significant are the responsibilities of the sponsor – the holder of the investigational new drug (IND) application or investigational device exemption (IDE). When the principal investigator is the sponsor (then called the sponsor-investigator, holding the IND or IDE), she has the same obligations as a pharmaceutical company would and must exercise energetic oversight and control of every aspect of the clinical investigation to remain compliant. The driving concern behind these requirements is safety: conducting monitoring of protocol adherence, data and safety monitoring, and registering a study with a publicly accessible website like Clinical Trials.[3] Explore regulatory information on the FDA website[4] and work with the IRB and any clinical research support services at your institution to understand and comply with these important rules, especially if you plan to serve as sponsor-investigator. The FDA also provides written guidance on a number of issues bearing on the conduct of clinical investigations. Remember that any time you are conducting research that involves a drug or device, the question should be raised as to whether an IND is required. Your IRB may ask you to ask the FDA for a determination, since the rules to determine IND status are not very explicit.

22.3   IRB Process

22.3.1   IRB Jurisdiction: Research with Human Subjects

If you are conducting research as part of your employment or as part of being a student of a university or hospital, chances are that your institution has a policy regarding human research, and which IRBs you can use to have your research approved by.

---

3 www.clinicaltrials.gov.
4 www.fda.gov.

22.3.1.1 Definitions

In many cases, what is considered research, and what is considered HSR is obvious, but at other times, it is a little gray. The following regulatory terms and definitions are key:

Research (from 45 CFR 46) is:

> a systematic investigation, including research development, testing and evalua-
> tion, designed to develop or contribute to generalizable knowledge. Activities
> which meet this definition constitute research for purposes of this policy, whether
> or not they are conducted or supported under a program which is considered
> research for other purposes. For example, some demonstration and service pro-
> grams may include research activities.

A *Human subject* (from 45 CFR 46) is:

> a living individual about whom an investigator (whether professional or student)
> conducting research obtains: (1) Data through intervention or interaction with
> the individual, or (2) Identifiable private information.

22.3.1.2 Some Items that May Not Be Under Your IRB's Jurisdiction

First, it is imperative that you check with your IRB to determine what local rules may be in place; however, in most institutions the following guidelines apply.

In an academic medical center or similar facilities, it is not unusual for unique and interesting clinical cases to be written up as *case reports* for publication in medical jour-nals or presentation at medical or scientific meetings. In general, the review of medical records for publication of "case reports" of up to *three to five* patients is NOT consid-ered human-subject research and does NOT typically require IRB review and approval because case reporting on a small series of patients does not involve the formulation of a research hypothesis that is subsequently investigated prospectively and systematically.

Many peer review journals require the submitter of the case report obtain informed consent from the patients, or have a letter from your IRB indicating that no such require-ment is needed.

22.3.1.3 Quality Improvement (QI) Non-Research versus Research

The regulations are not explicit, thus making the determination that a project is quality improvement (QI) and not research, is difficult and thus varies between institutions. Based on the definitions above, activities that are *not designed* to produce information that expands the knowledge base of a scientific discipline (or other scholarly field) do not constitute research. An operations or quality improvement activity does *not* constitute research if *both*:

- The activity *is designed* and/or implemented only for *internal* institutional purposes (i.e. its findings are intended to be used by and within that institution or by entities responsible for overseeing that institution).
- The activity *is not designed* to produce information that expands the knowledge base of a scientific discipline (or other scholarly field).

Many times, a quality improvement project is also a research project, but either can occur separately. Review your institution's policy on this topic, since conducting human research without proper approval can bring significant institutional, federal, and privacy (HIPAA) penalties.

22.3.2 Types of IRB Reviews

There are four types of IRB reviews that exist to guide investigators to conduct proper research. The best advice that we could give to the investigators before initiating a research project and submitting an IRB application is to familiarize yourself with your institution-specific IRB regulations and documentations to eliminate any unnecessary questions down the road.

22.3.2.1 Not Human Subjects Research

The first type of IRB review is "Not Human Subjects Research" (NHSR) designation. If your project does not fall into the classification of HSR as defined in 45 CFR 46, it can be granted the NHSR status. Although not required by federal regulations, most institutions do not allow investigators to establish this status themselves and designate the IRB or another office to perform this function. It is recommended that investigators submit an application for an official determination. Examples of this kind of project could include the use of cadaveric specimens or de-identified blood products from a blood bank.

22.3.2.2 Exempt Review

The second type of IRB review is an exempt review. A research project that fits one or more of the categories described in 45 CFR 46.101(b) is exempt from further IRB review (i.e. expedited or full board). Examples include observation of public behavior (not including children if the investigator participates); surveys collecting information that would not put the participant at risk of harm if a breach of confidentiality occurs; and use of secondary data that are publicly available and collected in such a way that the data cannot be traced back to the subject.

22.3.2.3 Expedited Review

The third and the most common type of IRB review is an expedited review. In order for a project to be considered under expedited review, it must meet the following criteria: poses *no more than minimal risk* to the human subjects, does not meet the criteria for an exemption, and meets at least one of nine categories listed under Office of Human Research Protection – Categories of Research.[5] This review does not apply to research that involves prisoners or "classified" research. Only one designated and qualified IRB member (often the chair) is required to review and approve HSR by expedited review. Though it refers to research that involves minimal risk, the IRB will examine this application according to the same approval criteria as it employs at a convened meeting. The most common documents included in the application are human subjects protocol, special populations information (i.e. research involving children, pregnant women, prisoners), informed consent/assent or a waiver of authorization and authorization materials, sponsor/funding information, data-collection instruments such as surveys, and any institution-specialized approvals with signatures. Examples of this research include

---

5 http://www.hhs.gov/ohrp/policy/expedited98.html.

the collection of limited amounts of blood samples from a healthy live subject, and retrospective collection of data that was previously recorded solely for non-research purposes. Some retrospective research may fall under the exempt category, so please check with your local IRB.

22.3.2.4 Full Board Review

The fourth type of IRB review is "full board" review, which occurs at a convened meeting of a quorum of the IRB. All research that involves any of the following conditions will need to go through the full IRB review: poses more than minimal risk to human subjects; does not meet the criteria for the expedited review; and involves new investigational drugs/devices. Disapproval of a study must be done by the full board under the regulations. The main difference between this type of review and the others is the number of IRB members required to review and vote on the protocol. There is usually a deadline for applications because the IRB needs adequate time to review each study before a meeting. The board pays special attention to the methodology, subjects selection, and risk and benefit ratio. This is why it would be very helpful to become familiar with your institution IRB procedures and deadlines. The application for the full IRB review is similar to the expedited one with addition of a few more documents such as sponsor's protocol, any NIH/FDA approved documents, and funding application submitted to the sponsor.

22.4 Investigator Responsibilities in Conducting Human Subjects Research

Conducting HSR is not a right, it is a privilege. Thus, as an investigator, you must know the rules, must understand the ethics, and must be aware of your responsibilities.

22.4.1 Training for Investigators

If you are starting out as a researcher at your institution, or have recently moved to another institution, it is in your best interest to set up time with your IRB to discuss training requirements. This training may not only include didactic courses, but is usually a way to receive some hands-on assistance from those who will be reviewing your research protocols.

The Health and Human Services HHS regulations for the protection of human subjects (45 CFR part 46) do not require investigators to obtain training in the protection of human subjects in research. However, an institution holding an OHRP-approved FWA is responsible for ensuring that its investigators conducting HHS-conducted or -supported HSR understand and act in accordance with the requirements of the HHS regulations for the protection of human subjects.

Therefore, as stated in the Terms of the FWA, OHRP strongly recommends that institutions and their designated IRBs establish training and oversight mechanisms (appropriate to the nature and volume of their research) to ensure that investigators maintain continuing knowledge of, and comply with, the following:

- Relevant ethical principles
- Relevant federal regulations

- Written IRB procedures
- OHRP guidance
- Other applicable guidance
- State and local laws
- Institutional policies for the protection of human subjects

Furthermore, OHRP recommends that investigators complete appropriate institutional educational training before conducting HSR.

In some cases, other federal requirements regarding training for investigators must be met, such as the National Institute of Health's (NIH) requirement for the training of key personnel in NIH-sponsored or -conducted HSR. The NIH provides an online training course on its website.[6] This provides a general overview of the items listed above, and is a requirement by NIH, for the grant awardee and his/her study staff to complete.

Even if your research is not funded, or funded by a nonfederal agency, most institutions, whether they are universities or private hospitals, will require some type of investigator and\or study staff education prior to the IRB reviewing your research study.

The majority of institutions utilize the Collaborative Institutional Training Initiative (CITI Program) at the University of Miami.[7] This program is a leading provider of research education content. Its web-based training materials serve millions of learners at academic institutions, government agencies, and commercial organizations in the United States and around the world.

The CITI course selections include

- Animal Care and Use (ACU)
- Biosafety and Biosecurity (BSS)
- Export Control (EC)
- Good Clinical Practice (GCP)
- Human Subjects Research (HSR)
- Information Privacy and Security (IPS)
- Responsible Conduct of Research (RCR)

Under HSR, the course options are

- Belmont Report and CITI Course Introduction
- History and Ethics of Human Subjects Research
- Basic IRB Regulations and Review Process
- Informed Consent
- Social and Behavioral Research (SBR) for Biomedical Researchers
- Records-Based Research
- Genetic Research in Human Populations
- Research with Protected Populations – Vulnerable Subjects: An Overview

It is best to contact your IRB or research office to determine if CITI classes are required, and if so, which classes. Also, your institution may have an agreement with CITI, to provide the courses at a discount price. Again, contacting your local IRB or research office is the key.

---

6 http://phrp.nihtraining.com/users/login.php.
7 https://www.citiprogram.org.

Most all "covered entities," or those facilities that bill Medicare or Medicaid, require its employees and faculty to complete some type of HIPAA training. In some cases, the institution or facility may have developed a HIPAA for researchers training that may be in addition to, or instead of, the standard HIPAA training.

Finally, your institution or facility may have developed a local training component to the researcher training requirement that is taken in addition to some of those already discussed. The purpose of the local training is to ensure that researchers are aware of any local research requirements, and some of the nuts-and-bolts of how to submit your protocol.

22.4.2 The Responsibilities of a Research Investigator

The principal investigator (PI) is responsible for all aspects of a research protocol:

- Select your research team wisely.
- Obtain IRB approval before conducting human subjects research.
- Delegate, do not abdicate, your responsibilities.
- Disclose any conflicts of interest; this may result in some disclosure of your conflict in your consent form, or other ways the conflict will be managed.
- If you do not have time to run your study correctly, do not become a PI.
- Protect human subject rights and welfare.
- Know where to go with questions about the regulations on human subjects research.

22.4.2.1 Informed Consent

Informed consent is a process, not a piece of paper. The following guidelines will help ensure that proper procedures are covered:

- Make sure your study subject is well informed and has time and the ability to ask questions.
- Provide all subjects a copy of the IRB approved informed consent, this is not an option, it does not need to be a signed copy, if they do not want it, they can discard it.
- Obtain consent prior to starting any research intervention, unless approved by your IRB.
- Only use most currently approved (by the IRB) version of your consent form.
- Keep all copies of signed consent forms.
- All revisions, no matter how minor, must be reviewed and approved by an IRB prior to making that change.
- Exception to that is if the change is if there is a need to eliminate potential harm to study subjects; if so, report to your IRB as soon as possible (usually within five days).
- Report any adverse events or unanticipated problems to your IRB per your IRB's policy. In many institutions, if the event is serious and unanticipated, the event must be reported within five working days of discover.
- Report all protocol deviations according to your local IRB. Deviations can be:
  - Regulatory compliance violations (i.e. you enrolled a minor and you were not approved to do so), or
  - Protocol deviations (i.e. subjects were supposed to receive a follow-up MRI and didn't).

- Close out studies properly. When your study is complete, or if you leave your current institution, it is your responsibility to close out your study correctly, submitting a closure report to the IRB, or transfer to a different PI (with IRB approval).

Conducting research with human subjects is a privilege. IRBs may impose their own corrective and preventive action plans for serious or continuing noncompliance, such as removal of the principal investigator, suspension or revocation of approval, or more frequent progress reports. The enforcement agencies can impose sanctions for serious noncompliance, including termination of support, repayment of grant funds, suspension from federal grant programs for a period of time, and even disbarment from participation in federal programs. The compliance teams at your institution are your partners in understanding and complying with a complex regulatory framework, so treat them as valued colleagues.

22.5 IRB Application

The IRB approval is a very important step that every investigator must take prior to initiating any research that involves human subjects. Even if the study can obtain an exempt status, an exemption IRB application must be submitted. In most cases, the IRB application will take time to be written. It requires a thorough and detailed explanation of the research protocol. Depending on the type of research that is done, different forms will need to be filled out. The most common ones are described in this section.

It is strongly advised to review your local IRB guidelines for completing the appropriate paperwork, because each IRB will have some requirements unique to their institution. Some institutions will provide orientation and/or assistance as well.

Remember, IRB was set in place to help conduct proper research, and not to be an obstacle. Its goal is to protect the subjects, researchers (you), and the institution from risks of unethical and noncompliant study conduct, so establish a good relationship with your IRB from the beginning by communicating and asking questions about your research project ahead of time.

Here is the list of the most common forms submitted in the IRB application[8]:

- *Human subjects protocol.* This is the core of the IRB application. This document requires information on the investigators, funding, location, special approvals, purpose, background including thorough literature review on the topic, study design, subjects, methodology, list of variables to be collected, risk and benefit discussion, data and safety monitoring plan, privacy and confidentiality discussion, and informed consent discussion.
- *Informed consent/assent.* This is one of the essential documents that guides the consent discussion and, if signed, records subjects' agreement to participate in the research study. Generally, informed consent forms should be written at a 6th–8th grade English reading level to be understandable to the average population. Children of certain developmental stages aligned with certain age ranges are considered to able to understand right from wrong and are able to assent for themselves along with the parental consent. Each IRB institution determines that age range. Therefore,

---

8 HHS IRB guidebook - http://www.hhs.gov/ohrp/archive/irb/irb_chapter3.html.

there are three common situations if a child is the subject in the study: (i) the child is too young/immature and only parental permission is required; (ii) both parental permission and child's assent are required; (iii) consent from both parent and child is required until the child reaches the age of majority.

- *HIPAA authorization.* This form must be signed by the subject/legal guardian to allow for use and disclosure of the health information for research purposes. Use of any of the 18 HIPAA identifiers in the research requires an authorization from the subject. This form is the last page of the informed consent form.

- *Waiver of HIPAA authorization and informed consent.* This document requests permission to use the PHI for research purposes only without obtaining consent from the subject to be in the study. To obtain this waiver, investigators must give solid and valid reasons for why the research cannot be practicably conducted without this waiver, how PHI will be protected from misuse and disclosure, and why consent and authorization cannot be practicably obtained.

- *Waiver of informed consent.* This document requests permission to collect information to which HIPAA does not pertain and an informed consent needs to be waived. To obtain this waiver, investigators must give solid and valid reasons for why the research cannot be practicably conducted without this waiver and why this research will not involve more than minimal risk and will not adversely harm the subjects.

- *Waiver of documentation of informed consent.* This document requests permission for subjects to give spoken consent to participation the study (without a signature or similar documentation). If this record is the only link between the subject identity and the study, and the primary risk of the study is a breach of confidentiality, then a waiver may be granted. Another common basis for the request is to accommodate cultural norms of the subject population.

- *Waiver of HIPAA authorization.* This document requests use of PHI only for screening and recruiting of potential participants. To obtain this waiver, investigators must give solid and valid reasons for why the research cannot be practicably conducted without use of PHI and this waiver.

- *Special populations review form: Pregnant women, children, prisoners.* When these vulnerable populations are considered for enrollment into a research study, the IRB must ensure that additional safeguards are in place prior to study approval. The IRB takes a deeper look at the risks, possible benefits, and ways in which these unique populations with limited autonomy may be guarded against coercion or undue pressure.

- *Sponsor's protocol/Funding.* If a study is sponsored, then include the sponsor's protocol and the funding application in the packet.

- *Study materials/Advertising.* If a research study involves a questionnaire, survey, advertising, or other study material that is given out to a patient or is used in the interaction with the patient, it must be submitted with the IRB application for approval as well.

- *Special approvals.* If the research project involved special topics or devices such as genetic testing, samples, or drugs, then a special approval form will need to be submitted with the application as well. Please check with your local IRB for institution-specific forms.

- *Protocol oversight review.* This form will vary from institution to institution and may go by a different name. Its role is to obtain departmental approval of the study prior to submitting to IRB. This form confirms the review of scientific integrity and credibility of study personnel.

- *Project revision/amendment*. Any changes to the study protocol must receive IRB approval *prior* to their implementation. This form notes any changes to the study protocol, design, personnel, consent forms, etc. If these changes affect consent/assent process, then new updated consent/assent forms must be submitted with the amendment for a review and approval.

- *Continuing review/Investigator's progress report*. This form is required for a study renewal to obtain a new IRB approval to continue conducting a research study. In most cases, this is submitted on annual basis but please check with your local IRB regarding your institution and your specific research study. One year is the longest approval period permissible under US federal regulations. With the renewal, you must update the IRB on the progress of enrollment and have an option to make any necessary changes to the study as well. Along with this form, you must submit any relevant study documents for approval as well. Examples of those can include updated consent/assent forms, any study materials, copies of all approved amendments since last approval date, and more.

22.6   Related Research Committees

Based on the type of research you wish to conduct, there could be several other committees that must provide approval prior to either IRB review, or your institutions local approval. Common committees include:

22.6.1   Human Use of Radioisotopes and Radiation Committee or Radiation Safety Committee

The Human Use of Radioisotopes and Radiation Committee must review all projects that involve the research of diagnostic X-rays, radioactive materials, or therapeutic radiation. Even a standard diagnostic test used more frequently than clinically indicated would constitute research with radiation. If this type of review is required, you cannot begin the project until you have both IRB and radioisotope/radiation approvals.

22.6.2   Institutional Biosafety Committee (IBC)

The Institutional Biosafety Committee (IBC) must review all projects that involve human gene therapy. A project that needs IBC review cannot begin until it has the approval of both the IBC and the IRB. Also, the IBC will review all revisions to human gene therapy projects to ensure that NIH guidelines are being met.

22.6.3   Other Potential Committees Could Include

- *Billing committee*. To ensure that research related tests and procedures are properly billed to your study and not an insurer.
- *Hospital committees*. May exist to ensure that clinical resources are not being used for research purposes that may include health care providers.

22.7 Publishing and the IRB

There are no specific federal regulations regarding publishing case reports. Rather, it is regulated at an institutional level and is subject to the local IRB determination of what approvals are required to publish case reports. With this in mind, make sure to contact your IRB office regarding the required documentation and HIPAA office regarding the privacy and informed consent requirements. The majority of medical journals now require proof of the IRB's approval and/or the subject's informed consent to publish the case report.

   In regard to publishing any other type of article, not just case report specific, most of the medical journals will require a statement within the manuscript confirming that, prior to commencing the study, an IRB approval was obtained. Check with the corresponding journal to which you are submitting the manuscript for more details.

22.8 Selected Frequently Asked Questions (FAQs)

For more information please visit OHRP FAQ[9]:

Q: What is an IRB?
A: Institutional Review Board for Human Use (IRB) is a committee established under federal regulations for the protection of human subjects in research (45 CFR 46). Its purpose is to help protect the rights and welfare of human participants in research.

Q: What is "research?"
A: *Research* is "a systematic investigation, including research development, testing and evaluation, designed to develop or contribute to generalizable knowledge" (45 CFR 46).

Q: What is a "human subject?"
A: *Human subject* is "a living individual about whom an investigator (whether professional or student) conducting research obtains (1) data through intervention or interaction with the individual, or (2) identifiable private information." (45 CFR 46).

Q: When do I need to submit an IRB application?
A: Prior to start of any research activities, including screening.

Q: Can I make any changes to the study without IRB approval?
A: No. You must obtain IRB approval first by submitting an IRB amendment.

Q: How long does my IRB approval last?
A: In most cases and institutions, up to one year. Always check with your local IRB for verification.

---

9 http://www.hhs.gov/ohrp/policy/faq/index.html.

Q: Does secondary use of an existing dataset require IRB approval?
A: It depends. Best practice is to ask your IRB. The scope of consent for the primary data collection, whether the data are de-identified, and possibly other factors should be considered.

Q: Conducting a survey, do I need an IRB approval?
A: It depends. Yes, if the study involves human subjects and research as defined above. Some surveys of adults may be eligible for exempt approval.

Q: What is assent?
A: Assent is agreement given by a child or other person lacking legal capacity to give consent. Affirmative assent should be sought whenever a subject is capable of giving it.

Q: Who are considered children for assent purposes?
A: The Code of Federal Regulations (45 CFR 46) defines children a "persons who have not attained the legal age for consent to treatments or procedures involved in the research, under the applicable law of the jurisdiction in which the research will be conducted." State and local laws may also define who is a child for purposes of research consent.

Q: When is informed consent not required by the IRB?
A: (1) If the IRB finds that the consent waiver criteria are met. (2) If the project does not involve "research" or "human subjects," the IRB will not review it, but seeking informed consent of some kind may still be appropriate under considerations of autonomy. (3) In exempt research, the Belmont Report ethical principles still apply and consent should be sought.

Q: When is a child capable of assent?
A: This depends on age, maturity, and psychological state of the child. Age ranges from IRB to IRB but can start around age 7.

Q: Can I get an exemption for a study involving children?
A: Yes, this is an option under categories 1, 4, 5 and 6 and under category 2 if "research involves educational tests or observations of public behavior when the investigator(s) do not participate in the activities being observed" (45 CFR 46.401(b)).

Q: Can I email a dataset that includes PHI to a colleague?
A: It depends at a minimum on whether the colleague is an IRB-approved member of the research team, whether the security settings on the email meet HIPAA Security Rule standards, and whether the dataset is emailed for a legitimate research purpose.

Q: What happens if my protocol expires?
A: You must stop all the work on the project and notify the IRB of the lapse of approval. You most likely will need to explain the lapse and the steps you will take to prevent such situations in the future in the Corrective Actions Plan. If ceasing study procedures would pose imminent risk to enrolled subjects, you should contact the IRB to arrange for continuing them on the treatment.

Q: Who can consent for a study subject if the subject is not of age, or is temporarily or permanently incapacitated?
A: Parent, Legally Authorized Representative, or Guardian

Q: What is considered "minimal risk" (for nonprisoner subjects)?
A: It is that the probability and magnitude of harm or discomfort anticipated in the research are not greater in and of themselves than those ordinarily encountered in daily life or during the performance of routine physical or psychological examinations or tests.

Q: What should I do when assuming the responsibility of an existing research study?
A: Obtain proper IRB training; Notify IRB to officially be added to the protocol by approved amendment; Read, understand, and fulfill the responsibilities of the principal investigator.

23

International Research with Human Subjects

Sarah B. Putney

Administrative Compliance Office, Internal Audit Division, Emory University, Atlanta, GA, USA

23.1 General Overview

Conducting research with human participants across countries and cultures is compli-cated in many ways – administratively, technically, and ethically. Challenges can arise from multiple sets of rules, ethical codes, and socio-cultural assumptions about individ-ual autonomy, consent, and gender roles. The economic inequalities between developed sponsoring countries and less-developed host countries often presents challenges to fairness in the distribution of burdens and benefits at the levels of investigators, institu-tions, participants, and their communities. Long distances make it difficult to maintain close oversight of scientific and technical aspects of a study. In all respects, the success of a transnational research project depends on mutually respectful, collaborative rela-tionships among investigators and between the research team and the host community. Building such relationships requires significant time and effort. All of this means that if you anticipate conducting research in a foreign country, it will require more time and patience than conducting a study in the United States, so plan accordingly.

The Belmont Report principles of autonomy, beneficence, and justice suggest that US-based researchers should avoid structuring research studies that leave study partic-ipants and host communities worse off than they were before the research study began. Thus, research collaborations across the divides between richer and resource-poorer research partners tend to involve what is known as capacity-building: the development of technical, informational, scientific, educational, and healthcare-delivery skills and functionalities of the host-partner, assisted by the sponsor-partner, with a view to even-tual self-sufficiency and sustainability. Granting agencies such as the Fogarty Interna-tional Center (FIC) of the National Institutes of Health expect such considerations to be incorporated into proposals for human subjects research with developing country institutions. The FIC and private, nonprofit organizations such as the Bill and Melinda Gates Foundation and the Clinton Foundation have supported capacity-building in the area of ethically sound healthcare research for many years. The rewards are by no means one way: collaborative, transnational research can provide some of the most memorable experiences of a career precisely because of the meeting of different cultures and soci-eties in the pursuit of common goals. The lessons to be learned and relationships formed from living and working in a foreign location with in-country partners can be priceless.

A Guide to the Scientific Career: Virtues, Communication, Research, and Academic Writing, First Edition.
Edited by Mohammadali M. Shoja, Anastasia Arynchyna, Marios Loukas, Anthony V. D'Antoni, Sandra M. Buerger, Marion Karl and R. Shane Tubbs.
© 2020 John Wiley & Sons, Inc. Published 2020 by John Wiley & Sons, Inc.

The primary issues facing a collaborative research study involving human subjects can be separated into three intertwined categories: the regulatory and legal frameworks; IRB oversight; and the ethical frameworks. Volumes could be written about each of these topics, but here we will point out the main issues of which you should be aware so that you can dig deeper into the study-specific contexts and troubleshoot the problems that may arise.

23.2   Regulatory and Legal Frameworks and IRB Oversight

In addition to understanding the US regulations that apply to a study, you should learn about the host country's regulatory framework and identify what laws and rules apply. It is a good idea to consult with your institutional legal counsel on this topic, who may, in turn, want to hire in-country counsel, whose fees your research budget may have to pay. An excellent online resource with which to start is the Office for Human Research Protection (OHRP)'s International Compilation of Human Research Standards. Some basic legal issues to cover include: age of majority/minority, as well as legal rights or restrictions on married and unmarried women, children, or other populations you plan to recruit, in the consent/assent process. In some cases, pregnant women's legal rights may be different, based on their age and marital status. You should be prepared to distinguish between legal guidelines and customary practice in obtaining and documenting informed consent; for example, it may be customary for a caretaking relative such as a grandparent, aunt, or uncle to give consent for a child, but if that relationship is not legally authoritative for consent, you should design the consent process to meet the legal standards and plan how to explain and facilitate this counter-customary practice.

23.2.1   IRB Registration and FederalWide Assurance

If the research project is supported by a US federal agency, the non-US institution collaborating with the US institution must obtain and meet the terms of a FederalWide Assurance (FWA) in order to conduct human subjects research with the US funds. OHRP provides instructions for registering IRBs and obtaining FWA. Be prepared to help your foreign-sited research partners to understand US requirements; you can start by identifying the appropriate contact people at the US agencies who can help answer their questions.

The FIC offers many resources for investigators and IRBs, including research ethics educational programs. The US Food and Drug Administration (FDA) is a very large agency and it may be more difficult to find the right person to help you there, but you should not hesitate to ask questions and keep trying until you find assistance. In case you get different answers from different agency representatives, keep notes about with whom you speak, and whenever possible, obtain answers in writing.

The non-US institution seeking an FWA to engage in research must ensure that it and the IRB(s) upon which it relies for review will, at a minimum, comply with one or more of the regulations and guidelines listed in Table 23.1. It is also possible that a US federal department or agency head may review procedures prescribed by the FWA applicant and find that they provide protections to study participants that are at least equivalent to

Table 23.1 Regulations and guidelines for non-US institutions seeking a FederalWide Assurance (FWA).

The Common Rule or Federal Policy for the Protection of Human Subjects
The US Food and Drug Administration regulations at 21 Code of Federal Regulations (CFR) parts 50 and 56
The current International Conference on Harmonization E-6 Guidelines for Good Clinical Practice
The current Council for International Organizations of Medical Sciences International Ethical Guidelines for Biomedical Research Involving Human Subjects
The current Canadian Tri-Council Policy Statement: Ethical Conduct for Research Involving Humans
The current Indian Council of Medical Research Ethical Guidelines for Biomedical Research on Human Subjects
Other standard(s) for the protection of human subjects recognized by US federal departments and agencies, which have adopted the US Federal Policy for the Protection of Human Subjects

those provided by the Common Rule. In this case, such procedures could be substituted for the Common Rule.

The institution and its IRB must have written policies and procedures meeting the terms of the FWA, including those ensuring prompt reporting to the IRB, appropriate institutional officials, the head of any US federal department or agency conducting or supporting the research (or designee), and OHRP of any unanticipated problems involving risks to subjects or others, serious or continuing noncompliance with the applicable US federal regulations or the requirements or determinations of the IRB(s) and suspension or termination of IRB approval. The IRB must conduct continuing review at least annually. The process of developing and implementing such policies and procedures materials may require guidance to be provided to the partner in-country institution. Note that the non-US institution must also register its IRB(s) with OHRP (if it has any). The registration and FWA processes are linked but separate. An institution must have an FWA in order to receive US Department of Health and Human Services (HHS) support for research involving human subjects. Each FWA must designate at least one IRB registered with OHRP. Before obtaining an FWA, an institution must either register its own IRB (internal IRB) or designate an already registered IRB operated by another organization (external IRB) after establishing a written agreement with that other organization.

23.2.2 FDA Considerations

If the research study involves an FDA-regulated product and/or is designed to collect data to support a marketing application for an investigational product or new use for an approved product, the FDA requires that the study be conducted according to the International Conference on Harmonisation (ICH) Good Clinical Practices (GCPs) standards. If the study will be FDA-regulated, it may be prudent to employ a contract research organization (CRO) that has experience dealing with international collaborative FDA-regulated research. Study-level monitoring, data and safety monitoring, and even consent monitoring will be extremely important activities in which the site principal investigator and the US principal investigator must stay engaged in order to fulfill FDA requirements, regardless of whether a CRO is involved in study conduct.

23.2.3 Privacy

The Health Insurance Portability and Accountability Act (HIPAA) Privacy and Security Rules may or may not apply to your international research project, depending on the covered or noncovered status of the US partner institution and, if it is covered entity, how it interprets HIPAA with respect to collection of individually identifiable health information outside the United States and subsequent transfer to the covered entity. Obviously, it is simpler if HIPAA does not apply, but on this issue the US investigator should make no assumptions and should consult her HIPAA privacy officer and/or institutional legal counsel. In the other direction, the host country may impose privacy and/or data security regulations on the study, which may impose higher standards than HIPAA.

23.2.4 Host-Country IRB

The question will quickly arise whether the study should be reviewed by the U.S.-based IRB (institutional, independent/for-profit, or central IRB), or the host country IRB using a reliance or authorization agreement, or both. Using both a US IRB and an in-country IRB allows you to obtain the best of both worlds in terms of familiarity with applicable US requirements and cultural mores and compliance oversight (note that outside of the United States, the term "IRB" is often replaced with "ethical review board" or another similar term meaning the same sort of thing). The host-country IRB will be especially important in assessing the comprehensibility of the informed consent documents and the feasibility of the consent process, and in identifying and resolving any unintended sociocultural conflicts, oversights, or discrepancies. Developing-country IRBs tend to have little administrative support and lower resources than US IRBs. You may become the go-between for your home IRB and the in-country IRB. It may be useful to facilitate conversations between their respective chairs or administrators on issues of concern to either group.

If the host IRB relies on the US IRB, consulting a community advisory board (CAB) is highly recommended to obtain the views of individuals familiar with the local research context. A CAB may be in place already or you may need to develop one in collaboration with the in-country institution. Even if the in-country IRB will conduct parallel review with the US IRB, the input of a CAB can be very valuable, if not always easy to accommodate.

23.3 Ethical Framework

Ethical issues in international human research tend to revolve around resource and power inequities between the US sponsor-partner and host-partner, between investigators and study participants, and between cultural assumptions about normative behavior. There is extensive discourse in the literature on each of the following examples of key issues:

- The standard of care available in the sponsor-partner country is frequently different and more effective than that available to the participant community in the host country. Which standard of care should govern the medical treatments offered to study participants who may suffer harm as a result of study participation–those available in the host country or those available in the sponsoring country? What are the sponsor-partner's ethical obligations to help the host-partner build capacity to deliver a higher standard of care?
- In communities where education and literacy are very limited and written documents tend to be regarded with anxiety and suspicion, the emphasis on reading a detailed consent form may pose barriers to participation as well as truly informed, comprehending, and voluntary consent. To what lengths should a study team go to ensure that the consent process is structured to overcome these challenges for each prospective subject?
- The identification and description of benefits of participation should be accomplished with thoughtful input from the host community. What is in it for each side? What is in it for the participants? What is in it for their communities?
- Who speaks for the host community? Who should be on a CAB? How much weight should their advice be given?
- What are fair and appropriate ways to compensate participants for their time and effort in the study? If other researchers working in the same community do not offer monetary compensation for participation in their studies, should you refrain from offering monetary compensation in order to avoid disrupting the local norm, which could cause other researchers to have to offer monetary compensation in order to compete?
- US concepts of individual autonomy tends to differ from the concept in traditional cultures in Asia, Latin America, and Africa because it gives little attention to the individual's identity as belonging to a family or community. How should the consent process be structured to meet US and host-partner standards if the US and in-country IRB disagree on the adequacy of an unmarried adult woman's autonomy in giving consent for her own participation in a study without the concurrence of her father or brother?
- What is a fair arrangement to offer study participants for access to proven, beneficial study treatment after the study ends? What will happen to them after that period ends? Where does the ethical obligation of the sponsor-partner toward study participants begin and end?
- The "therapeutic misconception" may occur when a prospective study participant perceives a guaranteed benefit from participation in research for lack of understanding the uncertainties inherent in research, the purpose of research being discovery rather than treatment, and/or an assumption that in a healthcare setting, any activities are designed to improve their health. How should the consent process be designed to minimize the therapeutic misconception?
- In societies where the social respect customarily granted to doctors and nurses has not been eroded as it has in the United States where a culture of medical malpractice litigation prevails, prospective study subjects may have difficulty accepting the idea

that participation in a healthcare research study may not be designed to benefit them directly, because they trust that the doctor's or nurse's intentions are aligned with a patient's best health interests. This can mean that they underappreciate the risks involved in participation and cannot provide comprehending consent. How should a study team deal with this dilemma?

There are no easy answers to these questions, which are a mere sampling of the kinds of issues that arise in international collaborative research. The process should be guided by the ethical principles or code(s) to which the reviewing IRBs subscribe. Usually, the questions boil down to a matter of balancing interests in a reasonable way that can be

Table 23.2 Selected resources containing recommendations and regulations for pragmatic and ethical issues involved in conducting international research with human subjects.

Entity	Website
Agencies	
Council for International Organizations of Medical Sciences (CIOMS)	http://www.cioms.ch
Fogarty International Center (FIC), National Institutes of Health	http://www.fic.nih.gov
International Council for Harmonisation of Technical Requirements for Pharmaceuticals for Human Use (ICH)	http://www.ich.org
National Bioethics Advisory Commission (1996–2001)	https://bioethicsarchive.georgetown.edu/nbac
Nuffield Council on Bioethics	http://nuffieldbioethics.org
Office for Civil Rights (OCR), US Department of Health and Human Services (HHS)	https://www.hhs.gov/ocr
Office for Human Research Protections (OHRP), US Department of Health and Human Services (HHS)	https://www.hhs.gov/ohrp
Presidential Commission for the Study of Bioethical Issues	https://bioethicsarchive.georgetown.edu/pcsbi/node/851.html
The Hastings Center	http://www.thehastingscenter.org
US Food and Drug Administration (FDA)	http://www.fda.gov
World Medical Association	http://www.wma.net
Journals	
American Journal of Bioethics	https://www.tandfonline.com/toc/uajb20/current
American Journal of Public Health	https://ajph.aphapublications.org
Journal of Empirical Research on Human Research Ethics	http://journals.sagepub.com/home/jre
Journal of Law, Medicine & Ethics	http://journals.sagepub.com/home/lme
Journal of the American Medical Association	http://jamanetwork.com/journals/jama
New England Journal of Medicine	http://www.nejm.org

articulated in the language of the applicable ethical framework. Often, investigators and IRBs must settle for a practical compromise or the best possible solution due to budgetary or structural restraints, but the process of arriving at such solutions is best served by employing an ethical lens to the discussions internal to the IRBs, and between the IRBs and investigators, with input from the community advisors and potential study participants.

There is by now a rich literature on research ethics built on theory, practice, and empirical studies. Table 23.2 is a limited list of selected resources that can be searched for more in-depth considerations of the pragmatic and ethical issues involved in conducting international research with human subjects.

Further Reading

Council for International Organizations of Medical Sciences (2016) *International Ethical Guidelines for Biomedical Research Involving Human Subjects* [WWW] Available from: http://www.cioms.ch/index.php/12-newsflash/400-cioms-inernational-ethical-guidelines [Accessed 4/15/2017].

Macklin, R. (2004). *Double Standards in Medical Research in Developing Countries.* Cambridge: Cambridge University Press.

National Bioethics Advisory Commission (2001) *Ethical and Policy Issues in International Research: Clinical Trials in Developing Countries* [WWW] Available from: https://bioethicsarchive.georgetown.edu/nbac/clinical/execsum.html [Accessed 4/15/2017].

Nuffield Council on Bioethics (2002) The Ethics of Research Related to Healthcare in Developing Countries [WWW] Available from: https://nuffieldbioethics.org/wp-content/uploads/2014/07/Ethics-of-research-related-to-healthcare-in-developing-countries-I.pdf [Accessed 4/15/2017].

Presidential Commission for the Study of Bioethical Issues (2011) *"Ethically Impossible" STD Research in Guatemala from 1946 to 1948* [WWW] Available from: http://bioethics.gov/sites/default/files/Ethically-Impossible_PCSBI.pdf [Accessed 4/15/2017].

Office for Civil Rights (2017) *HIPAA for professionals* [WWW] U.S. Department of Health and Human Services (HHS). Available from: https://www.hhs.gov/hipaa/for-professionals/index.html [Accessed 4/15/2017].

Office for Human Research Protections (2011) *Federalwide Assurance (FWA) for the Protection of Human Subjects* [WWW] U.S. Department of Health and Human Services (HHS). Available from: https://www.hhs.gov/ohrp/register-irbs-and-obtain-fwas/fwas/fwa-protection-of-human-subjecct/index.html [Accessed 4/15/2017].

Office for Human Research Protections (2016) *International Compilation of Human Research Standards* [WWW] U.S. Department of Health and Human Services (HHS). Available from: https://www.hhs.gov/ohrp/international/compilation-human-research-standards/index.html [Accessed 4/15/2017].

Shuster, E. (1997). Fifty years later: the significance of the Nuremberg code. *New England Journal of Medicine* 337 (20): 1436–1440.

US Food and Drug Administration (2013) *ICH guidance documents* [WWW]. Available from: https://www.fda.gov/scienceresearch/specialtopics/runningclinicaltrials/guidancesinformationsheetsandnotices/ucm219488.htm [Accessed 4/15/2017].

World Medical Association (2013) *Declaration of Helsinki–Ethical Principles for Medical Research on Human Subjects* [WWW] Available from: https://www.wma.net/policies-post/wma-declaration-of-helsinki-ethical-principles-for-medical-research-involving-human-subjects [Accessed 4/15/2017].

Section V

Research Grants and Proposals

24

Grants and Funding Sources

Jamie Dow and Chevis Shannon

Department of Neurosurgery, Vanderbilt University Medical School, Nashville, TN, USA

24.1 Introduction

Preparation, development, execution, and follow-through are required for any project to be successful. Obtaining funding for your research idea is no exception. This chapter provides a brief overview of the grant process, highlights some of the top grant funders, breaks down some of the most common types of grants, and provides you with a number of resources and tools to help you find your next funding opportunity.

So, what exactly is a grant? The grants process involves the distribution of funds to support specific projects that align with the goals, mission, or vision of the organization offering the funds. This is usually done via an application process, which can be an electronic or hard-copy/paper submission. Obtaining grant funding is a competitive process. To be successful, investigators must sell their research plan by clearly explaining their strategy for addressing specific criteria outlined in the funding organization's request for proposal (RFP). A competitive project should include innovative solutions that can be achieved within a reasonable time frame and budget.

24.2 Grant Life Cycle

To give you a better sense of the overall process and the main components of the grant process, we have mapped out a general life cycle of a grant (Figure 24.1). The easiest way to look at the grant life cycle is to divide it into two components, or stages. The pre-award stage consists of everything you need to do on the front end of the award (such as generating your idea, finding funding, developing your proposal, and submitting your application). The post-award stage involves starting your project, managing the funds, and providing final information to the funder.

First, you need to identify your project; once you have a project in mind you will need to find funding opportunities that align with your area of research. To begin your search, you may want to ask yourself the following questions:

- Are you seeking support for training, fellowship opportunities, travel grants, or novel research?

A Guide to the Scientific Career: Virtues, Communication, Research, and Academic Writing, First Edition.
Edited by Mohammadali M. Shoja, Anastasia Arynchyna, Marios Loukas, Anthony V. D'Antoni, Sandra M. Buerger, Marion Karl and R. Shane Tubbs.

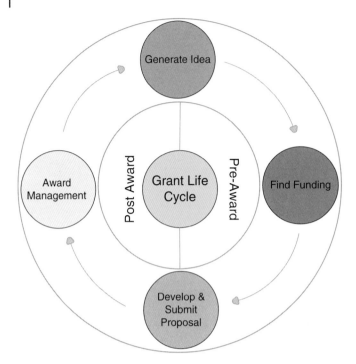

Figure 24.1 Grant life cycle.

- In what area of research does your project fall?
- Will you be working with animals or humans?
- What is your goal? What do you hope to achieve?

As you start to plan your application, these questions can help you narrow your focus to particular types of grants (such as an F or T grant) or specific types of grant funders like the National Institutes of Health (NIH) or a foundation. Finding funding opportunities is an ongoing process and there are numerous resources, which at times can be overwhelming. The hardest part of applying for a grant is doing the research to find appropriate funders for your project.

24.3 Types of Funding Organizations

One of the key components of obtaining grant funding is making sure that you are applying to a funder that is a match for your proposed project. Identifying funding organizations that align with your topic or area of research is crucial. The most common types of funding organizations include the following:

- *Government.* Federal, state, and local
- *Nonprofit organizations.* Foundations, associations, and other nonprofit organizations
- *For profit corporations.* Business and industry

24.3.1 Federal Funders

There are 26 federal grant-making agencies with over 900 funding programs that are divided into roughly 23 categories, ranging from agriculture to transportation. The central portal to find and apply for federal funding opportunities is Grants.gov.[1] You can search opportunities by using keywords or more specific criteria such as type of funding mechanism (i.e. grant, cooperative agreement, etc.), by agency, or area of interest; you do not have to register with the website to search for grant opportunities. By subscribing to grant notices, users can receive notifications of new grant opportunity postings and updates on Grant.gov.

24.3.2 State Agencies

Some federally funded programs are also applicable to individual states. In addition, each state typically has unique special-purpose grants funded by its legislature. Most state agencies have a section on their website where they will list all of the funding opportunities. They will fall into a number of categories including public and legal notices of grants available, funds available, public notices of RFPs or requests for information (RFI), or by notices of funds available (NOFA).

24.3.3 Foundations

Foundations are another great resource for searching out funding opportunities. Foundations usually provide very clear criteria in their RFPs or calls for proposals (CFPs), so make sure your topic aligns with their interests. Also something to keep in mind, most foundations restrict their grant-making to specific areas of study or specific geographic locations. Make sure you understand the scope and the limitations of the foundation. Thousands of private foundations, corporations, and associations dedicate their philanthropy to education-related objectives. Private funding sites often include a quick eligibility questionnaire and a preliminary proposal process to help you determine if your request matches their current priorities and guidelines.

Foundations are generally divided into two categories, public and private. Within each of the two categories, we can find a number of foundation types outlined in Table 24.1 that span locally, nationally, and internationally.

24.3.4 Business and Industry

When seeking corporate or industry funding in an academic medical center, first you need to review your institution's conflict of interest (COI) policy. Many institutions have implemented policies that prohibit or limit the acceptance of any on-site or off-site gifts or monetary awards from industry by faculty, staff, students, and trainees. Additionally, industry-funded projects can be supported either by an individual company or by a consortium. The following are some of the most common award mechanisms for these funders:

---

1 www.Grants.gov.

Table 24.1 Foundation types.

Foundation type	Potential limitations
National foundations	Typically highly competitive and are not limited to geographic location
Corporate foundations	Usually look for projects or innovations that benefit organizational processes or work flow
Family foundations	Often restricted in geographical area, fund projects in areas of family interest
Community foundations	Usually specific to a geographic area

- *Research contracts.* They define a scope of work carried out by the principle investigator (PI) and institution and may provide funder access to resulting inventions and research results within the project scope and term.
- *Clinical trials.* They involve testing of a funding organization's own drugs or devices using the funding organization's protocol.
- *Gifts.* They are donations to general or specific areas of campus activity that carry no obligations for deliverables or rights to results.
- *Restricted grants.* Funders may designate or "restrict" the use of their donations to a particular purpose or project.
- *Unrestricted grants.* Donated funds are made available for the investigator or institution to use toward any purpose.

As with any grant process make sure you work closely with your grants office or office of sponsored programs and understand your institution's COI policy. If you receive an industry award, make sure to review the funder's report, which is made available to the public. You want to make sure the information that the funding organization has provided regarding you, your award, and vested or ownership interests are accurate.

24.3.5 Show Me the Money

There are literally thousands of government and foundation grant funders to sort through and review. This can be a tedious and confusing process that requires expertise. This is why most institutions have on-staff grant expertise or, at the very least, contract this work out to grant writers with this skill set. Although we cannot provide you with an exhaustive list, Section 24.9 provides a compilation of government and foundation funders that are major contributors to research. Table 24.2 lists just a few of the major federal funding sources and foundations.

24.4 Types of Agreements

There are several types of funding awards, grants being the most common. Below we have listed a few of the most common types of agreements made between the institution/organization and funders. All types of funding awards come with specific requirements for tracking of funds and reporting of activities.

Table 24.2 Major funding sources.[a]

US government grant sources

National Science Foundation (NSF)

Agency for Healthcare Research and Quality (AHRQ)

National Science Foundation: Social Psychology Program

National Science Foundation: Searchable Database of Awards

National Institutes of Health Office of Extramural Grants (NIH)

OppNet: Basic Behavioral and Social Science Opportunity Network

Centers for Disease Control and Prevention (CDC)

Heath Resources and Services Administration (HRSA)

Department of Defense (DOD)

National Aeronautics and Space Administration (NASA)

Department of Education (DoEd)

Department of Energy (DOE)

National Endowment for the Humanities (NEH)

Department of Transportation (DOT)

Environmental Protection Agency (EPA)

Catalogue of Federal Protective Assistance (CFPA)

National Technical Information Services (NTISs)

Major grant making foundations

Alfred P. Sloan Foundation (science, technology, education, and other topics)

American Psychological Foundation (APA, a nonprofit organization)

Bill and Melinda Gates Foundation (fosters advances in health for those that need them most)

Brain Aneurysm Foundation

Brain and Behavior Research Foundation

Brain Tumor Foundation

Burroughs Wellcome Fund (biomedical sciences)

Carnegie Corporation of New York (education, peace, health, and other areas)

Charles A. Dana Foundation (health and education)

Commonwealth Fund (health policy, health reform, and healthcare delivery)

David and Lucile Packard Foundation (wide range of programs)

Foundation for Psychocultural Research (interdisciplinary research projects)

Ford Foundation (wide range of programs)

Elisabeth Glaser Pediatric AIDS Foundation

Gerber Foundation

James McDonnell Foundation (biomedical and behavioral sciences)

John D. and Catherine T. MacArthur Foundation (wide range of programs)

John Templeton Foundation (exploring "life's biggest questions")

Prevent Cancer Foundation

Robert Wood Johnson Foundation (health, healthcare, and substance abuse)

Rockefeller Brothers Fund (wide range of programs)

Thrasher Research Fund (medical research grants to improve the lives of children)

William T. Grant Foundation (research that improves the lives of young people)

W. K. Kellogg Foundation

---

a) For full table and more information, see http://www.socialpsychology.org/funding.htm.

- *Grant.* In short, a grant is simply a sum of money given by a funder to an individual or organization, for a particular purpose.
- *Cooperative agreement.* This is similar to that of a grant, but with a cooperative agreement the funder is actively involved in the development of the proposal and is involved in the research activities once the funds have been awarded.
- *Contract.* A contract is a written, legal agreement between an institution and an awarding agency or funder. It is important to note that there are only a few people that are able to sign a contact, and the PI is never one of them. Check with your contracts and grants office regarding signatory requirements.
- *Subaward agreement*: These subawards (or subcontracts) are contracts between the institution or organization and a subcontractor (most often, a company or another academic research institution), to conduct tasks as part of a research program.

24.5 Types of Grant Applications

There are two basic types of grant applications and proposals, namely, unsolicited and solicited:

- An *unsolicited proposal* is initiated by the applicant, and is based on your research interest. There are often no submission date requirements and only broad guidelines by the funder that must be followed. In this case, the funding entity did not announce or request proposals to be submitted.
- A *solicited proposal* is one that is submitted in response to a funder's RFP or request for applications (RFA). These proposals require that you submit a potential research project or training program directly related to the funding interests of the funding organization. The RFP or RFA generally includes standard terms, conditions, and assurances that the institution is asked to accept. Most solicited proposals can be found though the following announcement types:
  - Program announcement (PA)
  - Request for proposals (RFP)
  - Request for applications (RFA)
  - Request for contracts (RFCs)
  - Funding opportunity announcement (FOA)

24.6 Federal Grant Mechanisms

There are several federal mechanisms to explore when looking to support your research. These include research grants (R series), career awards (K series), fellowships and training grants (F and T series), and research program project and center grants (P series). For more specific and detailed information, visit the NIH website.[2]

What is the right mechanism for you? When selecting a federal grant at the NIH for example, you will need to take the following into consideration:

- *Your training and experience.* Where do you fall in the career timeline?
- What will you need in terms of resources and budget?
- What would you like to do and how much time do you need?

---

2 http://grants.nih.gov/grants/funding/funding_program.htm.

24.7   Career Timeline and NIH Mechanisms

Table 24.3 gives you an idea of what NIH awards and grant mechanisms best fit with a researcher's career timeline. These are just general suggestions to give you a starting point to begin looking. It is important to note that there are exceptions and opportunities that fall within in these categories.

Table 24.4 outlines NIH grant mechanisms for the T32 training and F32 fellowship grants, as they are the most common mechanisms for individuals in training positions.

24.8   Funding Cycle

There are three major grant submission cycles every year for the NIH. Table 24.5 outlines the submission dates. These dates do fluctuate by a week or two each year, so it is best to review the NIH website for exact dates.

Keep in mind that new grants can only be resubmitted once (for a total of two submissions). Make sure that when resubmitting your proposal, you address the reviewer comments. Take a moment, process the information, and address the areas where the reviewers have requested new or additional information. If the reviewers suggest that you include an experiment or additional analysis that you are not familiar with, take this opportunity to identify a collaborator.

24.9   Searchable Databases

When you are searching grant databases you should start with keywords that are specific to your field of interest, or target population (e.g. diabetes, liver transplant, veterans with trauma brain injuries, etc.). However, funding agencies may not list their funding priorities at this level of detail but instead may use broad categories (child health, healthcare, the elderly, education, etc.) It is important to use keywords, but be willing to expand

Table 24.3 Career timeline and relevant grant mechanisms.[a]

Career timeline	NIH mechanism
Medical school or graduate school	T32
Medical residency training or internship	T32
Post graduate medical education (fellowship) or post doctorate fellowship	F32
Junior faculty	K01, K08, K22, K23, R03, R21
Career	R01, R21, K24

a)   For NIH support by career stage–MD Track, see www.niaid
    .nih.gov/researchfunding/traincareer/pages/careermd.aspx.
For NIH support by career stage–PhD Track, see www.niaid.nih
.gov/researchfunding/traincareer/pages/careerphd.aspx.

Table 24.4 T32 and F32 funding mechanisms.

NIH mechanism	Mechanism details
T32	To enable institutions to make National Research Service Awards to individuals selected by them for predoctoral and postdoctoral research training in specified shortage areas. These grants develop or enhance research-training opportunities for early-stage investigators.

For more information related to T grants see the NIH website:
http://grants.nih.gov/training/T_Table.htm

F32	Applicants must have received the doctoral degree (PhD, MD, DO, DC, DDS, DVM, OD, DPM, ScD, EngD, DrPH, DNS, ND, PharmD, DSW, PsyD, or equivalent doctoral degree from an accredited domestic or foreign institution) by the beginning date of the proposed award. For applicants holding the PhD degree, this award is designed to provide support for advanced and specialized training in basic research, in basic research associated with clinical problems, or in clinical research. For applicants holding the MD or other clinical-professional degree, this program is intended to provide at least two years of rigorous basic or clinical research training.

For more information related to F grants, see the NIH website:
http://grants.nih.gov/training/F_files_nrsa.htm

Table 24.5 Funding cycle.

Submission cycle	Range of flexibility
February 1[a]	(±2 wk)
June 1[a]	(±2 wk)
October 1[a]	(±2 wk)
Grant resubmissions	Usually due 1 month later and fall in March, July, and November

a) Make sure to review the website for specific deadlines, as they vary for each mechanism and area of interest (see http://grants.nih.gov/grants/funding/submissionschedule.htm).

your search to more general topics, and think outside the box when reviewing funding options and priority areas. For example, the Department of Agriculture funds all types of research related to farming technologies and food sustainability. Nevertheless, they also fund research related to nutrition and cognitive development, malnutrition, and retention of school-age children, just to name a few. Remember, you are looking for the best fit between the objectives of your project and the priorities of the funding agencies. If you are not sure if your project fits, contact the funder and ask. It may be that they have never funded a project like yours but have an interest in doing so.

Take the time to review websites, guidelines, and RFPs to carefully understand the kinds of projects that these funders are interested in funding. Doing your homework and identifying potential funders whose mission aligns with your research will save you and your organization time. After locating potential funding sources, you may need to contact both the funding organization for more information to determine if submitting a proposal would be appropriate. Additionally, you should notify your institution to make sure your intuition can accept funds if awarded.

In reviewing information about funding agencies and grant programs, pay particular attention to deadlines and notification dates. You may want to consider these basic questions when searching for a funding organization:

- Does the funding organization or program support graduates or medical students, residents, or fellows?
- What limitations does the funder put on its programs?
  - Are there geographical restrictions?
  - Does the funding organization only support certain types of institutions or individuals?
  - How many years do you have to complete the project?
- Does the funder support ideas that are similar in size and scope to your proposed project?
- What research expenses are allowable to the budget?
  - Budget items such as salary and travel are not covered on certain grant mechanisms.
- What are the deadlines for applying?
- What method of application is required?
  - This information is important to your timeline as most institutions will require the full application to be complete and ready to submit several days prior to an electronic submission. If hard copies are required, you need to know how many copies and in what binding format are demanded. Have packaging materials readily available.

By keeping these points in mind, you can quickly narrow down the number of potential funders for your project and eliminate from consideration programs for which you are not eligible or not a strong candidate.

24.9.1 Grant Databases

Grant databases are a great way to begin searching for appropriate funding sources. They allow you to filter your search by a number of fields and categories to help you find a funding organization that aligns with your area of research. Listed below are a few of the most common resources:

- Pivot, formerly known as the Community of Science (CoS) (pivot.cos.com)
  - Pivot is the leading global resource for hard-to-find information critical to scientific research and other projects across all disciplines. Check with your institution for login or access information.
- The Grant Advisor
  - The Grant Advisor is a leading source of information on grant, research, and fellowship opportunities for US institutions of higher education and their faculty.

- Grants.gov (www.grants.gov)
  - Grants.gov is a free resource to find and apply for federal grants.
  - The US Department of Health and Human Services is the managing partner.
- Foundation Center (foundationcenter.org/)
  - Established in 1956 and today supported by close to 550 foundations, the Foundation Center is a free source of information about philanthropy worldwide.
- GrantsNet (www.grantsnet.org)
  - A free service that helps locate funding in the sciences and undergraduate science education.
    - o Provided by the American Association for the Advancement of Science (AAAS), Council on Undergraduate Research, GeoScience World, International Brain Research Organization, and the National Alliance for Hispanic Health.
- Grants Resource Center (GRC) (http://www.aascu.org/grc)
  - A searchable database of public and private funding sources.
  - Provided by the American Association of State Colleges and Universities (AASCU).
  - This database is designed exclusively for GRC-member institutions. Check with your institution for access or login information.

24.9.2 Useful Tools

If you are a student or in a training position in an academic setting, chances are, your institution has a website dedicated to its grants process. Make sure you check out your grants office or office of sponsored research website for more information and useful tips and information. In most cases, your institution will have its own processes and timelines for navigating the grant process. Table 24.6 lists some additional resources that may help you get started.

Most funding agencies have Rich Site Summary (RSS) feeds. Subscribing to these feeds allows an investigator to be current with announcements and notices for funding opportunities. Grants.gov, for example, has RSS feeds that will allow you to filter by a number of criteria such as specific federal agency or by category like health, humanities, or science and technology, just to name a few. As we become more dependent on real-time

Table 24.6 Additional tools and resources.

Helpful resources and tools	Website
NIAMS homepage	http://www.niams.nih.gov
NIH grants page	http://grants1.nih.gov/grants/oer.htm
NIH grant application basics	http://grants.nih.gov/grants/grant_basics.htm
Welcome Wagon	http://grants1.nih.gov/grants/funding/welcomewagon.htm
Getting started at the NIH	http://grants1.nih.gov/grants/useful_links.htm
Common grant terms	http://grants.nih.gov/grants/glossary.htm
The Grantsmanship Center	http://www.tgci.com/funding-sources

NIAMS, National Institute of Arthritis and Musculoskeletal and Skin Diseases. NIH, National Institute of Health.

information, a large number of agencies or funders have options that allow individuals to subscribe to listservs or follow them on social media outlets (i.e. Facebook and Twitter). Using these options can provide you with the most accurate and up-to-date funding and research announcements. *Grant opportunities can come up with very little notice, so it pays to frequently check the posting or subscribe to an RRS feed or listserv so you can be notified as soon as opportunities become available.*

It is extremely important to promote ongoing and positive relationships with funding agencies. This may lead to additional funding down the road. It is essential that you file the necessary paperwork, such as progress reports, in timely and professional manner. This reflects not only on you as the investigator but also on the institution that you represent.

24.10   Time Commitment and Infrastructure and Support

24.10.1   Time Commitment

Consider the amount of time it will take to complete each aspect of the application process and plan accordingly. It can take a significant amount of time to get organized, refine your ideas, collect preliminary data, write the grant application, obtain institutional approval for your budget, and approval for working with human subjects or animal subjects, etc. Develop a realistic timeline that includes draft application deadlines, and give yourself enough time to meet them. Begin early to define, organize, and plan your content. Note that according to the NIH, early means six to nine months before the deadline.

24.10.2   Infrastructure: Planning Within Your Organization

It is important to understand the infrastructure within your organization. Understating your organizations timelines, key players, and their process for sending a grant through the system can save you a lot of time and headache.

Make sure you visit the grants office or office of sponsored projects (OSP) website for more information. These resources will likely provide you with many of the answers to questions we have discussed in this chapter. Additionally, these sites will provide you with the internal grant processes and timelines that you will also need to be familiar with. Use these guidelines to help map out your own timeline.

- *Introduce yourself*. Meet with your office of sponsored research (or central grants support office) early in the process. They can help guide you through the application process and can inform you of any institutional deadlines you must meet.
- *Create your own timeline*. Make sure to account for unforeseen events such as equipment failures, delays, and sick, personal or vacation time. This will help to ensure you get your application to your office of sponsored research on time. This is especially important when collaborating investigators are involved. Your last-minute submission and request for information should not become anyone else's burden.
- *Know your resources*. Making yourself familiar with all of your institution's players, procedures, processes, and requirements will make for a less stressful process.

- *Identify a key contact*. Experienced staff at your institution can be very helpful. If at all possible, find someone at your institution that can assist you in understanding all the steps necessary to complete your application. This person may be in a central grants office, or it may be another investigator, a departmental administrator, etc.

24.11   Post Award

24.11.1   Award Negotiation, Management, and Setup

You have successfully submitted your grant and have received your award letter from the funder. Now what? This is the time where you will rely on your institution's experience and expertise in setting up and managing your grant. Funders will not only notify you in writing of your award, but they will also notify your OSP. OSP will need a great deal of information in order to set up your grant from both a financial and administrative standpoint. Most of what they will need will come to them from the funder via the written notification. However, they will also need to defer to you, the grantee, for information. We cannot emphasize enough how important it is to get to know the grants officer that has been assigned to your grant. Make yourself available and respond to questions in a timely manner. This individual can help you tremendously as you progress through your project and can provide you with needed financial information for progress reports. Your grants officer will also be your liaison with the funder should you need to adjust your budget or timeline during the project period. Most importantly, your grants officer will be the one to receive the check and work with financial management to set up your grant account.

You will likely be required to sign a contract with the funder. This contract, as with any typical contract, is signed by the institution designee, the funding organization, and often, you. It states that all parties agree to the terms of conducting the study for the amount of money awarded and requires that you adhere to all federal, institutional, and funding agency regulations with regards to human subjects, confidentiality, and ethical conduct.

As the grantee, you are not only responsible for conducting the study per the approved protocol but also for managing the day-to-day operations of the grant. You may have resources that will help you with the administrative management. However, it is your responsibility to have insight into how your grant is being managed and if the money is being allocated appropriately. The business officer/administrative director for your department can help facilitate reports and financial statements that you, as the grantee, can review. You will want to work directly with this individual to determine the most effective way for you to gain access to your accounts so that you can manage your funding. Your business officer will also be the one to facilitate payment requests for patient incentives, purchase equipment specifically budgeted for your project, and ensure that funds are going to your salary per the grant. The NIH has additional information about award management and setup at its website.[3]

---

3 http://grants.nih.gov/grants/managing_awards.htm.

24.11.2 Setting Up Your Site

While OSP and your business officer are setting up the administrative and financial portions of your grant, you should take the time to verify all other logistical requirements needed to successfully start your project on time. Make yourself a checklist of all the logistics related to enrolling, or acquiring samples, and detail the steps related to such tasks. If you need to order lab supplies, find out how long it takes to get those in, make a list, and go ahead and place an order. If you need consent forms, go ahead and print those, or verify your electronic version is usable if you are having patients read and sign electronically. Make any additional edits to the protocol, and have them approved by the institutional review board (IRB). Confirm that your project team has thoroughly reviewed the protocol and understands the aims and outcomes of the study.

One important logistic that many people fail to acknowledge is the challenge of working with individuals not in your sphere of influence or recruiting in an area that is not your usual environment. Make sure you meet, train, and include anyone who may be a part of your project. Get their buy-in before the study starts so they become stewards of your success versus barriers. Dirty data, poorly handled specimens, and lost recruits can all be avoided with a little footwork on the front end. Other important steps that you should consider when setting up your site include:

- Create your essential documents binder (required for all clinical studies). See Table 24.7 for a sample of an essential documents binder table of contents.
- Build your patient/specimen database.
- Complete any revisions needed on your study forms.

Table 24.7 Example of an essential documents binder table of contents.

ESSENTIAL DOCUMENTS BINDER Protocol #: Title of the study here

Required elements	
1	Protocol
2	Manual of operations (MOO)
3	Research staff roles and responsibilities
4	IRB application and correspondence
5	Regulatory documents
6	Patient study log
7	Screened patient documents
8	Data forms
9	Patient records and source documents
10	Monitoring reports
11	Study correspondence
12	Other correspondence/notes-to-file
13	Telephone/communications log

- Create and obtain IRB approval for any recruitment materials you may need.
- Create a study frequently asked questions (FAQ) sheet, with contact information on it for you and/or your research team.

This information can be handed out or hung in patient treatment areas. It will help trigger your colleagues to contact you with potential participants.

24.11.3   Progress Reports

Progress reports are an annual requirement for most (if not all) funders and usually due 30–45 days prior to the grant anniversary date. Although your progress report may contain other pieces of information requested by the funding agency, it will at least contain the following information:

- Target enrollment verification
- Inclusion enrollment report
- Any changes to your involvement of human subjects or use of animals
- Verification or changes to personnel originally listed on your grant
- Budget verification and any changes you are requesting to the budget

24.11.4   Award Close-out and Record Retention

Each funding agency and institution will have its own close-out protocols that must be followed. Although these processes differ, there are a few standard points to note:

- Any funds not used for the study must be returned to the funding agency.
- Close-out reports are due within 90 days of closing the grant (unless otherwise specified by the funder).
- Grantees must retain all study materials, records, financials, and reports for a minimum of three years. Some funders, depending on the type of study may require you to retain this information longer but will notify the grantee if an extension is required.

Further Reading

Heinze, T. (2008). How to sponsor ground-breaking research: a comparison of funding schemes. *Sci. Public Policy* 35 (5): 802–818.

Hyden, C., Escoffery, C., and Kenzig, M. (2015). Identifying and applying for professional development funding. *Health Promot. Pract.* 16 (4): 476–479.

Jacob, B.A. and Lefgren, L. (2011). The impact of research grant funding on scientific productivity. *J. Public Econ.* 95 (9–10): 1168–1177.

MacLean, M., Davies, C., Lewison, G. et al. (1998). Evaluating the research activity and impact of funding agencies. *Res. Eval.* 7 (1): 7–16.

Schroter, S., Groves, T., and Højgaard, L. (2010). Surveys of current status in biomedical science grant review: funding organisations' and grant reviewers' perspectives. *BMC Med.* 8: 62.

Van den Broeck, J., Brestoff, J.R., and Baum, M. (2013). Funding and stakeholder involvement. In: *Epidemiology: Principles and Practical Guidelines* (ed. J. Van den Broeck and J.R. Brestoff), 157–169. New York: Springer.

25

Essentials of Grant Writing and Proposal Development

Chevis N Shannon and Jamie Dow

Department of Neurosurgery, Vanderbilt University Medical School, Nashville, TN, USA

25.1   The Research Plan

Proposal writing is an art form, where the writer "sells" his/her research idea or outcome of interest. An excellent research plan is characterized by:

1. A compelling question
2. Clarity of thought and expression
3. A strong, testable hypotheses
4. Logical steps (aims) to answer the question
5. Rigorous experiments to answer the question

Having a well-thought-out, focused application will allow the reviewer, no matter how technical the grant, to appropriately score/rank your proposal. This score or ranking is how funding agencies of every kind determine what applications they are going to support and ultimately award. While "grantsmanship" can turn a mediocre application into a fundable grant, often the opposite is seen, where a great idea is lost in an unorganized plan, resulting in poor grantsmanship. When a great idea is lost, it is difficult for a group of peer-reviewers, board of directors, and/or community leaders to recognize the value of your idea and fund your proposal.

While government funding for medical research amounts to approximately 30–35% in the United States annually, the average rate of successfully funded grants from the National Institutes of Health (NIH) varies, depending on the branch of the NIH, grant type, and the number of applications submitted. For instance, first-time (original) submissions of research project grants were funded at a rate of 9.3% during the fiscal year 2013, while 31.5% of initial resubmissions were funded. In 2013, 2300 F32 postdoctoral fellowship training grants were submitted to the NIH with only 24% of those funded. Of those funded grants, only 7% were funded to trainees with an MD or MD/PhD.

25.1.1   Specific Aims and Hypotheses

The specific aims are the objectives of your research project; they describe what you want to accomplish, and are driven by the hypothesis you set out to test. This portion

A Guide to the Scientific Career: Virtues, Communication, Research, and Academic Writing, First Edition.
Edited by Mohammadali M. Shoja, Anastasia Arynchyna, Marios Loukas, Anthony V. D'Antoni, Sandra M. Buerger, Marion Karl and R. Shane Tubbs.
© 2020 John Wiley & Sons, Inc. Published 2020 by John Wiley & Sons, Inc.

of the research plan is one of the most important sections. It is the first encounter the reviewer will have with your ideas, and he/she often draws conclusions about the total application solely from your aims. As you work through developing specific aims, consider the following:

1. *First impressions count.* Focused aims and clear hypotheses capture and keep the attention of the reviewer.
2. *Focused aims lead to effective experiments or interventions.* Remember, if you do not know what you are selling, they cannot buy it.

How do you know if you have a good idea? First, do your homework – conduct a thorough literature search or attempt to find meta-analyses. Have a good understanding of where gaps in research exist. Second, ask yourself if your aims are clinically relevant. Will others find value in the work that you do? Will the work that you do change or affect healthcare? Third, seek the opinions of others and utilize their expertise. Be willing to accept critical feedback and incorporate changes. After you have clearly defined your aims you will want to include hypotheses that specifically test your aims. Note that a hypothesis is not always needed; however, objectives should always be included. A good hypothesis should be logical and based on data. It should be testable and focused. Keep in mind your research methods will relate directly to the aims you have described, so your hypotheses should inform your methodology.

A good rule of thumb and a format requirement for most government and foundation grants is to limit your specific aims to one page. Three aims is a good number to shoot for, but definitely no more than four. The more aims you have, the less focused you become and the less likely you will be able to successfully complete the project.

25.1.2 Background and Significance

The information you include in this section should be directly related to your field of research and should support the research you are proposing. This literature review is important, as it demonstrates to the reviewers your understanding of the field and provides the opportunity for you to reveal gaps or areas of discrepancies in the research that exists. According to the NIH, this section of your research plan should:

- Explain the importance of the problem or critical barrier to progress in the field that the proposed project addresses.
- Explain how the proposed project will improve scientific knowledge, technical capability, and/or clinical practice in one or more broad fields.
- Describe how the concepts, methods, technologies, treatments, services, or preventative interventions that drive this field will be changed if the proposed aims are achieved.

Although there is often not a page limit for this section, you do have a page limit for the research plan itself. Our suggestion is to limit your background and significance section to one or two pages leaving you plenty of pages to describe your approach, methodology, and statistical analysis.

25.1.3 Preliminary Studies/Innovation

Providing preliminary data helps build reviewers' confidence that you can handle the technologies, understand the methods, and interpret results. Preliminary data may consist of your own publications or publications of your collaborators. You may also include unpublished data from your own laboratory or from others you are collaborating with, or some combination of these:

- Preliminary data should support the hypothesis to be tested and the feasibility of the project.
- Explain how the preliminary results are valid and how early studies will be expanded in scope or size. Make sure you interpret results critically. Showing alternative meanings indicates that you've thought the problem through and will be able to meet future challenges.
- Explain how the application challenges and seeks to shift current research or clinical practice paradigms.
- Describe any novel theoretical concepts, approaches, or methodologies, instrumentation or intervention(s) to be developed or used, and any advantage over existing methodologies, instrumentation, or intervention(s).
- Explain any refinements, improvements, or new applications of theoretical concepts, approaches or methodologies, instrumentation, or interventions.

If the aims you are proposing involve innovative technology, therapies, or new populations for which supportive preliminary data do not exist, you should acknowledge this in your application. Discuss how the anticipated research you are proposing correlates to similar research previously conducted or published. Explain how the proposed project will impact the field with the potential for expanded research in the future. Provide insight regarding access to infrastructure or expertise that will lead to successful completion of your project.

25.1.4 Research Design and Methods

Describe the experimental design and procedures in detail and give a rationale for their use. Organize this section so each experiment/intervention is referenced in the same order as the corresponding specific aim. The writer wants to convince reviewers that the methods chosen are appropriate for each given aim and that the methodology is well established. If the methods proposed are innovative, make sure to describe any potential pitfalls you see and how you plan to address those as well as address the advantages of utilizing a new method.

Although charts, graphs, and photographs are being utilized in applications more, be careful to use sparingly and avoid their use when possible. The research plan should be self-contained; if you need a chart or graph to further clarify your methodology or experiment, make sure to include these in the application itself. Otherwise, use your appendices for graphics. You should never duplicate charts/graphs or pictures in both the body of the application and the appendix.

25.1.4.1 Approach

When you are discussing the approach, unless all aims are tied together, it is a good idea to discuss each approach in the same order to the corresponding aim. In this section of the application make sure you are organized and detailed in your thought process as much as you can. If the approach you are using is innovative, make note of this and clearly discuss the feasibility of the approach and how advantageous to your project the approach will be compared to an existing known approach.

Discuss the potential pitfalls that you may encounter with each approach. It is important that you communicate to the reviewer alternatives that will circumvent potential limitations and/or how you will handle each challenge that arises. Also, describe any experiments or materials that may be hazardous and clearly define the precautions you will take. This information lets the reviewer know that you have thought through your study design and approach and are prepared to address any hurdles that may arise. Make sure any proposed model or approach is appropriate to address the research questions and is highly relevant to the problem being modeled. Estimate how much you expect to accomplish each year of the grant and state any potential delays you anticipate.

As you work through the details of your approach and experiment or intervention, make sure to include publications that you and/or your collaborators have published relevant to your study or methodology. These manuscripts should go in the appendix as a reference for the reviewer.

When writing your research plan, *the key is details, details, details*. It is your responsibility to sell yourself and your proposed project. You want to convince reviewers that you are an expert in the field or that you are part of a collaborative team with expertise. Reviewers do not want vague descriptions of experiments and methods; they want the nitty-gritty details. Although not every reviewer will be an expert related to your specific subject matter, you must write your plan as if the reviewers are experts.

25.1.4.2   Statistical Analysis

First and foremost if you do not have expertise in statistical analysis, *involve your statistician on the front end!* This individual can help you determine the amount of data collection needed, sample size calculations, statistical methods, and appropriate analyses needed to employ, which at the same time may impress the reviewers. In this section, you want to include the description of your statistical methodology proposed and data variables of interest, as well as define the criteria for evaluation of a specific tests. For instance, will your study population follow a normal distribution? Is a sample size calculation required for your study design? Are you looking at inter-rater reliability, and what tests do you use for such analysis? You want to show reviewers that you or someone on your team is able to accurately interpret the results so that they may be disseminated.

25.1.4.3   Recruitment and Retention

The proposal should include a concrete plan for the recruitment and retention of subjects. Recruitment begins with the identification, targeting, and enlistment of participants (volunteer subjects or controls) for a research study. It involves providing information to the potential participants and generating their interest in the proposed study. There are two main goals of recruitment:

- Sample size and power requirements for your study
- Recruiting to adequately represent the target population

Recruitment is perhaps the most challenging part of a clinical research study. Subject recruitment depends on the type of study undertaken, collaboration with the clinicians, characteristics and preferences of the subjects, and the recruitment strategies employed. When recruiting you have to be careful that selection bias (resulting from the way study participants or samples are selected for the study or experiment) does not occur, however. It is important to be aware of the types of factors that greatly impact response rates:

- Greater age
- Male sex
- Non-white race
- Urban residence
- Low educational status
- Low family income
- Unemployed or low occupational status
- Recent illness or poor present health
- High use of medical care

Subject retention is a critical aspect of recruitment. Poor retention is costly both financially and in terms of time. When developing your research plan you should determine what participants will be eligible for enrollment and identify retention techniques that will be most effective. Some examples of retention techniques include email or text messages between visits, participant birthday postcards, participant-specific webpages that provide information about the study itself or the disease you are studying, or phone calls and letters reminding the participants about their upcoming visit.

Ineffective recruitment and retention can disrupt the timetable for a research project, reduce the ability to detect a difference in therapeutic studies, produce negative results, or make it impossible to prove your hypothesis.

25.1.4.4 Data Monitoring and Regulatory Requirements

With any clinical study conducted you must adhere to all regulatory requirements from your institution as well as any federal governing bodies. Institutional review boards (IRBs) often requires a study to have a data monitoring plan in place. The complexity of your data monitoring plan depends on the complexity and risk associated with the study you are conducting. There are three main types of data monitoring boards:

1. A data and safety monitoring board (DSMB) is an independent group of experts that makes recommendations to the study investigators. The primary responsibility of the DSMB is to evaluate the accumulated study data for participant safety, study conduct and progress, and, when appropriate, efficacy.
2. A clinical study oversight committee (CSOC) is an independent group of experts that makes recommendations to the study investigators on clinical studies not involving an intervention. The primary responsibility of this committee is also to evaluate human subject safety and evaluate study progress.
3. An independent safety monitor (ISM) is a qualified clinician with relevant expertise whose primary responsibility is to provide independent safety monitoring in a timely fashion. This is accomplished by evaluation of adverse events, immediately after they occur, with follow-up through resolution.

In the grant you will want to discuss your plans for data monitoring, even if all of the details have not been worked out. The funding agency wants to be assured that you have a mechanism in place to conduct data quality checks, to verify the study is being conducted per protocol approved by IRB, and that any safety issues are being reported to the appropriate regulatory bodies.

25.1.4.5 Collaboration

As part of the preaward planning process, you should consider what expertise you need on your team and who might serve as an effective collaborator. You can identify collaborators from your current pool of colleagues, mentors, and peers, or you may have a mentor that can suggest individuals that may serve as potential collaborators. The use of collaborators can range from someone serving as a consultant in an area you are less versed or participating as an active member on the research project (with salary support budgeted), to providing editorial support prior to submitting your application. Your collaborators may have varied backgrounds or may be affiliated with another institution making your project seem multidisciplinary or multi-institutional. Given the funding environment today, this becomes even more important to a funding agency. Consider the following scenarios:

- You might need a piece of equipment that your lab does not currently have and that you are unable to budget for. Collaborating with a colleague from another lab may allow you to utilize his/her equipment that will assure the funder you have the ability to run the experiments.
- Let's say you want to conduct a study that requires more subjects to be recruited into your cohort that you currently see in your clinic. Identifying a peer who might have more eligible subjects for enrollment would allow you to recruit from his/her clinic, making it a collaborative effort. This individual might not be listed on the budget but would provide a letter of support stating willingness to participant by contributing patients.
- You are in a clinic and have an idea for a study based on a patient you have recently treated. Sitting with a group of colleagues or peers to flesh out your ideas, focus your aims, and clearly define your hypothesis is another way to collaborate. This not only saves you a lot of time but can also facilitate long-term collaborative relationships that may pay off for you in the future.

25.2 Budget and Budget Justification

When developing your grant budget, the rule is: *Be realistic*! As a grant reviewer, nothing is more disappointing than to have a well-proposed project rejected because the applicant over- or underestimated the financial requirements to be successful. When drafting your budget, make sure you have several pieces of information at hand:

- Know your limits. Have a clear understanding of the maximum budget you are allowed to submit as well as what the funder will and will not fund.
- Know the indirect cost rate for your institution as well as the indirect cost rate that your funder is willing to pay.
- Know the fringe benefit rate for your institution; you will need this for your key personnel. Fringe benefits are what the employer provides above and beyond your salary, such as health insurance or vacation time. Note that faculty and staff fringe rates

are usually different, while neither postdoctoral fellows nor graduate students are assigned fringe rates.
- Have other rates handy such as mileage and per hour rates for contractors or consultants.

Whether you are writing a NIH grant or submitting a grant to a foundation, we recommend you utilize the budget forms used for NIH grants. These forms are a great tool to help you organize your thoughts related to itemizing your budget. Ultimately, you may be required to transfer your budget to the forms your potential funder supplies, but this is a great place to start. This form can be found on the NIH webpage.[1]

The budget divides costs into direct and indirect costs, as described here.

25.2.1 Direct Costs

Direct costs are those costs that can be easily identified and allocated directly to a sponsored activity. They include items such as:

Salaries, wages, and fringe benefits. When possible, identify the specific individuals that will hold the positions you are requesting. Although this isn't always possible, it is more difficult for reviewers to delete a named person and position versus a "to be named" place holder. Additionally, this shows the reviewers that you have identified to personnel resources needed to be successful. If year-to-year fluctuations occur in your budget, make sure to clearly address those in your budget justification. For instance, you may need a biostatistician on your study but not until year 2 or 3 because the data analysis phase may not start until then. Make sure, however, that all of these changes parallel the research plan and specific aims.

Consultants and travel. Read the budget guidelines carefully before including consultants and travel. Most funding agencies have a policy regarding these. If they are not clearly stated, contact the program officer or grants manager to get clarification. The budget justification serves to assure that you will use your money wisely. Putting money into restricted categories shows the reviewers that you are not familiar with the grant guidelines.

Supplies, services/materials, and equipment. Often, applicants confuse these categories and include items in the equipment category that should be itemized in the supplies/materials category. A good example of this is computer infrastructure or lab equipment. Equipment less than $5000 should go into the supplies/materials category unless otherwise noted by the funder. Equipment tends to be capital equipment where depreciation will be accounted for over time.

25.2.2 Indirect Costs

Indirect costs are overhead, or the portion of the cost of promoting and doing research that cannot be directly measured and attributed to a specific project. Examples of indirect costs include:

- Computing facilities and internet service
- Equipment use and depreciation

---

1 http://grants.nih.gov/grants/funding/phs398/phs398.html.

- Utilities
- Custodial services and clerical support

Indirect costs are not itemized individually in the budget. You will, however, have to be aware of the indirect cost rate that your institution and/or the funder require.

25.3 Grant Documents and Grant Formatting

25.3.1 Grant Documents

Most grants now require a biosketch. All federal funders and most international foundation funders will require that you use the NIH-formatted biosketch. A template and example can be found on the NIH website.[2]

Every grant requires that you provide letters of support. These letters should come from your mentor, supervisor, colleagues, and collaborators. The goals of a letter of support are to do the following:

- Specify what the consultant(s)/collaborator(s) will contribute to the research.
- Convince the reviewer that the consultant(s)/collaborator(s) will fulfill the request.
- Convey enthusiasm for the work.
- Lend credibility to your proposal.

It is highly recommended that you draft your own letters of support to send out. This will ensure that the letter contains the information you need and that you will get the letters back in a timely manner. There is no right or wrong way to write a letter of support as long as the letter demonstrates to the funder what each collaborator will contribute and what infrastructure and support you will receive from your mentor and/or department.

It is appropriate to give your letter writers two to three weeks to get the letter back to you. You may need to give them a little nudge if your deadline is getting close, but that is OK; they would nudge you… so nudge away; just remember to be professional. Here are a couple of tips that will help you remind your letter writers of the deadline approaching:

- Send a reminder email to all your letter writers (blind copied, of course). Attach the original draft letter and include your drop-dead dates. Be sure to thank them for their time and support.
- Call your letter writer or their assistant and offer to help get the letter finalized. Often, this will be prompt enough and you will have your letter within hours. Definitely keep any assistants or administrative support staff in the loop so they can also help as needed.
- If the letter writer is an internal person, make an appointment to meet with them, or stop by and extend your appreciation and gratitude for the support.

---

2 http://grants.nih.gov/grants/funding/2590/biosketchsample.pdf.

If all your efforts go unanswered, always make sure you have a backup. Better yet, request one or two extra letters of support than you originally needed to ensure you have the required number for your application.

25.3.2 Formatting Your Grant

Usually, the research plan is limited to 15 pages for government grants, 12 pages for international foundation and academic grants funded by the NIH (such as Clinical and Translational Science Award grants grants), and 5–7 pages for local or family foundation grants and internal academic mechanisms (funded by universities).

NIH has specific guidelines related to font type, font size, and line spacing:

- Arial, Helvetica, Palatino Linotype, or Georgia typeface, only.
- Black font color, and a font size of 11 points or larger. (A Symbol font may be used to insert Greek letters or special characters; the font size requirement still applies).
- Type density, including characters and spaces, must be no more than 15 characters per inch.
- Type may be no more than six lines per inch.
- Use standard paper size (8½″ × 11), Use at least one-half-inch margins (top, bottom, left, and right) for all pages.
- No information should appear in the margins.

Keep in mind, if you are submitting a grant to a funding agency and you are not provided with formatting requirements, you should always default to the NIH guidelines.

When writing, be careful about using jargon. The writer should be able to convey the research clearly and professionally, allowing both scientist and the public to comprehend. See the NIH writing tips in Table 25.1 for more information about lay language and sentence formatting.

25.4 Conclusions

As you finish your research plan, prior to putting your application together, make sure you have someone (preferably more than one individual) read your proposal thoroughly. Grammar, spelling, and ease of read are extremely important. Verify that your abstract does not omit any aspects that should be addressed per the grant guidelines, but also confirm that it highlights the innovation or impact of your project. Confirm that your proposal is written with the reader(s) in mind. Some will have scientific expertise but may not be an expert in your field. Some, however, will be reviewers from the community and must be able to understand your project to properly score it. Make sure your application meets *all* the requirements, including page limit, word count, font, etc. Verify that your ideas are clearly thought out and clearly stated. Keep in mind, less is more. Make your sentences shorter, do not use more words when less will do, and stay focused.

Table 25.1 NIH writing tips.

1. The instructions require that materials be organized in a particular format. Reviewers are accustomed to finding information in specific sections of the application. Organize your application to effortlessly guide reviewers through it. This creates an efficient evaluation process and saves reviewers from hunting for required information.
2. Think like a reviewer. A reviewer must often read 10–15 applications in great detail and form an opinion about each of them. Your application has a better chance at being successful if it is easy to read and follows the usual format. Make a good impression by submitting a clear, well-written, properly organized application.
3. Start with an outline following the suggested organization of the application.
4. Be organized and logical. The thought process of the application should be easy to follow. The parts of the application should fit together.
5. Write one sentence summarizing the topic sentence of each main section. Do the same for each main point in the outline.
6. Make one point in each paragraph. This is key for readability. Keep sentences to 20 words or less. Write simple, clear sentences.
7. Before you start writing the application, think about the budget and how it is related to your research plan. Remember that everything in the budget must be justified by the work you've proposed to do.
8. Be realistic. Don't propose more work than can be reasonably done during the proposed project period. Make sure that the personnel have appropriate scientific expertise and training. Make sure that the budget is reasonable and well-justified.
9. Be persuasive. Tell reviewers why testing your hypothesis is worth NIH's money, why you are the person to do it, and how your institution can give you the support you'll need to get it done. Include enough background information to enable an intelligent reader to understand your proposed work.
10. Although not a requirement for assignment purposes, a cover letter can help the Division of Receipt and Referral in the Center for Scientific Review assign your application for initial peer review and to an IC for possible funding.
11. Use the active, rather than passive, voice. For example, write "We will develop an experiment," not "An experiment will be developed."
12. Use a clear and concise writing style so that a nonexpert may understand the proposed research. Make your points as directly as possible. Use basic English, avoiding jargon or excessive language. Be consistent with terms, references, and writing style.
13. Spell out all acronyms on first reference.
14. Use subheadings, short paragraphs, and other techniques to make the application as easy to navigate as possible. Be specific and informative, and avoid redundancies.
15. Use diagrams, figures, and tables, including appropriate legends, to assist the reviewers to understand complex information. These should complement the text and be appropriately inserted.
16. Use bullets and numbered lists for effective organization. Indents and bold print add readability. Bolding highlights key concepts and allows reviewers to scan the pages and retrieve information quickly. Do not use headers or footers.
17. If writing is not your forte, seek help!

Source: Adapted from http://grants.nih.gov/grants/writing_application.htm.

Further Reading

Developing Your Budget (2012). Retrieved January 15, 2014 from http://www.emich.edu/ord/downloads/downloads_subd/training/developing_budgets/wrkshp_develBudget.pdf

Grants and Funding (2014). Retrieved January 23rd, 2014 from http://grants1.nih.gov/grants/grants_process.htm

How to write a research grant application (1999). Retrieved February 11th, 2014 from http://www.vanderbilt.edu/dsr/PDFs/howto.pdf

How to write a research grant application (2014). Retrieved January 23rd, 2014 from http://www.ninds.nih.gov/funding/write_grant_doc.htm

Section VI

Research Principles and Methods

26

Clinical Research Methods and Designs

Daxa M. Patel, Beverly C. Walters and James M. Markert

Department of Neurosurgery, University of Alabama at Birmingham, Birmingham, AL, USA

26.1   Introduction

In the era of evidence-based medicine, clinical studies are consulted to establish management strategies, including patient assessment tools, diagnostic tests, and therapeutic methods that produce the best possible health outcomes. Clinical studies can be analyzed from two perspectives, namely, structure and function. The structure of clinical studies focuses on individual identifiable elements of each study, including the hypothesis or question, design, protocol, study population, and sample size, as well as clinical measurements as qualifiers for the study or as study outcomes. An investigator should design the study in such a way that feasibility and efficiency are maximized. The function of clinical studies concentrates on how the study works and incorporates the ability to make inferences from the study data and permit generalization to overall populations. An investigator must attempt to minimize random and systematic errors to maximize the validity of these inferences (Hulley et al. 2013). These aspects of clinical studies are intricately linked and interdependent; they work in combination to provide principles to follow in day-to-day clinical decision-making. Nevertheless, it is important to look at them individually to achieve clarity and comprehension of a complex topic.

26.2   Structure of Clinical Studies

26.2.1   Research Question

The structure of a clinical study begins with the identification of the objectives that the study wishes to achieve and formulation of the research questions to be used as the guiding principle of the study. The first step is to focus on the unanswered questions in the context of current knowledge of information or to seek to refine the treatment package utilized in the clinical practice; this help outline the main objectives of the study. Although the clinical question might begin with a broad topic, it should be narrowed to a researchable issue that is focused and feasible at the same time (Thabane et al. 2009; Beckman and Cook 2007). Furthermore, it should also contribute to or impact on our current state of knowledge. Finally, the question should be typically written as a null hypothesis that is amenable to testing.

A Guide to the Scientific Career: Virtues, Communication, Research, and Academic Writing, First Edition.
Edited by Mohammadali M. Shoja, Anastasia Arynchyna, Marios Loukas, Anthony V. D'Antoni, Sandra M. Buerger, Marion Karl and R. Shane Tubbs.
© 2020 John Wiley & Sons, Inc. Published 2020 by John Wiley & Sons, Inc.

26.2.2 Background and Significance

Next, the background regarding the question of interest is researched to identify the significance of the proposed investigation. This should begin with a report of the previous literature that outlines the inadequacies and uncertainties that persist (Chow and Liu 2008; Grimes and Schulz 2002; Hulley et al. 2013). This outline will establish how the proposed research question will aid in resolution of the questions left unanswered by prior research and lead to a new or more complete understanding of the area under study.

26.2.3 Study Design

Third and most importantly, the design of the clinical study or how the study is structured is addressed. The clinical research design is a complex topic. Two broad categories of study designs are experimental studies and observational, or quasi-experimental, studies (Boet et al. 2012).

26.2.3.1 Observational Studies

The observational study gathers data in a protocolized manner featuring the careful collection of appropriate data for subsequent analysis. There is no experimental maneuver, i.e. the investigator does not have control over the assignment of treatments or other factors, and the combinations or experiments are typically occurring in normal clinical practice (Rosenbaum 2010). These studies are important in clinical settings where it is unethical or impossible to assign factors. They are occasionally the only available option for a study design of the research question. Although evidence from these studies is empirically weaker than experimental studies, it provides preliminary data for hypotheses in future studies and can help in refining the treatment paradigm for a given disorder (Levin 2005). Among the types of observational design study are case reports and case series, cross-sectional studies, ecological studies, case-control studies, and cohort studies (Table 26.1).

Case Reports and Case Series A case report is a description of a single case or clinical entity and typically includes the presentation, clinical course, diagnosis, prognosis, management, and outcome of a condition. It is anecdotal evidence. Due to its limited scope, a single case report provides narrow empirical data for the clinician (Nissen and Wynn 2014). Case reports, however, are a way of communicating to the medical community regarding diagnosis, treatment, and outcome of rare disease entities and may provide the foundation for research questions that would require further investigation (Vandenbroucke 2001). For information on how to write a case report, see Chapter 59.

Similarly, a case series is a descriptive, observational study of more than one case encountered over a period of time. The cases typically describe the manifestations, clinical course, diagnosis, prognosis, management, and outcome of a condition of interest. A case series is a common study type in the clinical literature, in particular for reports of a number of patients treated in a certain way as a way of introducing new treatment modalities, and sometimes of rare disease entities or unique operative techniques.

The cases in a case series provide weak empirical evidence due to lack of comparison with other treatments or with the natural history of a disease and may suffer from a

Table 26.1 Types of observational study designs.

Type of study	Case series and report	Ecological study	Cross-sectional	Case-control	Cohort study
Definition	One or more subjects' clinical course is reported.	Two groups and their associations are compared at one time point.	A group of subjects are examined at one point in time.	Two groups are selected based on the presence or absence of an outcome.	A group of subjects identified at the beginning and followed over time.
Advantages	Report of novel ideas Foundation for future hypotheses and stronger clinical studies	Inexpensive Investigation of rare disease Low time consumption	Broad scale of information Prevalence data Examination of multiple exposures and outcomes	Inexpensive Quick Investigation of rare disease, multiple exposures Investigation of long latent periods	Examination of rare exposures Measurement of incidence of disease (relative risk) Investigation of multiple outcomes and exposures
Disadvantages	Anecdotal evidence Empirically weak	Confounding	Recall bias Selection bias Confounding	Selection and recall bias Confounding Causation trap	Time consumption Expensive Changes in exposure and disease diagnosis criteria Loss to follow-up
Selection bias	NA	NA	++	+++	+
Recall bias	NA	NA	+++	+++	+
Loss to follow-up	NA	NA	NA	+	+++

(Continued)

Table 26.1 (Continued)

Type of study	Case series and report	Ecological study	Cross-sectional	Case-control	Cohort study
Confounding	NA	+++	++	++	+
Cost	+	+	++	++	+++
Length of time required	+	+	++	++	+++
Applications	Documentation of novel diseases, mechanisms, effects (adverse or beneficial)	Investigation of rare disease or cause Studying multiple exposures	Examination of multiple exposures and outcomes	Investigation of rare disease Studying multiple exposures Investigation of long latent periods	Investigation of rare cause Multiple outcomes Multiple exposures Measurement of time relationship Measurement of incidence Investigation of lengthy latent periods
Level of empirical evidence	I	II	II	II	II
Example	An investigator reports on several patients who have higher incidence of coronary artery disease (CAD) and correlating higher levels of low-density lipoprotein (LDL).	An investigator looks at CAD in two different counties and their LDL levels.	An investigator measures LDL in a group of patients and correlates results with CAD history.	An investigator examines patients with CAD (cases) and compares them with a group without CAD (controls), analyzing their LDL levels.	An investigator measures LDL cholesterol level in a group of subjects at baseline and then periodically examines them at follow-up visits to see if those with higher LDL will have increased CAD.

lack of accuracy of identified associations and conclusions (Chan and Bhandari 2011). They are, however, the only method of establishing the natural history of a disorder prior to implementing treatments aimed at changing patients' outcome. They are best used as a platform for hypothesis generation to be used for further experiments by stronger study designs. They are best regarded as a way that clinicians can inform each other and researchers regarding the course of diseases and novel therapeutics. Thus, they remain a cornerstone of medical progress, in spite of the fact that many, if not most, case series could and should be replaced by more robust clinical studies (Vandenbroucke 2001).

Ecological or Aggregate Studies The ecological or aggregate study is an observational study in which two groups and their exposure and outcome status are compared at one time point. Appropriately, these studies are also known as group-based studies, as they are conducted from the data collected at the level of the whole population. Typically, different population groups are compared and risk factors for a particular disease and disease prevalence are analyzed to identify associations. Ecological studies are appropriate study designs for examining health effects of environmental exposures and investigation of rare diseases or disease etiologies. However, like most observational studies, there are a few disadvantages to aggregate studies. Since all of the data and analysis are carried out at the aggregate level, relationships at the individual level are inferred. This may lead to ecological fallacies, where the interpretation of data is used to generate inferences about individuals from data obtained from the group to which these individuals belong, and is a tremendous disadvantage for the aggregate studies (Pearce 2000; Kramer 1983). Furthermore, multiple other uncontrolled variables, unless carefully sought out and controlled for, could distort the effect of identified associations in these group-based studies. Although ecological studies are relatively inexpensive, they provide weak empirical evidence due to the confounding and occurrences of the aforementioned ecological fallacies (Lasserre et al. 2000).

Cross-Sectional Studies The cross-sectional study is a descriptive study where a defined population of subjects is examined at one point in time. This type of studies, also known as prevalence study, usually aim at estimating the association of demographic factors and presence or absence of certain variables and an outcome. Cross-sectional studies, by definition, lack information on the sequence of events. Their main outcome focus is disease prevalence based on data collected from population databases or surveys (Levin 2006). These studies are best for examining multiple outcomes and exposures relationships and measuring the prevalence of health outcomes or determinants of health, rather than identifying causation. They are less expensive and time-consuming than cohort studies, but more than case series or ecological studies. However, several disadvantages exist, particularly opportunities for systematic error, including selection bias, recall bias, and confounding (Carneiro 2005). Despite these disadvantages, cross-sectional studies can provide significant information regarding prevalence of diseases and contribute to advancement and progress in medicine.

Case-Control Studies The case-control study is an observational study where two groups are selected based on presence or absence of an outcome, and retrospective analysis is conducted to assess exposure or treatment. The two groups consist of cases, who are patients with the presence of a certain outcome, and controls, who are patients without

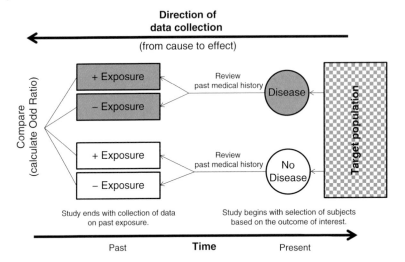

Figure 26.1 Case-control study design.

the outcome of interest (Figure 26.1). These groups are then assessed for a potential risk factor or treatment as causative or associated with the outcome. The former is inferred from the study's findings, whereas the latter is demonstrated. The common measure of association for case-control studies is the odds ratio of the outcome occurring in patients according to exposure or treatment. These studies are used as a preliminary investigation where little is known about the association between the risk factor and disease of interest, in particular for rare conditions or for risk factors with long induction periods. They may also be used as a preliminary study to a randomized controlled trial or used in place of a randomized controlled trial in circumstances where an experimental study will most likely, or could, never be done.

Although case control studies are quick and inexpensive, they may be prone to selection bias, recall bias, confounding, and causation trap. Like cross-sectional studies, case-control studies only provide inferential information on causation and on the chronology of disease and exposure; thus, they may fall into a causation trap, i.e. assuming a demonstrated correlation represents causation (Mann 2003). To minimize these systematic errors, careful selection of cases and controls, as well as limitation of confounding association between disease detection and risk exposure, is required (Lewallen and Courtright 1998). Case-control studies are a great resource for the study of rare diseases because the diseased are selected at the outset of the study, but their power of empirical evidence is limited by potential errors in their execution.

Cohort Studies A *cohort study* is a comparative longitudinal study where a group of subjects is identified at the beginning and followed over time. It is usually based on primary data from a follow-up period of a group in which subjects have had, have, or will have the exposure, or treatment, of interest to determine the association between that exposure/treatment and an outcome, or for identifying natural history or incidence. In the natural history paradigm, these studies begin with a group of individuals without disease that is followed prospectively, often for years, to assess the incidence of disease (Figure 26.2). These studies are largely about an analysis of

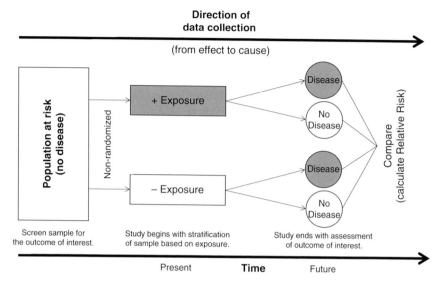

Figure 26.2 Prospective cohort study design.

risk factors and absolute risk of disease occurrence in the patients without the disease entity of investigation as compared to those who do develop the disease. They compare the relative risk of developing the disease between these two groups (Levin 2003). The advantages of the cohort study include examination of rare exposures, measurement of incidence of disease via relative risk, and investigation of multiple outcomes and exposures. The disadvantages include time consumption and cost, changes in exposure and disease diagnosis criteria, and differential loss to follow-up. Cohort studies are also used to identify comparative success or failure of various treatment modalities. They are quasi-experimental in that patients are not randomly assigned to treatments but may receive treatments according to clinician or healthcare site preference, or local "standards of care." Although cohort studies can be stronger than case-control studies when executed properly, they do not provide empirical evidence that is as strong as experimental studies (Foulkes 2008; Mann 2003).

26.2.3.2 Experimental Studies

The designs for experimental studies include parallel group trials (including randomized controlled trials), pre-post interventional trials, two-period crossover trials, cluster trials, and factorial trials. The *randomized controlled trial* is the experimental design that is the most rigorous, and therefore most robust, of all research designs, providing the "gold standard" against which all other designs are judged. Through experimental studies, the researcher takes on an active role of experimenter rather than being solely an observer. Experimental studies introduce an intervention and evaluate effects in real time (DerSimonian et al. 1982). They may, however, be prone to bias (systematic error) or random error (described in further detail below). Therefore, three hallmarks of experimental studies are adherence to a strict protocol of treatment or exposure, blinding of outcomes, and randomization of subjects into treatment or exposure arms.

Blinding is an effective method of eliminating systematic error and increasing the accuracy of endpoint measurement. Single-blind trials are those in which patients are unaware of their treatment; in double-blind trials, neither the patients nor the investigator are aware of the treatment assignment (Schulz and Grimes 2002). Randomization is a process by which patients are allocated to treatment groups, purely by chance using computer programs or random number tables or equivalent methods, thus minimizing random error. This process should equally distribute confounding variables between the allocated groups and help to prevent any manipulation by the investigators so that no prediction regarding outcomes can be made according the pattern of patient characteristics (Lachin et al. 1988). Currently, intention-to-treat analysis is a favored strategy for assessing outcomes in clinical trials. This paradigm states that patients must be analyzed and included in the data analysis in the group to which they were allocated, and not analyzed with the group that had the same treatment. Although somewhat counterintuitive, this is meant to maintain randomization's ability to minimize random error. Experimental studies essentially achieve symmetry of potential unknown factors and eliminate confounding through blinding, randomization, and, where appropriate, intention-to-treat analysis of subjects (Levin 2005). Further, they provide a stronger and more accurate basis for statistical analysis (Chow and Liu 2008).

The *pre- and post-intervention trial* (also known as a before-and-after study) is a prospective study in which subjects are enrolled into a trial and pre- and post-intervention data are collected and compared (DerSimonian et al. 1982). The subjects are not randomized, with all patients being treated with the same intervention, and pre- and post-groups might be matched and have similar characteristics. The evidence from a pre- and post-trial is useful, but is not a high level of evidence.

The *two-period crossover trial* is a prospective, randomized trial where subjects are allocated to two treatment groups and all patients are eventually exposed to both treatments. Individuals are randomly allocated to one of two treatment groups, specific data are collected about patient outcomes, the treatments are halted, and after a short period of time for washout, they are switched to the other treatment for the same time period (DerSimonian et al. 1982; Foulkes 2008). This method allows the study of both treatments in the same subject, and, in turn, eliminates confounding. However, this design is prone to carryover effects from the first treatment and may therefore influence the outcome of the second treatment (Hills and Armitage 2004). This trial is most suitable for conditions in which subjects return to an initial state during treatment withdrawal, for example, those involving pharmacological treatment. The crossover trial allows identification of treatment effect using a much smaller number of patients than the typical randomized controlled clinical trial.

The *cluster trial* is a prospective, randomized trial in which clusters of patients are randomly allocated to treatments (DerSimonian et al. 1982). In this trial, individual patients are not randomized. For example, one entire center of clinical care might be randomized to one treatment, while another hospital is randomized to another treatment (Murray et al. 2004). This trial is applicable when patients cannot be randomized individually. Unfortunately, this opens the cluster trial to confounding and, thus, decreases the level of empirical evidence obtained from it for improvement of medical practice (Foulkes 2008).

The *factorial trial* is a prospective, randomized trial where patients are randomized to two groups and allows the simultaneous evaluation of two or more treatments in each group. This trial is suitable for large trials to maximize return (DerSimonian et al. 1982; Foulkes 2008). However, this raises the possibility of the incompatibility of the combination of two treatments and introduces bias in the results (Brittain and Wittes 1989).

Among all of the clinical research designs, the randomized blinded trial is the best option, but observational or nonrandomized designs might be all that is feasible for some research investigations, as in the case of prohibitive costs and rarity of the condition under study.

26.3   Sample Size, Study Subjects, and Variables

After deciding on the clinical research design, sample size must be estimated. Generally, the sample size can be calculated based on the level of significance, power of the study, expected effect size, underlying event rate in the population, and the standard deviation in the population (Kadam and Bhalerao 2010). The level of significance chosen represents the potential for an *alpha error*, that is, concluding that a difference exists between treatments that isn't actually present. This is represented by the p value and it is typically set at $p < 0.05$ or $p < 0.01$, which is the probability of obtaining an outcome at least as extreme as the one that was actually observed by chance. The p value of 5% or 1% therefore signifies that the investigator is willing to accept 5% or 1% of false positive rate. Conversely, *beta error* is the risk of missing a difference between treatments that is actually present, but undetected by the study, or the false negative rate that the investigator is willing to accept. Most studies accept a 20% false negative rate, reflecting a *study power* of 80%. If a study has too few patients, it will fail due to insufficient power (Kadam and Bhalerao 2010). The effect size is the difference (*delta*) between the value of the outcome variable in the control group compared to the test group and is based on estimates from prior studies (Hislop et al. 2014). The underlying event rate in the population under study is the prevalence rate, which is also estimated from previously documented literature. The standard deviation of the study population is extremely important to note since it predicts the variability in the data and establishes variables that should be studied. All of the above factors influence sample size and must be calculated before initiating a study. Purely descriptive studies require sample size that produces acceptable narrow confidence intervals for mean, proportions, or other statistics that need to be calculated (Charan and Biswas 2013).

Additionally, the *study population* needs to be determined and described through the specification of inclusion and exclusion criteria. Also, the recruiting method to identify subjects must be decided. It is important to note the significance of a tradeoff of selecting patients from a subset of population versus the whole population: generalizability versus feasibility. Depending on the ultimate research hypothesis and question of study, appropriate study subjects must be chosen (Chow and Liu 2008; Foulkes 2008).

Next, the *measurements* – predictor, confounding, and outcome variables – need to be addressed. Determination of variables that need to be measured in a study depends on the design of the study. Observational studies examine several predictor or independent variables that affect the outcome or dependent variables. Most clinical trials investigate the effects of an intervention, which is a predictor variable that the investigator manipulates (Hartung and Touchette 2009). Also, this design, as described in detail above, includes minimization of confounding variables to identify the true association of intervention and outcome. All of these variables need to be identified prior to the onset of the study.

26.4 Functional Aspects of Clinical Studies

The ultimate goal of any research study is to provide conclusions that impact day-to-day clinical practice. Study findings are used to draw inferences about what happened in the study sample (internal validity) and then generalize those inferences to events in the world outside (external validity). Internal validity is the truth within a study. A study has internal validity if the study conclusions represent the truth for the subjects and results are not due to chance, bias, or interfered with by confounding (Hartung and Touchette 2009). The external validity is the truth beyond the study – generalizability. A study has external validity if the study conclusions represent the truth for the population of application. External validity can occur only if the study is internally valid and if the population of application is similar in characteristics to the study population (Jüni et al. 2001).

As indicated above, it is very important to limit those two major threats to internal and external validity – systematic (bias) and random error. Random error interferes with the clinical results due to chance and can distort results in either direction, usually unpredictably (Jüni et al. 2001). To minimize random error and increase precision, patients must be randomized and the sample size must be sufficient. Systematic error, or bias, threatens conclusions by distorting the results of the study in one direction. To minimize systematic error, mechanisms of patient selection must be carefully planned and specific protocol maneuvers must be established. Sampling or selection bias is an error in choosing the subjects under study or method of collecting samples. Specific focus toward study subject selection will limit this bias (Miller et al. 2001). Recall or responder bias results in differences in the recollections retrieved by study subjects regarding past events (Coughlin 1990). Ecological bias occurs when an association between group variables is mistakenly transferred to individual level (Kramer 1983; Pearce 2000). Measurement bias is an error that occurs due to the lack of blinding and measurement methods that are consistently different between the groups in the study. Confounding bias is an error due to the failure to account for the effect of variables that are related to both the factor being studied and the outcome and are not distributed equally between the groups under study. Confounding can be limited if the confounding variables are eliminated by the exclusion criteria or measured and included in the statistical models (Hulley et al. 2013). All of these biases can be limited if a study is well designed with specific focus on all of its structural parts, which allows the study's results to achieve validity (Jüni et al. 2001).

A properly designed randomized controlled trial results in Level I evidence (Force 1989). Following it are Level II studies – ranging from well-designed controlled trials without randomization, cohort or case-control analytical studies, and multiple time series designs with or without the intervention. Subsequent Level III evidence is based on noncomparative descriptive studies of clinical experience, or opinions of respected authorities, including reports of expert committees. With adequate attention to the structure and function of the study, and careful evaluation of its quality, all levels of evidence contribute to basic scientific knowledge and can impact clinical care.

26.5 Epilogue

Clinical research has its own role in the progress of medical science. Clinical studies should focus on the structural parts of a clinical study, including research question, significance and background, design, sample size, study subjects, and measurement approaches. With a structural focus on clinical studies, the functional aspects come alive with significant internal and external validity and adequate control over random and systematic errors. The combination of well-executed structure and function of these clinical studies permits discovery of new diseases, unexpected effects, and the study of novel mechanisms.

References

Beckman, T.J. and Cook, D.A. (2007). Developing scholarly projects in education: a primer for medical teachers. *Medical Teacher* 29 (2–3): 210–218. https://doi.org/10.1080/01421590701291469.

Boet, S., Sharma, S., Goldman, J., and Reeves, S. (2012). Review article: medical education research: an overview of methods. *Canadian Journal of Anaesthesia = Journal Canadien D'anesthésie* 59 (2): 159–170. https://doi.org/10.1007/s12630-011-9635-y.

Brittain, E. and Wittes, J. (1989). Factorial designs in clinical trials: the effects of non-compliance and subadditivity. *Statistics in Medicine* 8 (2): 161–171. https://doi.org/10.1002/sim.4780080204.

Carneiro, A.V. (2005). Types of clinical studies. III. Cross-sectional studies. *Revista Portuguesa de Cardiologia: Orgão Oficial Da Sociedade Portuguesa de Cardiologia = Portuguese Journal of Cardiology: An Official Journal of the Portuguese Society of Cardiology* 24 (10): 1281–1286.

Chan, K. and Bhandari, M. (2011). Three-minute critical appraisal of a case series article. *Indian Journal of Orthopaedics* 45 (2): 103–104. https://doi.org/10.4103/0019-5413.77126.

Charan, J. and Biswas, T. (2013). How to calculate sample size for different study designs in medical research? *Indian Journal of Psychological Medicine* 35 (2): 121–126. https://doi.org/10.4103/0253-7176.116232.

Chow, S.-C. and Liu, J.-P. (2008). *Design and Analysis of Clinical Trials: Concepts and Methodologies.* Wiley.

Coughlin, S.S. (1990). Recall bias in epidemiologic studies. *Journal of Clinical Epidemiology* 43 (1): 87–91. https://doi.org/10.1016/0895-4356(90)90060-3.

DerSimonian, R., Charette, L.J., McPeek, B. et al. (1982). Reporting on methods in clinical trials. *New England Journal of Medicine* 306 (22): 1332–1337. https://doi.org/10.1056/NEJM198206033062204.

Force, U. S. Preventive Services Task (1989). *Guide to Clinical Preventive Services.* DIANE Publishing.

Foulkes, Mary. 2008. "Study Designs, Objectives, and Hypotheses." http://ocw.jhsph.edu/courses/BiostatMedicalProductRegulation/biomed_lec2_foulkes.pdf.

Grimes, D.A. and Schulz, K.F. (2002). An overview of clinical research: the lay of the land. *The Lancet* 359 (9300): 57–61. https://doi.org/10.1016/S0140-6736(02)07283-5.

Hartung, D.M. and Touchette, D. (2009). Overview of clinical research design. *American Journal of Health-System Pharmacy: AJHP: Official Journal of the American Society of Health-System Pharmacists* 66 (4): 398–408. https://doi.org/10.2146/ajhp080300.

Hills, M. and Armitage, P. (2004). The two-period cross-over clinical trial. *British Journal of Clinical Pharmacology* 58 (7): S703–S716. https://doi.org/10.1111/j.1365-2125.2004.02275.x.

Hislop, J., Adewuyi, T.E., Vale, L.D. et al. (2014). Methods for specifying the target difference in a randomised controlled trial: the Difference ELicitation in TriAls (DELTA) systematic review. *PLoS Medicine* 11 (5): e1001645. https://doi.org/10.1371/journal.pmed.1001645.

Hulley, S.B., Cummings, S.R., Browner, W.S. et al. (2013). *Designing Clinical Research*. Lippincott Williams & Wilkins.

Jüni, P., Altman, D.G., and Egger, M. (2001). Systematic reviews in health care: assessing the quality of controlled clinical trials. *BMJ (Clinical Research Ed.)* 323 (7303): 42–46.

Kadam, P. and Bhalerao, S. (2010). Sample size calculation. *International Journal of Ayurveda Research* 1 (1): 55–57. https://doi.org/10.4103/0974-7788.59946.

Kramer, G.H. (1983). The ecological fallacy revisited: aggregate- versus individual-level findings on economics and elections, and sociotropic voting. *The American Political Science Review* 77 (1): 92. https://doi.org/10.2307/1956013.

Lachin, J.M., Matts, J.P., and Wei, L.J. (1988). Randomization in clinical trials: conclusions and recommendations. *Controlled Clinical Trials* 9 (4): 365–374. https://doi.org/10.1016/0197-2456(88)90049-9.

Lasserre, V., Guihenneuc-Jouyaux, C., and Richardson, S. (2000). Biases in ecological studies: utility of including within-area distribution of confounders. *Statistics in Medicine* 19 (1): 45–59.

Levin, K.A. (2003). Study design IV: cohort studies. *Evidence-Based Dentistry* 7 (2): 51–52. https://doi.org/10.1038/sj.ebd.6400407.

Levin, K.A. (2005). Study design I. *Evidence-Based Dentistry* 6 (3): 78–79. https://doi.org/10.1038/sj.ebd.6400355.

Levin, K.A. (2006). Study design III: cross-sectional studies. *Evidence-Based Dentistry* 7 (1): 24–25. https://doi.org/10.1038/sj.ebd.6400375.

Lewallen, S. and Courtright, P. (1998). Epidemiology in practice: case-control studies. *Community Eye Health* 11 (28): 57–58.

Mann, C.J. (2003). Observational research methods. Research design II: cohort, cross sectional, and case-control studies. *Emergency Medicine Journal: EMJ* 20 (1): 54–60.

Miller, K.D., Rahman, Z.U., and Sledge, G.W. Jr., (2001). Selection bias in clinical trials. *Breast Disease* 14 (1): 31–40.

Murray, D.M., Varnell, S.P., and Blitstein, J.L. (2004). Design and analysis of group-randomized trials: a review of recent methodological developments. *American Journal of Public Health* 94 (3): 423–432. https://doi.org/10.2105/AJPH.94.3.423.

Nissen, T. and Wynn, R. (2014). The clinical case report: a review of its merits and limitations. *BMC Research Notes* 7: 264. https://doi.org/10.1186/1756-0500-7-264.

Pearce, N. (2000). The ecological fallacy strikes back. *Journal of Epidemiology and Community Health* 54 (5): 326–327. https://doi.org/10.1136/jech.54.5.326.

Rosenbaum, P. (2010). *Design of Observational Studies*. Springer http://www.springer.com/statistics/statistical+theory+and+methods/book/978-1-4419-1212-1.

Schulz, K.F. and Grimes, D.A. (2002). Blinding in randomised trials: hiding who got what. *Lancet* 359 (9307): 696–700. https://doi.org/10.1016/S0140-6736(02)07816-9.

Thabane, L., Thomas, T., Ye, C. et al. (2009). Posing the research question: not so simple. *Canadian Journal of Anaesthesia = Journal Canadien D'anesthésie* 56 (1): 71–79. https://doi.org/10.1007/s12630-008-9007-4.

Vandenbroucke, J.P. (2001). In defense of case reports and case series. *Annals of Internal Medicine* 134 (4): 330–334. https://doi.org/10.7326/0003-4819-134-4-200102200-00017.

27

Retrospective Analysis from a Chart Review: A Step-by-Step Guide

Philipp Hendrix[1] and Christoph J. Griessenauer[2]

[1] Department of Neurosurgery, Geisinger, Danville, PA
[2] Research Institute of Neurointervention, Paracelsus Medical University, Salzburg, Austria

27.1 Stepwise Approach to Retrospective Studies

A retrospective study is a study that gathers information about a sample population regarding the occurrence of specific outcomes or phenomenon that have already taken place (Ott and Longnecker 2010; Kumar 2011). Students, junior residents, or novice clinician scientists working on their first clinical research project, e.g. a retrospective analysis from chart review data, frequently feel some discomfort due to a lack of experience with data gathering, statistical analysis, and manuscript writing. Here, we present a short and simple step-by-step guide on how to perform a retrospective analysis from the chart review (Table 27.1). This guide is intended to address frequent concerns and encourage readers to perform their first and subsequent retrospective analyses.

27.1.1 Define a Research Question

In clinical practice, *clinical questions* emerge regularly (e.g. whether one treatment is superior to another or which factors have a positive or negative influence on a patient's outcome, etc.). To answer a clinical question in a scientific manner, a *research question* needs to be formulated. Whereas clinical questions may be open-ended questions (e.g. what is the best surgical approach for tumor resection?), research questions have defined interrelated variables, usually referred to as independent and dependent variables (e.g. is the endoscopic surgical approach or open surgical approach [*independent variable is the type of surgical approach*] associated with a better neurologic outcome [*dependent variable is the neurologic outcome after surgery*]). To answer the research question correctly, it has to sustain statistical analysis. Therefore, the variables in the research question need to be measurable (e.g. surgery performed was endoscopic or open; or outcome was measured using an appropriate scale).

After the main research question has been defined, additional minor follow-up questions potentially associated with the problem may arise.

Draw a question-tree with the main research question as the trunk and follow-up questions as branches. This will be helpful to keep the research project organized and facilitate the preparation of the manuscript (see Section 27.1.3).

A Guide to the Scientific Career: Virtues, Communication, Research, and Academic Writing, First Edition.
Edited by Mohammadali M. Shoja, Anastasia Arynchyna, Marios Loukas, Anthony V. D'Antoni, Sandra M. Buerger, Marion Karl and R. Shane Tubbs.

Table 27.1 Checklist for retrospective studies.

	Project 1	Project 2
Topic area identified		
Clinical question formulated		
Research question formulated		
Literature searched and bibliography in software		
Literature reviewed		
Study design chosen		
Protocol written and reviewed		
IRB approval obtained		
Dataset designed		
Data gathered		
Data analyzed		
Statistics applied		
Paper written and reviewed by faculty or colleagues		
Paper submitted for journal review		

27.1.2 Search Literature, Create a Bibliography, and Review the Literature

A thorough review of the existing literature should be performed before any data are gathered. Prior studies may have already elaborated on certain aspects of the question-tree (i.e. the trunk or some of the branches), which may help reorganize the question-tree (prioritize questions as more or less important). It will also help decide which questions have already been answered and which warrant further investigation. As the literature is studied, the question-tree may change numerous times. Keep track of the articles that are pertinent to the study as they will serve as references for discussing the findings of your new retrospective study.

Critically appraise any articles that deal with the main research question (for more information on how to critically evaluate the clinical literature, see Chapter 48) and perform a thorough review of articles related to follow-up questions to identify additional variables that may contribute to a comprehensive investigation of the research question. Whereas some variables are considered standard (e.g. demographic information), other variable are unique to just one or a small number of articles and may warrant additional investigation. Moreover, the literature review will identify novel variables that have not been previously investigated.

Conclusively, the literature review reveals which variables have to be tested (e.g. demographic information, well-known predictors and outcome variables, etc.), which variables are equivocal (e.g. variables that are not yet established but have been mentioned in the literature), and variables that have not been tested and guarantee originality of the research project.

27.1.3 Reevaluate the Question-Tree and Variables

After completion of the literature review, reevaluate the question-tree. Reaffirm whether the main research question will extend the current knowledge in the research

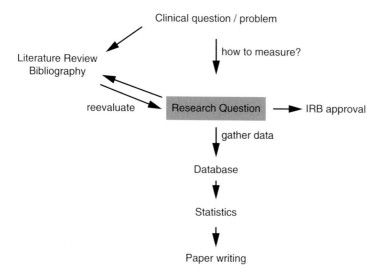

Figure 27.1 Research question in the retrospective studies. The ultimate success in conducting a retrospective study depends on the validity and importance of the research question posed.

area (Figure 27.1). To determine whether the research project will impact the current understanding of the problem, consider consulting a senior researcher.

27.1.4 Obtain Institutional Review Board Approval

Studies that involve human subjects are required to protect patients' right and welfare as their private data are explored and used for publication. It is the responsibility of a local ethics committee, commonly known as an institutional review board (IRB), to review, examine, and approve whether the planned study is in agreement with ethical standards and patients' interest. Hence, when considering a research project involving humans, check whether IRB approval needs to be obtained (see local IRB guidelines at your institution).

27.1.5 Design a Dataset and Gather the Data

The next step is to gather the required data to be able to answer the "research question." This requires endurance and a high level of accuracy to minimize the risk of having to return and gather additional data after the chart review is complete. We recommend investing sufficient time to prepare a database (e.g. with Microsoft Excel). When plotting the variables one should already think about how data can be grouped (e.g. age may be plotted as a ratio variable, i.e. age in years; additionally, patients may be grouped in below 50 years [e.g. designated as "0"] and 50 years old or older [e.g. designated as "1"]). Moreover, different grouping or coding of variables allows for comparison of subgroups in the statistical analysis and may therefore give additional information. However, coding of variables should follow a reasonable principle: either reaffirm an existing model (i.e. coding of variables similar to previous studies in the literature; e.g. patient age "0" for 0–30 years old, "1" for 31–60 years old, "2" for 61 years or older) or create a hypothesis and establish subgroups accordingly. One should mention this aspect explicitly in the methods section of the manuscript.

27.1.6 Perform Statistical Analysis

This step is often one of the most feared. We do not intend to go in any mathematical detail here as statistics are discussed in a separate section of this book (see Chapter 66). But we would like to stress two aspects: Before any statistic can be performed it is crucial to have well-defined, testable questions with related null and alternate hypotheses. Moreover, it is essential to have a well-prepared and organized dataset. This is an excellent preparation for the upcoming statistics part, as it allows us to directly copy and paste into a desired software for statistical analysis. Whether the statistical analysis is performed by oneself or whether it is supported by a trained statistician depends on individual knowledge and experience.

27.1.7 Prepare the Manuscript

Manuscript preparation is the final step. There is a separate section devoted to manuscript preparation in the book (see Section VIII). However, we would like to state a few aspects that shall help in the process. Preparation of the results section is fairly straightforward once the statistics are complete. Start with the presentation of the descriptive statistics (e.g. demographics, distribution of variables between groups, etc.) followed by inference statistics aimed at answering the research questions. In the results section, findings are presented without going into detail on how the results may be interpreted. Interpretation of the results belongs in the discussion section, where return to the question-tree may be of value. Discuss the findings in a logical manner and support the interpretation with the references added to the question-tree.

The method section is often underappreciated. However, this part of the paper outlines the quality of the presented results and the value of conclusions drawn from them. Mention the investigated patient sample and describe the research protocol: which data were collected and how were they collected. Explain how grouping and coding of research variables was undertaken. Always mention whether IRB approval/ethics committee review was necessary. Eventually prepare an additional section where the methods of statistical analysis are described: which software was used, which statistical tests have been performed, and especially for complex statistical analysis, bring up why certain tests have been chosen for the analysis. This contributes to the proper evaluation of statistics and, therefore, to the interpretation of the results and conclusions.

References

Kumar, R. (2011). *Research Methodology: A Step-by-Step Guide for Beginners*, 3e. Los Angeles: Sage.

Ott, R.L. and Longnecker, M.T. (2010). *An Introduction to Statistical Methods and Data Analysis*, 6e. Cengage Learning: Belmont.

28

Designing, Planning, and Conducting Clinical Trials

Ramón Rodrigo, Juan Guillermo Gormaz, Matías Libuy and Daniel Hasson

Molecular and Clinical Pharmacology Program, Institute of Biomedical Sciences, Faculty of Medicine, University of Chile, Santiago, Chile

28.1   Introduction

We are in an era where evidence-based medicine is the standard for developing new cures and improving on current clinical practices. Scientific research is a mainstay in the development and evaluation of new interventions used to prevent, detect, diagnose, treat, and manage illnesses in humans. For this purpose, clinical trials play a pivotal role in determining how well new therapeutic approaches work in patients. Although modern clinical trials have evolved from the eighteenth century, only in recent decades have they emerged as the preferred method for evaluating medical interventions (Day and Ederer 2004). Clinical trials are scientific investigations aimed to examine and evaluate the effectiveness but also the short-term safety of the new treatment. They apply to pharmacological, surgical, and clinical interventions, as well as diagnostic techniques used on human subjects. Clinical trials are performed on healthy human volunteers, as well as in patients with specific diagnoses. The participation of humans in clinical trials is permitted pending that all participants, or their legal representatives, understand the risks and benefits involved in the study and sign a written consent.

The first step in interventional clinical trial design is to develop a hypothesis related to the therapeutic intervention being studied. A trial protocol is then developed to test the hypothesis. The protocol introduces the scientific and methodological components of the trial, which should adhere to current standards of safety and ethics established in the World Medical Association (WMA) Declaration of Helsinki (1997).[1] The content put forth in a typical study protocol defines eligibility criteria, all involved treatment procedures, endpoints, or outcomes, randomization processes, and sample-size determination. It includes data analysis and interpretations as well.

The clinical trials for investigational new drugs are commonly conducted in four phases that produce different relevant information. Phase I investigates the metabolism of a drug, its pharmacodynamics and pharmacokinetics, and usually includes only healthy volunteers. Phase II is performed with small patient groups having a disease that is treatable using a standard dosage, evaluating side effects and treatment efficacy.

1 The revised Declarations of Helsinki can be found on the WMA website, https://www.wma.net.

A Guide to the Scientific Career: Virtues, Communication, Research, and Academic Writing, First Edition.
Edited by Mohammadali M. Shoja, Anastasia Arynchyna, Marios Loukas, Anthony V. D'Antoni, Sandra M. Buerger, Marion Karl and R. Shane Tubbs.
© 2020 John Wiley & Sons, Inc. Published 2020 by John Wiley & Sons, Inc.

Phase III provides evidence of the effectiveness of a treatment applied to large groups of patients. Phase IV studies – also called post-approval or post-marketing surveillance studies – are carried out after registration of a product to collect data on the long-term effects and rare adverse reactions. This chapter will provide an overview of the primary conventions used in designing, planning, and conducting clinical trials with human subjects.

28.2 Design of Controlled Clinical Trials

28.2.1 Definition and Design Basics

In this section, we will discuss the design basis of the randomized controlled trials (RCTs). The study protocols defined herein formulate the gold standard for establishing causation between an intervention and the primary endpoints of a study (Nallamothu et al. 2008). Despite the fact that RCTs have been traditionally reserved for pharmacological studies, these protocols have a wide range of applications in clinical medicine. The RCT research design must define eligibility criteria, random assignment practices, and masking techniques. It should also include robust endpoints and an evidence-based sample-size definition. The RCT could be defined as a prospective study that evaluates the effects of one or more interventions in comparison with a control group. RCTs are performed under the stringent control standards in order to minimize errors and the potential confounding variables, which may influence the outcome (Stanley 2007). These quality standards ensure a high degree of internal and external validity.

Internal validity refers to how well a protocol is performed, while external validity is related to ensuring a high degree of reproducibility of the results (Chalmers et al. 1981). Most RCTs are aimed at isolating a pathophysiological process, or a disease, and require stringent inclusion and exclusion criteria of the patient population ensuring a high degree of internal validity. However, this practice may undermine external validity of the trial as real-world patients usually have multiple comorbidities. Often in RCTs, as internal validity goes up, their external validity goes down. This is because the stringent inclusion and exclusion criteria often result in a homogenous sample population and not what is representative of the real world. For example, a trial on diabetes mellitus includes subjects who only suffer from diabetes and not from another condition like heart disease or hypercholesterolemia. This is certainly good for the internal validity of the study. Yet in the real world, many diabetic patients suffer from multiple conditions, which is why the external validity is lessened. External validity refers to generalizability of the trial results to subjects not included in the study, that is, the real-world population with the disease.

28.2.2 Ethical Considerations

Considering that human beings are the subject of study, ethical considerations must be placed as the first priority from start to end of the study design. A good study design focuses on safety and rights of the enrolled participants, as well as intervention efficacy established in the WMA Declaration of Helsinki (2005). To prove effectiveness of a new procedure, the innovative intervention should always be compared with the best treatment currently available. From a scientific point of view, it is only ethical to perform a

Table 28.1 Categorization by size of controlled clinical trials.

Study phase	Query	Aims	Study group	Size
I	Is it well-tolerated and safe?	Evaluate feasibility. Conduct preliminary dose-response analyses. Describe preliminary reports of side effects and toxicity.	A few healthy volunteers or patients with the disease, who are nonresponsive to the standard intervention	A few tens of patients
II	Does it have pharmacological applicability under controlled conditions?	Evaluate efficacy. Evaluate preliminary pharmacological effects. Evaluate short-term side-effect consequences.	A small amount of patients with the disease or at risk for developing the condition to be prevented	Several tens to a few hundred patients
III	Does it have pharmacological applicability under real or standard conditions?	Evaluate effectiveness. Evaluate pharmacological effects at standard conditions. Identify subgroups who may respond differently to the intervention.	Large amount of patients with the disease or at risk for developing the condition to be prevented	From several hundred to a few thousand patients
IV	Does it have long-term or other effects?	Evaluate long-term pharmacological properties. Evaluation of long-term safety and other potential effects. Identify potential interaction with other treatments.	Worldwide population with the disease or at risk for developing the condition to be prevented	Several tens of thousands of patients

clinical trial when there is genuine uncertainty in the scientific community about which intervention is more effective.

28.2.3 Categorization of Controlled Clinical Trials

28.2.3.1 Categorization by Sample Size
Categorization by sample size is commonly used in the clinical trials of the pharmaceutical industry. They are traditionally structured as four stages or phases, which are characterized by an increased complexity in objectives, logistics, and length (Table 28.1) (Wang et al. 2006; Brody 2011; Friedman et al. 2010; Gad 2009; Hackshaw 2009; Meinert 2012; Piantadosi 2005).

28.2.3.2 Categorization by Design
Categorization by design is used to describe how the participants are assigned to the intervention (Wang et al. 2006). The simplest RCT design is the parallel protocol, which

compares an innovative intervention to a standard procedure or placebo. The crossover design is structured to expose participants to all interventions at different times. In a two-intervention crossover design, for example, group A is first exposed to the innovative intervention and then to the standard procedure. Meanwhile, group B is first exposed to the standard procedure and then to the innovative intervention. In this case, each participant acts as their own control. Factorial designs are when participants have been randomized to more than one experimental unit, comparing all possible combinations of intervention and placebo groups. For example, a two-intervention factorial trial would comprise four groups total: (i) Exposition A + Placebo B; (ii) Exposition B + Placebo A; (iii) Exposition A + Exposition B; and (iv) Double Placebo.

28.2.4   Structuring of Controlled Clinical Trials

The RCT can be structured in three ways in terms of its clinical objective: a superiority study, an equivalence study, or a noninferiority study (Wang et al. 2006). A superiority study is structured to demonstrate that a novel intervention has a better performance than the control strategy (placebo or the current standard). Equivalence studies are structured to show that both interventions have the same performance. The aim of an equivalence RCT is to prove that potential differences between the innovative intervention and the current standard have no clinical significance. And a noninferiority trial is aimed to demonstrate that the novel intervention is not significantly worse than the current standard. It should be noted that in both equivalence and noninferiority studies, the innovative intervention can still be better than the control strategy; however, these trials are designed to show that the new treatment has a similar outcome to the standard, supporting the null hypothesis.

28.2.5   Endpoints

The endpoints of an RCT are the study outcomes, which determine if there is enough evidence to reject or accept the null statistical hypothesis of the study (Wang et al. 2006). Endpoints are used to quantitatively define whether the innovative intervention will be successful, and they are measured against the null hypothesis determined by the RCT design (Brody 2011; Friedman et al. 2010; Gad 2009; Hackshaw 2009; Meinert 2012; Piantadosi 2005). In a superiority study, accepting the null hypothesis implies that there is no statistically significant difference between both interventions in terms of the expected benefit. It is desirable that the endpoints are based on outcomes that are easy to measure, accurate, and precise, as well as sensitive and reproducible (Brody 2011). In most cases, the best approach is to follow standardized methodologies widely applied in clinical practice, as they ensure reliable outcomes that are easy to compare.

In an RCT, the categorization of endpoints must be performed in terms of the study objectives, which leads to the definition of primary and secondary endpoints. The general or main objective of the study is associated with primary endpoints, which are based on the outcomes that directly answer the question originating the clinical trial. In most cases, the primary endpoints are direct onset measurements of clinical events to be either reduced or avoided, e.g. reducing the occurrence of postoperative arrhythmias after cardiac surgery (Rodrigo et al. 2013). Secondary endpoints are related to other important objectives of the RCT. A secondary endpoint may be used to validate a biological mechanism deemed responsible for the potential efficacy of the innovative intervention. Or it could be used to prove the cost-effectiveness of the innovative strategy over the standard. In any case, the internal validity of the RCT will be determined largely by the use of endpoints to establish with certainty that the objectives of the intervention are being achieved.

28.2.6 Sample-Size Estimation

The sample size represents the amount of participants enrolled in a clinical trial. Sample sizes of RCTs are typically determined by fulfilling two major presumptions. First, the RCT presumes a difference in the primary endpoints between intervention and control groups to establish the study success (Wang et al. 2006). Second, the RCT presumes a degree of certainty required to establish that the difference is not due to chance, also known as the study power (Bangdiwala 2014).

Defining an optimal sample size is a key step of any RCT. Larger sample sizes reduce the chance of statistical errors and increase the probability of observing differences between the interventions. Smaller sample sizes, on the other hand, benefit the study as they simplify the design, are cost-effective, and shorten the trial length. Furthermore, a smaller sample size exposes fewer patients to the risks put forth by the clinical intervention. Today, due to the evolving sophistication of study designs, many researchers use specialized software to determine optimal sample size (Schulz and Grimes 2005). The underlying process remains the same: a measure of variance (e.g. standard deviation) must be inferred from a pilot study or previous studies, as well as the magnitude of the expected effect of the intervention (known as effect size). The study power and the level of significance provided by the sample size will be determined to best fit the study objectives. A basic formula for sample size calculation in a two-group clinical trial measuring a quantitative endpoint is shown below (Eng 2013):

$$N = \frac{4\sigma 2(z_{\text{crit}} + z_{\text{pwr}})^2}{D^2}$$

In this formula, N is the total sample size, σ is the measure of variance or standard deviation (assumed to be the same for both groups), Z_{crit} is a value dependent on the significance level, Z_{pwr} is dependent on the study power, and D is the minimum expected effect size. Both Z values are generated according to standard conversion tables.

28.2.7 Eligibility Criteria

Optimal eligibility criteria is one of the most important factors established in the RCT study design. This criteria allows those performing the trial to detect the expected difference between intervention and control groups, as well as to enhance the internal and external validity of the RCT (Wang et al. 2006; Brody 2011; Friedman et al. 2010; Gad 2009; Hackshaw 2009; Meinert 2012; Piantadosi 2005).

Eligibility criteria in RCTs are classified in two categories, exclusion and inclusion criteria. It is a *sine qua non* condition of all RCTs that enrolled participants meet all inclusion criteria and no exclusion criteria (which would disqualify them from the study). Exclusion criteria include any condition that either hinders the observation of endpoint differences between groups or makes the follow-up difficult. In addition, any condition incompatible with the intervention for clinical or ethical reasons should be listed among exclusion criteria (Nardini 2014). On the other hand, a suitable selection of inclusion criteria ensures optimal implementation of the protocol.

28.2.8 Random Assignment and Masking

Random assignment of enrolled participants to various RCT study groups is considered one of two important strategies used to reduce the confounding variables; the other strategy being to increase the sample size (Suresh 2011). The objective of randomization is to equally distribute known and unknown confounders among the intervention and control groups.

Masking, or blinding, obscures the identification of the study group assignment by either the researchers or the participants (Meinert 2012). Masking may be applied only to the subjects (a single-blind trial) or applied to the researchers as well (a double-blind trial). Masking is an essential method to reduce researcher and participant influence on RCT results, also known as observer bias. It has been reported that unmasked studies tend to favor the innovative intervention groups compared to the standard procedure or placebo groups (Schulz et al. 1995).

28.3 Project Management and Planning for Clinical Trial

28.3.1 Control and Uniformity Development

A robust trial design is essential to ensure a successful trial. It is not coincidence that efficacious trials often share similar characteristics (Pocock 1983; Wooding 1994; Knatterud et al. 1998). Most successful study designs are conceptually simple, address questions of clinical relevance where genuine uncertainties exist, avoid overly restrictive entry criteria (ensuring generalizability where appropriate), and avoid unnecessarily complex data requirements (Pocock 1983). Table 28.2 shows the key features of a trial design.

28.3.2 Informed Consent

Of major concern in clinical trial development is the ethical requirement to ensure the well-being of the study participants as written in the Declaration of Helsinki of 1964. Informed consent is the most important protection for subjects involved in a study. Yet, researchers can become trapped by common pitfalls such as including an excessive number of pages in the informed consent document, using different templates between research teams, or using a document that lacks structure or includes irrelevant information (Sterckx and Van Assche 2011). To combat these pitfalls, the Clinical Trial Taskforce in 2013 began encouraging the implementation of consistent templates for informed consent.

A proper template for informed consent should lead to the following advantages. First, a template would simplify both the editorial work and the evaluation process by the study's Ethics Committee. Second, it provides a standardization of project requirements for the review by research sponsors. Finally, an informed consent template would

Table 28.2 Key features in the trial design.

Action	Description	Main advice
Protocol development	The first step in setting up a clinical trial is to outline a trial protocol describing the basic details of the trial.	The protocol should be written according to good clinical practice (GCP) standards.
Sponsorship	Sponsorship can be formed as a financial or nonfinancial alliance with a person or institution.	Get a sponsor as soon as possible.
Trial file	A trial file is a document containing a record of notes generated during research.	Starting trial initiation can solve many problems during trial conduction.
Scientific consultation	Consult with appropriate academic partners about the consistency of the scientific proposal of the protocol.	This is the best moment to correct any mistakes present in the protocol.
Apply for funding	Stay alert about different kinds of funding and complete funding applications.	Remember to include trial management costs, statistician, managing product accountability, adverse event reporting, trial monitoring, data management, and archiving, etc.
Develop trial documentation	Include all necessary documents to develop the clinical trial, including adverse event reporting procedure. Additional work in the center of the clinical trial (ordering IMP supplies, data collection, data verification, etc.).	Check if the study protocol is following GCP standards.
Ethics approval	All clinical trials require a favorable opinion from the local ethics committee. The chief investigator is responsible for obtaining ethics approval.	The well-being of the study participants should always be a priority.

Table 28.3 General features of an informed consent.

1. Structured information with clear guiding principles

2. Understandable language: correct sentence structure, short sentences, absence of technical language, consistent use of identical terminology throughout the entire document, free of spelling errors

3. Sufficient organization and spacing of content throughout the document

streamline the process of informing and recruiting participants. Key requirements to be included in a requisite informed consent document are shown in Tables 28.3 and 28.4.

28.3.3 Logistics, Roles, and Organizational Structure

The organizational structure of a study must respond to the need to coordinate all parts of the project and maintain effective communication between both scientific

Table 28.4 Proposal of an informed consent structure.

1. Informed consent itself
2. Supplementary information about the objectives of the protocol that does not immediately inform the decision process

Table 28.5 Definition of roles.

Role	Responsibilities
General project coordinator	Supervision over complete project implementation, coordination of activities, on all levels of the project
Project implementation coordinator	To help with project management and make current decisions with cooperating entities; preparing documentation and documentation recording; maintain contact with prospective beneficiaries; and solving emerging problems
Scientific coordinator	Coordination of the research team meetings and ensuring the integrity of the trial protocol
Administrative coordinator	To act as a liaison between the administrative requirements of the scientific coordinator and the personnel instructed to respond to those petitions
Technical personnel	Assisting with sample processing procedures
Research team	Scientific staff, including academic researchers and postgraduate students
Monitoring personnel	Implementing evaluation tools in the enrollment process of patients
Financial specialist	Preparing financial reports; exercising control over current documentation preparation such as human resources, payments, and money transfers
Supplies specialist	To supervise, maintain, and secure the quality of supplies involved in the trial protocol
Auxiliary staff	To help with logistics support

and collaborative personnel. Defining the roles of members involved in the project should be part of the organizational scheme, guaranteeing productive relationships and contributing to the success of the trial (Hackshaw 2009). To maintain an organized study, members need to understand their responsibilities as well as their limitations, and should ideally incorporate the project goals into their practice.

Logistics and supply chain efficiency is vital in any clinical trial; the lack of resources compromises the execution of the trial procedure. Navigating the logistical hurdles of a study enables clinical trials to move efficiently through the various stages towards achieving a successful trial outcome (Mowery and Williams 1979).

The major roles of staff members involved in a research project are shown in Table 28.5, and a diagram proposing the organizational structure of a standard study is shown in Figures 28.1 and 28.2.

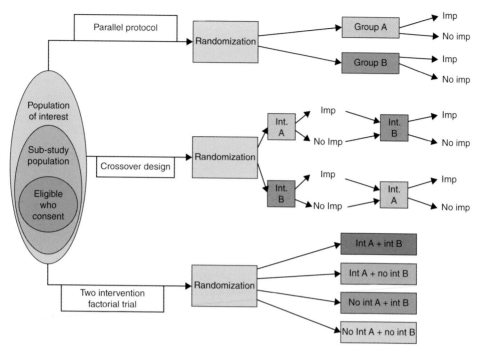

Figure 28.1 Categorization by study design: schematic representation of parallel protocol, two-intervention factorial trial, and crossover designs.

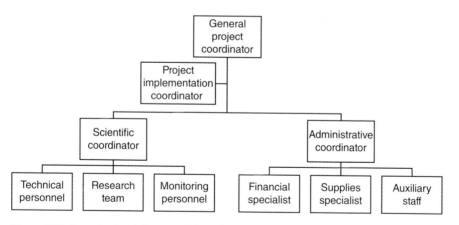

Figure 28.2 Organizational diagram of clinical trial staff.

28.4 Conduct of Clinical Trials

28.4.1 Best Practices

A basic tenet of a research study is that the methodological description of a clinical trial should make it completely reproducible by other research teams. In order to ensure

Table 28.6 Tasks to be performed when starting a clinical trial.

Tasks	Actions
Generating support documentation	Finishing the protocol
	Development of standard operating procedures (SOPs)
	Development of internal flowcharts
Pilot run	Personnel training
	Establishing definitive forms
	Establishing definitive procedures
Staff coordination	Role definition
	Establishing groups and subgroups
	Protocol training
Administrative tasks	Establishing internal and external accounting processes
	Definition of task supervision
	Establishing group and subgroup coordinators and preliminary meeting dates

this objective, and to promote ethical standards, international organisms have issued guidelines known as "best practices."

The International Conference on Harmonization developed the Good Clinical Practice (GCP) guidelines in 1997 using the Declaration of Helsinki as a foundation (European Medicines Agency 2002). Not to be misleading, the GCP guidelines apply to clinical research rather than clinical practice. The GCP guideline is a standard for designing, conducting, recording, and reporting trials that involve the participation of human subjects (Karlberg and Speers 2010; Australia Department of Health and Ageing, Therapeutic Goods Administration 2006). A comparable guideline on best practices used for clinical trials is the international standard ISO 14155 (Moenkemann et al. 2014). More recently, special guidelines have been generated specifically for data management practices, e.g. the Good Clinical Data Management Practices established by the Society for Clinical Data Management (2002).

There is a growing consensus that the conduct of clinical trials and their results should be made public regardless of whether the results are positive, neutral, or negative (Hudson and Collins 2015). Registering a clinical trial in an online database is now necessary for the eventual publication of the trial results in many reputable medical journals (Australia, Department of Health and Ageing, Therapeutic Goods Administration 2006).

28.4.2 Starting the Trial

Once the roles of the staff have been defined but before initiating the actual trial, extra steps must be taken to ensure a straightforward trial execution. These steps include generating support documents, conducting a pilot run, coordinating the staff, and other administrative tasks. See Table 28.6 for a more detailed outline of pretrial tasks.

28.4.3 Data Management

A determining factor in the success of any clinical trial is the quality of the data produced. As trials have increased in standardization and complexity, data management becomes more relevant. A key concept to understanding current data management is the trial file (TF). The TF is defined by the EU GCP directive as "essential documents, which enable both the conduct of a clinical trial and the quality of the data produced to be evaluated." The EU GCP further states that TFs may be stored as paper or electronic files.

Entry of primary data in clinical trials is known as the clinical record form (CRF). Key attributes for good CRFs are described by Machin (2004) as "pleasant to the eye, logical in layout, comprehensive yet focused on the key information required, easy to complete and easy to process." Data entry should be monitored through routine checks for missing data, including data that is only temporarily missing.

Today, many clinical trials store their trial data electronically; this is especially useful for multi-center trials. Specialized software tools for managing clinical trial data are now commercially available. It is of great importance to store TFs properly, whether in paper or electronic form, because regulatory requirements state that the TF must be "audit ready" at all times.

Proper data management systems should reveal whether the trial has missing data. In some instances, this data loss is not recoverable. Reasons for data loss include participants that have not attended all their sessions or who have withdrawn from the trial, or because the information has been simply mismanaged. A classification of the nature of the data loss was proposed by Rubin (1976); depending on the type of data loss, the research team can perform alternative data analysis techniques as described in Table 28.7.

28.4.4 Quality Control

When establishing quality control (QC) on systems of any nature, it should be understood that the process of exercising control over the system consumes limited resources. In clinical trials, this usually refers to the time required by personnel to monitor the QC process. In general, complete control is onerous and even unnecessary (Morganstein and Hansen 1990). Typically, key elements are identified that will undergo QC processes, including those that guarantee the integrity of the final results. Decisions on what parts of the trial are subject to QC should be made in a case-by-case basis. Priority elements to undergo QC in a trial include the process of screening prospective patients, the monitoring of the clinical condition of patients during the trial, and the monitoring of primary and secondary outcomes.

Many countries have instituted external QC processes in which a breach of imposed standards can mean the suspension of the trial (Australia, Department of Health and Ageing, Therapeutic Goods Administration 2006).

28.4.5 Trial Ending

Although most clinical trials are designed with a definitive end date, some trials may end prior to their proposed conclusion. The main reason for ending the trial early is because the preliminary results may be enough to convincingly answer the question posed by

Table 28.7 Classification of data loss and recommendations for obtaining unbiased estimates in missing data scenarios.

Type of data lost	Definition	Example	Actions to be taken
Missing completely at random (MCAR)	Missing data are not related to the hypothetical response value or to some or all of the observed variables.	Due to a protocol flaw, during the first two months of the trial, patient age was not recorded.	Participants with missing data can be excluded from the analysis.
Missing at random (MAR)	Missing data are not related to the hypothetical response value but are related to some or all of the observed variables.	The electronic file with the raw data on female participants of the trial was lost but not the data on male participants.	Analysis is of all available data without data replacement or replace missing values through "last observation carried forward" or "multiple imputation" methods.
Missing not at random (MNAR)	Missing data are related to the hypothetical response value.	Side effects of the drug being tested caused high rates of dropout among intervention-group participants.	No method will deliver unbiased data estimates.

the study. Another reason a trial is concluded prematurely is because the study group is exposed to an unacceptable risk. In both cases, the only ethically acceptable option is to close the trial to further patient entry (Jennison and Turnbull 1999).

Most modern trials are overseen by an external data monitoring committee (DMC), who performs the interim analyses on preliminary results. Interim monitoring allows DMCs to evaluate whether a divergence of event rates between groups is occurring. Usually the divergence has to be present for two consecutive interim analyses to be considered viable. However, interim analysis presents its own set of challenges, namely repeated statistical tests increase the type I error rate (Machin 2004). In order to solve this problem, many approaches have been proposed – group sequential methods being the most common among them. Jennison and Turnbull (1999) performed a thorough review of these methods.

28.5 Conclusions

The goal of this chapter is to introduce the reader to the principles of the scientific study design and the planning and conduction of clinical trials performed on human subjects. Ethical considerations are a priority from start to end, with an emphasis in subjects' rights and safety. A well-conceived informed consent form is a necessary standard. There are several techniques for calculating sample size, each with its own set of advantages and disadvantages. Correct randomization and sample-size estimation techniques allow researchers to limit confounding variables. Appropriate RCT design and application of GCPs ensures results, which can be trusted by both patients and the scientific community, and which can further be successfully applied to clinical settings. Modern data management usually involves handling of electronic records, which are being increasingly standardized. Data loss may occur during the trial and must be appropriately classified in order to best take advantage of the remaining available data.

References

Australia Department of Health and Ageing, Therapeutic Goods Administration (2006) *The Australian Clinical Trial Handbook*. Available from: http://www.tga.gov.au/pdf/clinical-trials-handbook.pdf [Accessed 15/04/14].

Bangdiwala, S. (2014). Power of statistical tests. *International Journal of Injury Control and Safety Promotion* 21 (1): 98–100.

Brody, T. (2011). *Clinical Trials, Study Design, Endpoints and Biomarkers, Drug Safety, and FDA and ICH Guidelines*, 1e. St. Louis: Academic Press Elsevier.

Chalmers, T.C., Smith, H. Jr., Blacburn, B. et al. (1981). A method for assessing the quality of a randomized control trial. *Controlled Clinical Trials* 2 (1): 31–49.

Day, S. and Ederer, F. (2004). Brief history of clinical trials. In: *Textbook of Clinical Trials*, 1e (ed. S. Day, S. Green and D. Machin). Chichester: Wiley.

Eng, J. (2013). Sample size estimation: how many individuals should be studied? *Radiology* 227 (2): 309–313.

European Medicines Agency (2002). *ICH Harmonised Tripartite Guideline E6: Note for Guidance on Good Clinical Practice (PMP/ICH/135/95)*, 13. London: European Medicines Agency.

Friedman, L.M., Furberg, C.D., and Demets, D. (2010). *Fundamentals of Clinical Trials*, 4e. New York: Springer.

Gad, S.C. (2009). *Clinical Trials Handbook*. Hoboken: Wiley.

Hackshaw, A. (2009). *A Concise Guide to Clinical Trials*. Chichester: Wiley-Blackwell.

Hudson, K.L. and Collins, F.S. (2015). Sharing and reporting the results of clinical trials. *JAMA* 313 (4): 355–356.

Jennison, C. and Turnbull, B.W. (1999). *Group Sequential Methods with Applications to Clinical Trials*. London: Chapman and Hall.

Karlberg, J.P. and Speers, M.A. (2010). *Reviewing Clinical Trials: A Guide for the Ethics Committee*. Hong Kong: Karlberg, Johan Petter Einar.

Knatterud, G.L., Rockhold, F.W., and George, S.L. (1998). Guidelines for quality assurance in multicenter trials: a position paper. *Controlled Clinical Trials* 19 (5): 477–493.

Machin, D. (2004). General issues. In: *Textbook of Clinical Trials*, 1e (ed. S. Day, S. Green and D. Machin), 11–45. Chichester: Wiley.

Meinert, C.L. (2012). *Clinical Trials: Design, Conduct and Analysis*, 2e. New York: Oxford University Press.

Moenkemann, H., Pilger, R., Reinecker, S. et al. (2014). Comprehensive and effective system of standard operating procedures for investigator-initiated trials. *Clinical Investigation* 4 (2): 115–123.

Morganstein, D. and Hansen, M. (1990). Survey operations processes: the key to quality improvement. In: *Data Quality Control: Theory and Pragmatics*, 1e (ed. G.E. Liepins and V.R.R. Uppuluri), 91–104. New York: CRC Press.

Mowery, R.L. and Williams, O.D. (1979). Aspects of clinic monitoring in large-scale multiclinic trials. *Clinical Pharmacology & Therapeutics* 25 (5 Pt 2): 717–719.

Nallamothu, B.K., Hayward, R.A., and Bates, E.R. (2008). Beyond the randomized clinical trial: the role of effectiveness studies in evaluating cardiovascular therapies. *Circulation* 118 (12): 1294–1303.

Nardini, C. (2014). The ethics of clinical trials. *Ecancermedicalscience* 8 (387): 1–9.

Piantadosi, S. (2005). *Clinical Trials: A Methodologic Perspective*, 2e. Hoboken: Wiley.

Pocock, S.J. (1983). *Clinical Trials: A Practical Approach*. Chichester: Wiley.

Rodrigo, R., Korantzopoulos, P., Cereceda, M. et al. (2013). A randomized controlled trial to prevent post-operative atrial fibrillation by antioxidant reinforcement. *Journal of the American College of Cardiology* 62 (16): 1457–1465.

Rubin, D.B. (1976). Inference and missing data. *Biometrika* 63 (3): 581–592.

Schulz, K. and Grimes, D. (2005). Sample size calculations in randomised trials: mandatory and mystical. *Lancet* 365 (9467): 1348–1353.

Schulz, K.F., Chalmers, T.C., Hayes, R.J. et al. (1995). Empirical evidence of bias. Dimensions of methodological quality associated with estimates of treatment effects in controlled trials. *Journal of the American Medical Association* 273 (5): 408–412.

Society for Clinical Data Management (2002) *Good Clinical Data Management Practices* [WWW] Society for Clinical Data Management. Available from: http://www.scdm.org/sitecore/content/be-bruga/scdm/Publications/gcdmp.aspx [Accessed 15/04/14]

Stanley, K. (2007). Design of randomized controlled trials. *Circulation* 115 (9): 1164–1169.

Sterckx, S. and Van Assche, K. (2011). The new Belgian law on biobanks: some comments from an ethical perspective. *Health Care Analysis* 19 (3): 247–258.

Suresh, K. (2011). An overview of randomization techniques: an unbiased assessment of outcome in clinical research. *Journal of Human Reproductive Sciences* 4 (1): 8–11.

Wang, D., Nitsch, D., and Bakhai, A. (2006). Randomized clinical trials. In: *Clinical Trials: A Practical Guide to Design, Analysis, and Reporting*, 1e (ed. D. Wang et al.), 1–14. London: Remedica.

Wooding, W.M. (1994). *Planning Pharmaceutical Clinical Trials*. New York: Wiley.

World Medical Association Declaracion of Helsinki (1997). Recommendations guiding physicians in biomedical research involving human subjects. *Journal of the American Medical Association* 277 (11): 925–926.

29

Animal Models in Science and Research

Ray Greek

Americans for Medical Advancement, Goleta, CA, USA

29.1 Overview of Using Animals in Science

Animals can be used in science in general, and research and testing in particular, in essentially nine different ways (Table 29.1) (Greek and Shanks 2009). Categories 3–9 are viable uses of animals in science and research as species differences are either insignificant or taken advantage of (Greek and Rice 2012). The use of animals for their predictive value for human response, such as is the case for categories 1 and 2 in Table 29.1, has been proven scientifically nonviable, however. By predictive value, I mean that there is a high probability that a response to a perturbation, such as a drug or a disease, that reveals itself in an animal model will also reveal itself in humans. This can be calculated as shown in Table 29.2.

There has long been an ethical controversy surrounding the use of animals in science in general and in testing and research in particular. Recently, however, this controversy has extended to whether animal models can scientifically fulfill the claims that have been made for them (Buzoni-Gatel et al. 2011; Committee on Applications of Toxicogenomic Technologies to Predictive Toxicology and Risk Assessment 2007; Hau 2003). The controversy revolves mainly around the claim that animal models have both high positive and negative predictive values for human responses to drugs and diseases. Empirical evidence suggests that animal models do not have this ability. For example, Seok et al. (2013) discovered that gene expression profiles for humans and mice suffering from inflammatory diseases such as sepsis, trauma, and burns differ considerably (see Figure 29.1). There was essentially no correlation between the species. Data like these explain why scores, if not hundreds, of drugs have been effective at curbing sepsis in animal models but none have been effective in humans (News 2012; Dolgin 2013).

Many studies preceding that of Seok et al. (2013) revealed very low predictive values for animal models (Litchfield 1962; Ennever et al. 1987; Igarashi et al. 1995; Mahmood 2000; Johnson et al. 2001; van Meer et al. 2012). Van Meer et al. (2012) retrospectively studied whether serious adverse drug reactions (SARs) in humans could have been identified using animal models prior to the drug being released. They evaluated drugs currently on the market and discovered that only 19% of 93 SARs were seen in animals. Van

A Guide to the Scientific Career: Virtues, Communication, Research, and Academic Writing, First Edition.
Edited by Mohammadali M. Shoja, Anastasia Arynchyna, Marios Loukas, Anthony V. D'Antoni, Sandra M. Buerger, Marion Karl and R. Shane Tubbs.

Table 29.1 Nine categories of animal use in science and research.

1	Animals are used as predictive models of humans for research into such diseases as cancer and AIDS.
2	Animals are used as predictive models of humans for testing drugs or other chemicals.
3	Animals are used as "spare parts" such as when a person receives an aortic valve from a pig.
4	Animals are used as bioreactors or factories, such as for the production of insulin or monoclonal antibodies, or to maintain the supply of a virus.
5	Animals and animal tissues are used to study basic physiological principles.
6	Animals are used in education to educate and train medical students and to teach basic principles of anatomy in high school biology classes.
7	Animals are used as a modality for ideas or as a heuristic device, which is a component of basic science research.
8	Animals are used in research designed to benefit other animals of the same species or breed.
9	Animals are used in research in order to gain knowledge for knowledge sake.

Source: Greek and Shanks 2009.

Table 29.2 Binary classification and formulas for calculating predictive values of modalities such as animal-based research.

		Gold standard (GS)	
		GS+	GS−
Test	T+	TP	FP
	T−	FN	TN

Sensitivity = TP/(TP + FN)
Specificity = TN/(FP + TN)
Positive predictive value = TP/(TP + FP)
Negative predictive value = TN/(FN + TN)

T− = Test negative; T+ = Test positive; FP = False positive; TP = True positive; FN = False negative; TN = True negative; GS− = Gold standard negative; GS+ = Gold standard positive.

Meer et al. (2012) accordingly stated, "The sensitivity of the animal studies for detecting SARs in humans was 19%." Litchfield (1962) studied rats, dogs, and humans in order to evaluate responses to six drugs. Only side effects that could be studied in animals were evaluated. The dog demonstrated a positive predictive value of 0.55 – in other words, slightly better than a coin toss. Approximately 1500 chemicals and drugs are known to produce congenital anomalies in animal models. Only about 40 are known to produce congenital anomalies in humans (Shepard and Lemire 2004). Figure 29.2 illustrates the random nature in correspondence in drug response between humans and animals. To all this must be added the fact that approximately 100 vaccines against HIV have been shown to be effective in animal models but zero have been effective in humans. Likewise, no drug exists for neuroprotection in humans, despite roughly 1000 doing so in animal models.

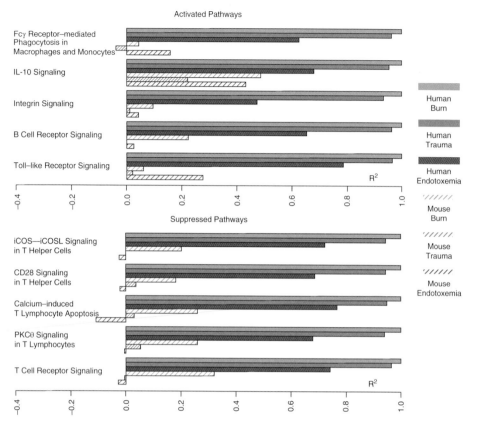

Figure 29.1 Pathway comparisons between the human burns, trauma, and endotoxin and mouse models (from Seok et al. 2013). Shown are bar graphs of Pearson correlations (R^2) for the most activated and suppressed pathways between the four model systems (human endotoxemia and the three murine models) vs. human trauma and burns. Negative correlations are shown as negative numbers ($-R^2$). Human burn is shown as the reference. In every pathway, human endotoxemia had much higher similarity to human injury than mouse models.

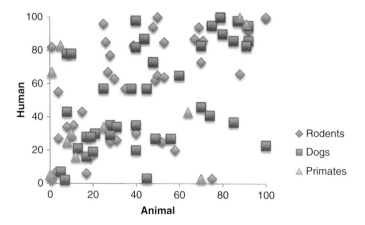

Figure 29.2 Variation in bioavailability among species. Graph by author using data from Sietsema (1989).

29.2   Evolved Complex Systems

In order to fully appreciate why animal models can successfully be used as bioreactors, to discover common properties of anatomy and physiology among mammals, as a source of spare parts for humans, to find potentially conserved processes among life, and so forth (see categories 3–9 in Table 29.1), we need to find a theory that accounts for the data (Greek and Rice 2012; Greek and Hansen 2013a,b). This requires some knowledge from one of the most ignored areas of science – philosophy of science.

Philosophy of science includes the study of critical thinking, defining and studying the foundations of science, defining and studying the axioms, postulates, and assumptions of science, the methodology used in and distinctive to science, the implications of science and specific science studies, along with the merit of science in general and specific areas in particular. Obviously, the philosophy of science is important in terms of doing science properly. Nevertheless, it is often ignored or even scorned by scientists who dismiss it as "mere philosophy."

Relevant to this discussion of animal models, philosophy of science asks how animal models are being used, what are the assumptions underlying their use, what is expected to come from such use, and whether their use is, in reality, consistent with knowledge from other areas of science (consilience) and with the reality of what society is actually obtaining from science. In terms of models, animal models can be categorized per Table 29.1 but also as causal analogical models (CAMs) or heuristic analogical models (HAMs) (LaFollette and Shanks 1993, 1995b, 1995a, 1996). Animals can be used as heuristic devices or HAMs, in other words as an aid to learning or discovery. This is how animal models have traditionally been used in basic research, and it is scientifically viable. The problem arises when animals are presented as CAMs, meaning that if X causes Y in a mouse, it will do the same in humans. This brings us back to the question: "Why can animals be used in categories 3–9 in Table 29.1 but not categories 1–2?" To answer this question, we must examine the fact that animals and humans are examples of evolved complex systems.

Systems can be divided into simple or complex. Simple systems include things like bicycles and a mass sliding down an inclined plane. Simple systems are nothing more than the sum of their parts, are able to be studied by reductionism alone, and are usually intuitive. A complex system is more than the sum of its parts. It is complicated, exhibits emergent phenomena, is highly dependent on initial conditions, interacts with its environment, has chaotic subsystems, is adaptive, has feedback loops, demonstrates nonlinearity, is composed of many components, some of which form modules, and is best described by partial differential equations. Furthermore, complex systems demonstrate a hierarchy of organization (Gell-Mann 1994; Goodwin 2001; Kauffman 1993; Kitano 2002a,b; Sole and Goodwin 2002; Csete and Doyle 2002; Lewin 1999; Morowitz 2002; Ottino 2004).

Complex systems are highly dependent on initial conditions, and evolution makes new species precisely by changing initial conditions in the form of:

- New genes
- New functions for old genes
- Different alleles
- Varying gene regulation and expression

- Presence of mutations such as
  - Copy number variants (CNVs)
  - Single nucleotide polymorphisms (SNPs)
  - Deletions
  - Duplications
- Modifier genes
- Pleiotropy
- Alternative splicing
- Different proteins or protein networks
- Horizontal gene transfer among others

Add to this list convergent evolution – the process by which unrelated or distantly related organisms evolve similar body forms or functions (Pianka 2015). One can see how the initial conditions of any organism might vary considerably from the ancestor organism.

Why are initial conditions important? Many people have heard the expression "a butterfly flaps its wings in Brazil and it rains in Kansas" (or something similar). The phrase dates back to Edward Lorenz (1963), who was running a computer simulation program for weather in the 1950s. One day when he wanted to rerun the program but was short on time, he shortened a variable from six decimal places to three. What he saw is illustrated in Figure 29.3. Note how the two lines begin almost identically but then diverge dramatically. Metaphorically speaking, this means a small change in initial conditions of the weather (like a butterfly flapping its wings in Brazil and influencing a weather front) could result in rain in Kansas on a day when it would have otherwise been sunny. Chaos is now considered a division of complexity studies.

The new field of medicine known as *personalized medicine* also illustrates the importance of changes in initial conditions. The recombination of genes due to sex can also result in changes in initial conditions, as can the environment. Monozygotic twins provide an excellent example of this. One monozygotic twin may suffer from a disease like multiples sclerosis or schizophrenia while the other does not – even though both were

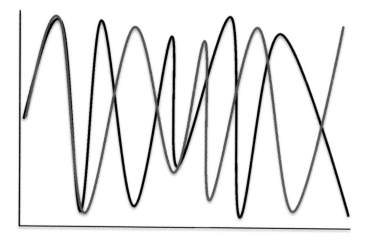

Figure 29.3 Importance of initial conditions.

raised in the same house, slept in the same bedroom, and sat side-by-side in the same classrooms. In other words, their environments were very close but not identical. However, a very small difference in the environment, even in utero, might have led to the methylation of a gene in one twin but not the other. This methylation could have turned on a gene that ultimately cause a disease. CNVs or other mutations could also account for the difference in susceptibility to disease (Maiti et al. 2011; Chapman and Hill 2012). Obviously very small changes in initial conditions can mean the difference between life and death.

Differences in disease susceptibility and drug response are also seen among ethnic groups and between the sexes (Klein and Huber 2010; Lyons et al. 2013; Dewland et al. 2013; Wilke and Dolan 2011; Xu et al. 2012). In the future, differences in what was supposedly the same disease will be identified through genotyping, and genotyping will allow physicians to match drug to disease (see Figure 29.4).

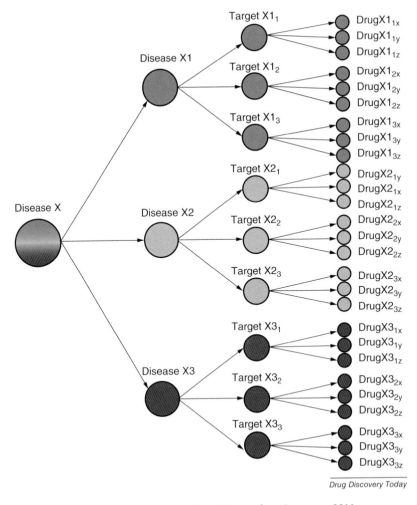

Drug Discovery Today

Figure 29.4 Drug development in the future. Source: from Jørgensen 2011.

29.3   Trans-Species Modeling Theory

I have summarized the above in TSMT (Greek and Hansen 2013b). TSMT states that:

> While trans-species extrapolation is possible when perturbations concern lower levels of organization or when studying morphology and function on the gross level, one evolved complex system will not be of predictive value for another when the perturbation affects higher levels of organization.

The combination of the theory of evolution and complexity theory,[1] along with the empirical data of animal modeling, allows us to understand why we can use animals as heuristic devices such as for basic research, defined as:

> Experimental or theoretical work undertaken primarily to acquire new knowledge of phenomena and observable facts without any particular application or use in view. It is usually undertaken by scientists who may set their own agenda and to a large extent organize their own work
> (Organization for Economic Cooperation and Development 1963).

TSMT also explains why perturbations effecting lower levels of organization can be discovered or demonstrated in animals or humans or perhaps even plants. The laws of physics and the principles of chemistry cross species lines without difficulty. But reactions to drugs and disease occur at higher levels of organization – hence, we should expect response to such perturbations to vary as initial conditions vary.

TSMT also explains why genetically modified animals have failed to be of predictive value for human response to drugs and disease (Geerts 2009; Liu et al. 2004; Mepham et al. 1998; Nijhout 2003). Genes are not pistons that can be replaced by a similar version provided it is for the same make and model automobile. The actions of genes vary with their environment, e.g. the presence of background or modifier genes.

29.4   Conclusion

Animal models can be used in scientific endeavors, but they have no predictive value for human response to drugs and disease. When communicating these concepts, one must categorize the uses of animal models, explain what predictive value is and how it is calculated, clarify the importance of relevant areas of evolutionary biology, describe what complex systems are and what their characteristics are, and explain why these are important when discussing animal models. Finally, one must present the empirical evidence illustrating the very low predictive values either calculated or inferred.

---

1 For the differences between a hypothesis and a theory, see American Association for the Advancement of Science 2011 and Committee on Revising Science and Creationism 2008.

References

American Association for the Advancement of Science (2011). *Q & A on Evolution and Intelligent Design*. American Association for the Advancement of Science, Last Modified 2011 Accessed August 10. http://www.aaas.org/news/press_room/evolution/qanda .shtml.

Buzoni-Gatel, D., Decelle, T., Hardy, P. et al. (2011). *Animal Models and Relevance/Predictivity: How to Better Leverage the Knowledge of the Veterinarian Field*. Fondation Mérieux Accessed October 6. http://www.fondation-merieux.org/ documents/conferences-and-events/2011/animal-models-and-relevance-predictivity- 10-12-october-2011-programme.pdf.

Chapman, S.J. and Hill, A.V.S. (2012). Human genetic susceptibility to infectious disease. *Nat. Rev. Genet.* 13 (3): 175–188.

Committee on Applications of Toxicogenomic Technologies to Predictive Toxicology and Risk Assessment, National Research Council (2007). *Applications of Toxicogenomic Technologies to Predictive Toxicology and Risk Assessment*. Washington, DC: National Academy of Sciences.

Committee on Revising Science and Creationism (2008). *Science, Evolution, and Creationism*. Washington, DC: National Academy of Sciences.

Csete, M.E. and Doyle, J.C. (2002). Reverse engineering of biological complexity. *Science* 295 (5560): 1664–1669.

Dewland, T.A., Olgin, J.E., Vittinghoff, E. et al. (2013). Incident atrial fibrillation among Asians, Hispanics, blacks, and whites. *Circulation* 128 (23): 2470–2477.

Dolgin, E. (2013). Animal rule for drug approval creates a jungle of confusion. *Nat. Med.* 19 (2): 118–119.

Ennever, F.K., Noonan, T.J., and Rosenkranz, H.S. (1987). The predictivity of animal bioassays and short-term genotoxicity tests for carcinogenicity and non-carcinogenicity to humans. *Mutagenesis* 2 (2): 73–78.

Geerts, H. (2009). Of mice and men: bridging the translational disconnect in CNS drug discovery. *CNS Drugs* 23 (11): 915–926.

Gell-Mann, M. (1994). *The Quark and the Jaguar: Adventures in the Simple and Complex*. Boston, MA: Little, Brown and Company.

Goodwin, B. (2001). *How the Leopard Changed its Spots: The Evolution of Complexity*. Princeton, NJ: Princeton University Press.

Greek, R. and Hansen, L.A. (2013a). The strengths and limits of animal models as illustrated by the discovery and development of antibacterials. *Biol. Syst. Open Access* 2 (2): 109.

Greek, R. and Hansen, L.A. (2013b). Questions regarding the predictive value of one evolved complex adaptive system for a second: exemplified by the SOD1 mouse. *Prog. Biophys. Mol. Biol.* 113 (2): 231–253.

Greek, R. and Rice, M.J. (2012). Animal models and conserved processes. *Theor. Biol. Med. Modell.* 9: 40.

Greek, R. and Shanks, N. (2009). *FAQs About the Use of Animals in Science: A Handbook for the Scientifically Perplexed*. Lanham, MD: University Press of America.

Hau, J. (2003). Animal Models. In: *Handbook of Laboratory Animal Science*, 2e (ed. J. Hau and G.K. van Hoosier Jr.,), 1–9. Boca Raton, FL: CRC Press.

Igarashi, T., Nakane, S., and Kitagawa, T. (1995). Predictability of clinical adverse reactions of drugs by general pharmacology studies. *J. Toxicol. Sci.* 20 (2): 77–92.

Johnson, J.I., Decker, S., Zaharevitz, D. et al. (2001). Relationships between drug activity in NCI preclinical in vitro and in vivo models and early clinical trials. *Br. J. Cancer* 84 (10): 1424–1431.

Jørgensen, J.T. (2011). A challenging drug development process in the era of personalized medicine. *Drug Discovery Today* 16 (19/20): 891–897.

Kauffman, S.A. (1993). *The Origins of Order: Self-Organization and Selection in Evolution*. Oxford University Press.

Kitano, H. (2002a). Computational systems biology. *Nature* 420 (6912): 206–210.

Kitano, H. (2002b). Systems biology: a brief overview. *Science* 295 (5560): 1662–1664.

Klein, S.L. and Huber, S. (2010). Sex differences in susceptibility to viral infection. In: *Sex Hormones and Immunity to Infection* (ed. S.L. Klein and C.W. Roberts), 93–122. Berlin: Springer-Verlag.

LaFollette, H. and Shanks, N. (1993). Animal models in biomedical research: some epistemological worries. *Public Aff. Q.* 7 (2): 113–130.

LaFollette, H. and Shanks, N. (1995a). Two models of models in biomedical research. *Philos. Q.* 45 (179): 141–160.

LaFollette, H. and Shanks, N. (1995b). Util-izing animals. *J. Appl. Philos.* 12 (1): 13–25.

LaFollette, H. and Shanks, N. (1996). *Brute Science: Dilemmas of Animal Experimentation*. London and New York: Routledge.

Lewin, R. (1999). *Complexity: Life at the Edge of Chaos*. Chicago, IL: The University of Chicago Press.

Litchfield, J.T. Jr., (1962). Symposium on clinical drug evaluation and human pharmacology. XVI. Evaluation of the safety of new drugs by means of tests in animals. *Clin. Pharmacol. Ther.* 3: 665–672.

Liu, Z., Maas, K., and Aune, T.M. (2004). Comparison of differentially expressed genes in T lymphocytes between human autoimmune disease and murine models of autoimmune disease. *Clin. Immunol.* 112 (3): 225–230.

Lorenz, E.N. (1963). Deterministic nonperiodic flow. *J. Atmos. Sci.* 20 (3): 130–141.

Lyons, M.R., Peterson, L.R., McGill, J.B. et al. (2013). Impact of sex on the heart's metabolic and functional responses to diabetic therapies. *Am. J. Physiol. – Heart and Circ. Physiol.* 305 (11): H1584–H1591.

Mahmood, I. (2000). Can absolute oral bioavailability in humans be predicted from animals? A comparison of allometry and different indirect methods. *Drug Metabol. Drug Interact.* 16 (2): 143–155.

Maiti, S., Kumar, K.H.B.G., Castellani, C.A. et al. (2011). Ontogenetic De novo copy number variations (CNVs) as a source of genetic individuality: studies on two families with MZD twins for schizophrenia. *PLoS One* 6 (3): e17125.

van Meer, P.J.K., Kooijman, M., Gispen-de Wied, C.C. et al. (2012). The ability of animal studies to detect serious post marketing adverse events is limited. *Regul. Toxicol. Pharm.* 64 (3): 345–349.

Mepham, T.B., Combes, R.D., Balls, M. et al. (1998). The use of transgenic animals in the European Union: the report and recommendations of ECVAM workshop 28. *Altern. Lab. Anim.* 26 (1): 21–43.

Morowitz, H.J. (2002). *The Emergence of Everything: How the World Became Complex*. Oxford: Oxford University Press.

(2012). Focus on sepsis. *Nat. Med.* 18 (7): 997.

Nijhout, H. (2003). The importance of context in genetics. *Am. Sci.* 91 (5): 416–423.

Organisation for Economic Cooperation and Development (1963). *The Measurement of Scientific and Technical Activities: Proposed Standard Practice for Surveys of Research and Development*. Paris.

Ottino, J.M. (2004). Engineering complex systems. *Nature* 427 (6973): 399.

Pianka, E.R. (2015). *Convergent Evolution*. University of Texas Accessed March 19. http://www.zo.utexas.edu/courses/THOC/Convergence.html.

Seok, J., Warren, H.S., Cuenca, A.G. et al. (2013). Genomic responses in mouse models poorly mimic human inflammatory diseases. *Proc. Natl. Acad. Sci. U.S.A.* 110 (9): 3507–3512.

Shepard, T.H. and Lemire, R.J. (2004). *Catalog of Teratogenic Agents*, 11e. Baltimore, MA: Johns Hopkins.

Sietsema, W.K. (1989). The absolute oral bioavailability of selected drugs. *Int. J. Clin. Pharmacol. Ther. Toxicol.* 27 (4): 179–211.

Sole, R. and Goodwin, B. (2002). *Signs of Life: How Complexity Pervades Biology*. Basic Books.

Wilke, R.A. and Dolan, M.E. (2011). Genetics and variable drug response. *JAMA* 306 (3): 306–307.

Xu, H., Cheng, C., Devidas, M. et al. (2012). ARID5B genetic polymorphisms contribute to racial disparities in the incidence and treatment outcome of childhood acute lymphoblastic Leukemia. *J. Clin. Oncol.* 30 (7): 751–757.

Further Reading

Abate-Shen, C. (2006). A new generation of mouse models of cancer for translational research. *Clin. Cancer Res.* 12 (18): 5274–5276.

APS. 2013. Why do scientists use animals in research?, Last Modified March 7, 2011. Accessed July 12. http://www.the-aps.org/mm/SciencePolicy/AnimalResearch/Publications/animals/quest1.html.

Arrowsmith, J. (2011a). Trial watch: phase II failures: 2008–2010. *Nat. Rev. Drug Discovery* 10 (5): 328–329.

Arrowsmith, J. (2011b). Trial watch: phase III and submission failures: 2008–2010. *Nat. Rev. Drug Discovery* 10 (2): 87–87.

Belmaker, R., Bersudsky, Y., and Agam, G. (2012). Individual differences and evidence-based psychopharmacology. *BMC Med.* 10 (1): 110.

Buckland, D. (2013). *The Extraordinary Secret of how Mice Are Curing Cancer*. Express, Last Modified June 9, 2013. Accessed July 31. www.express.co.uk/news/health/406168/The-extraordinary-secret-of-how-mice-are-curing-cancer.

Canto, J.G., Rogers, W.J., Goldberg, R.J. et al. (2012). Association of age and sex with Myocardial Infarction Symptom Presentation and in-hospital mortality. *JAMA* 307 (8): 813–822.

Chiou, W.L., Jeong, H.Y., Chung, S.M. et al. (2000). Evaluation of using dog as an animal model to study the fraction of oral dose absorbed of 43 drugs in humans. *Pharm. Res.* 17 (2): 135–140.

CSPI (2008). *Longer Tests on Lab Animals Urged for Potential Carcinogens*. CSPI Accessed November 17. http://www.cspinet.org/new/200811172.html.

Czyz, W., Morahan, J., Ebers, G. et al. (2012). Genetic, environmental and stochastic factors in monozygotic twin discordance with a focus on epigenetic differences. *BMC Med.* 10 (1): 93.

Edelstein, L.C., Simon, L.M., Montoya, R.T. et al. (2013). Racial differences in human platelet PAR4 reactivity reflect expression of PCTP and miR-376c. *Nat. Med.* 19 (3512): 1609–1616.

Enna, S.J. and Williams, M. (2009). Defining the role of pharmacology in the emerging world of translational research. *Adv. Pharmacol.* 57: 1–30.

FDA (2004). *Innovation or Stagnation? Challenge and Opportunity on the Critical Path to New Medical Products*. FDA, Last Modified 2004 Accessed November 25. http://www .fda.gov/downloads/ScienceResearch/SpecialTopics/CriticalPathInitiative/ CriticalPAthOpportunitiesReports/ucm113411.pdf.

Fletcher, A.P. (1978). Drug safety tests and subsequent clinical experience. *J. R. Soc. Med.* 71 (9): 693–696.

Fomchenko, E.I. and Holland, E.C. (2006). Mouse models of brain tumors and their applications in preclinical trials. *Clin. Cancer Res.* 12 (18): 5288–5297.

Gad, S.C. (ed.) (2007). *In Animal Models in Toxicology*. Boca Raton, FL: CRC Press.

Grass, G.M. and Sinko, P.J. (2002). Physiologically-based pharmacokinetic simulation modelling. *Adv. Drug Delivery Rev.* 54 (3): 433–451.

Greek, R. (2012). Animal models and the development of an HIV vaccine. *J. AIDS Clin. Res.* S8: 001.

Greek, R. and Menache, A. (2013). Systematic reviews of animal models: methodology versus epistemology. *Int. J. Med. Sci.* 10 (3): 206–221.

Greek, R. and Greek, J. (2010). Is the use of sentient animals in basic research justifiable? *Philos. Ethics Humanit. Med.* 5: 14.

Greek, R., Pippus, A., and Hansen, L.A. (2012). The Nuremberg code subverts human health and safety by requiring animal modeling. *BMC Med. Ethics* 13 (1): 16.

Greek, R. and Shanks, N. (2011). Complex systems, evolution, and animal models. *Stud. Hist. Philos. Biol. Biomed. Sci.* 42 (4): 542–544.

Greek, R. (2008). Letter. Dogs, Genes and Drugs. *Am. Sci.* 96 (1): 4.

Greek, R. (2012). Book review. Zoobiquity: what animals can teach us about health and the science of healing. *Animals* 2 (4): 559–563.

Greek, R. and Hansen, L.A. (2012). The development of deep brain stimulation for movement disorders. *J. Clin. Res. Bioethics* 3: https://doi.org/10.4172/2155-9627 .1000137, http://www.omicsonline.org/2155-9627/2155-9627-3-137.php?aid=9962.

Greek, R., Hansen, L.A., and Menache, A. (2011). An analysis of the Bateson review of research using nonhuman primates. *Med. Bioethics* 1 (1): 3–22.

Greek, R., Menache, A., and Rice, M.J. (2012). Animal models in an age of personalized medicine. *Pers. Med.* 9 (1): 47–64.

Greek, R., Shanks, N., and Rice, M.J. (2011). The history and implications of testing thalidomide on animals. *J. Philos. Sci. Law* 11: http://www6.miami.edu/ethics/jpsl/ archives/all/TestingThalidomide.html.

Hart, B.A., Abbott, D.H., Nakamura, K. et al. (2012). The marmoset monkey: a multi-purpose preclinical and translational model of human biology and disease. *Drug Discovery Today* 17 (21–22): 1160–1165.

Hughes, B. (2008). Industry concern over EU hepatotoxicity guidance. *Nat. Rev. Drug Discovery* 7 (9): 719–719.

Igarashi, T., Yabe, T., and Noda, K. (1996). Study design and statistical analysis of toxicokinetics: a report of JPMA investigation of case studies. *J. Toxicol. Sci.* 21 (5): 497–504.

Jankovic, J. and Noebels, J.L. (2005). Genetic mouse models of essential tremor: are they essential? *J. Clin. Invest.* 115 (3): 584–586.

Jones, R.C. and Greek, R. (2013). A review of the Institute of Medicine's analysis of using chimpanzees in biomedical research. *Sci. Eng. Ethics* 1–24.

Seok, J., Warren, H.S., Cuenca, A.G. et al. (2013). Genomic responses in mouse models poorly mimic human inflammatory diseases. *Proc. Natl. Acad. Sci. U.S.A.* 110 (9): 3507–3512.

Jura, J., Wegrzyn, P., and Koj, A. (2006). Regulatory mechanisms of gene expression: complexity with elements of deterministic chaos. *Acta Biochim. Pol.* 53 (1): 1–10.

Kola, I. and Landis, J. (2004). Can the pharmaceutical industry reduce attrition rates? *Nat. Rev. Drug Discovery* 3 (8): 711–715.

Kummar, S., Kinders, R., Rubinstein, L. et al. (2007). Compressing drug development timelines in oncology using phase '0' trials. *Nat. Rev. Cancer* 7 (2): 131–139.

LeCouter, J.E., Kablar, B., Whyte, P.F. et al. (1998). Strain-dependent embryonic lethality in mice lacking the retinoblastoma-related p130 gene. *Development* 125 (23): 4669–4679.

Lesko, L.J. and Woodcock, J. (2004). Translation of pharmacogenomics and pharmacogenetics: a regulatory perspective. *Nat. Rev. Drug Discovery* 3 (9): 763–769.

Lumley, C. (1990). Clinical toxicity: could it have been predicted? Premarketing experience. In: *Animal Toxicity Studies: Their Relevance for Man* (ed. C.E. Lumley and S.R. Walker), 49–56. London: Quay.

Lutz, D. (2011). *New Study Calls into Question Reliance on Animal Models in Cardiovascular Research*. Washington University, Last Modified August 3, 2011 Accessed May 31. http://news.wustl.edu/news/Pages/22540.aspx.

McArthur, R. (2011). Editorial: many are called yet few are chosen. Are neuropsychiatric clinical trials letting us down? *Drug Discovery Today* 16 (5/6): 173–175.

Miklos, G. and Gabor, L. (2005). The human cancer genome project--one more misstep in the war on cancer. *Nat. Biotechnol.* 23 (5): 535–537.

Morgan, P., Van Der Graaf, P.H., Arrowsmith, J. et al. (2012). Can the flow of medicines be improved? Fundamental pharmacokinetic and pharmacological principles toward improving phase II survival. *Drug Discovery Today* 17 (9/10): 419–424.

National Science Foundation (2011). *Press Release: Of Mice and Men*. NSF, Last Modified September 14. Accessed October 6. http://www.nsf.gov/news/news_summ.jsp?cntn_id=121653&org=NSF&from=news.

Ollikainen, M. and Craig, J.M. (2011). Epigenetic discordance at imprinting control regions in twins. *Epigenomics* 3 (3): 295–306. https://doi.org/10.2217/epi. 11.18.

Pearson, H. (2002). Surviving a knockout blow. *Nature* 415 (6867): 8–9.

Pinto, J.M., Schumm, L.P., Wroblewski, K.E. et al. (2013). Racial disparities in olfactory loss among older adults in the United States. *J. Gerontol. A Biol. Sci. Med. Sci.* https://doi.org/10.1093/gerona/glt063, http://biomedgerontology.oxfordjournals.org/content/early/2013/05/19/gerona.glt063.abstract.

Raineri, I., Carlson, E.J., Gacayan, R. et al. (2001). Strain-dependent high-level expression of a transgene for manganese superoxide dismutase is associated with growth retardation and decreased fertility. *Free Radic. Biol. Med.* 31 (8): 1018–1030.

Rohan, R.M., Fernandez, A., Udagawa, T. et al. (2000). Genetic heterogeneity of angiogenesis in mice. *FASEB J* 14 (7): 871–876.

Shanks, N. and Greek, R. (2009). *Animal Models in Light of Evolution*. Boca Raton, FL: Brown Walker.

Shanks, N., Greek, R., and Greek, J. (2009). Are animal models predictive for humans? *Philos. Ethics Humanit. Med.* 4 (1): 2.

Simon, V. (2005). Wanted: women in clinical trials. *Science* 308 (5728): 1517.

Smith, R.L. and Caldwell, J. (1977). Drug metabolism in non-human primates. In: *Drug Metabolism - from Microbe to Man* (ed. D.V. Parke and R.L. Smith), 331–356. London: Taylor & Francis.

Spriet-Pourra, C. and Auriche, M. (1994). *SCRIP Reports*. New York: PJB Publications.

Suter, K.E. (1990). What can be learned from case studies? The company approach. In: *Animal Toxicity Studies: Their Relevance for Man, Edited by CE Lumley and SW Walker* (ed. C.E. Lumley and S.W. Walker), 71–78. Lancaster: Quay.

Van Regenmortel, MHV and Marc, H.V. (2012). Basic Research in HIV vaccinology is hampered by reductionist thinking. *Front. Immunol.* 3: https://doi.org/10.3389/fimmu .2012.00194, http://www.frontiersin.org/Journal/Abstract.aspx?s=1247& name=immunotherapies_and_vaccines&ART_DOI=10.3389/fimmu.2012.00194.

Van Regenmortel, MHV, Marc, H.V., and Hull, D.L. (2002). *Promises and Limits of Reductionism in the Biomedical Sciences (Catalysts for Fine Chemical Synthesis)*. West Sussex: Wiley.

van Regenmortel, M.H.V. (1999). Molecular design versus empirical discovery in peptide-based vaccines. Coming to terms with fuzzy recognition sites and ill-defined structure–function relationships in immunology. *Vaccine* 18 (3–4): 216–221.

van Regenmortel, M.H.V. (2002a). Pitfalls of reductionism in immunology. In: *Promises and Limits of Reductionism in the Biomedical Sciences* (ed. M.H.V. van Regenmortel and D.L. Hull), 47–66. Chichester: Wiley.

Van Regenmortel, M.H.V. (2002b). Reductionism and the search for structure–function relationships in antibody molecules. *J. Mol. Recog. JMR* 15 (5): 240–247.

van Regenmortel, M.H.V. (2004a). Biological complexity emerges from the ashes of genetic reductionism. *J. Mol. Recog.* 17 (3): 145–148.

Van Regenmortel, M.H.V. (2004b). Reductionism and complexity in molecular biology. Scientists now have the tools to unravel biological complexity and overcome the limitations of reductionism. *EMBO Rep.* 5 (11): 1016–1020.

van Zutphen, L.F. (2000). Is there a need for animal models of human genetic disorders in the post-genome era? *Comp. Med.* 50 (1): 10–11.

Weaver, J.L., Staten, D., Swann, J. et al. (2003). Detection of systemic hypersensitivity to drugs using standard Guinea pig assays. *Toxicology* 193 (3): 203–217.

Willyard, C. (2009). HIV gender clues emerge. *Nat. Med.* 15 (8): 830.

Zhao, J., Goldberg, J., Bremner, J.D. et al. (2012). Global DNA methylation is associated with insulin resistance: a monozygotic twin study. *Diabetes* 61 (2): 542–546.

30

How to Identify a Timely and Relevant Topic for a Literature Review

Sanjay Patel, Jerzy Gielecki and Marios Loukas

Department of Anatomical Sciences, St. George's University School of Medicine, Grenada, West Indies

30.1 Introduction

A literature review summarizes the current literature on a topic into a concise and relevant manner. It provides insight into the primary literature, by interpreting the current knowledge and providing a direction for future research (Rapple 2011).

The first step to writing a literature review is to identify a topic. The topic and scope of the review should be stated early as it forms the framework that guides the rest of the review. When developing a topic for review, the academic writer should narrow the scope of the topic to a manageable size. A considerable amount of time is invested into writing a literature review; therefore, it is crucial to have a topic that is well considered and refined (Aveyard 2010; Dawidowicz 2010).

There is an ever-increasing number of scientific papers being published each year, and it is almost impossible for clinicians and academics to go through them all. Despite this, scholars are expected to stay informed on the advances in their field. A literature review is an excellent tool for scholars, as it allows them to review the current literature related to their field in a way that is systematic and efficient. Researchers who have an in-depth knowledge in their field and who have conducted many original studies are usually the ideal candidates to write literature reviews that will be read and be cited (Rapple 2011). The purpose of this chapter is to provide guidance on selecting a topic that is both relevant and timely.

30.2 Identifying a Relevant and Timely Topic

When choosing a topic for review, one must consider the intended audience. If the audience is mainly students, the writer must provide insight into the current views in the literature as well as giving historical background on the topic. Many students will also use the information as a steppingstone for finding relevant publications on a particular topic. On the other hand, the writer might summarize hundreds of articles into one manuscript for a different audience. This type of literature review will take more time and need a more critical perspective on the topic that will provide new insight into the

A Guide to the Scientific Career: Virtues, Communication, Research, and Academic Writing, First Edition.
Edited by Mohammadali M. Shoja, Anastasia Arynchyna, Marios Loukas, Anthony V. D'Antoni, Sandra M. Buerger, Marion Karl and R. Shane Tubbs.

field. In any case, the topic should be relevant for the reader whether it be students, scholars, or a boarder audience.

Some writers come across topics reviewing the current literature. If the current knowledge on a topic is widely debated, then it is usually an ideal topic for a review. After critically reading the literature, a researcher compares the strengths/weaknesses of various studies in the field and the degree to which other scholars have already addressed the topic. They can then find knowledge gaps that warrant further research into the topic (Timmins and McCabe 2005). This would provide justification for writing on a topic that will contribute meaningful reviews to the current field (Maier 2013). Literature reviews may also serve as a great avenue for graduate and postgraduate students to learn how to compose complex manuscripts under direct supervision (Aveyard 2010; Cronin et al. 2008; Pautasso 2013).

30.3 Narrowing the Topic

Many scholars will unknowingly choose a topic that is too broad. Authors need to be selective and try not to include every topic on a particular field. One of the objectives of a literature review is to be able to draw valid research-based conclusions. This task can become unfeasible if the scope is too large because a tremendous amount of work will be required to bridge knowledge gaps (Dawidowicz 2010). Most journals set a deadline for the completion of reviews and limit the number of citations and pages that can be used. Such restraints need to be considered when developing a topic for a literature review.

The best way to narrow down a topic is to brainstorm specific questions that can be asked about a narrow aspect related to the main field and to search the current literature to see whether there is enough information available to write a review. Creating a mind map with subtopics, along with notes on the strengths and weaknesses of each topic, would help a writer to select a topic that is focused and manageable (Aveyard 2010).

30.4 Literature Reviews and Clinical Practice

Another important source that researchers can refer to in choosing literature review topics is clinical practitioners. Systematic reviews, which are discussed in more detail in Chapter 32, play a vital role in guiding clinical practice. Most medical practitioners refer to systematic reviews on treatment options for evidence-based guidance on how best to treat their patients. Systematic reviews are used to compare different outcomes of possible interventions to find out which is most advantageous to the patient. If current medical protocols are not clear about which treatment options are ideal in a particular clinical scenario, then this would be ideal for a systematic review on that topic (Khan et al. 2011).

30.5 Conclusion

The task of finding a topic on which to write a literature review might seem daunting at first. However, by first deciding which type of review to write and developing an in-depth understanding on a particular topic, one will be able to choose and refine a topic that is relevant.

References

Aveyard, H. (2010). *Doing Literature Review in Health and Social Care: A Practical Guide*, 2e, 23–29. Maidenhead: McGraw-Hill/Open University Press.

Cronin, P., Ryan, F., and Coughlan, M. (2008). Undertaking a literature review: a step-by-step approach. *Br. J. Nurs.* 17: 38–43.

Dawidowicz, P. (2010). *Literature Reviews Made Easy: A Quick Guide to Success*, 7–11. Charlotte, NC: Information Age Pub.

Khan, K.S., Kunz, R., Kleijnen, J. et al. (2011). *Systematic Reviews to Support Evidence-Based Medicine: How to Review and Apply Findings of Healthcare Research*, 2e. London: Hodder Arnold.

Maier, H.R. (2013). What constitutes a good literature review and why does its quality matter? *Environ. Modell. Softw.* 43: 3–4.

Pautasso, M. (2013). Ten simple rules for writing a literature review. *PLoS Comput. Biol.* 9: e1003149.

Rapple, C. (2011). The role of the critical review article in alleviating information overload. *Annual Reviews.* Retrieved from http://www.annualreviews.org/page/infooverload.

Timmins, F. and McCabe, C. (2005). How to conduct an effective literature search. *Nurs. Stand.* 20: 41–47.

31

The Structure and Conduct of a Narrative Literature Review

Marco Pautasso

Forest Pathology and Dendrology, Institute for Integrative Biology, ETH Zurich, Zurich, Switzerland
Animal and Plant Health Unit, European Food Safety Authority (EFSA), Parma, Italy

31.1   Introduction

Literature reviews are increasingly needed in most scholarly disciplines (Colebunders et al. 2014) (Figure 31.1). Their perpetual necessity originates from the relentless growth in new scientific publications (Rapple 2011). Compared to 1991, in 2010, 4, 8, and 17 times more papers were indexed in Web of Science with the keywords "genetics," "immunology," and "neuroscience," respectively (Pautasso 2012). Given such rising mountains of papers, scientists can only examine in detail a minority of the papers relevant to their interests (Erren et al. 2009). Timely and thorough literature reviews thus often have a wide readership. Producing regular summaries of the recent literature is a service to one's own community (Ketcham and Crawford 2007). When they lead to new synthetic insights, literature reviews can be just as innovative as primary research papers, and should thus be just as important for career progression and peer recognition (Aveyard 2010; Hampton and Parker 2011).

The main aim of a literature review is to recapitulate the stand of knowledge on a particular issue. This is essential to avoid unnecessary and unaware duplication of previous research (Boote and Beile 2005). All researchers are theoretically in a position to write a literature review, because before starting a research project, it is important to do a survey of the literature so as to make sure that the research project has not already been done (Ioannidis et al. 2014). Since professionally conducted literature reviews are essential in the practice of science, guidelines about how to approach and carry out a literature review are needed (Steward 2004; McMenamin 2006; Bearfield and Eller 2007; Cronin et al. 2008).

This chapter aims to provide a brief guideline on how to conduct a narrative literature review, and is based on 10 simple rules that I learned working on about 30 literature reviews as a PhD and postdoctoral student (Pautasso 2013). Ideas and insights also come from coauthors and colleagues, as well as from feedback of reviewers, editors, and readers. Writing a literature review is a welcome break from the research and teaching routine, provides some historical perspective, and enables a meta-perspective on one's own research field. It is a recipe against excessive specialization in science, because it

A Guide to the Scientific Career: Virtues, Communication, Research, and Academic Writing, First Edition.
Edited by Mohammadali M. Shoja, Anastasia Arynchyna, Marios Loukas, Anthony V. D'Antoni, Sandra M. Buerger, Marion Karl and R. Shane Tubbs.
© 2020 John Wiley & Sons, Inc. Published 2020 by John Wiley & Sons, Inc.

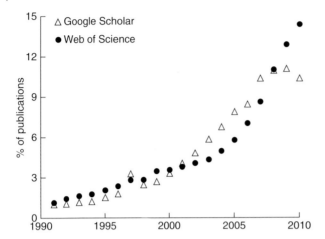

Figure 31.1 Proportional increase in the articles from Web of Science and Google Scholar mentioning "literature reviews" over a 20-year period (1991–2010), as of February 2014. If the amount of publications retrieved were evenly distributed through time, they would comprise about 5% of the materials published annually over this period. During the 1990s, the proportion of material retrieved annually was less than 3%. At the end of the 2000s, more than 10% of the whole literature corpus was retrieved each year. This means that about half of the literature reviews of this 20 year-period have been published in the last 5 years.

requires the ability to master multiple tasks such as finding and appraising relevant material, summarizing information from various sources, and identifying research gaps (Budgen and Brereton 2006).

31.2   Review Team

Readers of a literature review should gain a picture of (i) the major accomplishments in the reviewed field, (ii) the key issues of dispute, and (iii) the most pressing research gaps. You should thus be aiming to cover adequately all these three aspects in your review. Since it is difficult to be proficient on all these fronts, reviewing the literature is often the product of a set of complementary coauthors (Cheruvelil et al. 2014). Some researchers have expertise in the historical (or recent) development of a subject, some others find it easier to spot anomalies in the current research paradigm, and some have instead a talent at predicting which further work is going to help moving on a field. If your research team has exactly this sort of skills, then you are in a perfect position to achieve a well-rounded review of the literature.

Having a team of co-authors working on a literature review is useful to make the review succeed, because different people will tend to retrieve literature using different strategies. This will make it less likely that you will overlook relevant material. A team might be more suitable to assess the contribution of papers that have just been published, because their significance and impact on further research and society is still difficult to predict and might require some discussion. Writing a literature review on your own is possible, but multi-authored literature reviews are probably more fun, although they will require a coordinator to take charge of the distribution of tasks among co-authors.

31.3   Topic and Audience

Choosing a topic to review (and an audience) is essential (Maier 2013), and goes hand in hand with putting together a team to write the review. Only a well-defined issue will allow you to discard the vast amounts of publications that are not relevant to the chosen topic. The choice of a topic can be daunting because of the vast array of issues in contemporary science. A well-considered topic is both within your research interests and of concern to other researchers and practitioners.

Ideas for potential reviews may come serendipitously during desultory reading and informal discussions. If you have been asked to contribute to a literature review, the topic might have already been chosen, but you can still contribute to its refinement. Choosing a topic from papers providing lists of key research questions to be answered (Sutherland et al. 2011) can be risky because several other people might have had the same idea at the same time.

Together with choosing your topic, you should also identify a target audience. In many cases, the topic (e.g. networks in epidemiology) will logically define an audience (e.g. network epidemiologists), but that same topic may also be relevant to researchers in neighboring fields (e.g. network theory, epidemiology, etc.). It is important to define early on the criteria for exclusion of irrelevant papers. These criteria need then to be:

1. Described in the review to help define its scope;
2. Used during the literature search;
3. Refined when coming across publications, which defy the original criteria for inclusion.

31.4   Literature Search

Since you are already familiar with the chosen topic and audience, you already have a basis upon which to start the literature search. You can use the relevant papers you already have to do the following:

1. Identify relevant keywords and use them to search within various databases.
2. Check the list of references cited in those papers.
3. Track recent publications that cite those papers.

When searching and re-searching the literature, it is important to take a note of the search items, databases, and time limits you use, so that your search can be replicated at a later stage (Maggio et al. 2011). This is important both for yourself (and your co-authors) when having to revise or update the review and for other people who might wish to build on your review (or check what has been published on the topic since) (vom Brocke et al. 2009). Feedback from colleagues and librarians should be sought already at this stage, so as to choose a suitable set of keywords and databases (Tannery and Maggio 2012; Bramer et al. 2013).

Inevitably, some full texts of relevant papers will not be accessible immediately. Keep a list of them so as to be able to retrieve them later by emailing authors and/or colleagues at other institutions. The use of a reference management system (e.g. Mendeley, Papers2,

Qiqqa, Sente, Zotero, etc.) can help in achieving your literature review more efficiently. Care must be taken to search also for negative results (i.e. those reporting absence of a significant difference or association of interest), as these will tend to be underreported in the literature (Pautasso 2010). You might also consider searching for evidence in the gray literature (Yasin and Hasnain 2012).

During the literature search, you might come across a literature review on (or very related to) the issue you are planning to work on. This is because there are already various reviews of the literature on most topics (Bastian et al. 2010). Even if you come across a series of literature reviews covering your targeted issue and audience, I encourage you to carry on with your own literature review. Your review can stand out from the crowd of previous related reviews by discussing their standpoints, limits, and key findings. Moreover, past reviews need updating because of the novel studies that have unavoidably appeared since their publication.

31.5   Taking Notes

A good literature review is often the outcome of an iterative process between reading, writing, and learning (Boell and Cecez-Kecmanovic 2014). Reading the literature does not need to wait for the literature search to be finished. You can start perusing the papers you retrieve while you are still looking for relevant material. You can stumble upon useful additional publications while reading. However, it is definitely good practice reading the literature whilst at the same time taking notes. This way, you will avoid misremembering who wrote what and where, and forgetting the thoughts you had while reading each single paper. If you take notes of what you read, by the time you have read the literature you selected, you will already have a preliminary draft.

Although this first draft will still need much reorganization to achieve a text with a well-flowing story (Torraco 2005), rewriting, inserting, and removing parts of the text is less of a challenge than starting from scratch after having spent weeks reading papers without writing anything. Even before formally starting with literature search, it is useful to write down (i) useful findings and quotes, (ii) insights about how to structure the review, and (iii) ideas about what to write.

Take care while taking notes not to forget quotation marks if you are temporarily writing down entire sentences from the literature (i.e. verbalism). It is wise to then rephrase such excerpts with your own words in later drafts. Be careful to record references in the text from the very beginning in order to avoid mistakes in citations. This is important because wrongly attributed references can mislead many more readers if they are present in a literature review than in a research paper. The systematic use of referencing software is advisable, as it is likely to save you time.

31.6   Type of Review

Before starting to refine your first draft, it is time to decide which type of literature review is more appropriate. This choice depends on (i) the nature and amounts of the material found, (ii) the preferences of the target journal(s) and co-authors, and (iii) the time available to write the review.

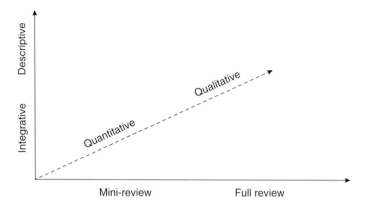

Figure 31.2 Types of literature review. Reviews can (i) be on the short (mini-reviews) or on the long side (full reviews); (ii) integrate the reviewed studies in a coherent story (integrative reviews) or focus on describing their findings and methodology (descriptive reviews); and (iii) use qualitative or quantitative approaches. The reviews can be narrative or systematic depending on the methods used to conduct them. Although literature reviews using systematic methods to search for relevant literature are supposed to yield more studies, reviews analyzing published results in a quantitative, systematic way (meta-analyses) end up in discarding many more studies than narrative reviews.

The choice is threefold between mini- and full reviews, descriptive and integrative reviews, and narrative vs. systematic reviews (Figure 31.2).

Whether to write a mini-review or a full review depends on the number of relevant studies found during the literature search. This, in turn, follows from the definition of the topic to review. A specific issue on which few researchers have already published will fit well with the limits in number of words and references typical of a mini-review. A broader topic with many ramifications is more suitable to a comprehensive literature review with hundreds of cited references. Mini-reviews have the advantage of conciseness, but they suffer from more rapidly becoming out of date. Most researchers are busy readers and will tend not to have time to read in detail long reviews, but full reviews are appropriate for topics that need more room for explanation of their subtleties, historical development, and likely further progress.

Authors of both mini- and full reviews need to decide whether to pursue a descriptive or an integrative review. Descriptive reviews summarize the methods, results, and conclusions of the reviewed material, whereas integrative reviews aim to find common ground in the cited studies (Khoo et al. 2011). It is possible to do both, although most reviews will tend to focus on one or the other approach.

Another choice exists between narrative and systematic reviews, depending on whether the focus is on qualitative or quantitative evidence (Dijkers et al. 2009). Typical narrative reviews are qualitative summaries of the relevant literature, whether this made use of statistical tests or not. Narrative reviews are able to combine studies that addressed different research questions and methodologies (Baumeister 2013). Systematic reviews aim instead to test a hypothesis based on the available qualitative and quantitative evidence, which are gathered using a predefined protocol to reduce bias (Rosenfeld 1996; Cook and West 2012). When systematic reviews use quantitative methods to analyze quantitative results, they are called meta-analyses. Narrative reviews may be strengthened by the inclusion of tables comparing quantitative results from various studies, and systematic reviews as well as meta-analyses benefit from a narrative description of their context. This chapter focuses on narrative reviews, whereas systematic reviews and meta-analyses are treated in detail in Chapters 67 and 68.

31.7   Balance

Literature reviews need to strike a balance between depth and breadth (Eco 1977; Hart 1998). Focusing on the chosen topic avoids covering materials that are only tangentially relevant, thus potentially losing readers along the way. Making the review of broad interest makes it instead potentially inviting to more readers, who might not have started reading the review if this had been too focused.

Finding this balance is one of the major challenges of writing a literature review. In order to stay focused, it is helpful to (i) avoid including material just for the sake of it – only cite relevant studies, (ii) strictly follow the criteria for inclusion agreed at the beginning of the review process, and (iii) only briefly discuss the wider implications of the reviewed topic for other disciplines.

Just as with research papers, reviews should not try to do too many things at once. Nonetheless, the slicing of literature reviews into least-publishable units should also be avoided (Bertamini and Munafò 2012). The need to keep a review of broad interest is less of a problem for interdisciplinary reviews, because the aim to establish connections between fields will automatically make the review interesting to researchers active in various areas (Wagner et al. 2011). If you are writing a review on, for example, how network approaches are used in modeling the spread of epidemics, you might need to include some materials from both parent fields, epidemiology and network theory. This will make the review of broad interest. To keep it focused, the review should only deal in detail with the studies at the interface between epidemiology and network science.

31.8   Criticism, Consistency, Objectivity

Reviewing the literature is not (yet) something that is commonly outsourced to a computer (Tsafnat et al. 2013; Bekhuis et al. 2014). A good literature review does not mechanically provide an annotated bibliography, but evaluates the significance of previous studies and points out their strengths and weaknesses (Carnwell and Daly 2001; Bolderston 2008). Not all publications are gold, as many researchers are under pressure to publish and some peer reviewers are less strict than others. Literature reviews should be a solution to the problem of the current scientific information overload, and not an aggravating factor.

A critical eye needs to be kept on your own draft too. Keep asking yourself:

1. Is what you are writing understandable?
2. Are you reporting the gist of the reviewed studies or unnecessary details?

One important quality of a good review is consistency in terms of content and language. Consistency does not imply that all reviewed studies should be allocated the same amount of words: a literature review provides a better service to the reader if it spends more time on landmark publications than on those without much substance. A multi-authored literature review is often weakly consistent, when different sections are written by different authors and nobody makes an attempt at harmonization. Be consistent throughout the review in the choice of, e.g. passive vs. active voice, present vs. past tense, and US vs. British spelling.

Literature review often involves a conflict of interest, because reviewers have typically published studies relevant to the review they are conducting. In fact, some people argue that you can only review the literature on a topic if you have done research on that topic. But it is a matter of debate how you can objectively review your own research (Logan et al. 2010). You may be particularly keen on your publications, so that you might tend to review your own findings in a rosy light. Or you may be rather dismissive of your own work, with the risk of giving a low profile to your involvement (if any) in the development of a research topic.

To avoid a review of the literature bordering on competitive self-denial or resembling an exercise in public relations, treat your own research as if it had been published by someone else. If your paper is relevant, include it succinctly. If it is irrelevant, dismiss it to the pile of papers possibly useful for another review. Is it a borderline case? Ask a colleague. If you do not feel to be up to the job of reporting objectively on your own relevant publications, a way out is to have your studies reviewed by a colleague, who might then point out some further studies, suggest additional ideas to be covered and help make the review more interesting. That is how teams writing a literature review are sometimes formed.

31.9 Structure

A good review is reminiscent of a well-baked gateau. It is neither rushed nor overdone, it provides both inspiration and energy, and it is ready just at the right time. Reviewing the literature is a balance between creative and systematic work, single-mindedness and encyclopedic knowledge, criticism and enthusiasm. Good literature reviews also need a sound structure.

The classic subdivision of research papers into introduction, methods, results, and discussion does not work with literature reviews, unless they are meta-analyses. Nonetheless, narrative reviews also benefit from starting with an introduction of the context and ending with a discussion of the main conclusions. A methodological section (or a few sentences), including information about how the literature was searched (such as information on databases, keywords, time limits, etc.), makes sense not just for systematic reviews (Roberts et al. 2006).

The main body of the review then needs to be organized to (i) make the review flow seamlessly, (ii) help the reader see the connections between the different sections, and (iii) balance the various parts so that each section is roughly of equivalent length. Interrelation between the components of a literature review can be represented visually with a conceptual scheme, e.g. using mind-mapping diagrams (see Figure 31.3 for an example). Such drawings – to be individualized for each review – can help recognize a better way to order the sections of a review than originally planned and are often instrumental in structuring the text of a literature review (Ridley 2008; Kelleher and Wagener 2011).

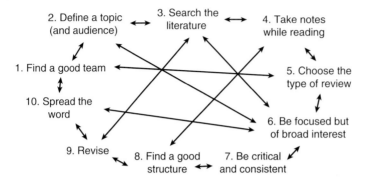

Figure 31.3 Conceptual diagram of the present chapter. Writing a literature review starts with building a team and choosing a suitable topic. Searching the literature and taking notes while reading are carried out alongside. Having chosen the type of literature review, the first draft needs to be kept focused and made interesting, with a critical eye and in a consistent way. The final draft has to be structured in a logical way and revised using feedback from a variety of sources. The process is often circular, because at the end of a literature review there are often more ideas for future reviews. Some of the links between nonconsecutive components of the review are shown (e.g. revising the review can benefit from an additional literature search).

31.10   Feedback and Revision

Comments from peer reviewers are generally very helpful in refining literature reviews (Oxman and Guyatt 1988). Peer reviewers with a fresh mind often spot key studies that had been overlooked, gaps in the argumentation, and inconsistencies. Feedback from colleagues unfamiliar with the reviewed topic is also fundamental to writing a good review, because it makes it clear whether the review will be understandable to the general reader. A diversity of views on the draft may be difficult to accommodate when revising the paper, but such a situation is better than no feedback at all. Science is organized skepticism, so (i) bear in mind that the average reader might be skeptical about what you write and (ii) do your best to make the review representative of the current scientific understanding of an issue (May 2011).

Feedback should be sought from colleagues before peer review too, thus enabling the commentator to provide advice on the content rather than the form. Particularly when the review is a collaboration of many co-authors, one more reading of the draft before submission is advisable. A last-minute amendment of misprints, logical leaps, and untidy sentences will make it more likely that peer reviewers will deliver a positive recommendation.

There is a multi-way relationship between research, readership, peer review, and literature reviews (Figure 31.4):

1. Authors of literature reviews are often asked to perform more peer reviews than usual.
2. Peer reviewers who do their job thoroughly are in a good position to write a literature review after having assessed a particular submission, because they have basically checked the main literature relevant to a specific research question.
3. Reviewing the literature is akin to performing a post hoc peer review, i.e. an assessment of papers after (instead of before) their publication.

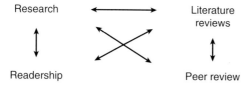

Figure 31.4 Interactions between research, readership, literature reviews, and peer review. Many readers are now overwhelmed by the amount of research and thus make use of literature reviews to keep up with new developments. New research is sometimes inspired by literature reviews, and literature reviews are, of course, based on published research. Both primary research and literature reviews are peer-reviewed, and peer reviewers rely on literature reviews for their reports. Traditionally, (anonymous) peer-review reports are not made available to readers, but this might change if open peer-review becomes popular.

31.11   Dissemination

Getting your review accepted is not the end of the story. Given the competition for attention from many hundreds of other literature reviews published every month, it is important to spread the word about your review in order for it to be widely read (Dijkers 2013). A variety of strategies can be used, such as publishing in open-access journals or in journals with optional open-access publication, posting an unformatted version of the accepted review on online institutional or personal repositories (e.g. academia.edu, Mendeley, ResearchGate, etc.), highlighting the review on social media (see Cocchio and Awad 2014), and presenting the review at conferences.

To make the community aware of your newly published literature review, it can be helpful to send the pdf to potentially interested colleagues, although this is sometimes regarded as too boastful. It can be argued that if a literature review is timely, well-organized, and in-depth, then it will reach its intended audience regardless of your efforts to disseminate it. However, it can also be argued that there is a responsibility to inform taxpayers about the findings from research projects funded with their money. If the journal that accepted the literature review is keen to publish a press release about it, you might wish to see whether this will multiply the number of potential readers compared to your other reviews without a press release (Morris et al. 2007). Also, your readers' feedback is valuable after publication, as it might help improve subsequent updates of the review and other literature reviews.

Acknowledgments

The author is grateful to O. Holdenrieder for insights and support, and to T. Matoni for helpful comments on a rough draft of this chapter.

References

Aveyard, H. (2010). *Doing a Literature Review in Health and Social Care: A Practical Guide.* Maidenhead, UK: McGraw-Hill.

Bastian, H., Glasziou, P., and Chalmers, I. (2010). Seventy-five trials and eleven systematic reviews a day: how will we ever keep up? *PLoS Medicine* 7: e1000326.

Baumeister, R.F. (2013). Writing a literature review. In: *The Portable Mentor – Expert Guide to a Successful Career in Psychology* (ed. M.F. Prinstein), 119–132. Berlin: Springer.

Bearfield, D.A. and Eller, W.S. (2007). Writing a literature review: the art of scientific literature. In: *Handbook of Research Methods in Public Administration* (ed. G.J. Miller and K. Yang), 61–72. Boca Raton: CRC Press.

Bekhuis, T., Tseytlin, E., Mitchell, K.J. et al. (2014). Feature engineering and a proposed decision-support system for systematic reviewers of medical evidence. *PLoS One* 9 (1): e86277.

Bertamini, M. and Munafò, M.R. (2012). Bite-size science and its undesired side effects. *Perspectives on Psychological Science* 7: 67–71.

Boell, S.K. and Cecez-Kecmanovic, D. (2014). A hermeneutic approach for conducting literature reviews and literature searches. *Communications of the Association for Information Systems* 34 (1): 12.

Bolderston, A. (2008). Writing an effective literature review. *Journal of Medical Imaging and Radiation Science* 39 (2): 86–92.

Boote, D.N. and Beile, P. (2005). Scholars before researchers: on the centrality of the dissertation literature review in research preparation. *Educational Researcher* 34: 3–15.

Bramer, W.M., Giustini, D., Kramer, B.M.R. et al. (2013). The comparative recall of Google Scholar versus PubMed in identical searches for biomedical systematic reviews: a review of searches used in systematic reviews. *Systematic Reviews* 2: 115.

Budgen D and Brereton P (2006) Performing systematic literature reviews in software engineering. Proceedings of the 28th International Conference i Software Engineering, ACM New York, NY, USA, pp. 1051–1052.

Carnwell, R. and Daly, W. (2001). Strategies for the construction of a critical review of the literature. *Nurse Education in Practice* 1: 57–63.

Cheruvelil, K.S., Soranno, P.A., Weathers, K.C. et al. (2014). Creating and maintaining high-performing collaborative research teams: the importance of diversity and interpersonal skills. *Frontiers in Ecology and the Environment* 12: 31–38.

Cocchio, C. and Awad, N. (2014). The scholarly merit of social media use among clinical faculty. *Journal of Pharmacy Technology* 30: 61–68.

Colebunders, R., Kenyon, C., and Rousseau, R. (2014). Increase in numbers and proportions of review articles in tropical medicine, infectious diseases, and oncology. *Journal of the Association for Information Science and Technology* 65: 201–205.

Cook, D.A. and West, C.P. (2012). Conducting systematic reviews in medical education: a stepwise approach. *Medical Education* 46: 943–952.

Cronin, P., Ryan, F., and Coughlan, M. (2008). Undertaking a literature review: a step-by-step approach. *British Journal of Nursing* 17 (1): 38–43.

Dijkers, M. (2013). Want your systematic review to be used by practitioners? Try this! *KT Update* 1 (4): 1–4.

Dijkers, M. and The Task Force on Systematic Reviews and Guidelines (2009). The value of "traditional" reviews in the era of systematic reviewing. *American Journal of Physical Medicine and Rehabilitation* 88: 423–430.

Eco, U. (1977). *Come si fa una tesi di laurea*. Milan, Italy: Bompiani.

Erren, T.C., Cullen, P., and Erren, M. (2009). How to surf today's information tsunami: on the craft of effective reading. *Medical Hypotheses* 73: 278–279.

Hampton, S.E. and Parker, J.N. (2011). Collaboration and productivity in scientific synthesis. *Bioscience* 61: 900–910.

Hart, C. (1998). *Doing a Literature Review: Releasing the Social Science Research Imagination*. London: SAGE.

Ioannidis, J.P.A., Greenland, S., Hlatky, M.A. et al. (2014). Increasing value and reducing waste in research design, conduct, and analysis. *The Lancet* 383: 166–175.

Kelleher, C. and Wagener, T. (2011). Ten guidelines for effective data visualization in scientific publications. *Environmental Modelling and Software* 26: 822–827.

Ketcham, C.M. and Crawford, J.M. (2007). The impact of review articles. *Laboratory Investigation* 87: 1174–1185.

Khoo, C.S.G., Na, J.C., and Jaidka, K. (2011). Analysis of the macro-level discourse structure of literature reviews. *Online Information Review* 35: 255–271.

Logan, D.W., Sandal, M., Gardner, P.P. et al. (2010). Ten simple rules for editing Wikipedia. *PLoS Computational Biology* 6: e1000941.

Maggio, L.A., Tannery, N.H., and Kanter, S.L. (2011). Reproducibility of literature search reporting in medical education reviews. *Academic Medicine* 86: 1049–1054.

Maier, H.R. (2013). What constitutes a good literature review and why does its quality matter? *Environmental Modelling and Software* 43: 3–4.

May, R.M. (2011). Science as organized scepticism. *Philosophical Transactions of the Royal Society A* 369: 4685–4689.

McMenamin, I. (2006). Process and text: teaching students to review the literature. *PS: Political Science and Politics* 39 (1): 133–135.

Morris, T., Harding, G., Miles, L. et al. (2007). Getting R&D results into the press. In: *Communicating European Research 2005* (ed. M. Claessens), 183–188. Berlin: Springer.

Oxman, A.D. and Guyatt, G.H. (1988). Guidelines for reading literature reviews. *Canadian Medical Association Journal* 138: 697–703.

Pautasso, M. (2010). Worsening file-drawer problem in the abstracts of natural, medical and social science databases. *Scientometrics* 85: 193–202.

Pautasso, M. (2012). Publication growth in biological sub-fields: patterns, predictability and sustainability. *Sustainability* 4: 3234–3247.

Pautasso, M. (2013). Ten simple rules for writing a literature review. *PLoS Computational Biology* 9 (7): e1003149.

Rapple, C. (2011) The role of the critical review article in alleviating information overload. Annual Reviews White Paper. Available: http://www.annualreviews.org/userimages/ ContentEditor/1300384004941/Annual_Reviews_WhitePaper_Web_2011.pdf. Accessed May 2013.

Ridley, D. (2008). *The Literature Review: A Step-By-Step Guide for Students*. London: SAGE.

Roberts, P.D., Stewart, G.B., and Pullin, A.S. (2006). Are review articles a reliable source of evidence to support conservation and environmental management? A comparison with medicine. *Biological Conservation* 132: 409–423.

Rosenfeld, R.M. (1996). How to systematically review the medical literature. *Otolaryngology–Head and Neck Surgery* 115: 53–63.

Steward, B. (2004). Writing a literature review. *British Journal of Occupational Therapy* 67 (11): 495–500.

Sutherland, W.J., Fleishman, E., Mascia, M.B. et al. (2011). Methods for collaboratively identifying research priorities and emerging issues in science and policy. *Methods in Ecology and Evolution* 2: 238–247.

Tannery, N.H. and Maggio, L.A. (2012). The role of medical librarians in medical education review articles. *Journal of the Medical Library Association* 100 (2): 142–144.

Torraco, R.J. (2005). Writing integrative literature reviews: guidelines and examples. *Human Resources Development Review* 4: 356–367.

Tsafnat, G., Dunn, A., Glasziou, P. et al. (2013). The automation of systematic reviews. *British Medical Journal* 346: f139.

vom Brocke, J., Simons, A., Niehaves, B. et al. (2009). Reconstructing the giant: on the importance of rigour in documenting the literature search process. In: *Proceedings of the ECIS 2009* (ed. S. Newell, E. Whitley, N. Pouloudi, et al.), 2206–2217.

Wagner, C.S., Roessner, J.D., Bobb, K. et al. (2011). Approaches to understanding and measuring interdisciplinary scientific research (IDR): a review of the literature. *Journal of Informetrics* 5: 14–26.

Yasin A and Hasnain MI (2012) On the quality of grey literature and its use in information synthesis during systematic literature reviews. Master Thesis, Blekinge Institute of Technology Karlskrona, Sweden.

32

A Guideline for Conducting Systematic Reviews

Paul Posadzki[1] and Edzard Ernst[2]

[1] The Centre for Public Health, Liverpool John Moores University, Liverpool, UK
[2] Department of Complementary Medicine, Peninsula Medical School, University of Exeter, Exeter, UK

32.1 Introduction

Systematic reviews (SRs) of therapeutic interventions, diagnostic methods, and prognostic or preventive measures help healthcare professionals in making informed decisions. SRs are intertwined with the concept of evidence-based medicine (EBM) aimed at integrating the best available evidence with clinical expertise and patient values into a coherent structure when making decisions about the care of individual patients or societies, or the delivery of health services (Sackett et al. 1996). This chapter is aimed at providing a brief guideline for conducting a high-quality SRs in healthcare. All 17 stages/phases of conducting SRs described here might be applicable to both qualitative and quantitative SRs.

32.2 Why Systematic Reviews?

The principle of an SR is to consider the totality of the existing evidence related to specific questions. SRs can minimize the confusion caused by contradictory findings, from single primary studies by minimizing random and selection biases. As the number of randomized controlled trials (RCTs) published each year grows exponentially (Figure 32.1), it is rarely possible for a clinician to know all the studies of a given question. An SR can assist in finding quick answers to complex problems. For instance, if 100 RCTs existed of the effects of therapy x, on an outcome y of which 50 were positive (indicating the intervention to be effective) and 50 were negative, the overall result might be seen as inconclusive. By including a rigorous evaluation of the reliability of the 100 primary studies, a SR might show what the most reliable type of evidence tells us. Similarly, SRs can highlight areas where the evidence is as yet insufficient for drawing firm conclusions and can thus guide future research activities.

SRs should adhere to a detailed protocol, which means that they have predefined hypotheses, aims and methods and frequently follow external guidelines, e.g. PRISMA (Moher et al. 2009). One key characteristic of an SR is the application of scientific/methodological rigor, transparency and objectivity during all stages of the

A Guide to the Scientific Career: Virtues, Communication, Research, and Academic Writing, First Edition.
Edited by Mohammadali M. Shoja, Anastasia Arynchyna, Marios Loukas, Anthony V. D'Antoni, Sandra M. Buerger, Marion Karl and R. Shane Tubbs.

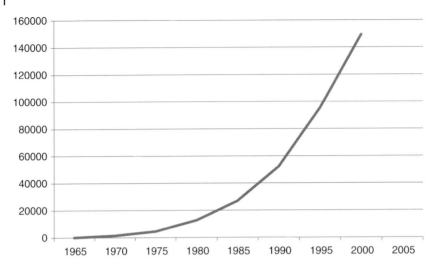

Figure 32.1 Number of Medline-listed randomized controlled trials published each year since 1964. Source: Adapted from Tsay and Yang 2005.

review process. If statistical pooling of the primary data is possible, SRs can employ meta-analytic techniques, which provide quantitative results based on the totality of the pooled data (see Chapters 67–68 for more details).

SRs have a clearly stated set of objectives with pre-defined eligibility criteria for primary studies, an explicit/reproducible methodology, a repeatable search strategy, validity assessments of the findings and a systematic presentation, and synthesis of the characteristics and findings of the included primary studies. None of these features apply to non-SRs, which are wide open to bias.

32.3 A Guideline for Conducting Systematic Reviews

The process of conducting a SR can be dissected into 17 distinct steps, which will be briefly discussed below.

32.3.1 Assembling a Team of Experts

Depending on the research subject, it might initially be wise to think about recruiting a team that covers all the expertise needed for the SR at hand. At the very minimum, the following types of expertise are needed for a high quality SR of a therapeutic intervention:

- An expert in the intervention that is being assessed
- An expert in the condition for which the intervention is employed
- An expert in conducting literature searches
- An expert in the methodology of SRs and/or meta-analyses

Occasionally, this expertise exists in one or two researchers but if that is not the case, it is advisable to gather more researchers participating in the planning, conducting, writing, and publishing of a SR.

32.3.2 Finding a Gap in the Evidence Base

The identification of a gap or deficit in the existing evidence base usually is a prime reason for considering an SR. There are several ways of finding such gaps. Initially, a reviewer should check International Prospective Register of Systematic Reviews (PROSPERO)[1] as it helps avoiding unplanned duplications; time, efforts, and money can be saved simply by searching this database. As the PROSPERO database is not complete, scoping searches are advisable for locating nonregistered SRs. A second option is conducting scoping searches in PUBMED,[2] or Excerpta Medica Database (EMBASE),[3] Cochrane Database of Systematic Reviews[4] or similar databases. A scoping search is useful for providing an overview of the literature on a given topic.

Occasionally, an update of an existing SR can be a good idea. The precondition is that a sizable amount of new data has become available, which renders the current SR obsolete. Such updates must apply the same methodology (search terms, eligibility criteria) as the original SR.

32.3.3 Asking a Focused Research Question/Choosing the Right Hypothesis

The research question of an SR may range considerably from addressing the effectiveness of a therapy, diagnostic methods, prognostic or preventive measures to their safety or cost-effectiveness. A good research question should be focused and must not be too general (e.g. how effective is complementary medicine?). If it is too specific, there is a danger that it turns out to be unanswerable because of the lack of primary studies addressing it. There are, of course, examples of "empty" SRs, which include no primary studies (Ernst 2009). Such SRs can nevertheless be of value in that they demonstrate that studies are needed for answering an important research question.

A concrete example might help understanding the formulation of a focused research question. For instance, if it relates to acupuncture as a treatment for nausea, one might progress from the general to the specific in the following fashion:

- Is acupuncture effective for nausea?
- Is needle acupuncture effective for nausea?
- Is needle acupuncture on pericardium 6 (P6) effective for nausea?
- Is needle acupuncture on P6 more effective than sham acupuncture for nausea?
- Is needle acupuncture on P6 more effective than treatment with a nonpenetrating sham-needle for nausea?
- Is needle acupuncture on P6 more effective than treatment with a nonpenetrating sham-needle for nausea caused by morning sickness?
- Is needle acupuncture on P6 more effective in the short-term than treatment with a nonpenetrating sham-needle for nausea caused by morning sickness?
- Is needle acupuncture on P6 more effective in the short-term than treatment with a nopenetrating sham-needle for nausea caused by morning sickness and quantified with Visual Analogue Scale (VAS)?

---

1 http://www.crd.york.ac.uk/PROSPERO.
2 http://www.ncbi.nlm.nih.gov/pubmed.
3 http://www.elsevier.com/online-tools/embase.
4 www.thecochranelibrary.com.

The degree of specification of the research question depends on a range of factors; as a rule of thumb, a research question should be as specific as necessary and as general as practical.

32.3.4   Clearly Defined Aims and Outcome Measures

Having identified gaps in the evidence base as well as a focused research questions that fits into the research agenda, the researcher must now formulate appropriate aims. These should be neither too specific nor too general and usually follow directly from a well-formulated research question (see above). An example of an SR's aims might include investigating the effectiveness of needle acupuncture vs. standard of care in breast cancer patients undergoing treatment for chemotherapy-induced nausea and vomiting.

The main outcome measure for a SR normally has to be orientated on the endpoints used in the primary studies. Often, it is advisable to use the outcome measure that is included in the majority of the trials. Whenever there is a choice, it might be wise to take an outcome measure that is adequately validated and generally accepted.

32.3.5   Choosing the Right Search Terms and Databases

The choice of the search terms for an SR will often automatically follow from the preceding steps of the SR development. For example, if a researcher conducts a SR on the effectiveness of reflexology, all relevant derivations of this word should be used such as reflex/zone therapy, reflexotherapy, foot massage, fussreflexzonen massage (the foreign language term); whereas irrelevant ones such as neuroreflexology should be avoided. A similar procedure needs to be followed regarding the specific condition, which the SR might be focused on. It is advisable to rely on the expertise of an information specialist or librarian at this stage, as the literature searches for a SR can be complex. A poor quality search runs the risk of missing relevant primary studies which, in turn, would invalidate the SR.

Choosing the right databases is equally important (Table 32.1). The choice obviously depends on the type of papers that are being searched for. If systematic reviewer needs

Table 32.1 The most commonly used databases in healthcare systematic reviews.

Name of the database
1. AMED (Allied and Complementary Medicine Database)
2. CINAHL (Cumulative Index to Nursing and Allied Health Literature)
3. Cochrane Library
4. EMBASE (Excerpta Medica Database)
5. ISI Web of Knowledge
6. ISI Web of Science
7. PEDro (Physiotherapy Evidence Database)
8. PsycINFO
9. PUBMED/MEDLINE
10. REHABDATA
11. Scopus

to include conference proceedings, ISI Web of Knowledge can be inspected, and if the subject area demands it, foreign language databases must be searched. For instance, if the topic of the SR is traditional Chinese medicine, it would seem essential that Asian databases are included; and in such a case, it is mandatory that the research team includes experts with relevant language skills.

32.3.6 Defining Eligibility Criteria

The definition of exclusion and inclusion criteria also follows from a well-focused research question. Often it is advisable to restrict them to only the highest-quality, double-blind, placebo-controlled RCTs. In other instances, it might be opportune to also include investigations with other study designs. The eligibility criteria determine many other details, for instance:

- The time span from which primary articles are drawn
- The languages in which they are published
- The exact nature of the experimental treatment
- The type of control treatments
- The exact nature of the medical condition(s)
- The type of patients included
- The type of concomitant therapies allowed

Once the eligibility criteria have been determined, it is essential that they are outlined in the final report of the SR in sufficient detail to render the exercise reproducible by the third parties.

32.3.7 Running the Searches

Now the research team must decide which interface(s) to use for running the searches. Each interface/database has its own strengths and weaknesses, which should be kept in mind. For instance, OVID and EBSCO are two examples of commonly used databases. EBSCO is limited as to the number of studies that can be exported at once. The strength of both OVID and EBSCO is their compatibility with ENDNOTE – the standard software tool for publishing and managing bibliographies, citations, and references on Windows and Macintosh platforms.

Running the searches would be straightforward if all interface/databases used the same truncations, wildcards, and proximity tools. However, this is often not the case, and systematic reviewers have to adapt and modify their search terms to fit in a particular interface/database. For instance, if a reviewer searches MEDLINE (via PUBMED) for the evidence regarding osteopathy, one should use wildcard symbol asterisk (*) in the term osteopath* to retrieve all the relevant citations on the topic. Adjacency/proximity operators allow a reviewer to search for two or more words in relation to one another. An example from EBSCO would be randomized "N4," which finds all the phrases related to RCTs either single blind, double blind, placebo controlled, multicenter, prospective, or feasibility, etc.

Frequently, researchers employ manual searches of reference lists of included studies to identify further relevant papers. Some SRs also include unpublished studies or the "gray literature," which is all material not listed in electronic databases but produced by government, academics, business, and/or industry.[5] All of these methods are designed to improve the chances of locating relevant studies.

32.3.8   Managing Abstracts and Coding

Once the searches are complete, one has to identify a sizable amount of hits, i.e. articles that might or might not fit the inclusion criteria and that are usually available as abstracts of the original papers. Usually hundreds, if not thousands, of abstracts are found, and it is now necessary to deal with them efficiently and transparently.

Managing abstracts involves reading and screening and is often the lengthiest part of the SR process. At this stage, the reviewers have to decide which of the retrieved articles *might* fit the eligibility criteria and which ones are *definitely* to be discarded. An effective way of doing this is through a coding system. It effectively means denoting or rating each and every abstract according to predefined criteria. Screening abstracts is usually done by two or three reviewers who work independently of one another to minimize the potential bias.

32.3.9   Retrieving Full-Text Versions

Retrieving full-text versions of all relevant papers inevitably requires access to publishers' resources and libraries. Often researchers rely on inter-library loans, or contact authors of the published studies directly requesting full-text version of an article; in addition, they can use the Google Search Engine. After retrieval of hard copies of the papers, the researchers have to read these articles in full to decide on the inclusion or exclusion of each article. This task is usually done by two independent reviewers. Discrepant opinions can later be settled through discussing the fine details and, if necessary, through arbitration by a third party.

32.3.10   Extracting Data

Data extraction is best done according to predefined criteria using a form that is usually designed already at the protocol stage, when the team decides which data/variables need extracting from the primary studies according to the research question(s) posed. For instance, if the effectiveness of an intervention is to be determined, the reviewers ought to extract as much details regarding its duration, frequency, or intensity as possible. It is equally important to use uniform terminology in this phase. For instance, it must be clear whether between-group or within-group differences were reported in RCTs, mean or median or percentage values were provided, or what kind of diagnostic tools were used to ascertain the presence or absence of a medical condition.

Different types of software are often used during the extraction phase. Some researchers use tables in Microsoft Word, others prefer Microsoft Access. All the information collected and processed ought to be checked, preferably by more than a one reviewer.

---

5 See www.greylit.org.

32.3.11 Critically Appraising the Quality of the Primary Data

Several tools for assessing the methodological quality and risk of bias of the primary studies are available. By and large, the choice depends on the type of primary data. For instance, adequate sequence generation, allocation concealment, patient/assessor blinding, addressing of incomplete data, selective outcome reporting alongside other sources of biases are relevant for RCTs but not necessarily for other study designs (Posadzki and Ernst 2011a,b). In case-control studies, adequate definition and representativeness of cases, selection and definition of controls; comparability of cases and controls on the basis of design and analysis; ascertainment of exposure or nonresponse rate are more important (Wang et al. 2013). Researchers sometimes use custom-made quality assessment tools (Posadzki et al. 2012), where no explicit criteria for methodological assessment exist (Sanderson et al. 2007).

Guidelines for reporting RCTs exist (e.g. CONSORT). Authors evaluating the quality of data/reporting might use such tools for critically assessing RCT quality.

32.3.12 Analyzing the Data

Quantitative data analysis (or meta-analysis) provides quantitative estimates e.g. odds ratio (OR), relative risk (RR), risk difference (RD), hazard ratio (HR), mean difference (MD), or standardized mean difference (SMD) with confidence intervals (CIs) or statistical significance (*P* value). Numbers can be pooled using fixed-effect or random-effect model. Type of data (i.e. dichotomous vs. continuous) and study design (i.e. cross-over vs. open label) play a key role in planning a quantitative data analysis.

When reviewers analyze quantitative data narratively, their job is to be as explicit as possible in describing and evaluating all important details of the primary studies, e.g. patient or population, intervention, exposure, or maneuver, comparison group, and outcomes. Sub-group analyses can often provide meaningful insights. These can involve, for instance, calculating the total number of patients who experienced benefit/improvement vs. the number of patients who did not; comparing the findings of high-quality versus low-quality studies; short-term vs. long-term improvements; and/or geographical location/cultural setting.

32.3.13 Discussing the Findings

The discussion section should put the results of the SR in a wider context. This might include a discussion of previously published work related to the same or a similar research question, a review of research on the mechanism of action of the intervention in question, a discussion of alternative treatments for the condition in question, a consideration of possible biases affecting the conclusion, a review of the limitations of the current SR, suggestions for further research, etc. In general, structured discussion sections "as a way of imposing discipline and banishing speculation" are preferred to aimless ramblings (Skelton and Edwards 2000).

32.3.14 Drawing Meaningful Conclusions

Theoretically, SRs should be objective and draw conclusions that are transparently based on the data available. Sadly, this is not always the case, and occasionally a considerable degree of confusion can emerge from two or more SRs of a similar topic. For example, two different research teams have recently published two SRs of the effectiveness of spinal manipulation for migraine, using similar eligibility criteria. Yet, they arrived at very different conclusions. One SR concluded that "current evidence does not support the use of spinal manipulations for the treatment for migraine headaches" (Posadzki and Ernst 2011a,b), whereas the other stated that "(…) spinal manipulative therapy might be equally efficient as propranolol and topiramate in the prophylactic management of migraine" (Chaibi et al. 2011). Such differences can usually be explained by the fact that the research questions posed were marginally different, which, as we have seen in the preceding discussions, determines much of the methodology of the SR. Faced with discrepant findings from two otherwise identical SRs, one can also ask which team of authors might have had a conflict of interest that, in turn, could have led to bias and a false positive or false negative overall result. In other words, SRs are designed to generate findings, which are as objective and free of bias as possible. However, this is not to say that reviewers are always sufficiently impartial and rigorous to guarantee a top-quality product.

32.3.15 Drafting the Research Paper

Drafting a research paper is frequently the pre-final stage of SR process. When writing such a paper, the researchers must ensure that the SR's format is tailored to the audience/readership; and that typically varies from one journal to another. The report must provide sufficient detail of all aspects of the SR. One should remember that the article should accurately describe the process of the SR in such a way that the reader can replicate the project without problems.

32.3.16 Assessing One's Own Performance

Each and every SR has its strengths and weaknesses. Nobody is in a better position to know them than the authors themselves. It is therefore not just good form but an essential step toward transparency to openly discuss the limitations of the SR. All research projects are merely human endeavors; they are thus never perfect. The frank disclosure of weaknesses and candid criticism of one's own work is thus far from shameful; on the contrary, it increases transparency and enables progress.

32.3.17 Publishing

The authors' conflicts of interest have the potential to introduce bias in the SR and thus distort the overall findings. For instance, it has been shown that studies whose authors have links to the food industry are five times more likely to report no association between consuming sugary drinks and weight gain than studies whose authors have no such conflicts of interest (Bes-Rastrollo et al. 2013). Bias may also arise from nonfinancial conflicts of interest such as strongly held beliefs, personal relationships, and desire for career advancement (Viswanathan et al. 2013). Therefore, in the interest of transparency, authors should always address all potential conflicts of interests and sources of funding.

32.4 Weaknesses of SRs

SRs are not without their weaknesses. Five are worth noting:

1. They collate heterogeneous primary studies that were not meant to be collated together.
2. They often fail to provide the specific answers clinicians need.
3. They are often based on poor primary research.
4. They often exclude relevant material published in languages other than English.
5. They get outdated as soon as new data become available.

32.5 Summary and Conclusions

SRs are critical evaluations of the totality of the evidence on a defined subject. They minimize random and selection biases, and represent the most reliable external evidence in health care. They are a valuable source of information for clinicians, patients, and decision makers looking after the health of individual patients or societies. SRs usually provide the most consistent information available but they are not fool-proof and do have limitations. Scientific rigor and integrity during all stages of the review process is the key to producing a high-quality SR.

References

Bes-Rastrollo, M., Schulze, M.B., Ruiz-Canela, M. et al. (2013). Financial conflicts of interest and reporting bias regarding the association between sugar-sweetened beverages and weight gain: a systematic review of systematic reviews. *PLoS Med.* 10 (12).

Chaibi, A., Tuchin, P.J., and Russell, M.B. (2011). Manual therapies for migraine: a systematic review. *J. Headache Pain* 12 (2): 127–133.

Ernst, E. (2009). Re: chiropractic for otitis? *Int. J. Clin. Pract.* 63 (9): 1393.

Moher, D., Liberati, A., Tetzlaff, J. et al. (2009). Preferred reporting items for systematic reviews and meta-analyses: the PRISMA statement. *BMJ* 339: b2535.

Posadzki, P. and Ernst, E. (2011a). Spinal manipulations for the treatment of migraine: a systematic review of randomized clinical trials. *Cephalalgia* 31 (8): 964–970.

Posadzki, P. and Ernst, E. (2011b). Yoga for asthma? A systematic review of randomized clinical trials. *J. Asthma* 48 (6): 632–639.

Posadzki, P., Alotaibi, A., and Ernst, E. (2012). Prevalence of use of complementary and alternative medicine (CAM) by physicians in the UK: a systematic review of surveys. *Clin. Med.* 12 (6): 505–512.

Sackett, D.L., Rosenberg, W.M.C., Gray, J.A.M. et al. (1996). Evidence based medicine: what it is and what it isn't – It's about integrating individual clinical expertise and the best external evidence. *Br. Med. J.* 312 (7023): 71–72.

Sanderson, S., Tatt, I.D., and Higgins, J.P. (2007). Tools for assessing quality and susceptibility to bias in observational studies in epidemiology: a systematic review and annotated bibliography. *Int. J. Epidemiol.* 36 (3): 666–676.

Skelton, J.R. and Edwards, S.J. (2000). The function of the discussion section in academic medical writing. *BMJ* 320 (7244): 1269–1270.

Tsay, M.Y. and Yang, Y.H. (2005). Bibliometric analysis of the literature of randomized controlled trials. *J. Med. Libr. Assoc.* 93 (4): 450–458.

Viswanathan, M., Carey, T.S., Belinson, S.E. et al. (2013). *Identifying and Managing Nonfinancial Conflicts of Interest for Systematic Reviews*. Methods Research Report. Rockville, MD: Agency for Healthcare Research and Quality.

Wang, Y., Yang, S., Song, F. et al. (2013). Hepatitis B virus status and the risk of pancreatic cancer: a meta-analysis. *Eur. J. Cancer Prev.* 22 (4): 328–334.

33

Clinical Management Guidelines

Vijay M. Ravindra, Walavan Sivakumar, Kristin L. Kraus, Jay K. Riva-Cambrin and John R.W. Kestle

Division of Pediatric Neurosurgery, Primary Children's Hospital, University of Utah, Salt Lake City, UT, USA

33.1 Introduction

Clinical healthcare guidelines are of great interest to practitioners, national organizations, professional societies, government entities, and researchers. The Institute of Medicine defines clinical guidelines as statements systematically developed to assist practitioners in decision making for specific clinical scenarios (Field and Lohr 1990). Guidelines can aid and instruct practitioners about diagnostic or screening studies, whether or not to provide medical or surgical treatment, and, if so, how they are to be delivered, and how long patients should stay in the hospital (Woolf et al. 1999). Clinical guidelines exist across every field of medicine (Woolf et al. 1999). In addition, guidelines have played an important role in health policy formation (Field and Lohr 1990; Browman et al. 2003) and have evolved to cover multiple topics, including research protocols, health promotion, and screening and diagnosis (Grimshaw and Russell 1993; Davis and Taylor-Vaisey 1997; Grol 2001). Clinical guidelines must be scrutinized carefully because the potential benefits are only as good as the quality of the guidelines themselves. The quality of guidelines can be extremely variable, and some fall short of basic standards (Shaneyfelt et al. 1999; Grilli et al. 2000; Burgers et al. 2004).

33.2 How Is a Clinical Guideline Created?

Clinical practice guidelines, or clinical practice parameters, are created to aid in decision-making. Specialties may organize their guidelines differently, but overall, the development, structure, and practice provisions that guidelines provide are uniform. The National Guideline Clearinghouse (NGC) is an initiative of the Agency for Healthcare Research and Quality (AHRQ) whose mission is to provide access to information on clinical practice guidelines and to encourage their creation, dissemination, and use in daily practice.[1] Development of a clinical guideline requires certain standards to

1 http://www.guideline.gov/about/index.aspx.

A Guide to the Scientific Career: Virtues, Communication, Research, and Academic Writing, First Edition.
Edited by Mohammadali M. Shoja, Anastasia Arynchyna, Marios Loukas, Anthony V. D'Antoni, Sandra M. Buerger, Marion Karl and R. Shane Tubbs.
© 2020 John Wiley & Sons, Inc. Published 2020 by John Wiley & Sons, Inc.

be met. It is important that readers of the guideline understand the process by which the guideline was created, how it was funded, and whether the experts involved in its preparation have any conflicts of interest. Authors of guidelines, typically panels of experts in the subject, may have conflicts of interest related to the subject matter, which should be clearly disclosed.

In 2011, the Institute of Medicine outlined the standards for developing trustworthy clinical practice guidelines. Guideline development groups should be composed of a variety of experts in the field and should be multidisciplinary and balanced to ensure the information delivered is uniform, complete, and reflects the most recent and pertinent evidence (Institute of Medicine 2011).

Typically, the professional organization representing a specialty will present guidelines prepared by a committee of experts, although in some cases, a national medical board or governing body may develop guidelines. Depending on which particular professional society or group prepares a set of guidelines, there may be different standards and methods for their creation. In neurological surgery, for example, the *Guideline Development Methodology*, a product of the American Association of Neurological Surgeons (AANS), Congress of Neurological Surgeons (CNS), and the AANS/CNS Joint Guideline Committee, is used as a specialty-wide guide in developing clinical guidelines.[2]

Guidelines are created in a systematic manner to ensure that they contain the most up-to-date and appropriate information for the clinical question at hand. The initial step in creating a guideline is identification of a clinical question that needs to be addressed. These may include topics of controversy, highly debated topics, and new issues that develop within a field. Furthermore, there must be a demand for a guideline to help practitioners and organizations address their concerns about the clinical question. The clinical question must be clear and must have enough literature to support recommendations made in the guideline.

After a clear clinical question has been formulated by the guidelines committee, the next important step in guideline creation is an extensive literature search on the subject. According to *Guideline Development Methodology*, the literature search should be extensive; at the very minimum, the search should encompass the available English-language literature for a length of time appropriate to the subject using the computerized database of the National Library of Medicine. The search should be broad, and the search terms should contain and reflect the clinical question as much as possible. Relevant abstracts are reviewed, and articles are selected for further review based on this process.

Papers should be organized by study type (therapy, diagnosis, clinical assessment). Each of the study types should then be analyzed using specific predetermined criteria. Therapy literature can be reported by multiple study designs. The strongest of these, if well designed and executed, is the randomized controlled trial. The results of randomized controlled trials and other less robust reports are evaluated to determine what consensus has been achieved regarding successful protocols and outcomes, and this information is used to formulate the proposed clinical practice guideline.

The developers should use systematic reviews to evaluate the clinical practice guideline and its intersection with clinical medicine and effectiveness. The next

---

2 https://www.cns.org/guidelines/guideline-procedures-policies/guideline-development-methodology.

Figure 33.1 Algorithm outlining the creation of clinical practice guidelines.

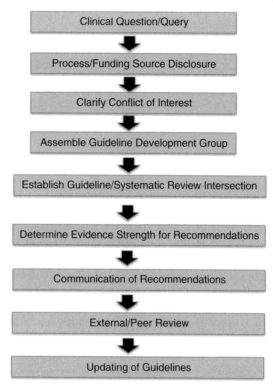

important step in creating a guideline is to establish foundations for the evidence presented and an explanation of the reasons underlying the recommendation. The strength of recommendations and levels of evidence play a large role in determining the legitimacy of clinical practice guidelines. Peer-reviewed, controlled trials, which have been previously published, level A recommendation, provide the best evidence. Results from peer-reviewed, epidemiological studies, level B recommendation, are the next best level of evidence. Case reports and small case series, level C recommendations, are the lowest recommendations that can be given (Embree 2000).

Succinct articulation of recommendations stating what the action is and under what circumstances it should be used should follow the grading or rating of the evidence available. Possibly the most important component of guideline preparation and creation is the external review process. External reviewers should be relevant stakeholders, other experts in the field, pertinent organizations or agencies, and others that may be affected by guideline implementation. Finally, updating of guidelines to reflect changes in the body of literature and use of new technology should occur on a regular basis (Figure 33.1).

33.3   Benefits and Limitations of Clinical Management Guidelines

Clinical practice guidelines are based on the most pertinent and strongest clinical evidence available. These protocols offer some assurance to the practitioner that they are following the best practices when treating a patient. Yet despite the value of these guidelines, the medical provider must consider all of the relevant data in the case at hand when determining the best course of action for the patient. Woolf et al. (1999) discussed the topic of clinical practice guidelines and the possible benefits, limitations, harms, and pitfalls that come with their widespread use.

33.3.1 Benefits

33.3.1.1 Benefit to Patients

The greatest potential benefit for patients from the use and adherence to clinical practice guidelines is improved healthcare outcomes. This hinges on the notion and practice that guidelines are created from the most up-to-date evidence-based information available. The implementation of practices that have proven benefit and, more importantly, the elimination of ineffective practices can significantly reduce morbidity and mortality and improve overall survival (Woolf et al. 1999). Guidelines also help to standardize the level of care that is received by patients, regardless of practitioner experience or location of care. Clinical guidelines can help inform patients and the general public of benefits and harms of certain treatment methods, procedures, or treatment algorithms (Entwistle et al. 1998), which empowers patients to make more informed decisions and requires that physicians keep up to date with current guidelines and literature (Woolf et al. 1999). Guidelines may call attention to under-recognized health problems, clinical services, and preventive interventions that may be out of reach of certain populations (Woolf et al. 1999).

33.3.1.2 Benefit to Healthcare Professionals

Clinical guidelines are created to improve the quality of decision-making. They are built to offer explicit recommendations, to challenge the practice patterns of physicians, to improve the consistency of care, and to provide authoritative recommendations that reassure practitioners about treatment policies and choices (Woolf et al. 1999). Guidelines that are constructed carefully should incorporate up-to-date scientific evidence, document the quality of the supporting data, and make succinct recommendations based on this information. They should alert clinicians to interventions unsupported by up-to-date literature, reinforce critical decision-making, and call to attention harmful practices (Woolf et al. 1999). Clinicians familiar with and incorporating clinical practice guideline recommendations into their practice may also benefit financially because external evaluation sources are more frequently using guideline adherence to create the metrics by which physicians are measured. Furthermore, clinical guidelines can also help produce personal progress and growth. They encourage clinicians to become familiar with new techniques as they are developed and their outcomes corroborated. The methods of guideline development that emphasize systematic reviews also focus attention on key research questions that must be answered to establish the effectiveness of an intervention, thus affording medical researchers the benefit of extrapolating on these areas to improve quality (Cook et al. 1997).

33.3.1.3 Benefit to Healthcare Systems

Clinical guidelines can also inspire and lead to change in the healthcare environment. Quality improvement is a topic that has gained widespread publicity over the recent years and has helped incite change within healthcare systems throughout the world. Development and implementation of clinical guidelines can inspire quality improvement endeavors, and healthcare systems have found that clinical guidelines may be effective in improving efficiency and optimizing value for money (Shapiro et al. 1993). Implementation of certain guidelines can streamline hospital procedures and even become a point of advertisement and improved public image. In fact, economic motives behind the implementation and creation of guidelines have led to a recent increase in their utilization (Woolf et al. 1999).

33.3.2 Limitations

The limitations of clinical guidelines lie in their own creation, which depends on the current best level of evidence on a certain topic. The problem inherently lies in the fact that clinical practice guidelines necessitate timely evidence presented in the literature. Often, problems or clinical questions may arise before useful evidence on the topic becomes available. When considering the construction of a guideline, one must first look at the quality of evidence, if any, that is available for that topic. Woolf et al. (1999) described three main reasons that guideline developers might not make the best decisions in certain clinical scenarios. The first is the lack of scientific evidence for many of the particular problems that are faced in a clinical setting. Guideline development groups often lack the time to thoroughly scrutinize every piece of literature available for a specific topic. The second reason is the inability to ignore opinions and clinical experience regardless of the evidence. Recommendations ultimately are at the mercy of the developer, which can make them subjective and open to interpretation and bias. The third reason is that the individual patient is not always at the forefront of thought and conversation during guideline development. Thinking of the greater good rather than individual needs to improve health systems, conserve costs, and improve overall healthcare can misplace the recommendations of the guideline developer (Woolf et al. 1999).

33.3.2.1 Harm to Patients

Flawed clinical guidelines can cause obvious harm to patients. Poorly constructed guidelines that do not factor in the best scientific evidence available, or that substitute anecdotal information for clinical evidence, can lead to suboptimal and even harmful practices (Woolf et al. 1999). Guidelines that are inflexible and rigid and leave little room for interpretation to the practitioner can also cause harm to patients. Blanket recommendations rather than a list of choices or options can ignore patient preferences and disrupt the autonomy of the individual practitioner (Woolf 1997). Clinical guidelines may also adversely affect public policy for patients. Recommendations against an intervention may lead to insurance withholding coverage for a necessary procedure. On the other hand, costly recommended interventions may drain money from other healthcare maintenance or preventive testing that is necessary and beneficial.

33.3.2.2 Harm to Healthcare Professionals

Guidelines that are poorly constructed may encourage clinicians to adopt or to continue to use ineffective, harmful, or wasteful interventions. Outdated guidelines and recommendations may perpetuate outdated practices and technology (Woolf et al. 1999). In addition, conflicting guidelines from different professional organizations may be difficult to sort through and might be frustrating to follow (Feder 1994). As mentioned previously, clinical guidelines may be used a benchmark against which practitioners are measured; however, poorly constructed guidelines may be used against practitioners by using any unclear language or case-specific variations to support litigation against them, as will be discussed below.

33.3.2.3 Potential Harm to Healthcare Systems

Healthcare systems can be harmed if guidelines suggest escalation of costly testing or treatment practices, compromise operating efficiency, or waste limited resources (Woolf et al. 1999). Guidelines may also harm the progress of medicine and medical investigation if further research is inappropriately discouraged (Woolf et al. 1999).

33.4 Medicolegal Implications of Clinical Guidelines

Although extensive evaluations of the effects of guidelines have shown that clinical practice guidelines do indeed improve the quality of patient care (Grimshaw and Russell 1993), the totality of the effect is often complex and difficult to evaluate. Patients, doctors, administrators, payers, and members of the medicolegal system all define quality and subsequently the effectiveness of guidelines differently (Woolf et al. 1999). These details are at the essence of how guidelines are used in medical litigation.

Assessing whether or not the physician has been clinically negligent is based largely on whether the care provided has fallen below the standard of care (Samanta et al. 2003). Although clinical guidelines are systematically developed to facilitate best practices, they do not necessarily equate to standard of care. The standard of care, the criterion used in medicolegal situations, may be well stated in clinical practice guidelines or may be nebulous. Thus, guidelines largely play a supplementary role to expert medical opinion (Hurwitz 1999). Despite the recent proliferation of clinical guidelines, there is no guarantee of consistent use in practice (Feder et al. 1999). In the United States, for example, the role of clinical guidelines in medical malpractice varies significantly (Samanta et al. 2003). In a study reviewing 259 claims of professional liability, only 17 (7%) cases utilized clinical guidelines, primarily by the plaintiffs (Hyams et al. 1995). Hyams and colleagues concluded that clinical guidelines had only a modest impact on medical malpractice cases in 1995. A study from the Netherlands found that guidelines were followed in only 55% of clinical decisions (Gill 2001). Explanations for this include provider disagreement with the listed guidelines, a lack of clear understanding of the guidelines, resource limitations preventing the utilization of these guidelines, and ignorance of the existence of the guidelines (Grimshaw and Russell 1993; Lugtenberg et al. 2009).

There has been an increase in utilization of clinical practice guidelines (Woolf et al. 1999) and the impact on physician practices has been variable; however, given the current healthcare climate in the United States, which places a greater emphasis on evidence-based medicine and quality of healthcare delivery, the use of guidelines will be tied to reimbursement and patient satisfaction. Medicolegal implications for guideline adherence and best practice standards are not far down the road.

33.5   Conclusions

The overall goal for the creation and implementation of clinical guidelines is to improve the quality of care that is delivered and received by patients throughout the healthcare system (Grimshaw and Russell 1993; Davies et al. 1994). An improvement in the quality of care is difficult to measure, however, because quality improvement is measured differently across disciplines.

Clinical management guidelines can provide information for practitioners to aid in clinical decision-making. The use of guidelines must be carefully assessed, however, as there are limitations in the formation and writing of management guidelines. Although guidelines are created to improve patient care, there are both advantages and disadvantages to the use of guidelines. For the patient, the benefits and limitations mostly affect healthcare outcome, while for clinicians the pros and cons of guideline adherence can extend into the medicolegal evaluation of a case or practice. For all of these reasons, the structure and content of guidelines should be continually updated, and practitioners utilizing the information provided should be aware of the methodology behind guideline creation and appropriately assess the suitability and pertinence of the guidelines for their own clinical practice.

The overall attitude about clinical guidelines varies from group to group. Guidelines that are created by government or paying enterprises may be resented by clinicians and patients, while specialists might think that those created by generalists are inadequate. Furthermore, "inflexible" rules – whether it is through clinical practice guidelines or other tools of healthcare provision – developed to govern a field that is nonuniform and is faced with unique challenges every day can be difficult to implement (Woolf et al. 1999). Review of medical malpractice cases also hint that clinical guidelines do not currently play a large role in the final ruling, whether they are used by practitioners to guide patient care or not; however, with the significant increase in the number of clinical guidelines being released and the general trend in medicine toward standardizing treatment protocols, it is anticipated that clinical guidelines will begin to play a larger role in the administration of patient care and the medical and legal evaluation of clinical outcomes.

References

Browman, G.P., Snider, A., and Ellis, P. (2003). Negotiating for change. The healthcare manager as catalyst for evidence-based practice: changing the healthcare environment and sharing experience. *HealthcarePapers* 3 (3): 10–22. Available at: http://www.ncbi .nlm.nih.gov/pubmed/12811083. [Accessed August 25, 2016].

Burgers, J.S. et al. (2004). International assessment of the quality of clinical practice guidelines in oncology using the appraisal of guidelines and research and evaluation instrument. *Journal of Clinical Oncology: Official Journal of the American Society of Clinical Oncology* 22 (10): 2000–2007. Available at: http://www.ncbi.nlm.nih.gov/ pubmed/15143093. [Accessed August 25, 2016].

Cook, D.J., Mulrow, C.D., and Haynes, R.B. (1997). Systematic reviews: synthesis of best evidence for clinical decisions. *Annals of Internal Medicine* 126 (5): 376–380. Available at: http://www.ncbi.nlm.nih.gov/pubmed/9054282. [Accessed August 25, 2016].

Davies, J. et al. (1994). Implementing clinical practice guidelines: can guidelines be used to improve clinical practice? *Effective Health Care Bulletin* 8 (1): 1–12.

Davis, D.A. and Taylor-Vaisey, A. (1997). Translating guidelines into practice. A systematic review of theoretic concepts, practical experience and research evidence in the adoption of clinical practice guidelines. *CMAJ: Canadian Medical Association Journal = Journal de l'Association Medicale Canadienne* 157 (4): 408–416. Available at: http://www.pubmedcentral.nih.gov/articlerender.fcgi?artid=1227916&tool=pmcentrez&rendertype=abstract. [Accessed August 25, 2016].

Embree, J. (2000). Writing clinical guidelines with evidence-based medicine. *The Canadian Journal of Infectious Diseases = Journal Canadien des Maladies Infectieuses* 11 (6): 289–290. Available at: http://www.pubmedcentral.nih.gov/articlerender.fcgi?artid=2094789&tool=pmcentrez&rendertype=abstract. [Accessed August 25, 2016].

Entwistle, V.A. et al. (1998). Developing information materials to present the findings of technology assessments to consumers. The experience of the NHS Centre for Reviews and Dissemination. *International Journal of Technology Assessment in Health Care* 14 (1): 47–70. Available at: http://www.ncbi.nlm.nih.gov/pubmed/9509795. [Accessed August 25, 2016].

Feder, G. (1994). Management of mild hypertension: which guidelenes to follow? *BMJ: British Medical Journal* 308 (6926): 470. Available at: http://www.ncbi.nlm.nih.gov/pmc/articles/PMC2539532. [Accessed August 25, 2016].

Feder, G. et al. (1999). Clinical guidelines: using clinical guidelines. *BMJ (Clinical Research Ed.)* 318 (7185): 728–730. Available at: http://www.pubmedcentral.nih.gov/articlerender.fcgi?artid=1115154&tool=pmcentrez&rendertype=abstract. [Accessed August 25, 2016].

Field, M. and Lohr, K. (eds.) (1990). *Clinical Practice Guidelines: Directions for a New Program*. Washington, DC: National Academy Press.

Gill, G. (2001). Going Dutch? How to make clinical guidelines work: an innovative report from Holland. *Clinical Medicine (London, England)* 1 (4): 307–308. Available at: http://www.ncbi.nlm.nih.gov/pubmed/11525579. [Accessed August 25, 2016].

Grilli, R. et al. (2000). Practice guidelines developed by specialty societies: the need for a critical appraisal. *Lancet (London, England)* 355 (9198): 103–106. Available at: http://www.ncbi.nlm.nih.gov/pubmed/10675167. [Accessed August 25, 2016].

Grimshaw, J.M. and Russell, I.T. (1993). Effect of clinical guidelines on medical practice: a systematic review of rigorous evaluations. *Lancet (London, England)* 342 (8883): 1317–1322. Available at: http://www.ncbi.nlm.nih.gov/pubmed/7901634. [Accessed August 25, 2016].

Grol, R. (2001). Successes and failures in the implementation of evidence-based guidelines for clinical practice. *Medical Care* 39 (8 Suppl 2): II46–II54. Available at: http://www.ncbi.nlm.nih.gov/pubmed/11583121. [Accessed August 22, 2016].

Hurwitz, B. (1999). Legal and political considerations of clinical practice guidelines. *BMJ (Clinical Research Ed.)* 318 (7184): 661–664. Available at: http://www.pubmedcentral.nih.gov/articlerender.fcgi?artid=1115098&tool=pmcentrez&rendertype=abstract. [Accessed August 25, 2016].

Hyams, A.L. et al. (1995). Practice guidelines and malpractice litigation: a two-way street. *Annals of Internal Medicine* 122 (6): 450–455. Available at: http://www.ncbi.nlm.nih.gov/pubmed/7856994. [Accessed August 25, 2016].

Institute of Medicine (2011). *Clinical Practice Guidelines We Can Trust: Standards for Developing Trustworthy Clinical Practice Guidelines (CPGs).* Washington, DC: National Academies of Sciences, Engineering, and Medicine.

Lugtenberg, M. et al. (2009). Why don't physicians adhere to guideline recommendations in practice? An analysis of barriers among Dutch general practitioners. *Implementation Science: IS* 4: 54. Available at: http://www.pubmedcentral.nih.gov/articlerender.fcgi?artid=2734568&tool=pmcentrez&rendertype=abstract. [Accessed August 25, 2016].

Samanta, A., Samanta, J., and Gunn, M. (2003). Legal considerations of clinical guidelines: will NICE make a difference? *Journal of the Royal Society of Medicine* 96 (3): 133–138. Available at: http://www.pubmedcentral.nih.gov/articlerender.fcgi?artid=539423&tool=pmcentrez&rendertype=abstract. [Accessed August 25, 2016].

Shaneyfelt, T.M., Mayo-Smith, M.F., and Rothwangl, J. (1999). Are guidelines following guidelines? The methodological quality of clinical practice guidelines in the peer-reviewed medical literature. *JAMA* 281 (20): 1900–1905. Available at: http://www.ncbi.nlm.nih.gov/pubmed/10349893. [Accessed August 25, 2016].

Shapiro, D.W., Lasker, R.D., Bindman, A.B. et al. (1993). Containing costs while improving quality of care: the role of profiling and practice guidelines. *Annual Review of Public Health* 14: 219–241. Available at: http://www.ncbi.nlm.nih.gov/pubmed/8323588. [Accessed August 25, 2016].

Woolf, S.H. (1997). Shared decision-making: the case for letting patients decide which choice is best. *The Journal of Family Practice* 45 (3): 205–208. Available at: http://www.ncbi.nlm.nih.gov/pubmed/9299998. [Accessed August 25, 2016].

Woolf, S.H. et al. (1999). Clinical guidelines: potential benefits, limitations, and harms of clinical guidelines. *BMJ (Clinical Research Ed.)* 318 (7182): 527–530. Available at: http://www.pubmedcentral.nih.gov/articlerender.fcgi?artid=1114973&tool=pmcentrez&rendertype=abstract. [Accessed August 25, 2016].

34

Why Is the History of Medicine and Biology Important?

Paul S. Agutter

Theoretical Medicine and Biology Group, Glossop, Derbyshire, United Kingdom

34.1   The Value of Knowing Our History

We're all busy—as clinicians, our priority is our patients; as researchers, it's our research—but we also have administrative duties, not to mention family and other responsibilities. The history of medicine and biology might be interesting, but can we afford to devote time to it when we face so many more pressing demands? If my answer weren't yes, I wouldn't be writing this chapter. I base my yes on personal experience, my own and my colleagues', but it needs to be explained to the reader.

Knowing something of the history of our field gives us three main advantages:

- It helps us to understand and respect the beliefs, methods, practices, and conventions of earlier times. Many of our ideas and techniques were born in those times, in cultures that differed (substantially or not) from our own. "Progress" in medicine and biology isn't just accretion of knowledge and technical innovation; it's a facet of general cultural change.
- We understand our discipline's past in a new and illuminating way, and we learn to perceive our cherished present-day beliefs and practices not as immutable certainties but as ephemera likely to be superseded in future generations.
- We gain a deeper appreciation of individuals who've contributed to the emergence of biology and medicine. We see how innovative and original they were; we recognize how often they were misled; and we learn that what our textbooks and teachers reveal about them is often selective, distorted, or even false. Some writings by great contributors of the past contain valuable insights or data that are overlooked in modern teaching, research, and practice.

34.2   An Illustration: Rudolf Virchow

I've stated these generalities with neither evidence nor illustration, so I'll devote the rest of the chapter to an exemplar: Rudolf Virchow (1821–1902). Biologists are likely to associate Virchow's name with the pronouncement "Every cell originates from another cell" (*omnis cellula* e *cellula*). Clinicians will associate it with "Virchow's triad," which is

A Guide to the Scientific Career: Virtues, Communication, Research, and Academic Writing, First Edition.
Edited by Mohammadali M. Shoja, Anastasia Arynchyna, Marios Loukas, Anthony V. D'Antoni, Sandra M. Buerger, Marion Karl and R. Shane Tubbs.

alleged to encapsulate the causes of deep venous thrombosis (DVT). Virchow was a prolific contributor to both medicine and biology, but as I'll reveal, those two associations are in different ways misleading.

Virchow passed his student years in Berlin during an exciting period in European history. The eighteenth century fashion for "experimental philosophy" had spread to the study of life, and even into political and religious affairs (Merz 1928; Temkin 1946). He was noted for his liberal ideology and his concern with public health, probably rooted in his government-sponsored investigation of the typhus outbreak in Upper Silesia in 1848. On the Berlin City Council, from 1859, he urged reforms in public hygiene, meat inspection, and hospital building. He was an avid social reformer and an implacable opponent of Bismarck, and exerted significant influence on German society in the late nineteenth century (Ackerknecht 1981). His conception of cellular pathology during his early career in Würtzburg became the backbone of his contribution to biology and medicine (Nuland 1988).

François-Vincent Raspail (1794–1878) had used the phrase *omnis cellula e cellula* as early as 1825 and – using the new achromatic microscope lenses – outlined what would later be called "cell theory." In 1827–1828 Raspail asserted that all animal and plant tissues consist of cells, and that disease processes arise from pathology at the cellular level (Weiner 1968). He thus anticipated Schwann and Schleiden by more than a decade and Virchow by almost three. Virchow probably acquired the principle, if not the wording, *omnis cellula e cellula* from the writings of John Goodsir (1814–1867), who first identified the cell as the "center of nutrition" and regarded Raspail highly (Turner 1868). Virchow dedicated his *Cellular Pathology* (Virchow 1858) to "John Goodsir FRS, (Curator of the surgical museum and) Professor of Anatomy in Edinburgh, the earliest and most acute observer of cell-life," but he never mentioned Raspail; Franco-German antagonism was rife during the nineteenth century. The philosopher Johann Gottfried Herder (1744–1803) had called upon his countrymen to end France's cultural domination of Europe and assert the superiority of German learning, not least in science. Virchow was later to be skeptical about Louis Pasteur's discoveries, while Pasteur (1822–1895) plagiarized experiments by Theodor Schwann (1810–1882) refuting spontaneous generation and published them as his own (cf. Harris 2002).

Prejudice and unauthorized borrowing notwithstanding, Virchow's *Cellular Pathology* – a series of lectures to clinicians, researchers, and interested laymen in Berlin during the mid-1850s – is a minor masterpiece. It contains a wealth of novel observations and incisive reasoning, and it transcends anything that Raspail, Goodsir, Matthias Jakob Schleiden (1804–1881), Schwann, or anyone else had written about cells. Despite its density of content, it's also a delight to read.

In the tenth lecture in *Cellular Pathology*, Virchow presented a drawing of a venous thrombus as he'd seen it under the microscope; he also attacked the account of venous thrombogenesis from his great predecessor Jean Cruveilhier (1791–1874), another illustration of his antagonism toward French science. He drew attention to the morphology of thrombi, and to the fact that all thrombi are anchored in valve pockets, which accounts for their appearing to arise in the middle of the vein lumen rather than on the wall (cf. Hunter 1793). The words *thrombosis* and *embolism* are Virchow's inventions (Virchow 1856), along with other medical terms now in common use such as *leukemia*. However, the crucial point here is his revised definition of *thrombus*. Prior to Virchow's writings, *thrombus* had been an ill-defined term often signifying a lump such as a superficial hematoma, but during the course of his research on thromboembolism, he found it necessary to distinguish between an intravascular coagulum and an ex vivo clot. He might have first used *thrombus* in this sense as early as 1847 (Malone and Agutter 2008). In a later publication (Virchow 1862), he spelled out the three-part distinction between a venous thrombus and an ex vivo clot:

1. The structure of the thrombus is in layers (i.e. the lines of Zahn).
2. The fibrin content is denser.
3. The population of colorless corpuscles is denser, and to a striking degree; these corpuscles were present in the blood from the beginning, and were separated from it with the fibrin.

In view of these observations by so eminent a pioneer of the study of thrombosis and embolism, it is surprising that the words *thrombus* and *clot* are treated as synonyms in many modern-day texts, and that the crucial role of valve pockets in venous thrombogenesis is widely overlooked. If clinicians and researchers could find time to read Virchow's work rather than (often misleading) accounts of it by other authors, their understanding of the nature and the etiology of venous thrombi would be enhanced.

A study of Virchow's writings would also preclude a serious misattribution: "Virchow's triad." This didn't appear in the medical literature until about 100 years after *Thrombose und Embolie* (Virchow 1846) was published, and there's nothing of the kind in Virchow's own writings (Brinkhous 1969; Brotman et al. 2004; Dickson 2004; Malone and Agutter 2008). Yet we find the following remarks – and many others of the kind – by leading authorities on DVT and thromboembolism:

- Virchow's triad of stasis, vessel injury, and hypercoagulability remains a valid explanation of the pathogenesis of thrombus formation (Peterson 1986).
- The cause of thrombosis is often unknown but is universally ascribed to part of Virchow's triad: stasis, hypercoagulability, and intimal injury (Burroughs 1999).
- The modern era of understanding the etiology of thrombosis began with the pathologist Virchow, who in the mid-1800s postulated three major causes of thrombosis: changes in the vessel wall, changes in the blood flow, and changes in the blood composition (Rosendaal 2005).

One can only suppose that Virchow, a habitually irascible man, would have taken each of these authors to task for so blatantly misrepresenting his work and attributing to him a more or less vacuous latter-day proposition. "Virchow's triad" is vague hand-waving. It masquerades as an etiological hypothesis but leads to no testable predictions (and is therefore unscientific by accepted criteria) and provides no useful guidance in respect to prophylaxis, management or therapy for thromboembolism. Instead, it obfuscates our understanding of the etiology of DVT, and, in particular, it deflects our attention from Virchow's real and profound contributions to the subject.

34.3 Conclusions

Using Virchow as exemplar, I have demonstrated that finding time to study the history of medicine and biology cannot only add to our understanding but also potentially redirect both research and practice in valuable ways. Moreover, it can prevent landmark contributions to the emergence of present-day biology and medicine from being mis-represented so that we're blinded to their true significance. I chose Virchow because he illustrates my case particularly well, but he isn't unique in this regard. The interested clinician, researcher, or student will discover comparable insights if she or he can find a few hours in an inevitably busy schedule to study salient historical contributions to her/his field of interest.

References

Ackerknecht, E.H. (1981). *Rudolf Virchow*. New York: Arno Press.

Brinkhous, K.M. (1969). The problem in perspective. In: *Thrombosis* (ed. S. Sherry), 37–48. Washington, DC: National Academy of Sciences.

Brotman, D.J., Deitcher, S.R., Lip, G.Y. et al. (2004). Virchow's triad revisited. *South. Med. J.* 97: 213–214.

Burroughs, K.E. (1999). New considerations in the diagnosis and therapy of deep vein thrombosis. *South. Med. J.* 92: 517–520.

Dickson, B.C. (2004). Venous thrombosis: on the history of Virchow's triad. *Univ. Toronto Med. J.* 81: 166–171.

Harris, H. (2002). *Things Come to Life: Spontaneous Generation Revisited*. Oxford: Oxford University Press, chapters 9–10.

Hunter, J. (1793). Observations on the inflammation of the internal coats of veins. *Trans. Soc. Improv. Med. Chir. Knowledge* 1: 18–41.

Malone, P.C. and Agutter, P.S. (2008). *The Aetiology of Deep Venous Thrombosis: A Critical, Historical and Epistemological Survey*. Dordrecht: Springer.

Merz, J.T. (1928). *A History of European Thought in the Nineteenth Century*, 3e. Edinburgh and London: Blackwell.

Nuland, S. (1988). *Doctors. The Fundamental Unit of Life, Sick Cells, Microscopes, and Rudolf Virchow*, 304–343. New York: Random House.

Peterson, C.W. (1986). Venous thrombosis: an overview. *Pharmacotherapy* 6: 12S–17S.

Rosendaal, F.R. (2005). Venous thrombosis: the role of genes, environment, and behavior. *Hematology Am. Soc. Hematol. Educ. Program* 1: 1–12.

Temkin, O. (1946). Materialism in French and German physiology of the early nineteenth century. *Bull. Hist. Med.* 20: 322–327.

Turner, W. (ed.) (1868). *Anatomical Memoirs of John Goodsir, F.R.S., Edited by W. Turner, with Memoir by H. Lonsdale*. Edinburgh: Edinburgh University Press.

Virchow, R.L.K. (1846–1856). *Thrombosis and Embolie*. In: *Klassiker der Medizin herausgegeben von Karl Sudhoff*; Leipzig, 1910; translated by Matzdorff, A.C. and Bell, W.R. (ed. J.A. Barth). Canton: Science History Publications, 1998.

Virchow, R.K.L. (1858). *Die Cellularpathologie in ihrer Begründung auf physiologische und pathologische Gewebelehre*. Berlin: A. Hirschwald.

Virchow, R.K.L. (1862). *Gessamelte Abhandlungen zur wissenschaftlichen Medizin von Rudolf Virchow*. Frankfurt am Main: Meidinger.

Weiner, D.B. (1968). *Raspail: Scientist and Reformer*. New York: Columbia University Press.

35

Historical Articles: A Methodology Guide

Anand N. Bosmia[1] and Mohammadali M. Shoja[2]

[1]Department of Psychiatry, LSU Health Sciences Center, Shreveport, LA, USA
[2]Division of General Surgery, University of Illinois at Chicago Metropolitan Group Hospitals, Chicago, IL, USA

The historical article is a project that enriches a researcher's perspective on the scientific enterprise. The odysseys of the individuals whose insights, mistakes, and collaborations with one another are intriguing to investigate, and the ancient tomes of the forefathers of science and medicine may contain pearls of wisdom that have not yet been revealed. Our objectives in this chapter are to provide general guidelines for writing historical articles, identify the categories of historical articles, and note any challenges unique to preparing specific types of historical articles within each category. We note that the list provided in this chapter regarding the types of historical articles is not exhaustive, but reflective of what we have encountered in our professional experience. Furthermore, articles that are very wide in scope likely are hybrids of the types of historical articles we note.

35.1 General Guidelines

35.1.1 Step 1: Identify the Subject

The first step is to identify the subject of your project. That subject could be a person, an event, or even an idea. Ask yourself why you are interested in the particular subject you have decided to write about. Are you curious to know more about the life of a person commemorated by a scientific eponym? Do you want to learn more about an ancient society's treatments of particular medical problems? Complementing the identification of the subject of the project is the identification of questions regarding your subject of interest that you are asking and aiming to answer. Doing so allows you to gauge the scope of the project.

35.1.2 Step 2: Identify a Mentor

The second step is to identify an individual who can serve as a mentor for your project. This step is critical for novice researchers. Of course, more than one person can function in this capacity. Identifying such a person can be done by searching the extant literature for publications related to your subject of interest, identifying individuals who

A Guide to the Scientific Career: Virtues, Communication, Research, and Academic Writing, First Edition.
Edited by Mohammadali M. Shoja, Anastasia Arynchyna, Marios Loukas, Anthony V. D'Antoni, Sandra M. Buerger, Marion Karl and R. Shane Tubbs.

have published extensively on your topic or related areas, and, if feasible, establishing a collaborative professional relationship with those individuals.

35.1.3   Step 3: List Potential Databases for References

The third step is to make a list of potential databases that you can examine for references on your topic of interest. Maintain this list for future projects so that you can determine which databases are more fruitful than others for you, with regard to finding pertinent references for your manuscript. We highly recommend that Google Books, PubMed, and the National Library of Medicine databases be researched for any project on the history of medicine. Several caveats must be mentioned regarding the navigation of those databases, covered in the following rules for navigating databases.

35.1.3.1   Rules for Navigating Databases

Rule 1: Always attempt to obtain the full text of a reference.   The abstract may omit important details regarding the article's contents and lead you to think erroneously that the article is less relevant to your project than it actually is. Simply because a link to a free copy of the full text is not available through one database does not mean that you must pay for the full text. For example, an article's full text might not be available through PubMed, but if you were to search for the same article on a computer belonging to your academic institution, the full text of the reference might be available through a database such as Ovid. Sometimes, Googling the full name of the reference in quotes will result in a hit consisting of a link to a full-text copy of the article.

Rule 2: Search for references using alternative spellings of keywords.   Terms pertinent to your subject of interest might be spelled differently among references, and thus your initial search of the databases might not be fruitful unless you alternate among the various spellings of those terms. For example, information on the Swiss anatomist Johann Conrad Brunner can be found under "Johann Konrad Brunner." Furthermore, Latinized or Anglicized versions of the names of prominent people outside of the European continent exist. For example, the medieval Arab physician Ibn Sina is also known as "Avicenna."

Rule 3: Verify the authenticity of non-peer-reviewed references.   You must always verify the authenticity of non-peer-reviewed references before citing them in the bibliography of your manuscript. For example, if you refer to a Wikipedia page, do not assume that all of the references cited on that page are authentic. Some information might have been obtained through online references that are no longer available and will not show if you click on the link to that reference on the respective Wikipedia page. Such a link is referred to as a "dead link." Although some peer-reviewed journals will accept Wikipedia pages as references listed in the bibliography of an article, the authors contend that investigating the references on which Wikipedia pages are based and citing those references upon the confirmation of their authenticity constitute better form in the world of academe.

Rule 4: Do not ignore suggestions for other references provided by databases.   When you research a particular database, be vigilant for suggestions for other references related to your subject of interest produced by that database. Do not be lackadaisical in examining additional references that the databases suggest. Such suggestions are based on your prior searches of those databases, and the references presented may provide more complete answers to the questions you are addressing in your manuscript. Those new references also can lead you to formulate more questions about your subject of interest and thereby broaden the scope of your manuscript or provide the basis for future projects. Furthermore, upon securing the full texts of the primary publications that you have identified in your initial search of the literature, examine the references cited by the primary publications because the references that informed those primary publications could contain additional information regarding your subject of interest. This practice can enable you to find more authoritative references on your subject of interest and potentially identify inconsistences within the literature about your subject of interest. Identifying such inconsistencies can prevent you from incorporating inaccurate information into your article.

35.1.4 Step 4: Find a Sample Article

The fourth step is to find a sample of the type of historical article that you are planning to write and using that sample to guide the development of an outline, which functions to guide your initial drafting of the article. If you have a mentor for your project, you can consult that mentor for this step. However, have your list of references prepared and ensure that you have read most, if not all, of your references before meeting with your mentor. This practice conveys commitment and interest in the project to your mentor, who subsequently is more likely to invest a greater amount of time and energy in guiding you. In addition, your mentor can recommend more references to you.

35.1.5 Step 5: Write the First Draft

The fifth step is to write the first draft of your manuscript according to your outline and revising the draft until you and your mentor are satisfied with the final product. While writing your article, be extremely vigilant about citing references so that you do not forget to give proper credit for a sentence or paragraph derived from one of your references and thereby risk committing plagiarism. All sentences or paragraphs that are cited verbatim from a reference must be included in quotes. It is important to distinguish which sentences or paragraphs are included in your article as written in the original reference and which have been paraphrased by you for the purpose of succinctness and narrative flow.

If you heavily rely on one or more references for information that you incorporate into your article, acknowledge those references in the abstract and/or introductory paragraph of your article. Doing so will convey to the reader your humility and gratitude for the work of your predecessors and colleagues more effectively than only consistently citing the references in the text of your article.

35.1.6 Step 6: Add Images to the Article

The sixth step is to incorporate images into your article. Images are an integral component of historical articles. Do not be apathetic toward this element of the historical article. Images can enhance the quality of your product tremendously because of their visual appeal. Just as graphs and diagrams complement discussions of data from clinical trials, images complement discussions of historical figures in science and texts of antiquity. We recommend the National Library of Medicine for students to peruse for images relevant to their research. The National Library of Medicine has a large collection of images, *Images from the History of Medicine. Wellcome Trust Collection* is another invaluable resource.

35.1.6.1 Copyrights

The issue of copyright is important to note. You must be sure that you have obtained all the required permissions from the copyright holder before incorporating an image into your article and submitting your manuscript to a journal. Some of the images from the National Library of Medicine are in the public domain. If an image is listed as being in the public domain, that image does not have a particular copyright holder and can be used freely. However, even for images in the public domain, mention in the figure legend the source from which you obtained that particular image and, if available, the creator of that image (e.g. the artist who painted a portrait of a historical figure, the photographer who took a picture of a historical figure, etc.). Such practice is a professional courtesy that gives the proper credit due to artists, photographers, librarians, and archivists who collaborate with scholars for the production of historical articles. Furthermore, do not hesitate to contact a librarian or archivist for assistance in clarifying the copyright status of an image or obtaining permission to utilize the image from the database over which he or she has stewardship.

35.1.7 Step 7: Submit the Manuscript

The seventh step is to find a suitable journal to which you can submit your manuscript, formatting the manuscript for that journal, and submitting the manuscript. There are peer-reviewed journals devoted to historical articles. However, many peer-reviewed journals that primarily publish scientific articles have sections devoted to historical articles. One such example is the journal *Clinical Anatomy*, which has a section devoted to historical articles titled "A Glimpse of Our Past." In short, do not limit yourself to journals devoted exclusively to historical articles when submitting your manuscript for consideration for publication. Cast a wide net, but do so by submitting your manuscript to one journal at a time. Simultaneous submission to multiple journals is considered to be unethical, and patience is of paramount importance as you persevere to find a proper home for your manuscript.

Formatting your manuscript for a particular journal can be a tedious task, but is critical because failure to do so will cause your manuscript to be rejected or its publication delayed. Furthermore, formatting the manuscript according to a journal's guidelines is a professional courtesy to the journal's editorial board and the referees who will review your manuscript. Peer-reviewed journals typically list author guidelines on their websites. The author guidelines provide instructions on how to format your manuscript. A particularly frustrating issue is formatting the bibliography. Sometimes, the author guidelines do not provide instructions on formatting specific types of references, most notably online references. An alternative approach is to download a sample article available on the journal's website and using that published article's bibliography as a guide for formatting your own.

35.2 Types of Historical Articles

35.2.1 Biographical Articles

Biographical articles are historical articles that focus on aspects of an individual's personal and/or professional life. Biographical articles can be divided into four types: classic biographical articles, obituaries, commemorations, and autobiographical articles.

35.2.1.1 Classic Biographical Articles

In a classic biographical article, the author aggregates various biographical data (e.g. date of birth and death, notable personal and professional relationships, time and location of professional training, etc.) on one or more historical figures. Classic biographical articles are frequently comprehensive in their discussion of an individual's life and work. An example is "Hieronymus Brunschwig (1450–1513): His Life and Contributions to Surgery" (Tubbs et al. 2012). This article relies heavily on a book discussing Brunschwig's life by Henry E. Sigerist (1946), and the utilization of this important work is noted in the article. The existence of this book did not make the authors' project on Brunschwig moot. Information about Brunschwig was collected from several other references to build upon Sigerist's book. An important lesson regarding biographical articles can be gleaned from this experience: The existence of a book or lengthy article on the life of a historical figure does not mean that you cannot write your own article about that historical figure.

Classic biographical articles do not have to be comprehensive as such, but alternatively might examine a certain period within that individual's life or a particular aspect of that individual's career. Classic biographical articles of such narrow scope are typically written when the extant literature contains numerous publications on a historical figure. However, the author may wish to study a specific aspect of that individual's personal life or professional career in greater detail than any other publication in the extant literature does. An example is "Michael Servetus (1511–1553): Physician and Heretic Who Described the Pulmonary Circulation" (Bosmia et al. 2013).

35.2.1.2 Obituaries

An obituary is typically written by an author who knew the deceased individual well (e.g. a family member or close colleague) shortly after that individual's death. An example of an obituary is "In Memoriam: John W. Kirklin. 1917–2004" (Cooley 2004).

35.2.1.3 Commemorations

A commemoration is written on the anniversary of an individual's birth or death or in the context of greater awareness among the scientific community or general public about an individual's work. The author may have been a close friend or associate of the deceased individual, but such a relationship is not a requirement to author such an article. Instead, the author can be a great admirer of that person without having known him or her personally. An example of a commemoration is "Alton Ochsner, MD, 1896–1981: Anti-Smoking Pioneer" (Blum 1999).

35.2.1.4 Autobiographical Articles

An autobiographical article is, of course, written by the author himself. Candidacy for authorship is evidently limited with this type of biographical article. Autobiographical articles typically serve to educate readers about an aspect of the author's areas of interest through the lens of his or her personal experiences and thus are not comprehensive in their discussion of the author's life and work. Such an objective is better met through a book (e.g. an autobiography) than through an article published in a peer-reviewed journal. An example of an autobiographical article is "Reflections on a Heart Surgery Career with Insights for Western-Trained Medical Specialists in Developing Countries" (Kabbani 2011).

35.2.2 Articles on *Materia Medica*

The term *materia medica* is a Latin medical term for the body of collected knowledge about the therapeutic properties of any substance used for healing. The term derives from the title of a work by the ancient Greek physician Pedanius Dioscorides published in the first century AD, *De materia medica*, which translates to English as "On medical matters" (Janick and Storlarczyk 2012). The preparation of articles that examine *materia medica* requires at least one author to be literate in both the language of the work to be translated and the vernacular language in which the translation will be written. The examination of *materia medica* by present-day scientists can inform and enhance their scientific and clinical investigations by using modern-day methods to examine the purportedly therapeutic properties of medicinal agents of antiquity. Various types of *materia medica* are currently being studied (e.g. Chinese, Greek, Indian, etc.), as the therapeutic potential of these ancient remedies can be elucidated using modern investigational techniques and subsequently utilized to create safer and more effective drugs. Articles on *materia medica* examine a single reference or multiple references. An example of an article on *materia medica* is "Scientific Evaluation of Medicinal Plants Used for the Treatment of Abnormal Uterine Bleeding by Avicenna." This article reviews medicinal plants mentioned by the philosopher and physician Avicenna (980–1037) in his book *Canon for Treatment of Abnormal Uterine Bleeding* (Mobli et al. 2015).

35.2.3 Translations

Just as with articles on *materia medica*, translations require the involvement of an individual who is literate in both the language of the work to be translated and the vernacular language in which the final product will be written. However, the primary aim of a translation is to discuss the contents of translated reference regardless of whether it contains information on *materia medica*. The manuscript should not consist solely of the translation of the reference in question. A discussion of the reference's historical context and the significance of the translation should be included in the final product. The significance of the translation might lie in it being the first translation of the reference in a particular language. The translation can have additional significance in that it corrects a misconception about a particular historical figure, brings to light a detail about a historical figure that previously had not been appreciated, or chronicles a historically significant moment not as well known among speakers of the vernacular language in which the translation is written. For example, in the article "Johann Conrad Brunner (1653–1727) and the First Description of Syringomyelia" (Bosmia et al. 2014), the authors provide the first translation of a case report by Johann Conrad Brunner, who is well known for his work in gastroenterology. This case report details his observation of syringomyelia, a potentially devastating neurological condition.

35.2.4 Articles on the History of Science

Articles on the history of science examine paradigm shifts in the history of science or the socioeconomic and cultural milieus in which a scientific concept was conceived, altered, sustained, or dismissed, and the reciprocal influences between a scientific concept and a particular milieu. Articles on the history of science can be divided into two types: cross-sectional historical articles and longitudinal historical articles.

35.2.4.1 Cross-Sectional Historical Articles

A cross-sectional historical article examines a single reference's discussion of a scientific concept and its application, and compares that primary reference's discussion of that specific subject with discussions of the same subject found in one or more references. An example of a cross-sectional historical article is "Vasovagal Syncope in the Canon of Avicenna: The First Mention of Carotid Artery Hypersensitivity" (Shoja et al. 2009).

35.2.4.2 Longitudinal Historical Articles

A longitudinal historical article examines the evolution, promotion, or dismissal of a scientific concept and the ramifications of the perceptions toward that scientific concept held by the scientific community and/or the public (e.g. the concept of a particular disease and its treatment) over a specified period of time. Such articles can be limited to a specific time period (e.g. the Renaissance) or be more exhaustive in their examination. Unlike cross-sectional articles, longitudinal articles do not make a particular reference's discussion of a scientific concept the focus of inquiry but focus on how various factors might have influenced a given scientific theory over a period of interest. Such factors include, but are not limited to, academic texts. An example of a longitudinal historical article is "A Historical Perspective: Infection from Cadaveric Dissection from the 18th to 20th Centuries" (Shoja et al. 2013).

35.2.5 Corrective Historical Articles

Corrective historical articles examine inconsistencies or suspected misconceptions in the secondary literature regarding the life and work of a historical figure or the history of a scientific theory or practice's inception, evolution, dismissal, or reception. The articles are very narrow in scope, but require an exhaustive and meticulous examination of multiple references to confirm any inconsistencies or misconceptions. The references that you conclude to be inaccurate should be identified as such in your manuscript. However, you must avoid framing your conclusions in a condemnatory context but instead promote them as constructive criticism. An example of a corrective historical article is "Karl Ewald Konstantin Hering (1834–1918), Heinrich Ewald Hering (1866–1948), and the Namesake for the Hering-Breuer Reflex" (Bosmia et al. 2015). In the article, the authors contend that multiple references mistakenly attribute the discovery of the Hering-Breuer Reflex to Heinrich Ewald Hering. In actuality, his father, Karl Ewald Konstantin Hering, is the individual after whom the Hering-Breuer reflex is named.

References

Blum, A. (1999). Alton Ochsner, MD, 1896–1981: anti-smoking pioneer. *The Ochsner Journal* 1 (3): 102–105.

Bosmia, A.N., Watanabe, K., Shoja, M. et al. (2013). Michael Servetus (1511–1553): physician and heretic who described the pulmonary circulation. *International Journal of Cardiology* 167 (2): 318–321.

Bosmia, A.N., Tubbs, R.I., Clapp, D.C. et al. (2014). Johann Conrad Brunner (1653–1727) and the first description of syringomyelia. *Child's Nervous System* 30 (2): 193–196.

Bosmia, AN, Gorjian, M, Binello, E, et al. 2015. Karl Ewald Konstantin Hering (1834–1918), Heinrich Ewald Hering (1866–1948), and the namesake for the Hering-Breuer reflex. Manuscript submitted for publication.

Cooley, D.A. (2004). In memoriam: John W. Kirklin (1917–2004). *Texas Heart Institute Journal* 31 (2): 113.

Janick, J. and Storlarczyk, J. (2012). Ancient Greek illustrated Dioscoridean herbals: origins and impact of the Juliana Anicia Codex and the Codex Neopolitanus. *Notulae Botanicae Horti Agrobotanici* 40 (1): 9–17.

Kabbani, S.S. (2011). Reflections on a heart surgery career with insights for Western-trained medical specialists in developing countries. *Texas Heart Institute Journal* 38 (4): 333–339.

Mobli, M., Qaraaty, M., Amin, G. et al. (2015). Scientific evaluation of medicinal plants used for the treatment of abnormal uterine bleeding by Avicenna. *Archives of Gynecology and Obstetrics* 292 (1): 21–35. Available from: http://link.springer.com/article/10.1007%2Fs00404-015-3629-x. [6 May 2015].

Shoja, M.M., Tubbs, R.S., Loukas, M. et al. (2009). Vasovagal syncope in the Canon of Avicenna: the first description of carotid artery hypersensitivity. *International Journal of Cardiology* 134 (3): 297–301.

Shoja, M.M., Benninger, B., Agutter, P. et al. (2013). A historical perspective: infection from cadaveric dissection from the 18th to 20th centuries. *Clinical Anatomy* 26 (2): 154–160.

Sigerist, H. (1946). *A Fifteenth Century Surgeon: Hieronymus Brunschwig and His Work*. New York: Ben Abramson.

Tubbs, R.S., Bosmia, A.N., Mortazavi, M.M. et al. (2012). Hieronymus Brunschwig (c. 1450–1513): his life and contributions to surgery. *Child's Nervous System* 28 (4): 629–632.

Section VII

Publication and Resources

36

An Introduction to Academic Publishing

Mohammadali M. Shoja[1] and R. Shane Tubbs[2]

[1] *Division of General Surgery, University of Illinois at Chicago Metropolitan Group Hospitals, Chicago, IL, USA*
[2] *Seattle Science Foundation, Seattle, WA, USA*

36.1 Introduction

A manuscript is often the most important scholarly output of a research endeavor in academia. Before submitting your manuscript for consideration for possible publication in a journal, follow this checklist:

1. Double-check the content to make sure that there is no inadvertent error.
2. Make sure all authors have read the manuscript and are in agreement with its content.
3. Prepare the manuscript and accompanying files to comply with the journal's requirements for new submissions.
4. Always keep a copy of the final manuscript and related dataset in a safe folder in your computer. Back up the files into an archive that can easily be restored any time in the future. Remember, your co-authors should have access to the files at their request.
5. Continuously update your co-authors on the status of the manuscript, during and after submission.

36.2 Manuscript Submission

During the submission process, almost all journals require a cover letter directed to the editor-in-chief, briefly signifying the strength of the manuscript, providing a list of suggested reviewers, a statement on the potential conflicts of interest in connection with the manuscript, and a statement confirming that the submitted manuscript is not simultaneously being considered for publication elsewhere. The corresponding author should sign the cover letter. Upon submission, some journals also require the authors to transfer the copyright of the work to the publisher, who will retain it until an editorial decision is reached and after the manuscript is accepted for publication. If the manuscript is rejected, the copyright automatically reverts to the authors.

After a manuscript is submitted for consideration for publication in a journal, one editor or a small group of editors assesses the manuscript to make sure that it is formatted properly, fits the scope of the journal and is sufficiently novel to warrant further

A Guide to the Scientific Career: Virtues, Communication, Research, and Academic Writing, First Edition.
Edited by Mohammadali M. Shoja, Anastasia Arynchyna, Marios Loukas, Anthony V. D'Antoni, Sandra M. Buerger, Marion Karl and R. Shane Tubbs.

evaluation. After initial approval by the editor-in-chief or his assigned co-editor, the manuscript is sent to external reviewers who are experts on the subject and have similar publication records. An editorial manager is usually assigned to communicate and coordinate among the authors, reviewers, and editors. In rare instances, the standards and merits of the submission might be high enough to convince the editor-in-chief that an internal review by him/her or his/her co-editors is sufficient; they might even accept the manuscript with no revision.

The process of finding eligible reviewers, inviting them and following up on their reviews is daunting. External reviewers usually receive no remuneration for their independent judgments, and their institutional credit for this voluntary and invisible contribution is often nil or minimal. Therefore, journal editors often invite many reviewers so that two or three independent reviews will ultimately be secured. Alternatively, they might ask the submitting authors to provide a list of potential reviewers, or even require them to do so. To make the process fair and balanced, the journals also allow the submitting authors to exclude particular reviewers; unfortunately, rivalry among researchers and scientists is not uncommon, and sometimes a researcher's manuscripts can be rejected because the reviewer, who publishes in the same field, wants to suppress competitors.

The issue of authors suggesting or excluding reviewers has been the subject of extensive debates in ethical fora. Some editors see it as a welcome opportunity for the authors to bring their works to publication provided that they can attract two expert reviewers to affirm the quality and merit of the work submitted. Others see it as an undesired source of additional conflict of interest whereby the review process is biased by any relationship between the authors and reviewers. In the extreme form of misconduct associated with suggesting reviewers, there have been instances in which authors created fake reviewers by inventing false identities and email accounts to which only they had access. They then could recommend the fake reviewers upon submitting their manuscript, and they would receive the journal's editorial invitation to "peer-review" their own manuscript. To make it seem natural, they would then recommend a few minor revisions. This type of irregular academic behavior is extremely dangerous to the integrity of science and makes the offending authors liable to legal action, institutional or international sanctions, and ultimately the permanent loss of academic position. For authors, it is therefore ethically and professionally necessary to disclose to the journal editors any past or recent relationship with the suggested/excluded reviewers, and the reason for suggesting or excluding them. This can be stated in the cover letter or in the designated forms used for electronic manuscript submission. Ultimately, it is the editor's decision whether to engage or exclude those reviewers and to judge whether the manuscript is compatible with the journal's scientific mission.

36.3 Peer-Review Process

The peer-review process takes various lengths of time. Not all reviewers can complete and submit their reports by the deadline set by the journal, and most journals will extend the deadline upon the reviewers' request. Depending on the style, quality, and subject of the submitted work, it can also take longer to secure reviewers because most reviewers' agreements to assess a manuscript are contingent on their interest in the subject matter. Not uncommonly, editors decide simply to reject a manuscript after multiple failed attempts to recruit reviewers, as the recurrent failures can be perceived as indicating a general lack of interest in the submission. The peer-review process is often blinded, with neither the reviewers nor the submitting authors made aware of one another's identities. Some journals disclose the identities of the submitting authors to the reviewers in order to help them to consider the academic backgrounds and past works of those authors in refining their decision, but it is very unusual for a scholarly journal to disclose the identities of the reviewers to the submitting authors.

Nowadays, almost all journals have electronic submission portals or accept submissions by email. Print submissions, such as those in the pre-internet era, are no longer favored or deemed practical. Every promising advance has its own shortcomings. The relative ease of electronic submission has fooled some novice researchers into submitting their works to multiple journals simultaneously. This is unethical and unprofessional and wastes the resources of the journals, editors, and reviewers. No manuscript should be submitted to more than one journal at a time for consideration for publication. If you desire to publish the whole or part of the same manuscript in different languages, the editors and publishers should be notified upon simultaneous submission or publication in a different language.

In addition to facilitating manuscript submission by the authors and handling by the editors, the electronic submission portal allows the authors to monitor the status of their manuscript during editorial process and peer review, to submit a revision if requested by the editors, and even to follow up on the post-acceptance stage of the manuscript.

36.3.1 Manuscript Revision and Resubmission

After a manuscript undergoes peer review and the reviewers' comments arrive in the editorial office, the editors evaluate the comments, taking into account the overall priority ranking and the scores on quality, novelty, and presentation of the manuscript. They then decide whether to return the manuscript to the authors for revision or to reject the submission. If there is any conflict among the reviewers' comments, the editor might seek another reviewer's feedback or just use his/her own judgment. In any case, the authors must be adequately advised on how to revise and resubmit their work. Requests for revision always come with a deadline, and if the authors cannot meet that deadline, they should notify the editors in advance and ask for an extension. Although a reasonable extension is often granted, a request for prolonged extension or failure to notify the editors of the delay in resubmission is likely to result in the manuscript being rejected or treated as a new submission. The revision should be submitted with a list of changes made in the manuscript – occasionally, the changes are highlighted within the manuscript – and responses to the queries and recommendations posed in peer review. This response-to-reviewers letter is sometimes called a rebuttal note. If a reviewer's recommendation is not addressed in the revision, it should be noted and explained in the rebuttal note. It is unprofessional to claim that a revision is complete without addressing an issue raised about the original manuscript.

After resubmission, the revised manuscript is assessed by the editors, who compare the revision with the original version and the reviewers' comments. If the editors find the revision satisfactory, they either accept it and send the manuscript for production, or send it to the original reviewers for approval, or even seek feedback from new reviewers.

If the revision does not satisfy the editors and reviewers, the manuscript is rejected or returned to the authors for another round of revision. Chapter 44 in this book addresses how to revise the manuscript and properly respond to reviewers.

36.4   After the Manuscript is Accepted for Publication

Some journals immediately publish the abstract and/or full version of the accepted manuscript on the journal website to make it available for viewing and downloading by the scientific community and public before copyediting and proofreading. They might then sequentially publish the copyedited version, uncorrected proof, and corrected proof before final publication in the journal issue.

36.4.1   Publishing Agreement and Other Pre-publication Forms

After the manuscript is accepted for publication, you must carefully read and sign the publishing agreement, usually including a copyright-transfer form and disclosure-of-conflict-of-interest form. At this stage, make sure that all authors' names are spelled accurately and that their highest academic degrees – if required by the journal – and affiliations are correct. Some require all contributing authors to sign the forms and agreement separately, while others require only the corresponding author to sign the forms on behalf of all co-authors. Any source of conflict of interest that might have affected the conduct of the study and manuscript preparation in any way must be disclosed fully and honestly. Pay particular attention to the terms of the copyright-transfer form. When this form is signed and sent to the publisher, the manuscript and all included images and illustration will be the property of publisher, who will maintain the right to distribute, sell the product, and grant permission for reproduction to a third party without consulting the authors. Most often, the copyright-transfer terms are negotiable, but such negotiations should be done as early as possible before the copyright-transfer form and publishing agreement are signed. Authors publishing in open-access journals retain copyright but allow unrestricted distribution of the content by the publisher and any third party. Anyhow, read the terms of the publishing agreement and copyright transfer carefully, as they differ among journals and can affect the visibility and impact of your work post-publication.

36.4.2   Copyediting

The next step is copyediting, in which the accepted manuscript is reviewed and edited by one of the journal editors or by a medical editor hired by the publisher to ascertain that the language, grammar, punctuation, and reference formatting are accurate and consistent. Some journals also interpose an extra step of scientific verification of the reported study and statistical data and accuracy of references before further processing. Copyediting helps to improve the quality of the manuscript and its presentation, and if any correction or clarification is deemed necessary by the copyeditor, a list of queries is sent to the authors to answer either before the proofreading stage or along with the galley proofs. It should be noted that the language and content of the manuscript are the full responsibility of the authors, not the publisher or the editors. The copyediting is only an additional step in the publishing process that most journals adopt in order to improve the quality of publication. Some publishers, especially novice open-access publishers, skip this stage to save money.

36.4.3 Typesetting, Page Proofs, and Proofreading

After copyediting, the manuscript is sent to a layout editor for typesetting. The layout editor puts it into the final publication format and creates the page proof or galley proof. This is the preliminary version of the article to be reviewed by authors, editors, and proofreaders before formal publication. It is often sent as a watermarked PDF, with unnumbered pages along with the queries posted by the copyeditors. The authors should read the page proofs carefully to make sure that meanings have not been inadvertently changed during copyediting and also that the organization of the proof and placements of images and tables are satisfactory. Extensive revisions at the proofreading stage are not allowed by many journals, and some charge a fee for extensive changes at this stage because they require double efforts to prepare a new proof. Editors can even deem the article unsuitable for publication if the proofreading changes significantly alter the manuscript. Most journals also forbid changes in the authors' list at this stage. The changes can be made by using annotation tools in the PDF to insert, delete, or replace any text or add a note; by printing the proof, making the changes and notes with a pen and then scanning and emailing the file back to the publisher; or by writing an itemized list of desired changes in the email. After the proofreading is completed and the authors have fully addressed the queries raised by the copyeditors, the publisher implements the changes and proceeds with publication.

36.5 Final Publication

There are differences among journals with respect to publishing schedules and processes. All print and most electronic journals have designated issues for publishing articles, usually 2–12 per annually, with additional supplemental issues. The issue to which an accepted article is to be assigned is decided mainly on the basis of when the proofreading is completed as well as the editor's priorities and perception of the article's fitness for a particular issue. Therefore, it can take weeks, months, or even a couple of years before the final article is published in a journal issue and given page numbers. To allow articles to be made available to readers ahead of print or final electronic publication in an issue, most journals publish them online as articles in press, whereby a reader can find the full article and download the corrected page proofs.

Some electronic journals publish final articles in an ongoing issue of the journal until the article limits are reached or until the end of the year or season to conclude the issue and start a new one. Thus, the article pages are either consecutively numbered or each article receives a unique ID number, and the pages are numbered starting from one. This practically removes any delay in final publication from the time the proofs are corrected.

37

Various Types of Scientific Articles

José Florencio F. Lapeña[1] *and Wilfred C.G. Peh*[2, 3]

[1] *Department of Otorhinolaryngology, Philippine General Hospital, University of the Philippines Manila, Manila, Philippines*
[2] *Department of Diagnostic Radiology, Khoo Teck Puat Hospital, Singapore*
[3] *Yong Loo Lin School of Medicine, The National University of Singapore, Singapore*

37.1   Introduction

What characterizes an article as scientific? Is it the subject matter, the approach used to research and gather content, the style of writing, the manuscript structure (i.e. IMRAD – Introduction, Materials and Methods, Results, and Discussion), or the process of being peer-reviewed? Perhaps it has to do with the nature of the publication (i.e. a peer-reviewed medical journal or an open-access publication)? There are in fact many types of articles that are considered scientific.

That said, it is useful for authors to understand the various kinds of scientific articles and their nuances before approaching the writing process. Ideally, authors should tailor their writing to fit the most appropriate article category so as to optimize their research reception and enhance their chances of publishing (Peh and Ng 2008).

There are numerous types of articles to be found in scientific journals, all of which contribute to the individual journal's scope (Table 37.1). Information on the format, style, and purpose of each type of article are usually detailed in a journal's instructions to authors or author guidelines. In this chapter, we will provide an overview of the major types of scientific articles. ur goal is to guide medical students, residents, and young researchers in selecting the most appropriate manner of communicating scientific material.

37.2   Primary or Original-Research Articles

Primary or original-research articles communicate knowledge arrived at or discovered by the author(s). They include theoretical articles, which present new or established abstract principles (e.g. mathematical modeling of biological and physiological processes), as well as observational and experimental research. Primary articles may be in the form of a short report, focusing on a single case, or in the form of a case series, presenting a series of cases encountered over a limited period of time. They may report

A Guide to the Scientific Career: Virtues, Communication, Research, and Academic Writing, First Edition.
Edited by Mohammadali M. Shoja, Anastasia Arynchyna, Marios Loukas, Anthony V. D'Antoni, Sandra M. Buerger, Marion Karl and R. Shane Tubbs.
© 2020 John Wiley & Sons, Inc. Published 2020 by John Wiley & Sons, Inc.

Table 37.1 Types of scientific articles.

Major type	Examples
Primary or original research articles	Randomized controlled trial
	Clinical trial
	Before-and-after study
	Cohort study
	Case-control study
	Cross-sectional survey
	Diagnostic test assessment
	Case report/case series
	Technical note
Secondary or review articles	Narrative review article
	Systematic review
	Meta-analysis
Special articles	Letters/correspondence
	Short communications
	Editorials/opinion
	Commentaries
	Pictorial essay
	Other special categories
Tertiary literature	Textbooks, handbooks, manuals
	Trade or professional publication articles
	Encyclopedias
Gray literature	Conference proceedings, posters, abstracts
	Government reports
	For-profit and nonprofit organization reports online forums
	Blogs, microblogs, tweetchats, and other social media

on clinical procedures, diagnostic, or therapeutic. Technical reports, including the evaluation of equipment, instrumentation, and technology, belong in this category. Original theses and dissertations are also counted among primary-research articles.

The original article is the most important type of scientific paper; it provides unique information based on an original-research design. The articles can be descriptive or analytical in nature, and report on studies that are retrospective (examining past outcomes) or prospective (looking for new outcomes). The format of the body of an original article usually follows the IMRAD structure. A structured abstract usually precedes the article. Examples include randomized controlled clinical trials, before-and-after studies, cohort studies, case-control studies, cross-sectional surveys, and diagnostic test assessments (see Table 37.1).

A case report describes a single study, documenting unique features discovered during the research process. Notable features can include previously unreported observations of a known disease; the unique use of imaging or diagnostic testing to reveal a disease; or an undocumented clinical condition, treatment, or complication. Case reports are usually short and focused with a limited number of figures and references. Their format generally includes a short abstract (or none at all), a brief introduction, an account of the case study, and a discussion giving context to the research findings. If relevant, a limited review of the literature might be included in a case report.

A technical note, or technical innovation, is an article describing surgical or medical techniques or procedures; new devices or technologies, including instrumentation and equipment; or modifications of existing techniques, procedures, devices, or technologies. The discussion is usually narrowed to a specific message, and there is often a prescribed limit to the number of figures and references. A technical note might be a condensed version of a more comprehensive intellectual property (IP) application or patent application. Technical notes usually consist of a short, unstructured abstract (or none at all), brief introduction, methods, results, and discussion. The methods and results sections can be combined under the heading of technique.

37.3   Secondary or Review Articles

Secondary or review articles expand on knowledge that has been previously communicated by others. That includes revisiting, reviewing, analyzing, or synthesizing existing research, and presenting it in a new light. Secondary articles might come in the form of a monograph (a detailed study of a single subject or aspect of a subject), a descriptive review, or commentary.

A narrative review article is an authoritative and comprehensive analysis of a specific topic that is more descriptive in nature than a systematic review. It does not introduce new data, but often leads to formulation of new hypotheses. A narrative review can include an unstructured abstract, an introduction that gives a detailed background of the topic, a body of content organized under subheadings, and a summary and conclusion with a large number of peer-reviewed references. Because narrative reviews are often solicited by the editors of a journal, they are sometimes referred to as "invited reviews."

Systematic reviews typically belong in the category of secondary articles as well. Although it could be argued that if new knowledge is arrived at in the process of conducting the analysis, it might qualify such reviews as original articles. Systematic reviews are usually characterized as either qualitative or quantitative (using meta-analysis), and attempt to reduce bias by addressing the methodological selection, assessing the quality of the study (critical appraisal), and analyzing the literature. A systematic review article includes an introduction, a description of the methodology employed in the review process, and the systematic presentation and synthesis of the findings. For additional information on systematic reviews and meta-analyses, see Chapters 67−68.

37.4   Special Articles

Other types of scientific articles include letters to the editor, correspondences, and short communications that might be primary or secondary in nature. Editorials, opinion articles, commentary, and perspective articles traditionally express the opinion and point of view of the author(s).

Letters and correspondences are usually short and can be written on any topic of interest to the journal reader, including feedback on previously published articles. These should be objective and constructive, and supported by a limited number of references. When a previously published article is the subject of a letter, the editor usually invites the author of the original article to submit a written response (author's reply to letter), which is published side-by-side with the letter to the editor. Letters can also be used to float new hypotheses and to draw readers' attention to important issues affecting clinical practices. In fact, the initial publication of the double-helical structure of DNA by Watson and Crick (1953) was a letter to the editor published in *Nature*.

Short scientific communications are usually quick synopses of preliminary results or initial research findings that are not yet ready for full presentation. Such communications are beneficial as they may stimulate discussion leading to additional insights and collaborative work. They can include both larger case series and small sample studies.

Editorials most often take the form of a short review or critique of an original article accepted for publication in the same journal issue. An editorial typically consists of a brief description of a subject that does not warrant full review, or is a follow-up to recent innovations or topics of interest to the journal readers. Editorials are so termed because they are usually written or assigned by the journal's editor.

Commentaries are short articles that describe an author's personal experience with a specific topic. The subject might be controversial and provides other existing viewpoints before presenting the author's perspective. Commentaries might be based on a current hot topic or commissioned by a journal to accompany an original paper on the same topic. Unlike a review article, the number of references and illustrations are limited to support the author's opinion. Commentaries usually include an unstructured abstract (or none at all), introduction, and content organized by subheadings.

A pictorial essay is a special type of teaching article that relies on quality visuals to communicate a current trend or message that is educational in nature. The text is limited in favor of many figures, with much of the message contained in the figure legends and captions. Emphasis is placed on the teaching value of the article, and the format can include an unstructured abstract (or none at all), a brief introduction, optional subheadings and discussion, and limited references.

Other categories of articles exist, which are usually meant to complement the mission and style of the individual journal contributing to the journal's character.

37.5 Tertiary Literature

Trade-publication articles, textbooks, handbooks, manuals, and encyclopedias comprise what is called tertiary literature. These build on knowledge that has been around for a long period of time and incorporate the secondary literature. In turn, tertiary literature may be used to synthesize primary literature.

37.6 Gray Literature

These are documents issued by research institutes, universities, government agencies, international, national and local authorities, or industrial firms outside of the traditional academic publishing and distribution channels (Feather and Sturges 2003). Conference proceedings, poster presentations, and abstracts are some examples that are not included in the formal categories of scientific literature previously discussed. Strictly speaking, they are not fully accepted as scientific papers. Hence, the term gray literature.

37.7 Conclusions

For authors, being familiar with the various types of scientific articles is part and parcel of the academic career. Authors should know the purpose and requirements for each article type, and submit their work in the most appropriate category in order to maximize their chances of getting published and viewed.

The future of scientific articles and their classifications remain to be seen. Our transition from the Gutenberg era to the age of electronic publishing is not the end of the story; it marks a new beginning of exciting and innovative forms and types of academic communication. The division between the traditional and newer forms of scientific publication is further blurred with future multistage open-review systems that allow public discussion and critique of initially published articles, which are accordingly revised until a "final" version is eventually published (Pöschl 2012).

References

Feather, J. and Sturges, P. (2003). *International Encyclopedia of Information and Library Science*, 2e. London: Routledge.

Peh, W.C.G. and Ng, K.H. (2008). Basic structure and types of scientific papers. *Singapore Medical Journal* 49 (7): 522–525.

Pöschl, U. (2012). Multi-stage open peer review: scientific evaluation integrating the strengths of traditional peer review with the virtues of transparency and self-regulation. *Frontiers of Computational Neuroscience* 6 (33): 1–16.

Watson, J.D. and Crick, F.H.C. (1953). Molecular structure of nucleic acids: a structure for deoxyribose nucleic acid. *Nature* 171 (4356): 737–738.

38

Authorship

Stephen W. Carmichael

Department of Anatomy, Mayo Clinic, Emeritus Center, Rochester, MN, USA

38.1   Era of Collaborative Research

We are in an era of collaborative research. Although it is not impossible for a single person to conduct a study and publish the results, it is becoming increasingly uncommon to do so. Working with colleagues who have complementary backgrounds and expertise offers many advantages. One potential pitfall of collaborative research is the issue of authorship.

38.2   Giving Credit to Collaborators

Authorship is used here as the determination of who will be included as a co-author vs. mentioned in the acknowledgments or not mentioned at all. Regarding who is listed in the acknowledgments, it has been my practice to be as inclusive as possible. All who helped with the project in any way, especially above and beyond their job description, should be acknowledged. That would include students who conducted a literature search that helped get the project started. An acknowledgment is always appreciated by students and colleagues; furthermore, acknowledging everyone who contributed to your project will make them more inclined to help the next time you ask!

It can be a fine line between who is listed as an author and who is acknowledged. The number of authors should be proportional to the expanse of the study. Obviously, a study that included many experiments probably will have several authors who all contributed significantly to the study and/or the manuscript. Having several authors for a short paper, such as a case report, might reflect poorly on the team of authors. It will be obvious to readers that not all of the authors made a substantive contribution.

38.3   Assigning Authorship

It has been my experience that the key to avoiding difficulties in assigning authorship is to address the matter during the genesis of the project – the earlier the better. As the study is being designed, just who is (and who is not) considered to become an author

A Guide to the Scientific Career: Virtues, Communication, Research, and Academic Writing, First Edition.
Edited by Mohammadali M. Shoja, Anastasia Arynchyna, Marios Loukas, Anthony V. D'Antoni, Sandra M. Buerger, Marion Karl and R. Shane Tubbs.
© 2020 John Wiley & Sons, Inc. Published 2020 by John Wiley & Sons, Inc.

must be agreed upon by all participating individuals. As the study progresses and it becomes advantageous to bring other people into the project, the original participants need to agree whether the additional person/people will become authors. By dealing with authorship issues as soon as they arise, you can avoid misunderstandings at the completion of the manuscript.

A difficult, however infrequent, situation that can occur is when a colleague who demands to be listed as an author has contributed little or nothing to the project. One tactic is to bring this up with your group and reach a consensus. The International Committee of Medical Journal Editors (ICMJEs)[1] defines an author as an individual who made a "substantive intellectual contribution to a published study. ..." (for information on available authorship guidelines, see Chapter 39). You and your team of authors must decide together whether the contribution of your colleague meets the qualifications; if not, then it can be presented to that colleague as a group decision. Having said that, internal politics in some cases will dictate that this person be included as an author.

Another point of contention that is likely to arise is the order of the authors. Obviously, it is most advantageous to be the first author listed among multiple authors; for example, a paper by Smith, Jones, and Brown (and any number of additional authors) will be most commonly cited as Smith et al. On the other hand, two authors can share recognition; a paper by Smith and Jones will be cited as such.

The first author is usually the person who originally proposed the study to a colleague or group of colleagues. In any event, that original group needs to agree who will be the first author. The selected person will be responsible not only for seeing the study through to completion, but usually also for the writing of the first draft, and ultimately submitting the manuscript. Of course he or she can, and probably should, delegate certain tasks to other members of the group.

A situation that adds to the difficulty of agreeing on the first-referenced author relates to some academic institutions only considering first-authored papers for promotion and/or tenure. One way to deal with the issue is to identify the person who needs to be first author for such reasons and have him/her take the leadership role expected for a first author.

Aside from identifying the first author, there is no hard and fast rule about the order of the other authors. It is my opinion that the last author is the overall leader of the group (perhaps a chairperson or section head) but who did not take the leadership role for this particular project. If two equally senior authors are appropriate to be listed last, those two individuals need to work it out. The other authors can be listed in the order of the magnitude of their participation in the study, with the penultimate author being the person who likely contributed the least to the study. Again, the key is to determine this early on. If tasks change during the course of the study, and all authors agree, the order of authors can be rearranged.

38.4   Corresponding Author

Another factor for the group to determine is who will be the corresponding author. This is the person who is identified at the beginning of the article for readers to contact with

---

1 The URL is www.icmje.org.

questions or matters related to the paper. The corresponding author assumes primary responsibility for responding to inquiries from readers and, when appropriate, drafting responses to Letters to the Editor that address the original article. The corresponding author is often the first or last author, but it could be any author who is very familiar with the work. Usually, the physical address and email address are given; it has been my practice to not include telephone numbers because I might be inconveniently contacted by somebody several time zones away.

38.5 Ethical Responsibility

The ICMJE stresses that every author shares responsibility for ethical integrity of the study and the published article. The committee states that "each author should have participated sufficiently in the work to take public responsibility for appropriate portions of the content." That applies, but is not limited to, the protection of human subjects and animals used in the research. For example, if cadaveric material is used in the study, the authors need to be trained in the proper use and treatment of cadavers and a statement acknowledging the contribution of the body donors should be part of the manuscript.

The accusation of plagiarism is another serious ethical issue and one that affects the reputation of all of the authors involved. The quote from the ICMJE website cited in the previous paragraph is a clearly worded statement that also relates back to the issue of who should be included as an author. This statement stresses the shared responsibility of authorship.

A difficult scenario occurs when a person who has been an integral part of the research team and who would be considered an author by standard criteria declines to sign off on the submission as leverage to have something changed in the paper. It must be considered how such a member of the team not being included as an author affects the remaining author(s), the missing author, and the integrity and weight of the article.

Another scenario is when a colleague wishes to include you as an author on a study to which you did not substantively contribute. This colleague is usually acting out of respect and/or friendship. It is my practice to politely but firmly decline the offer. I do not want my name on a study if I did not participate.

38.6 Conclusions

Collaborative research and publication is here to stay. Deal with authorship issues early in a project and be unambiguous. Your professional reputation is associated with every publication that has your name as an author.

Acknowledgments

The author thanks R. Shane Tubbs, PhD, and Kathy Meyerle, JD, for their very helpful suggestions. Also, the constructive comments of various reviewers of early drafts of this manuscript are deeply appreciated.

Further Reading

Dulhunty, J.M., Boots, R.J., Paratz, J.D. et al. (2011). Determining authorship in multicenter trials: a systematic review. *Acta Anaesthesiol. Scand.* 55 (9): 1037–1043.

Garcia, C.C., Martrucelli, C.R., Rossilho Mde, M. et al. (2010). Authorship for scientific papers: the new challenges. *Rev. Bras. Cir. Cardiovasc.* 25 (4): 559–567.

Heinrich, K.T. (1995). Co-authorship: turning pitfalls into pleasures. *Nurse Author Ed.* 5 (4): 1–3.

McCann, M. (2007). The challenges encountered during a collaborative research project. *J. Ren. Care.* 33 (3): 139–143.

Teixeira da Silva, J.A. and Dobránszki, J. (2016). Multiple authorship in scientific manuscripts: ethical challenges, ghost and guest/gift authorship, and the cultural/disciplinary perspective. *Sci. Eng. Ethics* 22 (5): 1457–1472.

Tornetta, P. III,, Siegel, J., McKay, P. et al. (2009). Authorship and ethical considerations in the conduct of observational studies. *J. Bone Joint Surg. Am.* 91 (Suppl 3): 61, 7.

White, A.H., Coudret, N.A., and Goodwin, C.S. (1998). From authorship to contributorship. Promoting integrity in research publication. *Nurse Educ.* 23 (6): 26–32.

39

Recognition, Reward, and Responsibility: Why the Authorship of Scientific Papers Matters

Elizabeth Wager

SideView, Princes Risborough, Buckinghamshire, UK

39.1 Why Does the Authorship of Scientific Papers Matter?

Why should we be concerned about getting authorship right? It might be argued that authorship abuse is a victimless crime, which has no impact on scientific progress or the reliability of the medical literature. But this is a poor argument. Authorship matters because the entire research and publication process relies on trust. If scientists or clinicians are prepared to lie about the people involved with a research project or a publication, why should we expect them to be any more honest about their findings?

Since authorship has become the currency of research credit, and the output of individual academics and entire departments is judged on their publication record, authorship abuse is not a victimless crime (Sheikh 2000). The career of a researcher wrongly denied authorship of a paper could suffer, while guest authors could receive undeserved credit and benefit materially. Therefore, understanding authorship is especially important for junior researchers.

There is anecdotal evidence that people who flout authorship conventions might also commit other forms of research or publication misconduct. If one views publication misconduct as a spectrum of offenses, one can see that if a senior person (e.g. a head of department) insists on being listed despite not contributing to the research, or omits a deserving junior researcher from an author list, this might be only one step away from appropriating results of a junior colleague and thus committing plagiarism.

Inappropriate authorship practices by senior researchers (such as demands for guest authorship) set a bad example and are likely to damage relationships among team members.

39.2 What Is Authorship?

Discussions about the importance of authorship presuppose that we understand what the authorship of scientific papers really means or, at the very least, that there are widely accepted definitions about who should (and should not) be listed as an author. Sadly, this is not the case.

A Guide to the Scientific Career: Virtues, Communication, Research, and Academic Writing, First Edition.
Edited by Mohammadali M. Shoja, Anastasia Arynchyna, Marios Loukas, Anthony V. D'Antoni, Sandra M. Buerger, Marion Karl and R. Shane Tubbs.
© 2020 John Wiley & Sons, Inc. Published 2020 by John Wiley & Sons, Inc.

Outside the world of science, an author is simply somebody who creates new written material. The authorship of a novel, poem, play, or newspaper article therefore lies with its creator and is usually simple to assign. Most works of literature and journalism are the work of just one person (ignoring, for the sake of argument, the influence of editors who generally revise or remove rather than create text). The concept of authorship is therefore intrinsically linked not only with the act of writing but also with creativity.

In the early days of scientific experiment, most researchers worked alone. It was therefore no more difficult to say that Isaac Newton (1643–1727) was the author of *Principia Mathematica* than to say that John Milton (1608–1674) wrote *Paradise Lost*. The early scientists were true authors in terms of both creativity and writing, although, interestingly, in the seventeenth century, many scientific works were published anonymously (Kronick 1990).

Nowadays, it is rare for research to be undertaken by a single person. Clinical research, in particular, is almost always a collaborative effort. Yet not everybody who contributes to the project will necessarily wield a pen, so a distinction starts to emerge between roles, and authorship starts to become dissociated with writing. When people have different roles, and not everybody is (or should) be involved in drafting the report, then systems are needed to determine which roles deserve to be recognized by authorship.

39.3   What Guidelines on Authorship Are Available?

A number of guidelines relating to authorship are available (Table 39.1). Most have been produced by groups of journal editors. However, individual journals often do not give

Table 39.1 Authorship guidelines.

Document	Produced by	Aimed at	Available from
Recommendations for the conduct, reporting, editing, and publication of scholarly work in medical journals	International Committee of Medical Journal Editors (ICMJE)	Researchers	www.icmje.org
Role of professional medical writers in developing peer-reviewed publications	European Medical Writers Association (EMWA)	Professional medical writers	www.emwa.org
What to do if you suspect ghost, guest, or gift authorship	Committee On Publication Ethics (COPE)	Journal editors	http://www .publicationethics.org
How to handle authorship disputes	Committee On Publication Ethics (COPE)	Researchers	http://www .publicationethics.org
White paper on promoting integrity in scientific journal publications	Council of Science Editors (CSEs)	Journal editors and researchers	http://www .councilscienceeditors.org

specific guidance, and there are no universal standards for those that do. A survey of 234 biomedical journals found that 41% gave no guidance about authorship, 29% based their instructions on the criteria of the International Committee of Medical Journal Editors (ICMJEs), while 14% proposed other criteria, and 14% stated only that all authors should approve the manuscript (Wager 2007).

The ICMJE recommendations are the most widely followed guidelines for medical publications. They were first developed in 1978 at a meeting in Vancouver (and are therefore sometimes referred to as the Vancouver style) (Huth and Case 2004) and have been periodically updated since then, with a major revision in 2013, including the criteria for authorship and a change in title from Uniform Requirements for Manuscripts to Recommendations for the Conduct, Reporting, Editing, and Publication of Scholarly Work in Medical Journals (International Committee of Medical Journal Editors n.d.).

Most guidelines on the authorship of scientific papers give greater weight to creative and intellectual aspects of research than to routine or technical contributions. Until 2001, the ICMJE guidelines stated that only people who had made significant contributions to the design of a study, or its analysis and interpretation, qualified as authors (Huth and Case 2004). However, the current version includes data acquisition as an activity that qualifies for authorship (but explicitly notes – unlike for the other activities – that data collection alone is not enough; the individual must also contribute to developing the manuscript and must be prepared to take accountability for the work).

Some journals expressly forbid acknowledging individuals such as secretaries or technicians. I have heard it argued that a person did not qualify for authorship "because they were only doing their job" and that because these people were being paid for their services, this was reward enough, and they should not be listed as authors (which struck me as strange, because, presumably, the principal investigators were also "only doing their job" and were certainly getting paid, yet this did not disqualify them from being listed).

Guidelines that assume investigators always report their own research are unhelpful when professional writers help to develop a manuscript but make no other contribution to the project. Failure to acknowledge professional writers turns them into ghostwriters, which is undesirable especially when it masks the involvement of a commercial sponsor (Wager et al. 2003).

39.4   Solutions to Authorship Problems

Since nobody (I hope) is suggesting that the best way to solve these problems is for every individual involved with a research project to take a creative role in reporting it (which would lead to papers being drafted by large committees), we need to devise a system for recognizing the various roles on the basis that either drafting or revising a manuscript is not the sole qualification for authorship. Table 39.2 summarizes a possible system, described in more detail here.

One solution is to agree on a writing group (ideally no more than about six people) who will do the writing. Publications can be written on behalf of a larger group who are acknowledged collectively. But this system again places greater emphasis on the creative act of writing than on other functions that might not be fair and might not inform readers fully about how or where the research was done. Why, for instance, should somebody who made a minor contribution to the study design, and recruited only a handful of

Table 39.2 Practical hints on avoiding authorship problems.

- Read the ICMJE criteria (and any guidelines from your institution and likely target journals) carefully, and make sure you understand them.
- Discuss authorship with your colleagues at the start of any research project or when planning other types of publication (e.g. review articles).
- At the start of any research project, get written agreement from all key players on how authorship will be determined (i.e., what criteria will be used) and how the order of authors will be determined (it can be helpful to refer to ICMJE and any relevant institutional guidelines or funder requirements).
- When you are ready to write up the research, refer to this agreement and determine who should be invited to join the writing group, making it clear that, to be listed, people must make an active contribution to developing the manuscript as well as having been involved in other aspects of the research.
- If team members disagree about who should be listed or the order of listing, refer to the agreement – or for projects without an authorship agreement, refer to relevant guidelines (most likely ICMJE, those from the target journal, your institution, and funder), and try to seek agreement.
- If team members cannot agree, consider appointing an independent arbiter (e.g., somebody knowledgeable about authorship who was not involved with the research) whose judgment everybody agrees to follow.

patients, deserve greater recognition than somebody who recruited dozens of patients and diligently recorded data from them, yet was not involved in designing the study?

In commercially funded studies, the protocol can be developed largely by company employees (especially if it the study is designed to meet the requirements of regulatory agencies), the data management and analysis will be done by another set of employees, and the report and publications might be drafted by an in-house or freelance medical writer. The roles of design, implementation, analysis, and reporting are therefore completely separated. The clinical investigators might be involved in interpreting the findings but, apart from this, they will not meet a strict interpretation of the ICMJE authorship criteria. In a large, multicenter study, the selection of a writing group (who get the chance to interpret the findings and to contribute to the publication) can therefore seem rather arbitrary.

Another problem with the traditional system of authorship listing is that it leaves editors and readers in the dark about who did what. Traditional lists do not reveal which (if any) system has been used to allocate authorship, and it is therefore impossible for anybody outside the group to determine whether the right people have been credited with, and should take responsibility for, the work.

Almost inevitably (except in the tiniest of studies involving one or two people), the contributions of different individuals will vary in terms of their nature and size. The conventional list of authors tells readers nothing about the different types of roles (e.g. design, analysis, or reporting), but does it indicate the relative size of each individual's contribution? Paradoxically, although reward systems are based on the assumption that authors are listed in decreasing order of their contribution, this convention has virtually never been codified. In a survey of over 200 journals' instructions, only two gave any advice about the order of author listing or suggested criteria for determining it (Wager 2007). The ICMJE guidelines are also silent about how to determine the order of authors.

Even if we accept the premise that authors are listed simply in decreasing order of the size of their contribution to the project (and this is, by no means, universally true), how can we quantify different types of contribution? Equations and mathematical formulae have been proposed for solving this puzzle, but those assume that the authors agree on the relative importance of different activities (in effect, to say that one hour designing the study is equivalent to four hours taking measurements, or whatever) (Merz et al. 1997; Hunt 1991). But as lack of agreement about the relative values of different roles lies at the heart of most authorship disagreements, that approach is unlikely to succeed.

Several studies in different countries have shown that the ICMJE authorship criteria are, by no means, universally respected. The proportion of scientists who disagreed that all three of the ICMJE criteria should be met was 77% in a study of 39 French researchers (who had all acted as principal investigators in clinical trials) (Pignatelli et al. 2005) and 62% in a survey of 66 British scientists (Bhopal et al. 1997).

Authorship problems are common. Among the 39 French investigators interviewed by Pignatelli et al (2005), 41% considered they had been unfairly left off author lists, and 62% had discovered that they were an author only after publication. In a study of 77 Indian researchers at a teaching hospital, 39% reported having had a conflict over authorship (Dhaliwal et al. 2006).

It appears, then, that although many journals encourage the use of the ICMJE criteria, many researchers do not agree. Also, the criteria give no guidance about how to determine the order in which authors are listed. Several journals have therefore adopted a different approach. Instead of treating a research paper like a novel, with a clearly defined author (or authors), they treat it like film credits and list contributors so that readers can see who did what (Rennie 1997; Smith 1997). Instead of having to guess what contribution each individual has made, this is explicitly stated. Contributorship has several advantages over conventional author lists. It is more transparent and makes it easier to spot "guests" (who did not do enough to merit being on the list) and "ghost" roles (i.e., contributions that have not been acknowledged). For example, if nobody is listed as having drafted the manuscript, an editor can ask about this. Although the ICMJE encourages journals to develop a contributorship system, only about 10% of biomedical journals list contributors (Wager 2007).

However, contributorship is not a panacea. It does not solve the problem of the order in which authors are listed, although it does provide information so that editors and readers can judge the contributions. It also does not prevent dishonesty. Authors who are prepared to flout conventions (e.g. by demanding to be guest authors although this is clearly outlawed by the ICMJE guidelines) are probably also prepared to lie when it comes to describing their contribution to a project.

Editors from the *Croatian Medical Journal* showed that, as in other types of questionnaires, people might give "socially desirable" rather than truthful responses (Marusic et al. 2006; Bates et al. 2004). They set up a randomized trial and compared responses when individuals described their contribution in their own words, and without any guidance, or when they completed a checklist supplied with a reminder of the ICMJE criteria. Those given the copy of the ICMJE criteria and asked to tick boxes were more likely to indicate that they fulfilled the authorship criteria than those left to describe their contribution in their own words.

Another problem of both the traditional authorship and the contributorship systems is whether responsibility can be divided between co-authors/contributors. The ICMJE criteria state that each author should agree "to be accountable for all aspects of the work in ensuring that questions related to the accuracy or integrity of any part of the work are appropriately investigated and resolved. In addition to being accountable for the parts of the work he or she has done, an author should be able to identify which co-authors are responsible for specific other parts of the work. In addition, authors should have confidence in the integrity of the contributions of their co-authors" (International Committee of Medical Journal Editors n.d.). That suggests that authors need not be held accountable for the entire project. In a multicenter study, especially one spanning several continents, is it reasonable to blame all authors if misconduct is discovered at one center or is committed by a single author? Can a statistician be held responsible for inadequate patient consent procedures or a clinician be expected to defend the statistical techniques used? Some journals require that at least one contributor is named as the "guarantor." The BMJ (formerly *British Medical Journal*) defines guarantors as people "who take responsibility for the integrity of the work as a whole, from inception to published article."

Following disclosure of research fraud, co-authors have sometimes tried to distance themselves from the offense. In some cases, journal retractions have indicated that only some authors are to blame (Anon 2008). Yet in other cases, when serious misconduct has been discovered, all authors have been held culpable (Dyer 1995).

One of the problems with editors setting authorship criteria is that they are rarely in a position to be able to verify whether they have been followed or to adjudicate disputes. The ICMJE notes that "it is not the role of journal editors to determine who qualifies or does not qualify for authorship or to arbitrate authorship conflicts." Committee on Publication Ethic (COPE) has produced guidance on what editors should do when faced with authorship disputes or requests to change the list of authors, but emphasizes that editors should rely on the contributors' institutions to resolve the conflict.[1] Similarly, the US Office of Research Integrity will not investigate authorship disputes but refers those to the researchers' institutions.

Local knowledge of the specific research project, and who was involved in reporting it, are needed to resolve disagreements about authorship. Many disputes could be avoided if institutions had clear authorship policies, promoted them to all researchers and checked to ensure that the policies were followed. For multicenter projects, the listing and acknowledgment criteria should be agreed at the outset, in writing, by all researchers (Albert and Wager 2003). At the start of a project, it might not be possible to name the eventual authors (since personnel can change), but it is helpful to identify the roles that will qualify for membership of a writing team, or for listing as an author or contributor. As ICMJE recommends, "The group ideally should decide who will be an author before the work is started" (International Committee of Medical Journal Editors n.d.).

While it might be desirable to have universal criteria for how researchers are recognized in publications, local and specific knowledge about the relative importance of different roles for different types of studies are also important. If responsibility for authorship policies were devolved to the researchers' institutions or employers, rather than directed by journal editors, it might initially result in variation between institutions but probably no worse than at present. In time, policies would probably be harmonized, as researchers move between institutions. Although it is generally helpful to separate rule makers from enforcers in judicial systems, where detailed knowledge is required to verify statements about who did what, this is probably impractical. Institutions could

---

1 www.publicationethics.org

provide checks and balances to minimize unfair practices or abuse, for example, by establishing an independent ombudsman to arbitrate in disputes between institutions or those that cannot be resolved at the departmental level. Having institutional systems for agreeing and enforcing authorship policies might seem to add a layer of bureaucracy for researchers, but if policies are well designed and criteria agreed at the start of projects, disputes should be rare and, when they do arise, should be able to be resolved quickly. Some institutions produce detailed and helpful policies on authorship and procedures for handling disputes (for example, see University of Melbourne 2015).

In many countries, research output is largely measured by authorship of peer-reviewed publications, so academic departments should have an interest in ensuring that contributions are fairly reflected. Vouching for the accuracy and completeness of a contributor list could become a routine procedure.

Journal editors might continue to have a role in suggesting criteria for the authorship of non-research publications such as editorials and nonsystematic review articles. While such publications might not be so crucial in assessing the productivity of departments, it is still important for readers to know who was involved in their development. In particular, readers and editors deserve to know about any possible conflicts of interest. Some journals now have a transparency policy covering such articles – for example, the BMJ asks who had the idea for an article, whether the named authors received assistance in writing it, were paid to write it or have links with organizations that might benefit from its publication (BMJ n.d.).

39.5   Conclusion

Although the authorship criteria proposed by the ICMJE are the most widely promoted by medical journals, they are not universally accepted. While most journals retain the traditional system of listing authors, some have replaced (or augmented) this with a list of individuals' contributions to the research and/or the publication. Neither system determines the order in which authors/contributors are listed, although it is generally assumed that individuals are listed in decreasing order of their contribution. However, such a convention is virtually never explained, and importance is sometimes attached to being the last or corresponding author. Listing contributors does, however, increase transparency, provides more information for editors and readers, and might help detect (and therefore prevent) problems such as guest and ghost authorship.

Journal editors can do little to ensure that authorship policies have been followed. They should therefore consider encouraging institutions and employers to set and enforce policies on listing and acknowledgement. If listing and acknowledgement criteria are agreed, in writing, between all parties, at the start of every project, the incidence of authorship disputes should be reduced.

Acknowledgment

This article is an updated version of *Maturitas* 2008 **62**:109–112 and is republished with permission.

References

Albert, T. and Wager, E. (2003). *How to Handle Authorship Disputes: A Guide for New Researchers*. COPE Report. Committee on Publication Ethics www.publicationethics.org.

Anon. Retraction of Matsuyama et al., "Discoidin domain receptor 1 contributes to eosinophil survival in an NF-kappa B-dependent manner in Churg-Strauss syndrome", Blood 109:22–30 (Jan 2007) (2008). *Blood* 111: 2537.

Bates, T., Anic, A., Marusic, M., and Marusic, A. (2004). Authorship criteria and disclosure of contributions. *JAMA* 292: 86–88.

Bhopal, R., Rankin, J., McColl, E. et al. (1997). The vexed question of authorship: views of researchers in a British medical faculty. *BMJ* 314: 1009–1012.

BMJ 2017, *Transparency policy*. Available from: http://resources.bmj.com/bmj/authors/ editorial-policies/transparency-policy. [Accessed 6 April 2017].

Dhaliwal, U., Singh, N., and Bhatia, A. (2006). Awareness of authorship criteria and conflict: survey in a medical institution in India. *Medscape General Medicine* 8: 52.

Dyer, O. (1995). Consultant struck off for fraudulent claims. *BMJ* 310: 1554–1555.

Hunt, R. (1991). Trying an authorship index. *Nature* 352: 187.

Huth, E. and Case, K. (2004). The URM: twenty-five years old. *Science Editor* 27: 17–21.

International Committee of Medical Journal Editors n.d., *Recommendations for the conduct, reporting, editing, and publication of scholarly work in medical journals*. Available from: http://www.icmje.org/recommendations. [Accessed 6 April 2017].

Kronick, D.A. (1990). Peer review in 18th-century scientific journalism. *JAMA* 263: 1321–1322.

Marusic, A., Bates, T., Anic, A. et al. (2006). How the structure of contribution disclosure statements affects validity of authorship: a randomized study in a general medical journal. *Current Medical Research and Opinion* 22: 1035–1044.

Merz, J.F., Gorton, G.E., Cho, M. et al. (1997). Calculating coauthors' contributions. *Lancet* 350: 1558.

Pignatelli, B., Maisonneuve, H., and Chapuis, F. (2005). Authorship ignorance: views of researchers in French clinical settings. *Journal of Medical Ethics* 31: 578–581.

Rennie, D. (1997). When authorship fails. A proposal to make contributors accountable. *Journal of the American Medical Association* 278: 579–585.

Sheikh, A. (2000). Publication ethics and the research assessment exercise: reflections on the troubled question of authorship. *Journal of Medical Ethics* 26: 422–426.

Smith, R. (1997). Authorship: time for a paradigm shift? *BMJ* 314: 992.

University of Melbourne 2015, *Authorship Policy*. Available from: http://policy.unimelb.edu .au/MPF1181. [Accessed 10 March 2014].

Wager, E. (2007). Do medical journals provide clear and consistent guidelines on authorship? *Medscape General Medicine* 9: 16.

Wager, E., Field, E.A., and Grossman, L. (2003). Good publication practice for pharmaceutical companies. *Current Medical Research and Opinion* 19: 149–154.

40

Biomedical Journals: Scientific Quality, Reputation, and Impact Factor

Katherine G. Akers

Shiffman Medical Library, Wayne State University, Detroit, MI, USA

40.1 Introduction

Despite the tens of thousands of scientific and medical journals currently in existence, the most important research tends to be published in a relatively small number of journals (Garfield 1996; Ioannidis 2006). The reputation of these high-ranking journals is a culmination of several factors, including the prestige of their editors or affiliated organizations and associations, the rigor of their peer-review processes, the newsworthiness of their articles, and the influence of their articles on scientific thought or clinical practice. However, because such factors are subjective in nature, the ranking of journals is more easily accomplished using objective measures of how frequently their articles are cited in other journal articles.

40.2 The Journal Impact Factor

The impact factor, first released in the 1970s, is the most prominent quantitative measure of journal impact (Garfield 2006). Journals with higher impact factors publish articles that are cited a greater number of times, on average, than journals with lower impact factors. A journal's impact factor is calculated by dividing the number of citations in a given year to articles published in the preceding two years by the total number of citable articles published in those two years. For example, if a journal's 2018 impact factor is 4, it means that its articles published in 2016 and 2017 were cited four times, on average, in 2018.

Journal impact factors are calculated and reported annually through Clarivate Analytics' Journal Citation Reports database.[1] Journal Citation Reports does not include all scientific journals; instead, only leading journals are included. To receive an impact factor, a journal must meet several criteria, including its ability to publish issues on time,

1 Clarivate Analytics' Journal Citation Reports database, http://thomsonreuters.com/journal-citation-reports. Access to the Journal Citation Reports database is typically provided through a library subscription. To look up a journal impact factor, access the Journal Citation Reports database through your library webpage or contact your librarian.

A Guide to the Scientific Career: Virtues, Communication, Research, and Academic Writing, First Edition.
Edited by Mohammadali M. Shoja, Anastasia Arynchyna, Marios Loukas, Anthony V. D'Antoni, Sandra M. Buerger, Marion Karl and R. Shane Tubbs.

to apply adequate peer review, to publish articles that are valuable to particular research areas, and to demonstrate international diversity among authors and editorial boards.[2] Thus, a journal must undergo a lengthy evaluation process before being accepted into the Journal Citation Reports database and accrue three years of publication and citation data before receiving an impact factor. Thereafter, journals often prominently post their most recent impact factor as a badge of honor on their websites. As of 2018, Journal Citation Reports included over 11,500 journals.[3]

The impact factor has been heavily criticized for many reasons, including errors in the underlying citation data, the mathematics and lack of transparency of its calculation, its bias toward English-language publications, its focus on only short-term article impact, and the possibility that journals can increase their impact factors by publishing more review articles (which tend to be cited more frequently than research articles) or by promoting journal self-citations (Dong et al. 2005; Rossner et al. 2007; Seglen 1997a,b). In particular, attempting to deliberately inflate journal editors' or publishers' impact factors by encouraging authors to cite articles published in the same journal is considered an unethical practice, with offending journals being banned from the Journal Citation Reports database (Van Noorden 2013). Furthermore, impact factors do not always correlate with how experts in particular research areas rank journals in terms of their overall contribution to the field (Serenko and Dohan 2011). Nonetheless, because the impact factor is a systematic and objective measure that is easy to access, simple to use, and updated annually, it remains the most widely recognized index of journal quality.

The changing nature of how researchers access journal articles, however, might be diminishing the importance of the impact factor. Since the 1990s, researchers increasingly find individual articles by searching through online databases rather than by reading physical copies of journals. Across this time period, there has been a declining association between the impact factor of a journal and the number of times its articles are cited, suggesting that today's researchers are more likely to judge articles based on their own merits more than the reputation of the journal in which they are published (Lozano et al. 2012).

40.3 Alternatives to the Journal Impact Factor

Due to some perceived inadequacies of the journal impact factor, several alternative citation-based journal metrics have been developed (Table 40.1). In addition to the impact factor, which measures journal impact across a two-year time span, Journal Citation Reports also contains measures of more immediate and more prolonged journal impact. Its Immediacy Index and the five-year impact factor are similar to the original impact factor except that it considers the number of citations to articles during the year in which they are published or across a five-year time span, respectively. Journals with a high Immediacy Index tend to publish articles that are urgent or in "hot" research areas, whereas journals with high five-year impact factors tend to publish articles that have more sustained impact on the scientific community.

---

2 The Clarivate Analytics Journal Selection Process, https://clarivate.com/essays/journal-selection-process.
3 A list of all journals with impact factors can be found by selecting "View all journals" in the Journal Citation Reports database.

Table 40.1 Journal impact factor and alternative metrics.

Journal metric	Description	Access
Impact factor	Average article impact over 2 years	http://thomsonreuters.com/journal-citation-reports
Immediacy index	Average article impact over 1 year	
5-yr impact factor	Average article impact over 5 years	
Eigenfactor score	Overall journal impact over 5 years	http://www.eigenfactor.org
Article influence score	Average article impact over 5 years	
SCImago journal rank	Overall journal impact over 3 years	http://www.scimagojr.com
Journal h-index	Overall journal impact over 3 years	
Journal h-index	Overall journal impact over 5 years	http://scholar.google.com/intl/en-US/scholar/metrics.html

Two other alternatives to the journal impact factor, the Eigenfactor score and the SCImago Journal Rank (SJR) indicator, are based on the PageRank algorithm used by Google Search, which weights links from more important websites more heavily than links from less important websites. Thus, whereas the impact factor considers only the number of citations and assumes that all citations are of equal value, the Eigenfactor score (Bergstrom et al. 2008) and SJR indicator (Gonzalez-Pereira, et al. 2009) consider the structure of the citation network and weight citations from higher-ranked journals more heavily than citations from lower-ranked journals. Both alternative measures are indices of the total importance of a journal to the scientific community, but they differ in their sources of citation data (Elkins et al. 2010). The Eigenfactor score is based on citation data from the Journal Citation Reports database, whereas the SJR indicator is based on citation data from Elsevier's Scopus database, which includes a larger collection of journals published in a greater variety of languages.

Another Eigenfactor metric, the Article Influence score, is similar to the journal impact factor in that it is a reflection of the average impact of articles published in a journal. Different from the impact factor, however, the Article Influence score is normalized to the score of the mean article in the Journal Citation Reports database. For example, if a journal's Article Influence score is two, it means that its articles are, on average, twice as influential as the mean article.

Yet another alternative measure of journal impact is the journal h-index (Braun et al. 2006), which was adapted from the researcher h-index originally designed to assess the impact of particular scientists (Hirsch 2005). If a journal's h-index is 20 for a specific time span, it means that it published 20 articles with at least 20 citations each during that time span. A primary advantage of the journal h-index is that it minimizes the undue influence of a small number of highly cited articles on the average assessment of journal impact. SCImago reports journal h-indices over a three-year period for all journals in the

Scopus database, and Google Scholar Metrics reports *h*-indices over a five-year period for the top journals in several fields.

40.4   Caveats to Interpreting the Journal Impact Factor

While journal impact factors can be informative and beneficial, they also need to be viewed with caution in some instances.

40.4.1   Journal Impact Factors Should Not Be Compared Across Disciplines

Different disciplines and research areas vary in their citation practices (e.g. the number of citations included in reference lists) and the time lag between article submission and publication. As such variations can influence the magnitude of journal impact factors, impact factors should not be used to compare journals across disciplines, and differences in impact factors among research areas should not be taken to indicate that some research areas are more important than others. In general, journals in the life sciences tend to have the highest impact factors compared to other fields, although substantial differences in impact factors can also exist among sub-fields within a discipline. For instance, within the field of biomedicine, the average 2004 impact factor for neuroscience journals was 3.252, whereas the average 2004 impact factor for ophthalmology journals was 1.905 (Althouse et al. 2009). Furthermore, because clinical medicine journals draw heavily on basic science literature but not vice versa, basic science journals tend to have higher impact factors than clinical journals (Seglen 1997a,b), which, unfortunately, could be a contributor to the defunding of clinical research in university settings (Brown 2007).

40.4.2   Journal Impact Factors Should Not Be Used to Evaluate Individual Journal Articles

It has been estimated that 20% of articles published in a given journal account for 80% of that journal's citations (Garfield 2006), suggesting that most articles published in a particular journal are not cited as frequently as might be predicted by the journal's impact factor. Because publishing an article in a journal with a high impact factor does not guarantee that the article will be frequently cited, measures of journal impact should not be used to assess the impact of individual journal articles. Rather, the impact of individual journal articles is more appropriately judged directly by the number of times they are cited or by alternative metrics (altmetrics) such as number of downloads or mentions in social media (Priem et al. 2012). Furthermore, although it is often assumed that studies of higher scientific quality are published in higher-impact journals and accrue more citations, journal rankings and article citation counts are not always associated with quality-related aspects of individual articles, such as the presence of an explicit hypothesis statement, specifications of sample size, or errors in the reporting of statistical analysis (Callaham et al. 2002; Nieminen et al. 2005; Brembs et al. 2013). It is also interesting that journals with higher impact factors tend to publish more articles that are retracted due to suspected fraud or error, which could be due to either the greater scrutiny of articles published in top journals or an incentive for researchers to act unethically (i.e. to manipulate or falsify data) or cut corners in an effort to land publications in prestigious venues (Fang et al. 2013).

40.4.3   Journal Impact Factors Should Not Be Used to Evaluate Researchers

Much like individual journal articles, it is also the case that researchers should not be evaluated based on the impact factors of the journals in which they publish, although that has become common practice among hiring panels, tenure, and promotion committees, and granting agencies (Monastersky 2005). In some countries, including Turkey, China, and Korea, scientists receive cash bonuses from the government for publishing in highly ranked journals (Franzon et al. 2011). However, the practice of rewarding researchers based on the impact factors of the journals in which they publish can serve to skew or hinder scientific progress. Instead of being encouraged to explore their own research interests, researchers might be encouraged to pursue research questions that are more likely to be published in top journals, even though such research questions might not ultimately be as important to society as less popular avenues of inquiry that are disseminated through less prestigious journals. Also, a focus on impact factors can lead many researchers to "aim high" when trying to publish their articles – initially submitting manuscripts to high-impact journals followed by submissions to successively lower-impact journals upon manuscript rejection, resulting in scientists spending more time trying to publish their research than embarking on new research projects and ultimately delaying the communication of scientific findings. For this reason, a group of prominent journal editors, scientists, and directors of academic societies and funding agencies issued the San Francisco Declaration on Research Assessment[4] in 2012, which urges institutions to stop using journal impact factors in funding, appointment, and promotion considerations. Instead of indices designed to measure the impact of journals, the performance of researchers is more appropriately judged by their number of publications (i.e. a measure of research productivity), number of citations (i.e. a measure of research impact), or *h*-index (i.e. a combined measure of both productivity and impact; Hirsch 2005), as more fully described in Chapter 9.

40.5   Conclusions

The impact factor is an objective citation-based measure used to rank journals by the magnitude of their impact on the scientific community. Although it has been heavily criticized, spurring the development of alternative and complementary indices, the impact factor remains the most widely recognized measure of journal impact. When deciding where to submit their manuscripts, researchers can compare the impact factors of different journals to predict how much attention their articles might receive after publication. However, it is important to realize that the impact factor was designed to measure the impact of journals and not the impact of particular research areas, individual journal articles, or researchers themselves. Such misuse of the impact factor can have unfortunate consequences on scientific progress. Therefore, instead of becoming preoccupied with journal impact factors or succumbing to "impact factor mania" (Alberts

---

4 San Francisco Declaration on Research Assessment, http://am.ascb.org/dora.

2013), scientific and medical researchers should consider the impact factor as only one indicator of journal reputation and seek to value research quality and the advancement of knowledge over publishing in prestigious journals (Casadevall and Fang 2014).

References

Alberts, B. (2013). Impact factor distortions. *Science* 340: 787.

Althouse, B.M., West, J.D., Bergstrom, C.T. et al. (2009). Differences in impact factor across fields and over time. *Journal of the American Society for Information Science and Technology* 60: 27–34.

Bergstrom, C.T., West, J.D., and Wiseman, M.A. (2008). The Eigenfactor metrics. *Journal of Neuroscience* 28: 11433–11434.

Braun, T., Glanzel, W., and Schubert, A. (2006). A Hirsch-type index for journals. *Scientometrics* 69: 169–173.

Brembs, B., Button, K., and Munafo, M. (2013). Deep impact: unintended consequences of journal rank. *Frontiers in Human Neuroscience* 7: 291.

Brown, H. (2007). How impact factors changed medical publishing—and science. *BMJ* 334: 561–564.

Callaham, M., Wears, R.L., and Weber, E. (2002). Journal prestige, publication bias, and other characteristics associated with citation of published studies in peer-reviewed journals. *The Journal of the American Medical Association* 287: 2847–2850.

Casadevall, A. and Fang, F.C. (2014). Causes for the persistence of impact factor mania. *mBio* 5: e00064–e00014.

Dong, P., Loh, M., and Mondry, A. (2005). The "impact factor" revisited. *Biomedical Digital Libraries* 2: 7.

Elkins, M.R., Maher, C.G., Herbert, R.D. et al. (2010). Correlation between the journal impact factor and three other journal citation indices. *Scientometrics* 85: 81–93.

F.C. Fang, R.G. Steen, A. Casadevall. (2013). Misconduct accounts for the majority of retracted scientific publications. *Proceedings of the National Academy of Sciences of the United States of America*, 110: 17028–17033.

Franzon, C., Scellato, G., and Stephan, P. (2011). Changing incentives to publish. *Science* 333: 702–703.

Garfield, E. (1996). The significant scientific literature appears in a small core of journals. *The Scientist* 10: 13–15.

Garfield, E. (2006). The history and meaning of the journal impact factor. *The Journal of the American Medical Association* 295: 90–93.

B. Gonzalez-Pereira, V. Guerrero-Bote, F. Moya-Anegon. (2009) The SJR indicator: A new indicator of journals' scientific prestige. *arXiv*, http://arxiv.org/abs/0912.4141

J.E. Hirsch. (2005). An index to quantify an individual's scientific research output. *Proceedings of the National Academy of Sciences of the United States of America*, 102: 16569–16572.

Ioannidis, J.P.A. (2006). Concentration of the most-cited papers in the scientific literature: analysis of journal ecosystems. *PLoS One* 1: e5.

Lozano, G.A., Lariviere, V., and Gingras, Y. (2012). The weakening relationship between the impact factor and papers' citations in the digital age. *Journal of the American Society for Information Science and Technology* 63: 2140–2145.

Monastersky, R. (2005). The number that's devouring science. *The Chronicle of Higher Education* 52: A12.

Nieminen, P., Carpenter, J., Rucker, G. et al. (2005). The relationship between quality of research and citation frequency. *BMC Medical Research Methodology* 6: 42.

Priem, J., Groth, P., and Taraborelli, D. (2012). The altmetrics collection. *PLoS One* 7: e48753.

Rossner, M., Van Epps, H., and Hill, E. (2007). Show me the data. *The Journal of Cell Biology* 179: 1091–1092.

Seglen, P.O. (1997a). Why the impact factor of journals should not be used for evaluating research. *BMJ* 314: 498–502.

Seglen, P.O. (1997b). Citations and journal impact factors: questionable indicators of research quality. *Allergy* 52: 1050–1056.

Serenko, A. and Dohan, M. (2011). Comparing the expert survey and citation impact journal ranking methods: example from the field of artificial intelligence. *Journal of Infometrics* 5: 629–648.

R. Van Noorden. (2013). New record: 66 journals banned for boosting impact factor with self-citations. *Nature News Blog*, http://blogs.nature.com/news/2013/06/new-record-66-journals-banned-for-boosting-impact-factor-with-self-citations.html.

41

Scholarly Open-Access Publishing
Jeffrey Beall

Retired from Jeffrey Beall Auraria Library, University of Colorado Denver, Denver, CO, USA

41.1 Introduction

Although the first efforts at promoting open access (OA) to scholarly literature began around 1993 (Laakso et al. 2011), the earliest organized OA compact was the Budapest Open Access Initiative (BOAI) signed in 2002 (Budapest Open Access Initiative 2002). The 16 signers attended a George Soros-sponsored meeting in Budapest to create and sign the document. The initiative was partly a response to the serials crisis, which refers to a time when journal subscription prices were increasing greater than the rate of inflation, a situation that forced many academic libraries to cancel many needed academic journal subscriptions.[1]

The BOAI represents the formal beginnings of a social movement that continues to this day. After BOAI, similar initiatives soon followed, including the Berlin Declaration on Open Access to Knowledge in the Sciences and Humanities (2003) and the Bethesda Statement on Open Access Publishing (2003). Significantly, the goal of all three initiatives – universal OA – is yet to be achieved, more than a decade after their signing. That is to say, there are still many journals that publish using the traditional, subscription model.

The official BOAI definition of OA is broad:

> By 'open access' to this literature, we mean its free availability on the public internet, permitting any users to read, download, copy, distribute, print, search, or link to the full texts of these articles, crawl them for indexing, pass them as data to software, or use them for any other lawful purpose, without financial, legal, or technical barriers other than those inseparable from gaining access to the internet itself (Budapest Open Access Initiative 2002).

Thus, OA by this definition is much more than just "ocular" access to documents. It also includes being able to reproduce them, provided proper attribution is made. The definition skips mention of any commercial reuse rights, for this has been contentious. Typically, European OA activists prefer a broader waiver of rights that includes

---

1 For a brief and well-written history of scholarly open-access publishing, see Liesegang (2013).

A Guide to the Scientific Career: Virtues, Communication, Research, and Academic Writing, First Edition.
Edited by Mohammadali M. Shoja, Anastasia Arynchyna, Marios Loukas, Anthony V. D'Antoni, Sandra M. Buerger, Marion Karl and R. Shane Tubbs.
© 2020 John Wiley & Sons, Inc. Published 2020 by John Wiley & Sons, Inc.

Table 41.1 Publishing models and fees.

Publishing model	Fees charged to author	Fees charged to reader (subscriber)
Subscription	No[a]	Yes
Gold	Yes	No
Green	No	No
Platinum	No	No

a) Some subscription journals, especially journals published by scholarly societies, may impose "page charges" on their authors. These fees are usually nominal and may cover additional services such as color illustrations and the like.

commercial reuse, but North Americans have tended to limit the re-publication of OA materials to noncommercial uses (Beall 2013).

Today, in addition to the traditional or subscription-based model, many recognize three models of scholarly OA publishing, gold, green, and platinum. Table 41.1 summarizes the payments on the author and reader side for the various scholarly publishing models currently in use for journals. The subscription model is financed by subscription fees levied on libraries or other organizations and on individual subscribers. Additionally, online versions of subscription journals are for sale individually to internet users. The great advantage of the subscription model is that there is almost never a publishing charge to authors.

Gold open access is free to anyone with internet access, and the model is financed by a charge levied on authors upon acceptance of a manuscript. The advantage of this model is that articles are free to everyone. The gold (author-pays) OA model favors hard sciences, where grant money is often used to pay for the article-processing charges (APCs) that OA publishers impose upon acceptance of a scholarly manuscript. Typically, research in the humanities, the arts, and sometimes the social sciences is completed without grant funding, so research in these areas published in gold OA journals often requires that authors pay the APCs out of their personal funds. In a few instances, universities have set up funds to help their faculty members pay for these APCs, sparing them the burden of having to pay out of their own pocket. But these university funds have created some political problems at the universities. When the funds are limited, then committees must be formed to determine which researchers will be supported. This means that faculty compete with each other for the funds and that new bureaucracies must be created to administer the funds and their distributions.

The subscription-based publishing model almost never requires monetary transactions between authors and publishers, and this is turning out to be a great strength of the traditional system (still, some noncommercial publishers sometimes impose moderate page charges on authors to help subsidize publishing costs). Overall, however, the implementation of new scholarly publishing distribution models that require monetary transactions between scholarly authors and scholarly publishers has introduced many new problems into scholarly communication. For research paid for by grant-funding agencies, the new practice of paying APCs from grant money means that less money is going toward research, because a proportion of the money is used to pay for publishing.

Collectively, this means that millions of dollars once spent directly on research is now being paid to OA publishers. In this way, open access publishing has merely shifted the financing of scholarly publishing from library subscription revenue to researchers themselves.

Green open access is also called the author self-archiving model of OA publishing. In this model, authors publish in a non-open-access subscription journal and then make a post-print of the article available in an OA repository. A post-print is the author's last version of an article before it is sent off to the journal publisher for copyediting and publishing. Thus, most post-prints in repositories are usually Microsoft Word documents. Only in very rare cases do publishers grant authors the right to self-archive the final PDF version of the article. Green OA has two substantial advantages, for it allows authors to publish in high-quality and established subscription journals and at the same time makes their work available for free in OA repositories. There are generally two types of repositories, institutional (such as a university repository) and disciplinary (such as arXiv, organized by the particle physics community).

In green OA, authors transfer copyright of their work to the publisher, so permission is needed before authors can self-archive a post-print of the article in the repository. A database called SHERPA/RoMEO[2] makes available a list, by journal, of the standard copyright terms, letting authors know what self-archiving rights they have for their post-prints. Additionally, it is often possible for authors to negotiate terms of the author agreement before signing. A freely available document called the *SPARC Author Addendum to Publication Agreement*[3] provides boilerplate language that authors can use when negotiating. Finally, some publishers impose embargoes on author self-archiving. For example, a particular journal or publisher might allow author-self archiving of a post-print in an OA repository, but not until six months have passed after its original publication date. The length of embargoes varies by journal and by publisher, and the embargoes serve to prevent subscription cancellations.

Platinum OA is similar to gold OA except that no fee is charged to the authors upon acceptance of the manuscript. Instead, the publisher covers the publication costs. Thus, nonprofit organizations, including universities, associations, and the like, often publish scholarly journals using the platinum OA model. The model has the advantage of being free to readers and free to authors. Also, removing the practice of requiring monetary transactions between authors and publishers removes any potential conflicts of interests on the part of the publisher.

41.2   Monitoring Scholarly Open-Access Publishing

The Open Access Scholarly Publishers' Association (OASPA) is the main industry association for scholarly OA publishers. OASPA (pronounced oh-ASS-puh) is an international trade organization that publishes a members' code of conduct. OASPA is not a regulatory association. Overall, only a minority of OA publishers are members, but most of the larger OA publishers belong to OASPA. Its membership includes traditional

---

2 The SHERPA/RoMEO database is available at: www.sherpa.ac.uk/romeo.
3 The SPARC Author Addendum to Publication Agreement is available at: http://www.sparc.arl.org/resources/authors/addendum-2007.

publishers whose journal portfolios also include OA journals, i.e. publishers whose journal portfolios include both OA and subscription journals.

There are additional industry associations that OA publishers may belong to. One is the Society for Scholarly Publishing, and its membership includes mostly traditional scholarly publishers, but it also includes some OA publishers as well. Another organization worth mentioning is COPE, the Committee on Publication Ethics. Membership includes individual journals and publishers. Its website makes freely available many helpful resources for scholarly publishers, including workflows for dealing with research misconduct and a forum for its members to discuss individual cases of research misconduct such as plagiarism.

To gain credibility, some questionable OA publishers will claim on their websites to be members of COPE, even when they are not. Moreover, they will claim that they follow all the COPE guidelines when they in fact do not. Some abstracting and indexing services require that journals maintain ethical statements on their websites in order to be eligible to be considered for indexing. To meet this requirement, predatory publishers will simply copy portions of some COPE statements. See Section 41.3 for a description of predatory publishers.

Companies that assign journal metrics play a quasi-regulatory role for scholarly journals. For example, the company Clarivate Analytics markets a product called Journal Citation Reports, otherwise known as the impact factor. Because the impact factor metric reflects on the scholarly community's overall average rate of citations to articles in a journal, this metric reflects the perceived importance or value of the given journal in the scholarly community. Other journal-level metrics are emerging; one is called the Eigenfactor (West et al. 2010). Also, Google Scholar calculates and makes available some journal-level metrics. One of these is the h-5 index, which the database defines as "the h-index for articles published in the last 5 complete years. It is the largest number h such that h articles published in 2008–2012 have at least h citations each."[4] To access this data in Google Scholar, click on the link at the top called "Metrics," then enter a journal title in the search box.

There is a trend away from journal-level rankings toward more granular ratings, specifically at the article level. Some are experimenting with alternative metrics, also called *altmetrics*.[5] Some examples of altmetrics are the number of downloads a particular article has achieved, or the number of mentions on social media. Despite intense interest, the scholarly community has not yet accepted such measurements as valid, especially in terms of tenure and promotion.

I have also noted a number of bogus metrics ratings used by predatory publishers. Unscrupulous OA journals and publishers are in need of metrics that will attract article submissions. A number of firms have emerged that are filling this need, creating bogus measures that generally contain the term *impact* (Beall 2013).

41.3 Predatory Publishers

As described earlier, the gold OA model of scholarly publishing is one in which the author is charged a fee upon acceptance of the manuscript. This fee creates a conflict of

---

4 Google Scholar metrics are available at: https://scholar.google.com/intl/en/scholar/metrics.html#metrics.
5 For a description of the Altmetric Attention Score and donut, see www.altmetric.com.

interest for the publisher, because the more papers it accepts, the more money it makes. This conflict of interest leads in some cases to a lax or nonexistent peer review because the publishers may want to accept more papers and therefore earn more income from the APCs. Thus, predatory publishers are operationally defined as publishers that abuse the gold (author-pays) model of scholarly publishing for their own profit. While the gold OA model has been in use for many years, predatory publishers did not start to appear until around 2009, and their number has been growing at a rapid pace since that time. There is a low cost barrier to starting up a new publishing operation using the gold OA model in the advanced era of electronic publishing. It is not uncommon for predatory publishers to launch a fleet of 25 or more journals in a single day. These launches are usually accompanied by massive spam campaigns for editorial board members and article submissions. Very often, predatory publishing operations are the effort of a single individual, and these publishers frequently operate out of a dwelling.

Predatory publishers aim to appear legitimate. That is to say, their business model is to make scholars think that they are legitimate publishers so that they can attract articles and the accompanying (APCs) from researchers submitting their work. To appear legitimate, predatory publishers use tricks and deception. A common trick is to hide the publisher's true location and give the appearance that the publisher is based in a Western country rather than, for example, South Asia. Predatory publishers create journal titles that mimic established and prestigious journals, and they use unwarranted geographical terms in journal titles.

Unlike traditional, subscription journals, which must cater to readers or face losing subscriptions, OA journals focus on keeping the authors – their customers – happy, not the readers. Thus, they are naturally author-centric rather than reader-centric. The authors paying the APCs are their customers; the readers are of secondary importance. Predatory publishers and journals represent the extreme of this focus, concentrating their efforts on serving authors at the expense of readers. A fast and easy peer review process is desired by some authors, and predatory publishers accommodate this need. A high-quality peer review process often takes time, so speeding up this process may inadvertently reduce the quality.

Predatory publishers use spam to solicit scholarly manuscripts for their journals, and they have perfected their use of this tool. I have documented that some predatory publishers send out spam to researchers, soliciting submissions but failing to mention the required APCs. Then, they quickly publish the papers and send an unexpected invoice to the authors. Surprised, the authors do not know what to do. They may try to withdraw their article, but the publisher may refuse and state that the article cannot be withdrawn because it has already been assigned a DOI (digital object identifier). Though this claim is false, it is still used to pressure authors to pay the fee.

Among those most vulnerable to manipulation by predatory publishers are junior faculty, graduate students, and postdocs. Predatory publishers know that they have the least experience in academic publishing and are most vulnerable to exploitation. One very effective strategy involves personalized spam emails. The predatory publishers send a spam email to a researcher, praising an earlier work and inviting another. For researchers unaccustomed to getting praise or positive feedback on their published works, this method is very effective for it is pandering.

Most predatory publishers are based in South Asia, particularly India and Pakistan. According to one study of predatory journals, India is "the world's largest base for open-access publishing" (Bohannon 2013, p. 64). Other common headquarters locations for predatory publishers include Nigeria, England, the Province of Ontario, and the United States. However, most of the publishers in Western Countries are run by expatriates, and they often market their services to those in their home countries. Many institutions in Asia give greater credit to scholars who publish in "foreign journals," so the publishers have been set up in Western countries to accommodate this need. This explains why the terms *international*, *global*, *world*, and *universal* are so common in predatory journal titles. Moreover, many questionable publishers based in non-Western countries lie about their true locations, feigning addresses in the United States, Britain, Canada, or Australia. They do this using the address of a friend or relative who lives in the Western country, or they use the addresses of mail-forwarding services based in Western countries.

Predatory publishers hurt science and hurt science communication in many ways (Beall 2016). Most upstart OA publishers appear to be owned by a single person hoping to get in on the easy profits to be made by questionable publishers. The barrier to entry is low – all one needs is a website, and an email address, and some unique journal titles. Most of these smaller publishers devote no resources to what is called *digital preservation*, which is the process of backing up information to ensure against its loss. Most of these publishers just keep a single copy of every file on the web server that distributes or serves their files, despite there being some very low-cost options for digital preservation run by some nonprofit organizations that provide standardized backups. Some predatory publishers have disappeared from the internet, taking all their published content with them. Thus, all the scholars who paid to have their articles made available OA forever have seen their content disappear within only a few years.

Many predatory publishers also offer conferences as a side business, or in some cases, conference organizers publish a few journals as an outlet for the presentations made at the meetings they organize. These conferences are sometimes predatory in many of the same ways that make predatory journals predatory. For example, scholars regularly receive spam emails promoting questionable scientific conferences. There are no published criteria for evaluating questionable conferences, but the ones associated with known predatory publishers ought to be avoided. Often, these conferences are held in tourist cities such as Las Vegas, Nevada, and Orlando, Florida. Though they claim to have a peer review, virtually everyone who pays the registration fees gets his or her paper accepted for presentation at the conference. Sometimes these conferences bundle conference registration and APCs into a single price, allowing researchers to claim credit for both a conference presentation and a journal publication.

Predatory journals are allowing the publishing of much questionable science. The role of peer review is to function as a gatekeeper for science, only allowing honest science to be published in scholarly journals that bear science's stamp of approval. But because many predatory publishers conduct a perfunctory or sham peer review (Beall 2017), much pseudoscience is being published as if it were authentic science. The separation of authentic science and pseudoscience is called *demarcation*, and predatory publishers are perhaps the biggest threat to demarcation ever known. Examining predatory publishers' journals, it is not hard to find papers that offer nonscientific views of the etiology of autism, climate change, and the existence of alien beings.

Overall, predatory publishers are staining scholarly communication, poisoning the work of those promoting scholarly OA publishing, and blurring the line that separates authentic science from pseudoscience. Because they often operate internationally, because they often hide their true locations and identities, and because they benefit from the privilege of freedom of the press, there is very little action that can be taken to regulate or suppress predatory publishers. The best antidote is education; scholars must learn about predatory publishers and how to recognize and avoid them.

41.4 Discovering Open-Access Publications

The most comprehensive list of scholarly OA journals is DOAJ, the Directory of Open-Access Journals.[6] However, some find the list too comprehensive, meaning that its coverage is too broad and includes predatory journals. A sting operation conducted by journalist and scientist John Bohannon (2013) found that 45% of a sample of journals listed on DOAJ accepted a bogus scientific paper for publication. Reacting to criticism it received after news of the sting was published, DOAJ removed some journals from its central index.

Google Scholar indexes most OA journals, but unfortunately, it widely indexes predatory journals as well, so the resource must be used with caution. I am frequently asked for a list of "good" OA journals to publish in, but no such white list exists. Still, an adequate starting point might be the membership list of the OASPA. Also, the publisher Elsevier makes available a product called Elsevier Journal Finder[7] that allows a user to enter in the title and abstract of a scholarly manuscript and receive suggestions for appropriate journals to submit the work to. The system allows for limiting by OA journals. Of course, the search results are limited to Elsevier journals.

41.5 The Advantages and Disadvantages of Publishing in Open-Access Journals

Earlier, I discussed some of the advantages of scholarly open-access publishing, the chief advantage being that it makes research freely available to anyone with internet access. Scholars in developing countries are often cited as the beneficiaries of OA, but this is not exactly true. Most of the large scholarly publishers are members of organizations that distribute published journals for free in lower-income countries. Research4Life is the umbrella term for four programs that distribute research to developing countries for free or at a reduced cost. The programs include HINARI (Health InterNetwork Access to Research Initiative), AGORA (Access to Global Online Research in Agriculture), OARE (Online Access to Research in the Environment), and ARDI (Access to Research for Development and Innovation).

Universal open access would make such programs unnecessary. Gold open access, however, requires that authors finance scholarly publishing, and scholars in poorer

---

6 The Directory of Open Access Journals is located at http://www.doaj.org.
7 Elsevier Journal Finder is available at: http://journalfinder.elsevier.com.

countries generally lack the funds to pay the APCs. However, most scholarly OA publishers waive or greatly discount APCs for authors from lower-income countries. The ones who get squeezed in the middle are authors from middle-income countries, who are not eligible for APC waivers or discounts, and their institutions generally lack funds to subsidize their APCs. Thus, scholars from middle-income countries should consider platinum OA publishing, which does not impose author fees. However, because they do not impose such fees, platinum OA journals are more competitive and have lower acceptance rates. Also, the amount of an individual journal's APC is beginning to reflect the journal's status, with higher-quality journals charging more. For example, gold OA journals with impact factors are able to demand higher APCs.

Earlier in this chapter, I gave the standard description of green OA. One of the tacit goals of this OA model is that authors will become so accustomed to self-archiving their works, they eventually will not even bother to send them to journals; instead they will just self-publish them, removing publishers from the scholarly communication equation altogether. One of the weaknesses of the green model – author self-archiving – is that it sometimes does not function in the way for which it was developed. Once a paper is published in a journal, an author feels a sense of completion, and most neglect to upload the post-print to a repository. This reluctance is why we hear about OA mandates; in order to get the green OA model to work with any measure of success, scholars must be mandated into practicing the model, a practice that is disrespectful of individual liberties.

Those publishing their work in borderline, questionable, or predatory journals risk their work being stained merely by being published in the same journal as junk science. A pseudoscience article in a journal that falls next to a legitimate article in the same journal's table of contents will certainly not help the reputation of the good article. In fact, the opposite will happen. The juxtaposition of a poor article will hurt the credibility of the good one. For this reason alone, authors should always submit their work to the best possible journals.

41.6 The Future of Open-Access Publishing

Many who describe the future have a financial interest in seeing that their vision is realized, and the visions of others not realized. That is to say, company executives, for example, tend to describe a future in which their companies' products will be essential. Thus, talking about the future of scholarly publishing in general and OA publishing in particular has become hackneyed. Open-access advocates all see an OA future. Consultants all see a future where their individual specialties will be essential and something worthy of charging for. The future will probably see a mixture of scholarly distributions models.

I think everyone involved in scholarly communication should step back and not promote a particular model of scholarly publishing and instead should disinterestedly examine all potential models and support the one or ones that best serve the needs of scholarly communication. No one knows what that is at this point. Fortunately, we are seeing much experimentation involving different funding models for OA publishing, and we will all be able to learn from the successes and failures of these experiments. It is clear that there is a lot of money to be made in scholarly publishing, and that is why the stakes are so high and why so many are keenly invested in its future.

I think disappearing content will be a big theme of OA publishing's future. There is an abundance of one-man operations in place now – publishers that do not back up their content but are operated by sole-proprietors, and when these sole proprietors grow tired of the business or when they pass on, there will be a lot of content whose accessibility will be jeopardized. Much scholarly content will surely disappear forever – content that was paid for by author fees with the understanding, often tacit, that it would be published in perpetuity.

41.7   Why Open-Access Publications Are Gaining in Appreciation

Libraries like OA because it enables them to provide scholarly content to their users at no cost. At the same time, academic library scholarly instruction programs are increasingly teaching college freshmen how to critically analyze information, a skill that was not needed so much in terms of scholarly publishing in years past. When the subscription model dominated, scholarly journals played a strong validation role. Because print journals have limited space, they could only afford to accept and publish the very best articles. This limitation meant that print journals validated research and published only the very best of it. In the online world, this space limitation is gone, and the journals can publish volumes of unlimited size. Because of this lack of space restriction, many journals, especially the predatory journals, publish articles that are unworthy of being called science.

Competition will affect gold OA publishers in several ways. As mentioned earlier, OA journals with favorable, documented metrics such as valid impact factors will be able to charge higher APCs. At the other end of the spectrum will be discount OA journals, charging very low APCs as a way of generating business.

While libraries will increasingly save money as more scholarly content becomes freely available, universities, on the other hand, will see higher expenditures as they will increasingly be called to pay for and subsidize their faculty members' APCs, i.e. those faculty who lack grants to pay to publish their research. University support for APCs will bring new and onerous bureaucracy to universities. Committees and programs and departments and assistant vice chancellors for this and that will be tasked with administering APC funds, and politics will increase. Certainly, the funds will be limited, so faculty will have to compete for them. We will see faculty in gender studies departments competing for scarce funds with faculty in chemistry, civil engineering, and geology. This again is another area in which imposing financial transaction costs between publishers and scholarly authors creates many more new problems than it solves.

41.8 How to Recognize Predatory Journals

Many scholars read and write reviews of new books in their fields. My website content is tantamount to book reviews, only in this case the works being reviewed are not books but journals and publishers. Librarians have traditionally reviewed books and published these reviews in library journals, whose audiences are other librarians. These reviews have helped in library collection development – that is, deciding which books to buy. Now in the context of scholarly OA publishers, the reviews help libraries decide which journals and publishers to include in their online discovery tools, such as online catalogs, and they help academic librarians deal with questions from patrons asking for help in deciding which journals to submit a paper to, an increasingly common question at academic library reference desks.

Attempts have been made to characterize, develop and validate criteria for determining predatory open-access publishers (Tsuyuki et al. 2017; Gonzalez et al. 2018; Olivarez et al., 2018). A periodically updated criteria that I have introduced, broadly known as "Beall's Criteria" (Table 41.2), recognizes COPE's "Code of Conduct for Journal Publishers" and "Principles of Transparency of Best Practice in Scholarly Journals" as foundational documents. This "Criteria" is currently in its third edition and contains

Table 41.2 Characteristics of predatory journals and publishers (a truncated version of Beall's Criteria).

General	• Scope of the journal is inappropriately broad or does not align with its content. • Name and/or design of the journal mimics that of established publishers. • Name of the journal does not accurately reflect its origin (e.g., publisher, editor, and/or affiliates do not relate to Canada, despite being named a "Canadian" journal). • Ownership, management, and/or editorial teams of the journal are unclear, vague, or misleading. • Vague contact information, often unclear or misrepresentation of the journal's headquarters (e.g. through the use of addresses that are actually mail drops in the United Kingdom or United States). • Journal articles and/or websites contain serious formatting or editing errors or do not follow industry standards. • Journal is not included and/or fakes are included in journal whitelists (e.g. Directory of Open Access Journals, MEDLINE). • Journal uses falsified publishing standards and bibliometrics (e.g., International Standard Serial Numbers, impact factors). Beware of large improvements in impact factors over a short period of time.
For academics	• Publisher sends spam emails to solicit article submissions, peer review, and/or conference attendance (often with adulatory language about your previous work). • Publisher advertises promises of rapid publication and/or unusually quick peer review that is ultimately of low quality or is false. • Deception about publication fees and opportunities to edit your work once submitted. • Does not openly discuss or relinquishes copyright privileges regarding your work.
For clinicians	• Scientific evidence and discussion presented in the journal are consistently of low quality, due to a poor peer-review process.

Source: Tsuyuki et al. 2017, reproduced with permission.

27 items divided into four categories, namely, editor and staff, business management, integrity, and other (Olivarez et al. 2018).

I recommend that all scholars become experts on the journals in their fields. Read many articles from different journals in your discipline. This familiarization will help you learn about the various journals and their strengths and weaknesses. Subscribe to table of contents alerts, usually sent via email, to the journals you find most connected to your particular research agenda. Building on this, also consult with senior colleagues in your field about the top journals. Tenure criteria often list preferred journals, so pay careful attention to tenure criteria documents as you proceed on tenure track.

Above all, resist the temptation to purchase the goods that predatory publishers offer: a quick and easy acceptance of your work. Scholarly publishing is not a fast process. A publication in a poor journal early in one's career can do much damage over the length of the career. The scholarly publishing process, done correctly, takes time, and peer review – in addition to functioning as a quality-control mechanism for science – can also greatly help you improve your submitted papers. That is to say, the advice that peer reviewers give you can make your scientific article much better than it otherwise might have been.

Good scholarly publishers add great value to research. They provide expert copy-editing, ethics management, copyright protection, and services for improving datasets and figures. Their effective management of the peer review process is invaluable. Never underestimate the value that a high-quality scholarly publisher can add to your published research. For more details on how to find a suitable journal for your manuscript, see Chapter 42.

References

Beall, J. (2013). Avoiding the peril of publishing qualitative research in predatory journals. *Journal of Ethnographic & Qualitative Research* 8 (1): 1–12.

Beall, J. (2017). Predatory journals threaten the quality of published medical research. *Journal of Orthopaedic and Sports Physical Therapy* 47 (1): 3–5.

Beall, J. (2016). Medical publishing and the threat of predatory journals. *International Journal of Women's Dermatology* 2 (4): 115–116.

Berlin Declaration on Open Access to Knowledge in the Sciences and Humanities (2003) Available from: http://oa.mpg.de/lang/en-uk/berlin-prozess/berliner-erklarung [Accessed: 03/19/14].

Bethesda Statement on Open Access Publishing (2003) [WWW] Available from: http://legacy.earlham.edu/~peters/fos/bethesda.htm [Accessed: 03/19/14].

Bohannon, J. (2013). Who's afraid of peer review? *Science* 342 (6154): 60–65.

Budapest Open Access Initiative. (2002) Available from: http://www.budapestopenaccessinitiative.org [Accessed: 03/19/14].

Gonzalez, J., Bridgeman, M.B., and Hermes-DeSantis, E.R. (2018). Differentiating predatory scholarship: best practices in scholarly publication. *International Journal of Pharmacy Practice* 26 (1): 73–76.

Laakso, M., Welling, P., Bukvova, H. et al. (2011). The development of open access journal publishing from 1993 to 2009. *PLoS ONE* 6 (6): e20961.

Liesgang, T. (2013). The continued movement for open access to peer-reviewed literature. *American Journal of Ophthalmology* 156: 423–432.

Olivarez, J.D., Bales, S., Sare, L. et al. (2018). Format aside: applying Beall's Criteria to assess the predatory nature of both OA and non-OA library and information science journals. *College and Research Libraries* 79 (1): 52–67.

Tsuyuki, R.T., Al Hamarneh, Y.N., Bermingham, M. et al. (2017). Predatory publishers: Implications for pharmacy practice and practitioners. *Canadian Pharmacists Journal* 150 (5): 274–275.

West, J.D., Bergstrom, T.C., and Bergstrom, C.T. (2010). The Eigenfactor metrics™: a network approach to assessing scholarly journals. *College and Research Libraries* 71: 236–244.

42

How to Find a Suitable Journal for Your Manuscript

Mohammadali M. Shoja[1], Thomas P. Walker[2] and Stephen W. Carmichael[3]

[1] Division of General Surgery, University of Illinois at Chicago Metropolitan Group Hospitals, Chicago, IL, USA
[2] Holy Spirit Library, Cabrini University, Radnor, PA, USA
[3] Emeritus Center, Mayo Clinic, Rochester, MN, USA

42.1   Introduction

Writing manuscripts in any field is an opportunity for the researcher to communicate his ideas and work with the outside world. The scholarly journals serve to not only publish a researcher's work but also to certify the scientific validity and significance of the work (Babor et al. 2008). When looking for the most suitable journal to submit your manuscript, several objectives should be kept in mind: (i) the article should fit with the journal's intended scope and readership, (ii) the journal editors should be capable of providing appropriate and constructive comments (through engagement of experienced reviewers) to help authors in improving the manuscript, (iii) the journal should have a timely editorial process and publication schedule so that the manuscript undergoes peer review in the least amount of time and could be made available to the readers soon after being accepted for publication, and (iv) the journal should reach out to the potential audience of the work to maximize the exposure, and should have a publishing mechanism that ensures the published articles will remain available for use by the scholars anytime in future (Callaham 2002; Kloner et al. 2013; Thompson et al. 2007). The importance of the latter cannot be overstated, especially in the current advanced electronic age where a growing number of new online titles are created by various publishers every year. The first step in selecting a suitable journal is to exclude from your list those pseudo-scholarly journals whose prime aim is to charge an article processing charge (APC) through a faked or inadequate editorial process. Chapter 41 carefully addresses this matter, and for the remainder of this chapter, we assume that you have already excluded the pseudo-scholarly journals from your list for potential publications.

Both seasoned and novice researchers should make themselves familiar with the journals in their field on an ongoing basis. While novice writers may not be aware of many publication opportunities available for them, the seasoned writers show a tendency to publish their works in the same journals, overlooking the new publication opportunities

A Guide to the Scientific Career: Virtues, Communication, Research, and Academic Writing, First Edition.
Edited by Mohammadali M. Shoja, Anastasia Arynchyna, Marios Loukas, Anthony V. D'Antoni, Sandra M. Buerger, Marion Karl and R. Shane Tubbs.

(Conte 2017). By continually answering these questions, you can develop an updated list of potential journals to consider for submitting your current and future works:

- Which journals are being published in my field?
- Which journals publish in my field of interest?
- Which journals are suitable for my manuscripts?

42.2 Initially Deciding on Which Journal to Publish

Scientific researchers generally consider many factors in which to publish. Is the focus of the journal similar to the main theme of the manuscript? Is the manuscript focused toward basic science or clinical journals? Aiming for journals with the highest possible visibility and quality, while keeping in mind rapid publishing in some lower-visible journals, is a practical approach.

Researchers may want to consider searching their institutional library resources and other national and international databases such as PubMed, Science Citation Indexes, etc. to determine which journals publish in their fields of interest. In choosing the journal to publish in, a researcher will want to consider the journal impact factor (JIF), devised by Eugene Garfield, the founder of the Institute for Scientific Information, which is now part of Thomson Scientific, a large internationally known US-based publisher. Many journals post their annual impact factor (IF) on their websites. But, many other factors contribute to a successful publication, and in the following section we will consider them one by one.

42.3 Factors to Consider in Selecting a Suitable Journal

42.3.1 Audience

When looking for the most suitable journal to which you will first submit your manuscript, the key factor to address is the issue of who is the appropriate audience to read your article and if the journal attracts the intended range of readership. This might be obvious for a seasoned scientist. If you are new to the field, you will need to identify experienced people in the field and ask them what journals they read. If your senior colleagues or mentors are cooperative, briefly describe your study and ask them if it would be appropriate for the journals they read. This will go a long way toward finding a suitable journal.

Journals can be categorized into four classes based on their intended audience (Figure 42.1). Multidisciplinary journals that publish articles from a variety of disciplines (such as Nature and Science) usually have a higher circulation (prints or electronic article downloads), impact factor and prestige compared to the disciplinary, specialty, and constricted journals. Disciplinary journals deal with a specific discipline such as medicine, surgery, or biology. Next to disciplinary journals are specialty and constricted journals that respectively focus on a particular field (such as oncology, cardiology, neurology, etc.) and a narrow topic (such as melanoma, congenital heart

Figure 42.1 Journal categories based on the intended audience.

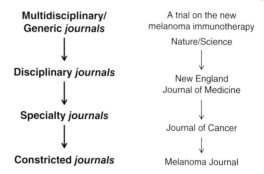

disease, Alzheimer's disease, etc.). Moving down this ladder, the JIF usually comes down, suggesting that it may be easier to publish in them. Although constricted journals have a narrow reader base, some have relatively high impact factor comparable to the counterpart specialty and even disciplinary journals, and it should not be assumed that publishing in them is universally easier than the rest of journal categories. When choosing the right journal, the authors may consider candidates from each of these categories that offer a unique and often overlapping reader base. An important consideration here is that as you go up this ladder from constricted to more broadly oriented disciplinary and multidisciplinary journals, the manuscript should be prepared and written in a language that is comprehensible to audience not specialized on the topic (Babor et al. 2008).

Factors increasing the journal exposure are electronic access, indexing and abstracting services, open-access, and English language publishing. While print circulation was the sole estimate of a journal's reader base in the pre-internet era, nowadays the most reliable indicator of a journal visibility and readership is the number of article view and downloads. Additional markers of greater exposure are library subscriptions and the number of copies distributed to the members of a learned society or professional organization (Babor et al. 2008). Most electronic journals report their average monthly download or the total number of view and pdf downloads for each article. The power of open-access publishing is the greater visibility and accessibility of the published contents it offers to the readers. Access to traditional, subscription-based journal articles is limited to member institutions and individuals or requires a payment varying between 5 and 100 USD for one-time article download. Some subscription-based journals offer alternative ways to improve visibility of the content by having optional open-access publishing choice, allowing open-access to some or all contents after a period of time post-publication or allowing the authors to self-archive their accepted or published versions either before or after press (Babor et al. 2008). Print-only publications can only publish a set number of pages each year. Space limitation is no longer an issue for electronic publications.

42.3.2   Fitness Factor: Journal Scope and Aim

Most journals publish their Aims and Scope online or in the first or the last issue of a volume. Again, determine if these are consistent with what you plan to submit. This will lead to a very short list of journals that are appropriate and desirable for your manuscript. Pay attention to the restrictions that the journals impose on the article types, field or topics, and word counts. For example, some journals do not consider case reports for publication; others might only consider review articles or accept original research articles that are likely to attract a wide audience versus a specific group of readers.

Assuming that a submission has scientific merit that warrants publication, the most common cause of manuscript rejection by the journal – in our experience – is not that the standards of the submitted work is lower than the standards of the journal, but instead it is due to a lack of fitness between the submitted content and journal's mission encompassing its aim, scope, and publishing trend and priorities.

42.3.3   Journal's Publishing Trend and Priorities

Most journals tend to publish predominantly in certain areas within their scope. Browsing the journal's archive of the last few years (particularly the most recent issues) would give you a better idea on the trend of publications in a given journal. For example, comparing the recent content of two nephrology journals, you may find out that one journal predominantly publishes articles related to hemodialysis or epidemiology of kidney diseases, while the other prioritizes laboratory or experimental studies related to pathophysiology or mechanism of diseases.

Babor et al. (2008) refer to this publishing trend as the general culture of the journal, tradition or its aspiration. While seasoned writers may be familiar with current trends of the journals in their fields, the novice writer will most likely need to consult peers or mentors in their chosen field for advice. Note that while a journal's mission statement and aim and scope is a good starting point in scrutinizing the journals, most journals do not specify their publishing trend and priorities, despite the fact that such trend still exists because of the interest of the individual journal editors or affiliated organizations.

Once a short list of potential journals has been identified, access recent issues of those journals to scrutinize the publishing trends and priorities. These will almost certainly be available online through your institutional library. Read the tables of contents carefully and select a few published articles that are similar in scope and theme to your study. Read those articles carefully to get a clear idea of what is covered and in what detail. A decision then needs to be made if these articles are consistent with your envisioned manuscript. If the journal has published articles in the past three to five years that are similar to the theme of your manuscript, the chances are high that the journal editors will welcome your submission. Alternatively, you can search the title, keywords, and/or abstract of your manuscript in citation databases or matching software (Table 42.1) to find the journals with a history of publishing similar topics.

42.3.4   Journal Impact, Prestige, and Reputation

Do not be intimidated by the reputation of a journal as being too prestigious. The prestige and reputation are difficult to be quantified, but several parameters have been suggested as surrogate markers of journal importance to the scientific community. Although it is hard to tell what contributes to a journal's prestige, the track record of published articles in the recent and far past, journal's editorial board, and its affiliated professional organizations or learned societies are among the elements that bring reputation and prestige.

Table 42.2 highlights several markers of the journal's significance to the scientific community. One of the most popular markers is standard JIF. Although there is not a general agreement that prestige and IF are related, many academicians relate

Table 42.1 Recourse for locating the scholarly biomedical journals.

Cabell's International journal directory	An independent platform that provides up-to-date details about academic journals to help researchers evaluate and select the most appropriate outlets for their work. Two of the reported metrics are Cabell's Classification Index© and Difficulty of Acceptance©.	https://www.cabells.com
Embase	Database contains published biomedical and pharmacological articles from over 8000 journals.	https://www.embase.com
InCites™ Journal Citation Reports– Thomson Reuters	Can sort journals from various fields based on the number of citations, impact factor and Eigenfactor score.	https://jcr.incites .thomsonreuters.com
Jane: Journal/Author Name Estimator	Jane compares the manuscript title and/or abstract to articles in Medline to find the best matching journals, authors, or articles. The results are sorted by confidence score.	http://jane.biosemantics .org
National Library of Medicine catalog	Contains the list of journals currently indexed in PubMed/MEDLINE.	https://www.ncbi.nlm.nih .gov/nlmcatalog/journals
Scopus	An abstract and citation database, which provides references cited in the entries and citations received by the articles.	https://www.scopus.com
Web of Science	Three important databases are Science Citation Index Expanded, Social Sciences Citation Index, and Arts and Humanities Citation Index.	https://www .webofknowledge.com

prestige to IF. Supposedly, the higher the IF, the more prestigious the journal, the more difficult is to have a manuscript accepted, and the more likely a published article to get cited post-publication. Comparing journals belonging to different specialties and fields by impact factor can be misleading. Journals from rapidly growing fields and those receiving a greater share of grant funds receive more citation than less rapidly growing fields (Babor et al. 2008). Articles from certain specialties, such as anatomical science, pathology, and medical humanities, are often used for educational purposes and are less likely to be cited by the subsequent research in a short term.

The Science Citation Index database, which is used for gathering citation data for calculating IF, only represents <4% of all journals, and the citations made in the remaining 96% of journals do not contribute to IF calculations (Kurmis 2003). The fields with higher share of Science Citation Index database, therefore, are given advantages in IF calculations.

Furthermore, the citation and referencing behavior among fields also varies to an extent, as some fields encourage an exhaustive reference list in the published articles (Kurmis 2003; Babor et al. 2008). The number of citations that an article receives is not necessarily an index of reporting quality and methodology (West and McIlwaine 2002; Nieminen et al. 2006). Therefore, IF cannot be used as a *sole* factor in scrutinizing a

Table 42.2 Journal metrics.

Parameter	Operational definition/characteristics
Acceptance rate	Accessible through journal's homepage or Cabell's directories of publishing opportunities.
Standard journal impact factor (IF)	The 2018 impact factor is calculated by taking the number of citations made in 2018 to articles published in the journal in the preceding two years (2016 and 2017) and dividing by the total number of that journal's articles in these two years.
	Web of Science is used to gather citation data.
	Standard journal impact factor (JIF) reflects recent impact of the journal and may be calculated with or without excluding self-citations.
	The five-year impact factors is also calculated, and is more reflective of journal's long-term impact.
Immediacy index	The average number of times a journal article is cited during the year of publication (Oosthuizen and Fenton 2014).
Cited half-life	It is the median age of the citations received by a journal during the JCR year.
	A 2018 cited half-life of four years means that half of all the cited articles were published before 1 January 2015 and half were published afterward.
Article influence score (AI)	The AI score is the measure of average influence of each article in the journal; the scores are normalized so that the articles in the JCR database have a mean AI score of 1.00 (Bergstrom et al. 2008).
	A score greater than 1.00 indicates that journal articles have above-average influence. A score less than 1.00 indicates that journal articles have below-average influence.
	Web of Science is used to gather citation data.
Eigenfactor score (EF)	It is calculated from the ratio of the number of times articles from the journal published in the last five complete calendar years have been cited in succeeding year to the total number of published articles in the five-year period, while taking into account the weight of each citation.
	It ranks the journal within the network of journals and gives more weight to the citations from highly ranked journals.
	It uses Eigenfactor algorithm (based on eigenvector centrality measures) to rank the journals, which is similar to the approach that Google's PageRank algorithm uses to rank webpages (Bergstrom et al. 2008).
	It is not influenced by the journal self-citation, as citations from one article in a journal to another article from the same journal are removed.
	The sum of Eigenfactor scores of all journals listed in Thomson's Journal Citation Reports (JCR) is 100.
	It is a measure of the journal's total importance within the network of journals; if a journal doubles in size while maintaining the same quality of articles, its EF is expected to double (Bergstrom et al. 2008).
	Web of Science is used to gather citation data.
Impact Per Publication (IPP)	2018 IPP is calculated by the number of citations in 2018 to articles published between 2015 and 2017 and dividing that by the total number of articles published between 2011 and 2013.
	Scopus is used to gather citation data from the articles, conference papers, and review articles.

Table 42.2 (Continued)

Parameter	Operational definition/characteristics
Not cited (%)	Manuscripts published in the last three complete calendar years that have never been cited (Oosthuizen and Fenton 2014).
Journal *h*-index and *h*-median	The *h*-index is the largest number *h* such that at least *h* articles in that journal were cited at least *h* times each.
	The *h*-core is the set of journal articles contributing toward journal *h*-index.
	The journal *h*-median is the median of the citation counts for the articles in the journal *h*-core.
Google Scholar metrics	They include journal *h*5-index, *h*5-core, and *h*5-median.
	They are the *h*-index, *h*-core, and *h*-median of the journal based on the articles published in the last five complete calendar years.
	Google Scholar is used to gather citation data.

journal value and influence and overemphasizing the journal IF can be unknowingly deceitful. Other factors such as efficiency of editorial process, quality of peer review, and timeliness of publication often compete with the journal IF (Babor et al. 2008).

Some authors chose to submit their manuscripts to the highest impact journal in the field, and if it is rejected, then they move down the rank of impact factors to submit to the next journal. It is very important to objectively and impartially scrutinize the quality and merits of your envisioned manuscript and match it to a suitable journal, instead of just relying on the JIF (Conte 2017). If the subject of manuscript is too narrow, the broadly oriented journal might not have editors or reviewers familiar with the topic, thereby reducing the chance of being published in that journal (Babor et al. 2008). One should expect to have a manuscript rejected by a very high-profile journal. This has the advantage of having your manuscript reviewed by well-qualified reviewers. The comments from these reviewers can be used to improve your manuscript before submitting it to another journal where it is more likely to be accepted. Just be sure to carefully follow the Instructions to Authors (or an equivalent document) for the journal. It is common that the format of the references varies among different journals, and you need to conform to the preferred format for that journal.

42.3.5 Expert Opinions

Ask your mentors, colleagues, and co-authors about their experience with different journals in the field. A seasoned author who has previously published in the same topic of your manuscript will have some good insight into the various journals in the field and their publishing trends.

42.3.6 Editorial Office Standards and Efficiency

It is important to take into account efficiency of the operations of the editorial office. The first approach in evaluating this is reviewing the Instructions to Authors and determining if they are readily accessible, and if they clearly explain the methods for manuscript submission, journal style format, and how the manuscript submission process will be handled. Journals with an online format are more likely to include an electronic manuscript submission system. This system is an important feature that allows rapid management of manuscripts, provides a record, and allows authors to track their manuscript through the review process.

42.3.7 Time, Publishing, and Distribution Factors

The style in which the publisher presents and promotes the author's manuscript has a direct impact on readership. Publishers prepare galley proofs of the material to be published to authors once it is accepted for publication. The galley proofs display the layout and typeface to be published, and display figures and graphs, which were included in the submission. The publisher will then wait for confirmation from the authors if the galley proofs are acceptable. The communication between authors and publisher are important to ensure the manuscript is presented in an attractive format to both parties, as well, as to the reader. Chapter 36 details the academic publishing process.

Timelessness with the editorial process is important. Factors to consider are the lag time to final publication, whether the journal conducts an independent copyediting of the contents, and if it will allow its readers electronic pre-publishing access to the accepted manuscript and/or proof on its website before it goes to press. Some journals reflect on the average time to editorial decision and average time from manuscript acceptance to publication. These timings may vary unpredictably due to the availability of peer reviewers, total number of article accepted and those in queue for press, and the number of journal issues published each year. One advantage of open-access and electronic publishing versus traditional and print publishing is that the lag time is often minimal with the former, and some electronic journals have ongoing issues in which the accepted manuscripts are consecutively added immediately after being accepted.

Open-access journals are usually electronic only. While traditional subscription-based journals have both electronic and print versions, a growing number of them every year turn into electronic-only journals for market demands in the internet era, improved efficiency and cost-effectiveness. The availability of the chosen journal, through subscription format, select retail stores, print format, and electronic access are key to its influence and visibility.

42.3.8 Costs

Some journals charge a submission fee, or an acceptance fee for manuscripts. Other journals also charge a "per-published" page fee or impose fee for additional pages or word counts beyond a certain limit. In either case, this usually influences authors' decisions if they will submit their manuscript to the journal. Costs may also incur when publishing in color figures or figures of scientific importance. If the authors decline the option of color printing upon manuscript submission or after it is accepted for publication, the journals often publish the color images in the online versions of the articles, and print them in black and white, without imposing any fee on the authors. Some publishers impose a discounted rate for online-only color publishing. The journal's color publishing policies are very important factors to consider when planning submitting a manuscript. A black-and-white production of the key images of the manuscript that should have been ideally produced in color may undermine the quality of the final publication. In case of gold open-access, publishers charge an APC varying from couple hundreds to several thousands of dollars or charge a membership fee from the institution or the authors, plus a content fee to the authors for maintaining the article on their webpage. Therefore, submission of manuscripts to certain journals can be ruled out completely because of restrictions in the personal or university budget.

42.3.9 Difficulty of Acceptance

A journal's acceptance rate is another important consideration, and closely reflects how difficult is to publish in that journal. Most journals report their preceding year acceptance rate in their homepage. Cabell's International provides a difficulty of acceptance metric, which is based on the discipline-specific, z-score transformed distribution of the average number of times an article from a top-performing institution publishes in a given journal. This distribution is then used to categorize the journal into one of three difficulty classifications, namely, rigorous (top 10%), significantly difficult, and difficult.

An article acceptance depends on many factors, including the number of annual submissions to the journal, available journal space and publishing schedule (the number of regular issues, supplements, and special issues published per year), and scientific quality of the submission in comparison to the other submissions that journal has received in a same period of time. Articles that are long, poorly formatted, badly written, or not following the journal's guideline for manuscript preparation are likely to lose the competition, even if they have sufficient scientific merits.

42.3.10 Abstract and Indexing Services

Instead of navigating to individual publication or journal websites, researchers nowadays widely use the national and international databases to search for, recruit, and access content. Therefore, to maximize exposure, you must ensure that the selected journal provides abstract and indexing services to its content. There are several databases and indexing platforms that register the article metadata, and each of them has its own technical requirements and inclusion criteria that publishers and journals must follow and satisfy for inclusion. It is necessary to find the registries and databases that are popular and commonly used in your discipline and field of research (e.g. PubMed for medicine and life science articles, Web of Science's Science Citation Index Expanded [SCIE], Social Sciences Citation Index [SSCI], and Arts and Humanities Citation Index [A&HCI] for related disciplines), and make sure that your article metadata will be indexed in them after publication. Consult the journal homepage to find the abstract and indexing services that it provides.

42.3.11 Language Factor

Non-native English scientists tend to publish in English (Di Bitetti and Ferreras 2017). Several studies have shown that English-language articles receive a higher number of citations than non-English-language articles (Di Bitetti and Ferreras 2017; Poomkottayil et al. 2011). This is because English is considered the universal language of science, economy, and politics, and therefore, English-language articles receive a greater exposure and are more likely to make an impact on the community. A small fraction of scientific articles is published in other languages such as German, French, Japanese, or Chinese, for example. This fraction constitutes 3% of natural science, 5% of social science, and 27% of arts and humanities literature indexed in Web of Science databases (Liu 2017). Considering the high volume of scientific publications, these fractions still amount to a large number of articles each year. Non-English-language journals add cultural diversity to the scientific community and usually target the local audience rather than international community (Babor et al. 2008). They also tend to have a higher acceptance rate and a lower impact factor. Most often, their focus is on the specific topics of interest to the region and the country of origin.

Publishing a scientific article in a non-English journal can improve its local impact (such as affecting the regional policy making) despite reducing the global visibility of the article. When it comes to deciding whether to publish in English or other languages, you should carefully weigh the advantages and disadvantages of each option, and chose one while consulting with your co-investigators and mentors. With only local audience in mind and in a region where English is not the primary language, publishing in a non-English-language journal is a valid option.

When publishing in a non-English journal, it is ideal that the journal at least publishes the abstract of the articles in English for indexing purposes. Some non-English journals require the authors to provide the English-language abstract. Some even publish English articles on a regular basis (Babor et al. 2008). You may also consider publishing your paper in two or more languages; but in such case, the publishers and editors of each journal should be made aware of this reproduction and approve it. The newly published article with different language should also clearly cite the previously published versions. Publishing the same article in different languages without taking these necessary steps is a form of academic misconduct.

42.3.12   Other Factors

Submission process is another consideration to take into account. Most journals have an electronic submission portal or accept new submissions by email. The electronic submission portal allows the authors to monitor the status of the submission through the editorial and peer-review process. Sustained availability of the publication to readers is another critical issue, as you may wish not to publish in a journal whose full-text articles is not available online in either PDF or HTML versions or those journals that do not have preservation plans, especially if they are published online and open-access when for any reasons the publisher fails to maintain the content online any time after publication. PubMed Central (PMC) is the National Library of Medicine (NLM's) official archive of electronic journals and digitalized print collections, which aims at permanently preserving the library materials pertinent to medicine in accord with NLM Act. Alternative preservation plans that journals are recommended to observe are LOCKSS (Lots of Copies Keep Stuff Safe) Program, under the auspices of Stanford University, and Portico, supported by Ithaka, a nonprofit organization. Indexing in Medline requires journals to have a preservation plan. The importance of verifying the preservation plan of the journal before submission cannot be overstated.

42.4 Recommended Approach

Table 42.3 outlines a recommended approach for submitting authors' work. We firmly advise against the "trial-and-error" method, in which the authors submit their manuscript to a popular journal or a journal with highest IF without conducting an in-depth analysis and then wait for the journal to process it. This trial-and-error approach is a kind of gambling and wastes a lot of time and resources from both authors and editors.

42.5 Contacting the Editorial Office

Journal editors are often open to authors contacting them or the editorial office to explain their study, and get answers about the journal. In contacting the editorial office, it is suggested that the corresponding author write a brief introduction explaining the subject of the paper to the editor. The introduction should be short and concise, and should be preferentially directed at the editor-in-chief. It is also good to enclose the abstract or the full text of manuscript along with the letter. Politely request that the editor indicate whether there is interest in considering the manuscript. The editor usually responds within a few days to a few weeks with advice on whether to submit or not. It is not uncommon for the editor to provide constructive feedback to the authors in order to improve the manuscript.

You may contact the editor in circumstances where you are uncertain about suitability of the journal for your submission, the journal has not published a similar article in the recent years, or if multiple attempts submitting to similar journals has previously failed. Our group had once prepared a manuscript. Several journals rejected the manuscript on ground that it does not fit with their scope or readership. We contacted

Table 42.3 Recommended approach for submitting manuscripts.

1	Discuss the choice of journals with your co-authors, ideally even before writing the manuscript.
2	Select up to five journals and list them in the order of suitability for submission.
3	Submit to the first journal in your list. Before submission, make sure the manuscript meets the journal's guidelines. Remember, it is an unethical practice to simultaneously submit a manuscript to multiple journals for consideration for publication. Almost all journals require a cover letter along with submission that includes a statement indicating that the manuscript is not under consideration for publication elsewhere.
4	If the first effort is not successful, then work your way down the short list you already prepared. Be sure to revise your manuscript at each step, taking into account the constructive comments offered by the reviewers. Again, check that the format of your manuscript and the references fit the journal where you are resubmitting your work.
5	If repeated efforts are not successful, then you will need to take a hard look at your study. Is there a fundamental flaw, such as a small sample size? If so, then the study may need to be expanded or redesigned to provide more meaningful results.

the editor of another journal in our publication shortlist regarding the suitability of potential submission. We wrote in the following way:

> We have recently prepared a paper. [Title and a brief description inserted here]. I was wondering if you are interested in considering this manuscript for peer review and possible publication. I am attaching the manuscript file for your kind considerations. If you find it suitable for your journal, I will be happy to format and formally submit it for further processing.

The response that we received from the editor was encouraging and accompanied by few constructive comments aimed at improving the manuscript prior to submission:

> Many thanks for this inquiry. The article you attached does seem of potential interest as a contribution to our journal, although we would suggest some modifications before formal submission to our journal.

42.6 Acceptable Format for Manuscripts

Ensure your research manuscript meets the journals guidelines prior to submission. Usually most manuscripts include a title page including all authors, affiliations, and their contact information. The manuscript should also include an abstract, to be around 250 words. The introduction defines the scope of the manuscript and provides key background information including the rationale for the study.

For original research articles, the introduction should only be about two to three paragraphs. Methods included in this type of manuscripts should be written very clearly so that reasonable scientists in your field will be able to repeat the same studies. List your results as the primary end point in the study, followed by any secondary end points, along with unexpected findings. In most cases, tables and graphs are preferred to help explain the results of your study. Finally, scientists who write manuscripts with any end results relate those results to other studies represented in the scientific literature. Before submission, proofread your manuscript and allow your fellow cooperating scientists to also proofread it, as the more "eyes" that proofread for possible errors, the better.

42.7 Handling a Negative Editorial Decision

Becoming an experienced researcher and writer takes much perseverance. The key to getting your paper accepted by a reputable journal also involves persistence as well as communication with your peers. In any attempt, remember that you should never take a rejection personally, but instead should take it as a learning experience. As rejections occur, work down your list of selected journals and modify your manuscript based on editors' comments and resubmit your work until it is finally accepted.

If repeated efforts are not successful, then you will need to take a hard look at your study. Is there a fundamental flaw such as a small sample size? If so, then the study might need to be expanded or redesigned to provide more meaningful results. Here, the comments of the reviewers should serve to guide you. If your study yields negative results, then it might be difficult to get it published. If this is the case, it is important that you do not take the rejection personally. If publishing the results of scientific studies was easy, then everybody would be doing it! Not all scientists get published.

42.8 The Possible Outcomes of Submitting to an Unsuitable Journal

1. You might receive an outright rejection without peer review. Some journals convey such negative decision on the same day of submission, but it may take days or weeks for others to respond.
2. Your submission might undergo peer review. But because the journal is unable to recruit appropriate and expert reviewers, the peer-review process will be longer than expected. You may receive unfair or unsound criticism from the editors or reviewers.
3. Your manuscript could get rejected after peer review on the ground that it is not of high enough priority for publication, compared to the other submissions that the journal receives.
4. It may take an unreasonable amount of time to get the manuscript published and make it available for the readers. For those articles that are "time-sensitive" or introduce a new treatment modality, an innovative technique, or a revolutionary concept, this can be detrimental.
5. Submitting to a wrong choice of journal ultimately results in the loss of time, efforts, motivation, and money on behalf of the authors and even the journal editors and publishers.

References

Babor, T.F., Morisano, D., Stenius, K. et al. (2008). How to choose a journal: scientific and practical considerations. In: *Publishing Addiction Science: A Guide for the Perplexed*, 2e (ed. T.F. Babor, K. Stenius, S. Savva, et al.), 12–35. Essex: International Society of Addiction Journal Editors.

Bergstrom, C.T., West, J.D., and Wiseman, M.A. (2008). The Eigenfactor metrics. *The Journal of Neuroscience* 28 (45): 11433–11434.

Callaham, M. (2002). Journal prestige, publication bias, and other characteristics associated with citation of published studies in peer-reviewed journals. *Journal of the American Medical Association* 287 (21): 2847–2850.

Conte S. 2017. Choosing the Right Journal for Your Research [WWW] American Journal Experts. Available from: http://www.aje.com/en/arc/choosing-right-journal-your-research. [Accessed 5/6/2017].

Di Bitetti, M.S. and Ferreras, J.A. (2017). Publish (in English) or perish: the effect on citation rate of using languages other than English in scientific publications. *Ambio* 46 (1): 121–127.

Kloner, R. (2013). A brief review of useful tips for publishing your scientific manuscript. *Journal of Cardiovascular Pharmacology and Therapeutics* 18 (3): 197–198.

Kurmis, A.P. (2003). Understanding the limitations of the journal impact factor. *Journal of Bone and Joint Surgery* 85-A (12): 2449–2454.

Liu, W. (2017). The changing role of non-English papers in scholarly communication: evidence from web of Science's three journal citation indexes. *Learned Publishing* 30 (2): 115–123.

Nieminen, P., Carpenter, J., Rucker, G. et al. (2006). The relationship between quality of research and citation frequency. *BMC Medical Research Methodology* 6: 42.

Oosthuizen, J.C. and Fenton, J.E. (2014). Alternatives to the impact factor. *Surgeon* 12 (5): 239–243.

Poomkottayil, D., Bornstein, M., and Sendi, P. (2011). Lost in translation: the impact of publication language on citation frequency in the scientific dental literature. *Swiss Med Wkly* 141: w13148.

Thompson, P.J. (2007). How to choose the right journal for your manuscript. *Chest* 132 (3): 1073.

West, R. and McIlwaine, A. (2002). What do citation counts count for in the field of addiction? An empirical evaluation of citation counts and their link with peer ratings of quality. *Addiction* 97 (5): 501–504.

43

Scientific Peer Review

Christoph J. Griessenauer[1] and Michelle K. Roach[2]

[1] Department of Neurosurgery, Geisinger, Danville, PA
[2] Research Institute of Neurointervention, Paracelsus Medical University, Salzburg, Austria

43.1 Introduction

Peer review is defined as the evaluation of a body of work by experts in the same subject field as the author (peers). The principles of peer review are utilized for a variety of purposes. Professional peer review is performed in numerous fields such as medicine, law, and engineering, and focuses on performance evaluation, quality improvement, establishment of standards, and provision of certification. In medical peer review, also commonly referred to as clinical peer review, a professional body determines how a healthcare provider's competence or professional conduct affects clinical privileges or eligibility for membership in a professional society. The primary objective is to set standards and to guarantee the highest quality of medical care as well as patient safety. As long as the professional peer-review body and members or staff of the body comply with these standards, they are protected from litigation under most federal and state laws (American Medical Association 2015). In academics, promotion of faculty and tenure are frequently determined by peer review.

The focus of this chapter is on scholarly peer review, meaning an author's scholarly work is subject to the scrutiny of experts in the same field. This may determine whether a manuscript is considered suitable for publication in a scientific journal, or if an application for a research grant will be awarded by a funding agency.

43.2 History of Peer Review

The first documented description of a peer-review process is in a book titled *Ethics of the Physician* by Ishap bin Ali Al Rahwi (854–931) of Al-Raha, Syria. The book encourages a physician to keep a duplicate medical record that can be examined by a local council of physicians if there is suspicion for malpractice (Spier 2002). After the invention of the printing press by Johannes Guttenberg (1400–1468), there was a sudden ability to distribute information to the public at a larger scale. There was also a push to regulate that information. This regulation, or review, even had consequences for the works of distinguished scientists such as Nicolaus Copernicus (1473–1543) and Galileo Galilei

A Guide to the Scientific Career: Virtues, Communication, Research, and Academic Writing, First Edition.
Edited by Mohammadali M. Shoja, Anastasia Arynchyna, Marios Loukas, Anthony V. D'Antoni, Sandra M. Buerger, Marion Karl and R. Shane Tubbs.
© 2020 John Wiley & Sons, Inc. Published 2020 by John Wiley & Sons, Inc.

(1564–1642) (cf. Spier 2002). It was Francis Bacon's (1561–1626) work *Novum Organum* in 1620 that inspired English scholars to engage in informal scientific discussions that resembled an early stage of peer review.

The Royal Society of London for Improving Natural Knowledge (commonly known as the Royal Society), founded in 1660, emerged out of this movement and established the journal *Philosophical Transactions* in 1665. The editor of the journal had the ultimate decision-making power in the determination of what would be published. In 1731 the Royal Society of Edinburgh was established, and each member had knowledge on a particular subject matter. The members reviewed submitted work for publication (Nielsen 2009). A similar strategy was adopted by the Royal Society of London for Improving Natural Knowledge in 1752, when the society assumed responsibility of the journal (Spier 2002). Increasing diversity and specialization of the works submitted required the recruitment of experts outside the society.

The journals *Science* and *JAMA* (Journal of the American Medical Association) did not allow outside experts for review until 1940 (Spier 2002). With the invention of the photocopier in 1959, replication of manuscripts for peer review was simplified. Also, around that time the number of scientists interested in publication increased significantly, creating competition for publication in many scientific journals (Spier 2002). With the introduction of the computer and the internet, the peer-review process evolved to its modern-day state.

43.3 Process

The process of peer review serves as a universal tool to assure quality, originality, and relevance of scholarly contributions, and was introduced in its current form by the prestigious journal *Nature* in 1967 (cf. Nature 2017). A journal's response to submitted work includes rejected, accepted pending revisions, or accepted without reservations. The peer-review procedure is usually initiated by the editor of the journal. The editor then identifies a variable number of qualified reviewers with expertise in the field, establishes contact via email or web-based processing systems, and invites them to review the submission.

Some journals give the author the option to suggest or reject specific reviewers. An open review means the author and reviewers know each other's entity. In anonymous review, currently the preferred method, the authors' and reviewers' identities are not disclosed to each other. Should the reviewers accept the invitation, they are then expected to perform an unbiased review in timely fashion and return their comments to the editor. Reviewers are also expected to share any conflicts of interest with the editor and decline the invitation if such conflict is significant.

Based on the individual reviews, the agreement between reviewers, and his own impression, the editor decides whether a manuscript is suitable for publication with or without revisions, or if rejection is warranted. If there is substantial disagreement between the reviewers, the editor may consider the recruitment of additional reviewers. The editor's decision is eventually communicated to the author and is accompanied by the reviewer's comments. The editor may elect to withhold individual reviewer comments from the author if there is serious concern about impartiality of the review. If the editor decides that a paper is accepted pending revision, the revised version may be returned to the reviewers for additional review. Through this process, the reviewer's role is entirely advisory. Select journals publish reviewer's comments in addition to the accepted manuscript.

43.4 Criticism

Although peer review plays an important role in academic practice, it has been criticized for numerous reasons. Editors stand between the author and the readership, function as gatekeepers, and hinder an unobstructed exchange between the two parties.

The process of peer review was not adopted until the second half of the twentieth century, contrary to the popular assumption that it began much earlier in the scientific community. As an example, Albert Einstein (1879–1955) published more than 300 manuscripts throughout his career and only a single paper was ever subject to peer review. That paper received a negative review, which prompted Einstein to accuse the editor of the journal of unauthorized showing of his work prior to publication and to withdraw the paper (Nielsen 2009). Before peer review became common practice, the editor was the sole deciding factor in the publication of articles. The rejection rate was low because editors were under pressure to fill their journal and the time to publication was short. Some societies even granted their members a right to publication (Nielsen 2009). However, with a massive increase in the number of scientists in the twentieth century, new technologies such as the photocopier and later the computer and the internet, and increasing importance of publications for advancement in academia, peer review evolved into a filtering mechanism.

One major criticism of the peer-review process is that it can fail and fundamental errors in a manuscript may go unrecognized by the author, editor, and reviewers. A systematic analysis of editorial peer reviews concluded that it was largely untested and the effects remained uncertain (Jefferson et al. 2002). Without any reliable evidence showing a significant impact of peer review, the question remains whether peer review helps with verifying the validity of scientific works, filters important scientific studies, or suppresses progress and innovation (Nielsen 2009).

There are numerous examples of peer-review failure. In 2000 and 2001, a German physicist published numerous articles in prestigious journals such as *Nature* and *Science* on breakthroughs in organic superconductivity, papers that were later found to be fraudulent and were retracted. Peer-review failures have occurred in some of the most prestigious journals (Goodstein 2002). In a study by the *British Medical Journal*, eight deliberate errors were inserted into a manuscript accepted for publication and sent to 420 potential reviewers. The reviewers detected a mean of two errors, and no reviewer caught more than five of the eight errors. Some of the reviewers were blinded to the author's identity and affiliation and were less likely to recommend rejection than reviewers who were unaware of the authors' identities (Godlee et al. 1998).

In an attempt to minimize the publication of fraudulent papers, certain journals require raw data be made available to editors and reviewers to independently analyze and confirm the author's interpretation of the results. If, however, a flawed manuscript does get published, a corrective letter may be published or the article may be retracted.

Another criticism of peer review is that innovating ideas are suppressed. The editor and reviewers are part of an elite community that may be negatively biased against a manuscript if it contradicts their personal views or does not follow mainstream theories. Successful scientists are more likely to be invited as reviewers; thus, manuscripts that harmonize with such experts' opinions are more likely to be published. Revolutionary work may suffer. David Horrobin documents some remarkable discoveries that almost fell victim to peer review (Horrobin 1990). The list includes George Zweig's discovery of quarks, rejected by *Physical Review Letters* and later published in a CERN report, or Krebs's work on the citric acid cycle, rejected by *Nature* and later published in *Experientia* (Horrobin 1990; Nielsen 2009).

In addition, the peer-review process is frequently lengthy. Since reviewers are generally not compensated for their efforts, they may have little incentive to perform a timely and thorough review. For peer review to function properly, the reviewer's interest to engage is indispensable. Peer review is just one aspect of the foundation of scientific knowledge, and should be perceived as an avenue to improve and expand on the author's work. However, it is in no way a perfect process and is not assumed to be an ultimate measure of a scholarly work or a victimless evaluation process.

References

American Medical Association, 2015. Medical Peer Review. Available at: URL: http://www.ama-assn.org/ama/pub/physician-resources/legal-topics/medical-peer-review.page? Accessed 1 September 2014.

Godlee, F., Gale, C.R., and Martyn, C.N. (1998). Effect on the quality of peer review of blinding reviewers and asking them to sign their reports: a randomized controlled trial. *JAMA* 280: 237–240.

Goodstein, D. 2002. In the matter of J Hendrik Schön. Physics World, 2002. Available at: URL: http://physicsworld.com/cws/article/print/11352. Accessed 5 March 2017.

Horrobin, D.F. (1990). The philosophical basis of peer review and the suppression of innovation. *JAMA* 263: 1438–1441.

Jefferson, T., Alderson, P., Wager, E. et al. (2002). Effects of editorial peer review: a systematic review. *JAMA* 287: 2784–2786.

Nature, 2017. History of the Journal, Nature: Available at: URL: http://www.nature.com/nature/history/timeline_1960s.html. Accessed 5 March 2017.

Nielsen M. Three myths about scientific peer review. Michael Nielsen's blog, 2009. Available at: URL: http://michaelnielsen.org/blog/three-myths-about-scientific-peer-review. Accessed 5 March 2017.

Spier, R. (2002). The history of the peer-review process. *Trends Biotechnol.* 20: 357–358.

44

How to Reply to Editors and Reviewers

Paul Tremblay

Sidney Druskin Memorial Library, New York College of Podiatric Medicine, New York, NY, USA

44.1   You Receive the Decision Letter...

You have been toiling for months and maybe years in order to conduct clinical trials, manage human subjects, have your study protocol approved by the appropriate review boards, sweat blood on the methodology, calculate and compare data, draft findings and conclusions, craft a groundbreaking study that your colleagues and friends proofread and that you submitted to a prestigious academic journal. You know your perfect manuscript will surely be deftly read and approved by a group of like-minded peer reviewers. Once emailed, you sit back, relax and wait; after all the bulk of the work is done. After a few more weeks, an email from the editor is in your inbox and it cannot be but a complete and glorious acceptance missive.

Falling into shock in your chair, you realize this editor (whom you now deeply loath) has not accepted your manuscript. Indeed, based on recommendations from the reviewers (most likely two or three), the editor informs you that your findings are in doubt, the methodology needs explanations, or your topic and conclusion are not novel enough.

After tossing the uncorked bottle of wine bought for the occasion, you get back on the computer, punching at the keys – not to implement the modifications suggested by the reviewers but to tell the editor what you really think of the journal.

Publishing is a very professional experience that can turn deeply emotional and irrational at times, especially when your submission is either dismissed or returned with an "unacceptable in its present form" letter (Seibert 2006). How should one react in this situation? A rejection or demands for significant modifications may understandably trigger frustration and the unfortunate knee-jerk reaction of strongly defending one's creation, but this behavior will almost inevitably lead to misunderstanding and bridge-burning. One word of advice: After you threw a fit and promised to never write another word in your life or for "that" journal, calm down and read the reviewers' comments very carefully. They are helping you to produce a better paper, so get on it!

Anyone choosing to submit a manuscript to a refereed journal should "develop a thick hide" (Cummings and Rivara 2002) and should not interpret comments merely as negative feedback but constructive criticism (Seibert 2006). It is understandable "to feel personally rebuffed and to feel like giving up" (Browner 1998) but this is only the beginning

A Guide to the Scientific Career: Virtues, Communication, Research, and Academic Writing, First Edition.
Edited by Mohammadali M. Shoja, Anastasia Arynchyna, Marios Loukas, Anthony V. D'Antoni, Sandra M. Buerger, Marion Karl and R. Shane Tubbs.

of the process; do not abandon all hope! This chapter will reach out to researchers and get them to think and act clearly about replying to those dreaded letters.

On their way to replying to editors or reviewers (also called referees), researchers have to discard two major misconceptions:

a) Reviewers see eye-to-eye with them and will love their manuscript. As discussed in the previous scenario, this is mostly a fairytale, which we need to clarify; however, authors and reviewers do share something in common, as will be explained later.
b) Most of the work is performed during the initial research process and the writing of the original manuscript.

We will return and expound on both of these misconceptions.

44.2   Context

Researchers publish in order to contribute to their respective professions and may be granted tenure due to their publication records. Reviewers and editors are gatekeepers who ensure that a manuscript conforms to a journal's scope, policies and general mission of forwarding a given discipline. They are also aware of "what they want in their journals" (Browner 1998).

All of the stakeholders (researchers, editors, and reviewers) have to adhere to the journal's terms, usually posted online or in print as "authors' submission guidelines," which run from the mundane (formatting and style, maximum number of words, indentations, paragraph and font, as well as electronic submissions rules) to the very spirit of the publication (its scope and scientific research expectations). Therefore, a journal's narrow scope is not all a prospective author should mind before writing and submitting a manuscript; to paraphrase many of those guidelines, publications rules and policies infer a level of originality as well as stringent methodology coupled with findings compatible with the researcher's own procedures and approach.

The ability, or lack thereof, to reproduce a researcher's findings have often led to brilliant careers or humiliating exposures, as per the famous cloning scandal (Hwang et al. 2004; Kennedy 2006; Minkel 2007). In 2004 Dr. Hwang, renowned stem cells and cloning researcher, submitted manuscripts to the journal *Science*. The groundbreaking studies on human cloning were peer-reviewed and published in 2004 and 2005. The inability to reproduce any of the experiments, added to Dr. Hwang's own confession to unethical scientific behavior (e.g. lying about the provenance of the embryos), forced an embarrassing retraction from *Science* in 2006 and Dr. Hwang and his team's eventual dismissal from their academic appointments. The episode raised a few issues, one of which was the timeliness of the substantiation and confirmation of Dr. Hwang's claim: The journal did publish the studies before any of the experiments could be properly duplicated. But the scientific community excels in policing itself, so to speak, and the claim was so spectacular and unheard of that many researchers and laboratories felt under pressure to test the findings, evidently unsuccessfully. It was only a matter of time before the studies (and the researchers) were discredited.

Researchers may consult the Uniform Requirements for Manuscripts Submitted to Biomedical Journals (just known as UR) to familiarize themselves with the basic requisite of authorship.[1] The principles of scientific writing are clarity, simplicity, and understandability: "Editors want and expect from the authors of submitted paper excitement and surprise of something new, importance of the issue, originality of the research data, relevance to the audience of the journal, as well as true, clearly and engagingly written" text (Donev 2013).

Reviewers are also expected to follow ethical and professional standards; the Committee on Publication Ethics (COPE) issues an Ethical Guidelines for Peer-Reviewers which notes, among other things, that a reviewer ensures "the integrity of the scholarly record" (Committee on Publication Ethics 2012). The same document stipulates unambiguously that peer reviewers should "be objective and constructive in their reviews, refraining from being hostile or inflammatory and from making libelous or derogatory personal comments." COPE is clear as to the absolute avoidance of bias and pursuit of fairness and confidentiality. Editors and peer reviewers are also bound by guidelines found in UR (ICMJE 2014a,b).

In other words, reviewers are not professionals who happen to have a minute or two to gloss over a text and stamp "OK" or "Bad" in the margins. They are dedicated and seasoned researchers who thoroughly understand and master their discipline (they may have written the book, as the saying goes). They are uncompensated scholars reviewing manuscripts to judge their validity, authenticity, and originality as well as their suitability and affinity for the readership of a given journal. Reviewers answer to editorial boards of journals who have very rigorous criteria leading to high rejection rates; some journals typically turn down 80% or 90% of manuscripts submitted to them, and reviewers are instructed to be strict and, to be honest, ruthless when judging a manuscript. Journals expect theory, design, and writing expertise, and any of these skills lacking may be ground for rejection (Daft 1995).

In a nutshell, researchers need journals to disseminate their work and journals need researchers to fulfill their missions. Misconception number 1 is actually half true: in a way, researchers, editors, and reviewers do see things eye to eye: they all want sound and robust science to see the light of day and contribute to the literature of a given discipline. When replying to the editors and addressing reviewers' concerns, researchers have to keep this fact in mind (Day and Gastel 2011). The reviewers' responsibility is "to show how the author might better have done his or her identified research" (Romanelli 1995).

There is, however, a lack of transparency in the peer-review paradigm, hence the shortage of mentoring: "The exchanges between authors and reviewers are almost entirely hidden from public view, so social learning cannot take place in the profession" (Pondy 1995). Many reviewers themselves admit that their own perspectives come from "experience and not from professional training" (Rousseau 1995). The point is that not much has been written or is solidly codified about the process; however, the peer-review system has been developed and evolved since the end of the eighteenth century, and it is not the purview of this chapter to discuss the ongoing debates concerning its fairness, open-access publishing, publishers charging authors a fee for publishing, etc. Readers would be much better served by exploring the literature on the topic (Jefferson et al. 2002; Schroter et al. 2008) and by reading relevant chapters in this book. The norms are unlikely to change anytime soon, and serious researchers will have to submit manuscripts to peer-reviewed journals for many years to come (Day and Gastel 2011).

---

1 See International Committee of Medical Journal Editors (ICMJE), www.icmje.org.

44.3   Types of Decision Letters and How to Answer

There are no "industry standards" as to the "acceptance/revision/rejection" letters; many journals do not even have specific procedures. Some journals may post some timetable for peer-review and editorial decision-making, but are by no means bound to it (Browner 1998), so the decision process is usually left to editors who try everything they can to rein in every party involved to adhere to a due process (Samet 1999). You will most likely receive a letter or email from the editor stating the decision based on reviewers' recommendations and rationales written in a more or less narrative style, along with bulleted details. Sometimes the editor's letter accompanies the reviewers' comments separately. In the email, there may be different attachments.

Reviewers write their comments and recommendations to the editors and researchers. There is a correspondence between editors and reviewers that researchers are not privy to. The editor reaches a decision, contacts the researchers, and shares the reviewers' comments; the editor may act as a facilitator in case of disagreement between reviewers or contradicting instructions. It is left to the editor to "communicate the likelihood of successful revision, assuming that revision is the decision" (Romanelli 1995).

For all of the circumstances we will discuss here (even the outright rejection), there are five fundamental principles to the reply process (Table 44.1):

1) Always reply, even before resubmitting, to thank the editor and reviewers.
2) Reply to the editor who, as previously stated, reaches the final decision based on reviewers' comments and recommendations. Your reply should include your feedback on reviewers' comments.
3) Always reply politely, thank them (editor and reviewer) profusely and, maybe later after the fog of ire has settled, ask for point-by-point clarifications (unless you are clear on the concepts and wish to start working on the revised manuscript). At all times remain in good terms with the editor and reviewers – you may encounter them again! Do not forget that you are all professionals, so communicate with them as such.
4) Attempt to identify the major concern in order to frame your reply satisfactorily; the reviews may be a few pages long, but what are the *major* issues? The novelty? The methodology? (Khanam 2012). Of course there may be many concerns, but try to detect the common point of dissension in the reviews.
5) Address *each* and *every* comment with facts and evidence (Seibert 2006; Williams 2004). Whether you agree with the comments is irrelevant as long as you address each and every concern. One author uses a courtroom analogy and even advocates for a hypothetical reviewer acting as defense attorney for the lonely researcher (Pondy 1995). The point is that all parties have to present facts to prove their case. Consider yourself a lawyer who has to submit evidence to a jury (they are sometimes called jury of reviewers for good reasons!); you cannot win your case with opinions, arrogance, or raw feelings (Day and Gastel 2011).

The one kind of letter we will not discuss here is the rare and much-coveted acceptance document, or a word from the editor and reviewers merely asking for esthetic or formatting revisions. Congratulations. Still, be gracious and humbly thank them, asking whatever questions you have in mind, revise the text, uncork the wine, and celebrate.

For the rest of us, what come after manuscript submissions are rejection or revision letters. They come in several forms, depending on the magnitude of revisions requested.

Table 44.1 Basic principles to replying to decision letters.

Basic principles	Decision letters		
	Acceptance	Rejection	Revise and resubmit
Always reply	Yes, and it should be easy!	Reply, thank them for their work, and enquire about their concerns.	Thank them in your initial reply and enquire about their stated comments and concerns.
Reply to the editor	Thank the editor who took the final decision, but do not forget the reviewers!	Tell the editor and the reviewers that you are disappointed but wish to keep collaborating with them; Ask for some advice.	Reviewers commented and recommended, the editor took the decision. Address the editor and each and every reviewer's comments.
Reply politely	You won the lottery, so dance around the stadium on your own time and be polite and gracious in your reply.	Do not forget that they worked hard, so do not be rude! You may meet them again.	There is a lot of work ahead, so start by being polite.
Identify major concerns	No needs.	Ask them what were the major reasons for the rejection. It would help you toward an eventual resubmission.	Identify the major concern(s) and ask questions about them, if need be.
Address each concern with facts and evidence	If they ask you to revise minor formatting things, just do it.	Do not give up! Based on their comments and concerns, rewrite the study (for the same journal or another publication).	Comply by revising according to their demands, one point at a time, or make a case against one issue (or some issues) with evidence.

They sometimes overlap, but they range from the complete rejection to a request to resubmit after reviewers' comments and points have been addressed.

44.3.1 Rejection

There are many rationales for rejecting a manuscript: the study not contributing to the extant literature, the incompatibility with the journal's scope, or a major flaw in the methodology of the study. If the letter is specific as to the lack of originality or novelty, or an incompatibility of the manuscript with the publication mission and scope, it may be time to thank the editor and reviewers for their hard work and submit elsewhere; still, review what they say, as their comments will assist you in drafting a better manuscript.

In case of a lack of novelty (or allegedly not contributing to the field), should you insist on being published by that particular journal and feel that you may surmount the hardship of redoing the entire work, you can eventually submit a major revision to the same journal. If you do, it is a good idea to send along with your resubmission their original comments and how you addressed them. If you resubmit months later (maybe a year or more), chances are you will deal with a different editor and, in all likelihood, different reviewers. The manuscripts are usually blind reviewed (authors are unknown to reviewers) so it might not matter. Take for granted that the persons are the same as per the first submission; showing them goodwill and dedication to the journal's mission is good public relations. Your resubmission will also be treated as an entirely new submission.

However, in the case of a manuscript falling outside the scope of a journal, it is a bit more complicated. In general, no one should submit a juvenile myoclonic epilepsy review article to the *Journal of Trauma*, so get an idea which journal(s) are appropriate for your research. Some journals have wider scopes by their nature, like the *Journal of the American Medical Association* (JAMA) or the *American Journal of Medicine*, or, relatively speaking, the *Journal of Cancer*. Others display their scope in their very title: *Nature Clinical Practice Gastroenterology and Hepatology*, for instance, or *Expert Opinion on Drug Metabolism and Toxicology*, *PLoS Neglected Tropical Diseases*. Read their author guidelines, the publication's scope and mission (do they publish exclusively groundbreaking research?). It might be optimistically possible to tailor your manuscript to their scope. For instance, the epilepsy article to the *Journal of Trauma* might be considered if the researcher pinpointed a correlation between seizure and head trauma. For the most part, however, it would be advisable to submit elsewhere. The reader would be much better served by relevant chapters in this book.

Another rationale for rejecting a manuscript may be, in the reviewers' opinion, faulty methodology or erratic conclusion. Reviewers may deem the data, methodology, or conclusion so farfetched or unfounded that they do not suggest a resubmission. Reviewers are usually quite specific as to such alleged errors. Read their comments multiple times, and after an initial reply stating your gratitude and a promise to carefully consider their comments, do just that: Verify the validity of the comments (they usually are sound and based on facts), study and address them point by point, and consider redoing some of the studies, surveys, or experiments. It will be up to you to decide if such an endeavor is worth it.

Should you decide that one or more criticisms are unfounded, or that the referees have failed to make a connection somewhere, contact the editor and politely rebuke with facts and details. Did the reviewers miss something due to your text not being clear enough? Brevity is essential, but remember that this is not a literary or poetry contest, so crossing your t's and dotting your i's – that is, making sure all the details are covered – may be worth your while. "Authors become supremely involved in the internal logic of their manuscripts" (Daft 1995) and if your text is not clear to either editors or reviewers, it will not be clear to general readers (Day and Gastel 2011). The decision letters indicating the lack of clarity in the submission are very frequent and probably constitute the bulk of rejections, so brace yourself. The reviewers, who most likely spent days pouring over your text and tables and experiments, dissecting and analyzing your findings, conclusion, etc. are in fact on their way to help you think clearly about your project. Whether you resubmit elsewhere or to the same journal, their initial comments will have improved your new manuscript (Day and Gastel 2011; Seibert 2006; Williams 2004).

A "cold" reject is the editor stating the reviewers have raised concerns but they do not specify any, nor do they offer help. This is the letter with a "don't call us, we'll call you" subtext. It is a rare occurrence, but such letters are usually a sign to submit elsewhere. Remain on good terms, thank them, and respond by asking them to elaborate; after all you want to know what went wrong and what was missing, in their opinion. Stress that you need the information in order to submit to another journal (Browner 1998). You could ask them to reconsider their decision, but always come up with facts as to why you think they should (Browner 1998).

44.3.2   Revise and Resubmit (R&R)

The R&R letter is very common, and this is why it is worth our while to dwell on the topic here. This kind of communication spans from minor revisions to nearly complete rewriting, critiquing the methodology, the findings, or the conclusion. If you are on the receiving end of an implicit or explicit R&R letter, be happy, as it is a reprieve of sorts: it was not a rejection letter, and it "gets your foot in the door" (Cummings and Rivara 2002). Seibert (2006) relates his journey through the peer-review process for one of his co-authored manuscripts and remembers how some of the referees' comments were "fairly devastating," and how he and his co-authors built on the critiques to draft a much better paper. An R&R letter once criticized the abstract provided by the author as being misleading and unsuitable to the article. Once the abstract was modified, the rest of the process went smoothly and the manuscript was eventually published. Critiques are usually directed at more fundamental aspects of the research and modifications requested are much more time consuming and involving.

Remember the second misconception? Authors usually agree on one aspect of the writing process: replying to reviewers and rewriting the manuscript may take more time than drafting the original research project; "responding to an R&R is a lot of work" (Seibert 2006). Most experienced researchers do expect to spend more time on the rewriting phase than on the initial stage.

There are as many hybrids of an R&R letter as you have submissions, but for all intents and purposes, you have two main kinds: editor will publish if revised, or will reconsider if revised. Sometimes the letter may be hard to interpret (Williams 2004): "We cannot accept in the present form" sounds deadly, but it might mean the reviewers are ready to reconsider if, and only if, they list specific comments and requests and you address their concerns. This is when you realize they have spent more than a few minutes gliding over the text; in many cases, they deconstruct the manuscript, enumerating issues and concerns point by point for pages (Williams 2004). "Publish if modified" is your cue to implement changes if your revision is deemed acceptable. The "reconsider" letter is an invitation to implement changes and resubmit without any kind of commitment from the editor to publish.

Your immediate reply should include a warm thank you and a promise to get back to them promptly. Once done, review their reviews (literally) and their comments, they are probably right and you are getting an education money cannot buy (Day and Gastel 2011). It will be a "process of give and take" between you and the reviewers (Williams 2004), but your manuscript will be improved. Remain professional: they are criticizing your manuscript, not you. It has been said that roughly 75% of all resubmitted manuscripts are eventually accepted (Henson 1995).

Reply to the editor and address reviewers' comments in bullet format, as it make the response letter easy to follow through. For instance, reviewer #3 has an issue bulleted as "d" about a perceived lack of background data connected to one of your control groups; replicate the comment in your reply and follow it up with your specific response:

> as to Reviewer #3's concern "d", we thank the reviewer for bringing up this concern. It has been addressed in our revision as per page 7, paragraph 4…

Your submission is not the only manuscript they are reviewing, so they certainly appreciate your time in numbering and matching questions and answers, if the reviewers' comments are not bulleted in the decision letter you have received (Cummings and Rivara 2002). This may take more than a day or two, so you will need preparation to comprehend, study, and respond to the reviewer's comments, especially if you collaborated on the manuscript. Was one of your arguments flawed or unexplained? Do the reviewers (or one of them) question the composition of your control group or your statistical methods? Are the reviewers puzzled with your initial presumption? Is the design of your experiment weak? Identify the concern(s) and frame your answer.

Are you going to agree with every single item, concern, and criticism? Of course not, although you might be right at times. Reviewers come from this very planet; hence, they err like the rest of us. This is when the rules of *answering politely* and *replying with facts* are very much relevant. Should you disagree with the reviewers (or just one of them) or maybe a comment from the editor, you may reply with, for example:

> Thank you for pointing out this fact; however, it is our belief that we had already addressed the issue in the fourth paragraph…

Just make sure it *is* in the fourth paragraph, as you claim. You do not want to ruin your chances by promising a revision in your comments, which cannot be found in your text. Doing so may irreparably damage the trust between you and the reviewers, to say nothing of the editor. "There should be reasoned disagreement and not just disagreement" (Samet 1999). All authors agree that you should be as organized as possible: "Your reply should be scientific and systematic" (Williams 2004).

Of course, there are the difficult cases, such as reviewers not agreeing with each other. Make a choice, just point it out in your reply, justify your selection, and explain you were indeed given conflicting instructions (Cummings and Rivara 2002; Seibert 2006). Sometimes you may feel the reviewers are simply rude. Retain the high ground and keep your cool (do not reply right away) and, again, pepper your answer with facts and evidence. Editors and reviewers have to adhere to rules of civility as defined by COPE, and most will "try to be directive but not proscriptive," but you will encounter many styles "ranging from sincere and helpful to arrogant and biting" (Samet 1999). I reiterate this advice: never take it personally. Editors generally act as arbitrators, even guiding researchers as to the direction of revisions, hence alleviating confusion.

A rather common challenging situation is when your conclusion is welcomed by this comment: "So what?" accompanied by explanations or simply without any further comments. The reviewer may be right in pointing out the lack of novelty of a seemingly anticlimactic finale. Do not reply with this: "We appreciate the fact that reviewer #2 could not properly read a shopping list if it were not accompanied by its Cliff Note version…" Write the last sentence on a separate sheet of paper and then toss it. Feel better? Good. Use the same introductory words:

> We appreciate reviewer #2's comments about the apparent lack of originality but our conclusion is supported by the findings and are interestingly different from [or concurrent with] Smith & Jones 2009…

Of course, the reviewer might be right, so after requesting more information from the editor and reviewers, have your manuscript (and the reviews) read by a colleague who may come with an interesting point of view: maybe there is something you forgot to mention, or you can come up with a new approach: "…a concise logical argument supported by relevant empirical findings will be your best reply" (Seibert 2006).

A very delicate scenario: You researched your topic systematically, all possible instances of important books or articles on the topic have been uncovered, commented upon, and listed in your literature review. You feel confident in your work, until reviewer #3 politely (or rudely) points out that you omitted to take into account Jeffrey in her 1992 seminal approach on this theory or that treatment. There is usually no easy way out of this one. You thank them and redo it. If you feel that "Jeffrey's" contribution is irrelevant to your topic, prove it. But all in all, you do have to incorporate and rewrite. At the other end of the spectrum, Samet (1999) warns against adding too many references, as he remembers a manuscript citing "100 references by the end of the introduction." Make sure you use your references and citations appropriately and fittingly. Balance is very important.

How should you respond to a request to cut 20–30% of your text, after which they may reconsider? There are many valid reasons behind such a request: too much chatter and deleting some paragraphs or graphs may help drive the point, etc. (Cummings and Rivara 2002; Williams 2004). It may be an issue of space in the journal. This being said, researchers are at the heartbreaking end of that demand, and the question is, what do we do? Again, have your manuscript read by colleagues, as there may be an easy way to remove background discussion or shorten a methodology. Editors and reviewers may provide you with some clues as to what could potentially be deleted. If you are unwilling to comply, explain yourself with evidence, but at the end of the day, they may decide against publishing your study.

44.4 After You Resubmitted

Once you addressed all concerns, does it mean your manuscript will be automatically published, even if the editor said it would? After resubmitting, your manuscript will go through another round, starting with the editor who may or may not consult with the same reviewers as to the validity and completeness of the revisions. Even if satisfied, the editor is by no means bound to publish your manuscript. A number of variables have to be taken into account. Is there enough space in this issue of the journal? Are you too late anyway? The editor-in-chief has to weight in and may veto the manuscript. More revisions may be necessary. Do not gloat until you have a written (print or electronic) confirmation: then and only then will your manuscript will be published. The only exception to this is when there is evidence of fraud, plagiarism, or other misdeeds in the manuscript detected prior to publication.

44.5 Conclusion

There are many diplomatic ways you can reply to the criticisms. Some authors have even come up with excellent, ready-made introductory remarks (Williams 2004). The point is to learn from the remarks, comments, and concerns listed by the editors and the reviewers. The next step is to reply, implement whichever modifications are necessary, and submit to either the same journal or to another, more suitable journal. But researchers would do best to submit their very best effort in their initial manuscript, as reviewers will not rewrite your manuscript for you: "The art of the reply is critical for a diamond in the rough but will not transmogrify a lump of coal" (Seibert 2006). Researchers may have to redo and rewrite entire projects: "sometimes a design has flaws that cannot be corrected" (Daft 1995).

Another last word: Keep in mind that this is your study and your name(s) will be at the top of the published article, so should you feel very strongly about some reviewers' comments, and after you responded to editors and reviewers you think the text would not reflect you, it may be time to consider other venues for your study (Cummings and Rivara 2002). Rousseau (1995) quotes a reviewer who once wrote to a researcher, "It's not our purpose to beat you into submission." If you feel that the integrity of the paper is at stake and that "accommodating means saying what the author really does not believe in," then it is time to reconsider (Rousseau 1995). Once researchers accept the idea that editors and reviewers are not the enemy but, to a certain degree, associates who will assist you into improving your study and get it published, your writing career will take off.

References

Browner, W.S. (1998). *Publishing and Presenting Clinical Research*. Philadelphia, PA: Lippincott Williams & WIlkins.

Committee on Publication Ethics (COPE). (2012). Guidelines of the committee on publication ethics. Retrieved March 3, 2014, from http://publicationethics.org/resources/guidelines

Cummings, P. and Rivara, F.P. (2002). Responding to reviewers' comments on submitted articles. *Archives of Pediatrics and Adolescent Medicine* 156 (2): 105–107.

Daft, R.L. (1995). Why I recommended that your manuscript be rejected and what you can do about it. In: *Publishing in the Organizational Sciences*, 2e (ed. L.L. Cummings and P.J. Frost), 164. Thousand Oaks, CA: Sage Publications.

Day, R.A. and Gastel, B. (2011). *How to Write and Publish a Scientific Paper*, 7e. Santa Barbara, CA: Greenwood.

Donev, D. (2013). Principles and ethics in scientific communication in biomedicine. *Acta Informatica Medica* 21 (4): 228–233. https://doi.org/10.5455/aim.2013.21.228-233.

Henson, K.T. (1995). *The Art of Writing for Publication*. Boston, MA: Allyn and Bacon.

Hwang, W.S., Ruy, Y.J., Park, J.H. et al. (2004). Evidence of a pluripotent human embryonic stem cell line derived from a cloned blastocyst. *Science* 303 (5664): 1669.

International Committee of Medical Journal Editors (ICMJE). (2014a). International committee of medical journal editors: Responsibilities in the submission and peer-review process. Retrieved February 28, 2014, from http://www.icmje.org/

recommendations/browse/roles-and-responsibilities/responsibilities-in-the-submission-and-peer-peview-process.html.

International Committee of Medical Journal Editors (ICMJE)(2014b).International committee of medical journal editors; ICMJE homepage. Retrieved February 28, 2014, from http://www.icmje.org

Jefferson, T., Alderson, P., Wager, E. et al. (2002). Effects of editorial peer review; a systematic review. *The Journal of the American Medical Association* 287 (21): 2784.

Kennedy, D. (2006). Editorial retraction. *Science* 311 (5759): 335.

Khanam, S. (2012). Do's and don'ts for responding to peer reviewer comments. Retrieved March 3, 2014, from http://www.editage.com/insights/do%E2%80%99s-and-don%E2%80%99ts-for-responding-to-peer-reviewer-comments

Minkel, J. (2007). Korean cloned human cells were product of "virgin birth". Retrieved March 3, 2014, from http://www.scientificamerican.com/article/korean-cloned-human-cells

Pondy, L.R. (1995). The reviewer as defense attorney. In: *Publishing in the Organizational Sciences*, 2e (ed. L.L. Cummings and P.J. Frost), 183. Thousand Oaks, CA: Sage Publications.

Romanelli, E. (1995). Becoming a reviewer: lessons somewhat painfully learned. In: *Publishing in the Organizational Sciences*, 2e (ed. L.L. Cummings and P.J. Frost), 195. Thousand Oaks, CA: Sage Publications.

Rousseau, D.M. (1995). Publishing from a reviewer's perspective. In: *Publishing in the Organizational Sciences*, 2e (ed. L.L. Cummings and P.J. Frost), 151. Thousand Oaks, CA: Sage Publication.

Samet, J.M. (1999). Dear author – advice from a retiring editor. *American Journal of Epidemiology* 150 (5): 433–436.

Schroter, S., Black, N., Evans, S. et al. (2008). What errors do peer reviewers detect, and does training improve their ability to detect them? *Journal of the Royal Society of Medicine* 101 (10): 507.

Seibert, S.E. (2006). Anatomy of an R&R (or, reviewers are an author's best friends...). *Academy of Management Journal* 49 (2): 203.

Williams, H. (2004). How to reply to referees' comments when submitting manuscripts for publication. *Journal of the American Academy of Dermatology* 51 (1): 79.

45

Causes of Manuscript Rejection and How to Handle a Rejected Manuscript

Sung Deuk Kim, Matusz Petru, Jerzy Gielecki and Marios Loukas

Department of Anatomical Sciences, St. George's University School of Medicine, Grenada, West Indies

45.1 Introduction

The rejection of a manuscript can be a great source of discouragement for academic writers. However, every author who has submitted any number of manuscripts experiences the rejection of a manuscript. The revision and resubmission of a manuscript is very common in the academic world (Peh and Ng 2009; Kotsis and Chung 2014). It is always wise for the author to respect the comments of the editor and the reviewers of the manuscript. Rather than being disappointed, the author needs to recognize why the manuscript was rejected and what can be done to improve it. Do not take reviewer comments personally. Once the rejected manuscript is revised and resubmitted, it will have a better chance of being published.

Authors should be aware that the revision and subsequent resubmission of a rejected manuscript might not result in acceptance of their paper. Even if it is the case, authors usually benefits from the process by analyzing the reasons behind the rejection of the manuscript, and this can enhance the paper for potential future publication.

45.2 Dealing with the Common Causes of Rejection

Lack of sufficient novelty is often cited as the most common reason for manuscript rejection (Wyness et al. 2009). Chapters 38 and 44 touch on some of the potential strategies that can be taken in managing this situation. This chapter will present some of the most common remediable causes of manuscript rejection and how to handle the rejected manuscript.

45.2.1 Journal's Specific Requirements Are Not Followed

Each journal has its own specific requirements for publishing articles. For instance, the journal might set a maximum word limit for a case report, or it might have a strict format on how to structure a review article. A manuscript can be rejected simply because the author did not follow the guidelines set by the journal (Pierson 2004). This is an issue of formality rather than context of the manuscript. It is essential that the author is familiar with and sticks to the specific requirements of the journal.

A Guide to the Scientific Career: Virtues, Communication, Research, and Academic Writing, First Edition.
Edited by Mohammadali M. Shoja, Anastasia Arynchyna, Marios Loukas, Anthony V. D'Antoni, Sandra M. Buerger, Marion Karl and R. Shane Tubbs.
© 2020 John Wiley & Sons, Inc. Published 2020 by John Wiley & Sons, Inc.

45.2.2 Author Fails to Revise and Resubmit a Manuscript

As mentioned at the beginning of this chapter, the revision and resubmission of a rejected manuscript is common. However, the author might become frustrated or lose confidence about his/her chances for publication. The discouragement might prevent the author from revising and resubmitting the manuscript. Also, the author might disagree with the comments made by the editor and the reviewers so that the author does not want to follow their suggestions. However, it is better to appreciate the suggestions made by the editor and the reviewers, since they are the experts in this field, and to review and make these suggestions (Pierson 2004). Also, when resubmitting a rejected manuscript, the author should include a letter to the editor that addresses every point that the editor and the reviewers mentioned, alongside the corrections and revisions (Pierson 2004).

45.2.3 Subject Matter Is Outside Journal's Scope

The author should be aware of what journal would be the best fit for his manuscript (Peh and Ng 2009). If the manuscript is not within the scope of the journal, or if it inappropriate for the journal's audience, the manuscript will usually be returned immediately (Day and Gastel 2006). The author should also pay attention to who is the main audience of the journal before submitting the manuscript. One study by Kotsis and Chung (2014) showed that 76% of manuscripts rejected from one journal were eventually published by different journals in about two years. This shows how important it is to choose the appropriate journal for publication in the first place.

45.2.4 The Manuscript Is Poorly Written

As far as scientific writing is concerned, the simplest and most direct statements are always preferred (Pierson 2004). Reviewers usually do not receive monetary compensation for reviewing manuscripts, and they are usually busy with their own jobs and family during the day. Therefore, in reality, reviewers frequently review manuscripts at night when they are most tired. Therefore, the manuscript should have good flow and concise sentences so that the reviewers can understand the manuscript easily without having to read it two or three times. The author should keep in mind that a simple, concise, and concrete sentence is much better than a long, confusing, and run-on sentence. A poorly written manuscript can also stem from the author's lack of skill of writing a manuscript. It is important to use correct grammar, spelling, and punctuation. In fact, misspelling one word can hurt the credibility of the entire manuscript. For those whose primary language is not English, more attention is required. It is always a good idea for multiple native English speakers to read the manuscript before submission. Also, it is beneficial to utilize software that checks grammar and spelling. Some available software includes WhiteSmoke, Grammarly, Writer's Workbench, Microsoft Word, CorrectEnglish, WordPerfect, and Grammar Expert Plus.

45.2.5 The Study Design Has a Serious Flaw

The study design should be developed before writing a manuscript. It is the framework of the entire research. Conducting experiments before even designing a study would result in a rejection of manuscript, since the study design is usually poor (Kapil 2013). If the study design has a serious flaw, it is very hard to fix it later, since experiments and data collection have already been done (Pierson 2004). If the experiments and data collection are repeated or altered, the study design needs to be rewritten before repeating the experiments and data collection. However, it is best to avoid having a serious flaw in the study design. It is advisable that the author put enough effort and takes suggestions from mentors or peers to ensure that the study is designed well.

45.2.6 The Manuscript Has an Inadequate Description of the Methods

The methods section in a manuscript should explain how the data are collected. It is certainly an essential part of the manuscript, and an inadequate description of the methods will not permit the reader to understand how the author gathered the data. The methods of the manuscript must be explained comprehensively with concise and concrete sentences. Furthermore, the method section should be detailed enough so that the reader can repeat the study, if necessary (Pierson 2004).

45.2.7 The Results Are Not Correctly Interpreted

The results can be incorrectly interpreted (Gupta et al. 2006). For instance, the author may want a particular result despite the fact that the data do not support it. Authors should never allow their own biases to appear in their work. Also, the results can be misinterpreted due to the inexperience of the author. It is necessary to be equipped with good scientific/medical knowledge to avoid this problem. Furthermore, if the results contain statistical analysis, it must be accurate. The manuscript can be rejected outright if inappropriate or incorrect statistical analysis is presented (Kotsis and Chung 2014).

45.2.8 Plagiarism or Duplication Is Involved in the Manuscript

This can be avoided by diligently citing references and paraphrasing whenever possible. The author should run the manuscript on software (e.g. Anti-Plagiarism, DupliChecker, PaperRater, Plagiarism Detector, Turnitin, etc.) that checks plagiarism before submission (Dogra 2011). Also, remember that duplication of the author's own previous work can be a reason for the rejection of a manuscript (Peh and Ng 2009). Therefore, caution is necessary to avoid the issues of plagiarism or duplication.

45.3 Summary

As a summary for this chapter, the checklist for resubmitting a revised manuscript is provided in Table 45.1.

45.4 Conclusions

As mentioned at the beginning of this chapter, manuscript rejection is commonplace. Whenever possible, revise and resubmit the rejected manuscript for publication. If the

Table 45.1 Checklist for resubmitting the revised manuscript.

Check the journal's specific formatting requirements.
Check the scope and main audience of the journal to decide whether the revised manuscript should be submitted to another journal.
Check whether the revised manuscript is written with simple, concise, and direct sentences and with correct use of English.
Check the validity of the study design.
Check the details of the methods.
Check the interpretation of the results.
Check for plagiarism and duplication.
Get the revised manuscript reviewed by colleagues and independent experts.
Give at least a few days' interval between each revision of the manuscript to revise.
Include a letter to the editor that addresses every point raised in the rejection letter.

editor is satisfied with the revision, there is a good chance that the manuscript will get published. The editor might reject the manuscript again, but it will be worthwhile for the author to keep revising and resubmitting the manuscript until it gets published.

References

Day, R.A. and Gastel, B. (2006). *How to Write and Publish a Scientific Paper*, 6e, 117–129. Westport, CT: Greenwood Press.

Dogra, S. (2011). Why your manuscript was rejected and how to prevent it? *Indian J. Dermatol. Venereol. Leprol.* 77: 123–127.

Gupta, P., Kaur, G., Sharma, B. et al. (2006). What is submitted and what gets accepted in Indian Pediatrics: analysis of submissions, review process, decision making, and criteria for rejection. *Indian Pediatr.* 43: 479–489.

Kapil, A. (2013). Rejection of a manuscript. *Indian J. Med. Microbiol.* 31: 329–330.

Kotsis, S.V. and Chung, K.C. (2014). Manuscript rejection: how to submit a revision and tips on being a good peer reviewer. *Plast. Reconstr. Surg.* 133: 958–964.

Peh, W.C.G. and Ng, K.H. (2009). Dealing with returned manuscripts. *Singapore Med. J.* 50: 1050–1053.

Pierson, D. (2004). The top 10 reasons why manuscripts are not accepted for publication. *Respir. Care* 49: 1246–1252.

Wyness, T., McGhee, C., and Patel, D. (2009). Manuscript rejection in ophthalmology and visual science journals: identifying and avoiding the common pitfalls. *Clin. Exp. Ophthalmol.* 37: 864–867.

46

Resources and Databases
Koichi Watanabe

Department of Anatomy, Kurume University School of Medicine, Fukuoka, Japan

46.1 Reference Search Using Textbooks

Writing a medical article often requires the author to obtain all relevant knowledge about their field of study from a diverse set of resources. This will help to establish a broad knowledge base from which to begin researching. Although textbooks do not provide the latest information, and several years may have passed since they were published, they contain standard knowledge that every researcher in a given field should know. That is to say, textbooks remain a relevant and important resource even when performing current research. Textbooks range in scope, and it is important to select a variety of texts for your research.

Some textbooks provide a general understanding of the medical disciplines (surgery, internal medicine, pediatrics, emergency medicine, etc.); other textbooks are more narrowly focused on the nuances associated with specific medical specialties (thoracic surgery, vascular surgery, neurosurgery, etc.). Textbook editors often demonstrate different areas of specialization as well, lending a valuable point of view on a specific subject. It is thus an important practice to consider the editorial team when selecting the most appropriate texts for your research. Including the right textbooks in your preliminary research efforts will make writing a manuscript easier and more straightforward. Furthermore, textbooks typically include a bibliography section, which can serve as a source for additional references. The majority of articles listed in a textbook bibliography have significantly contributed to medical practices or theories, and have already been determined as valid resources.

46.2 Reference Search Using Websites

The internet is a relatively open source of information on any topic you want to know about. And this seemingly infinite source of knowledge can be accessed instantly at the strike of a key or click of a mouse. With its vast and immediate collection of electronic media, the internet is often attributed with the decline of traditional paper media such as print newspapers, magazines, and books.

A Guide to the Scientific Career: Virtues, Communication, Research, and Academic Writing, First Edition.
Edited by Mohammadali M. Shoja, Anastasia Arynchyna, Marios Loukas, Anthony V. D'Antoni, Sandra M. Buerger, Marion Karl and R. Shane Tubbs.

The internet has substantially affected the process of medical writing. Before the era of internet, authors had to collect references by reading as many related articles as possible and identifying relevant citations from the reference list in the article. The traditional method for gathering references required time and good research practices in order to compile a substantial collection of articles. The results were often limited, and important articles were sometimes missed.

Using appropriate internet search tools, relevant articles can be obtained quickly, easily, and accurately. That said, it requires much more vigilance and a keen ability to discern valid resources on a platform where about anything – qualified or not – may be published. The rise of databases has made it much easier to access a curated collection of articles, and has become a mainstay of internet research. Medical databases (PubMed,[1] Google Scholar,[2] Web of Science,[3] Scopus,[4] etc.) allow one to search for quality articles based on specific search engine keywords.

46.2.1 Medical Databases

Many medical databases are currently available on the internet. Each database has distinctive characteristics with respect to its search algorithms, number of articles, and range of publication dates. Thus, the search results for a given keyword may differ in terms of what factors are given priority in sorting the retrieved entries. You should understand the core characteristics of each database in order to obtain better search results. Medline, the predecessor of PubMed, is the first internet-based medical database. It was introduced by the United States National Library of Medicine (NLM) in 1971. Various databases were subsequently developed and introduced different search features. Today, the internet databases most frequently used for scientific research are PubMed, Web of Science, Scopus, and Google Scholar. A comparison of the features of each database is shown in Table 46.1.

Of the four representative databases, PubMed and Google Scholar can be accessed free of charge, whereas Web of Science and Scopus charge a user fee. However, Web of Science and Scopus may be used free of charge by researchers whose affiliated institution is a registered patron. Free databases such as PubMed and Google Scholar are not any less accurate, and the quality of search results is comparable to databases that charge. In fact, most researchers regularly use these free databases as a primary resource for collecting valid references.

46.2.2 Keywords

Before searching for articles in databases, prepare appropriate keywords and search terms. Keywords should accurately describe the research topic of interest. Suitable keywords include names of organs and other anatomical structures, diseases and conditions, therapies and treatments, experimental methods, and publication information, such as the names of authors, journals, article titles, and publication years. Some databases use a controlled vocabulary that unifies different words with the same

---

1 Available from https://www.ncbi.nlm.nih.gov/pubmed.
2 Available from https://scholar.google.com.
3 Available from https://webofknowledge.com.
4 Available from www.scopus.com.

Table 46.1 Comparison of each major medical database.

	PubMed	Google Scholar	Scopus	Web of Science
Provider	National Library of Medicine (NLM)	Google, Inc.	Elsevier	Thomson Reuters
Numbers of journals	25 000	Not mentioned	21 000	12 000
Covered period	1950–present	Not mentioned	1966–present	1900–present
Covered sources	Journals	Journals, theses, books, abstracts, court opinions	Journals, trade publications, books	Journals, conference proceedings
Order of retrieved entries (search result)	Sort by date (newest to oldest)	Sort by relevance and publication date	Sort by latest date, oldest date, times cited, relevance, or alphabetical order	Sort by latest date, times cited, or relevance

meaning into one search term. The best example of a controlled vocabulary is the list of terms defined under Medical Subject Headings (MeSH).[5] The MeSH vocabulary was created by the NLM and was initially used to index articles for Medline and PubMed.

How does MeSH work? Let's take, for example, a search for articles on lung cancer. If the words "lung cancer" are entered into PubMed, the search results may not be adequate; important articles may not be retrieved because they use the synonymous terms – "pulmonary cancer," "pulmonary carcinoma," or "carcinoma of the lung" – instead. A search using MeSH vocabulary would retrieve entries that use synonymous terms as well as "lung cancer." The appropriate MeSH term related to this search is "lung neoplasms." MeSH terms to retrieve the best results for this search are shown in Figure 46.1.

You can also search for MeSH terms on the PubMed website by changing the keyword input box from "PubMed" to "MeSH" in the drop-down menu. When a keyword is than entered, the exact MeSH term will be provided, e.g. "lung neoplasms" for the keyword "lung cancer." Note that many special or unusual medical terms are not supported by the MeSH vocabulary, and some databases such as Google Scholar are not necessarily compatible with MeSH terms. In these cases, relevant articles can be searched and retrieved

Figure 46.1 Use of controlled vocabulary. MeSH is the National Library of Medicine's controlled vocabulary thesaurus.

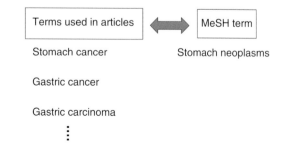

---

5 Available from http://www.ncbi.nlm.nih.gov/mesh.

Figure 46.2 Boolean logic.

A AND B   A OR B   A NOT B

by inputting each synonymous term or by using database search operators described in the next section.

46.2.3   Search Operators and Advanced Search

When a keyword is entered into a database and many entries are retrieved, it is practical to narrow them down. In contrast, when less than 10 entries are retrieved and there is a possibility of retrieval failure (synonymous terms are not retrieved), the search must be broadened. In both cases, better search results will be obtained if you combine multiple keywords using database search operators. Operators are words inserted between search terms in order to refine or broaden the search. In PubMed, these operators are known as Boolean operators and include "AND," "OR," and "NOT."

The concept of Boolean logic is illustrated in Figure 46.2. For two keywords, A and B, the use of each operator is as follows. "AND" is used to combine the search terms A and B. A search for "A AND B" will retrieve only the overlapping area between A and B. The results are narrow and contain both A and B. Let's take, for example, a search for the survival rate of stomach cancer using MeSH terms "Stomach Neoplasms" and "Survival Rate." The search formula using Boolean operators thus becomes "'Stomach Neoplasms' [MeSH] AND 'Survival Rate' [MeSH]."

"OR" is used to retrieve either of two terms. "A OR B" retrieves the entire area of A and B. The results are broad and contain either A or B. This operator is most useful when searching keywords that are not supported by MeSH terms; better results can be obtained by combining synonymous terms using "OR."

"NOT" is used to eliminate something from the results. "A NOT B" retrieves the residual area of A with omission of the overlapping area of A and B. For example, to investigate stomach cancers excluding adenocarcinoma, the search formula using Boolean operators becomes "'Stomach Neoplasms' [MeSH] NOT 'Adenocarcinoma' [MeSH]." While Boolean operators are specific to PubMed, Google Scholar has similar search operators, albeit with slightly different applications. Table 46.2 compares the search operators of PubMed and Google Scholar.

These databases also offer other ways to perform advanced searches. Some support phrase-searching or coinciding word searches, which use a wildcard character (usually an asterisk or percent sign) to perform a search on a truncated word. The PubMed advanced search function allows building a custom operation formula by entering keywords into the search box and selecting the appropriate keyword fields (MeSH, Author, Journal, Title, All Fields) and search operators (AND, OR, NOT). In Google Scholar, an arrow icon at the right edge of the search box accesses the advanced search menu. The results from an advanced search are more accurate than those retrieved in a basic search. Databases often have online tutorials and it is strongly recommended that the users view them to learn of how they work. The PubMed tutorials include videos that allow the user to master the search process within one hour.[6]

---

6 Available from https://www.nlm.nih.gov/bsd/disted/pubmedtutorial/cover.html.

Table 46.2 Comparison of search operators in PubMed and Google Scholar.

	PubMed	Example	Google Scholar	Example
AND	A AND B	"stomach cancer" AND adenocarcinoma	A B	"stomach cancer" adenocarcinoma
OR	A OR B	"stomach cancer" OR "gastric cancer"	A OR B	stomach cancer OR "gastric cancer"
NOT	A NOT B	"stomach cancer" NOT adenocarcinoma	A -B	"stomach cancer" -adenocarcinoma
Phrase search [a]	"A B"	"stomach cancer"	"A B"	"stomach cancer"
Truncated or wildcard search terms [b]	A*	stom*		
Exact term search [c]	+A	+ Cozaar		

a) Use quotation marks to search for phrases.
b) Both the asterisk (*) and the percent sign (%) can be used as wildcard characters to truncate search terms. This retrieves entries with all words that contain the letters entered. A minimum of three letters is required plus the wildcard.
c) When you need to search an exact word or phrase, put a + sign in front of the search terms.

46.2.4 Retrieved Entries or Search Results

No two databases will retrieve the same set of entries, even when the same keyword is used to perform the search. One of the main reason for these differences is the order in which the databases sort and display the results. For example, PubMed displays results from newest to oldest. Google Scholar, on the other hand, delivers results in order of relevance (typically the more an article has been cited) or date of publication. Other reasons for a variance in search results depends on the number of sources covered by a particular database or how far back in time the entries were published.

PubMed covers approximately 25 000 journals published since 1950. The data displayed in the search results includes the article title, authors, and journal information (journal title, issue, volume, pages, and published year). A link to the abstract is also included in the retrieved data and is displayed in the results list. The search results are able to be sent via text file, clipboard, email, etc. PubMed is updated continuously and sorts results by date; this is extremely useful when searching for the most recent articles on a given topic.

Although the search algorithm is uncertain, Google Scholar is most useful for general keyword searches. The most distinctive feature of this database is that the covered sources not only include journal articles but also book chapters. The search results will retrieve data on the article or chapter title, the authors, and publication information. The abstract is not included in Google Scholar, but the publisher's website or PubMed can be accessed from the Google Scholar results, and, thus, the article's abstract. Each article retrieved through Google Scholar also provides data on the number of citations it has received, and it allows you to link to related articles or create your own citation in the form of MLA, APA, or Chicago style.

Web of Science, which charges a user fee, covers more than 12,000 journals since the year 1900, which is the widest covered publication period among all databases.

This database contains carefully selected core journals in various areas and provides high-quality, accurate information. Scopus, which also charges a user fee, covers more than 21 000 titles (journals, trade publications, and books) since the year 1966. This database features useful functions such as an analysis of the search results and the ability to download the full text.

Optimal search results can be obtained by understanding the characteristic features of each database. Discerning the difference in databases is key to performing the most effective search and retrieving the most relevant articles.

46.3 Other Sources

A vast amount of information comprising both specialized and unspecialized knowledge is available from internet-based encyclopedias such as Wikipedia, internet dictionaries, and a variety of websites that individuals or organizations have created for the public to use free of charge. Keywords are entered into the search engine, and content relevant to your search will immediately appear. This rapid data acquisition is extremely useful and allows researchers to complete research for an article in no time. However, the objectivity of such information should be considered. There are big differences between the information obtained from a general internet search of unknown sources and that obtained from searching a database of published articles found in authentic medical journals.

Medical journal articles are generally published only after passing through the peer-review process. Peer review is an evaluation performed by one or more independent third-party reviewers, usually professionals in the field. After a manuscript is submitted to the journal or publisher, it is sent to reviewers who evaluate the full manuscript content for accuracy, relevance, and quality of research methods. The manuscript is usually published after revision according to the reviewers' comments. The peer-review system is important, as it upholds the objectivity of published articles. In contrast, most of the information flooding the internet is published without having passed through any sort of screening process. Thus, such information has a risk of being incorrect or biased. Before collecting information obtained from a general internet search, make sure the source and the accuracy of the information is thoroughly assessed.

Further Reading

van Etten, F. and Deurenberg, R.J. (2009). *A Practical Guide to PubMed*, 2e. Houten: Springer.

Falagas, M.E., Pitsouni, E.I., Malietzis, G.A. et al. (2008). Comparison of PubMed, Scopus, web of science, and Google scholar: strengths and weaknesses. *The FASEB Journal* 22 (2): 338–342.

Giustini, D. and Barsky, E. (2005). A look at Google scholar, PubMed, and Scirus: comparisons and recommendations. *Journal of the Canadian Health Libraries Association* 26 (3): 85–89.

Gralla, P. (2006). *Google Search and Tools in a Snap.* Indianapolis: Sams Publishing.

Greenhalgh, T. (1997). How to read a paper. The Medline database. *BMJ: British Medical Journal* 315 (7101): 180.

Katcher, B.S. (2006). *MEDLINE: A Guide to Effective Searching in PubMed and Other Interfaces.* San Francisco, CA: Ashbury Press.

47

Research: A Construct Defined by Context

Michael Dieter

College of Applied Health Sciences, Department of Biomedical and Health Information Sciences, University of Illinois at Chicago, Chicago, IL, USA

47.1 Introduction

Research can be defined from a number of perspectives. Your involvement as a potential medical researcher necessitates a holistic evaluation of research that includes considering research methods and the rationales for research significance. The goal of this chapter is to provide a broad perspective for thinking about research and research resources in order to guide reflective research planning intended to produce new knowledge or revise existing knowledge. A systems perspective allows us to define research as a self-critical form of systematic inquiry (Stenhouse 1981).

Research viewed as self-critical systematic inquiry consists of transformations from abstract ideas to decisions and outcomes, creating a set of circumstances that collectively define a context for research in terms of *premise*, *process*, and *content*. Your goal as a medical researcher is to perform and produce research grounded in claims and explanations that are both *credible and trustworthy*; claims convincingly supported by logic and evidence. This goal necessitates ongoing evaluation of the research context and its resources as fundamental to good research planning and practice. The synthesis of Research literacy (RL) with systems thinking provides clarity toward achieving the goal.

47.2 Research Literacy: A Mindset and Skillset for Self-Critical Systematic Inquiry

RL has been characterized as both as a societal education and medical education issue (Nolan and Behi 1995; Goodman et al. 2010; des Anges Cruser et al. 2012; Exner 2014) and as a key factor driving evidence-based medical practice and advancing societal health (Brody et al. 2012; Finch 2007). The construct of a framework for organizing literacies provides a useful way to understand how multiple forms of literacy contribute overall to an overarching concept of literacy (Phillips and Alston 2013). This construct can also be applied to RL, consistent with a systems approach to research where RL serves as a system for conceiving, planning, practicing, and evaluating research and research resources.

A Guide to the Scientific Career: Virtues, Communication, Research, and Academic Writing, First Edition.
Edited by Mohammadali M. Shoja, Anastasia Arynchyna, Marios Loukas, Anthony V. D'Antoni, Sandra M. Buerger, Marion Karl and R. Shane Tubbs.

The framework model supplied by Phillips and Alston (2013) provides the initial construct for thinking about RL as a configuration of constituent literacies. A fundamental RL model construct would include:

- *Functional literacy.* The basic ability to communicate and understand meaning using language and other semiotic forms, such as images (Kirsch and Guthrie 1977; White 2011)
- *Information literacy.* The ability to recognize an information deficit and articulate it as a searchable query posed to information systems; evaluate information retrieval selectively; and apply it appropriately (Eisenberg 2010; Bruce 2011)
- *Technological literacy.* The ability to utilize the affordances of information and communications technology (ICT) as a means for applying functional and information literacies (Compton et al. 2011; Ingerman and Collier-Reed 2011)
- *Methodological literacy.* The ability to selectively evaluate and apply research methodologies and their inherent research methods appropriately (Steinerová 2013; Poth 2014).
- *Ethical literacy.* The ability to understand and apply ethical principles appropriately to research (Folinsbee 2009; Whitmarsh 2009)
- *Critical literacy.* The ability to question the current state of affairs in order to plan and achieve progressive improvement (McDaniel 2004; Luke 2012).

Specific to *medical* RL, one can also consider the role and contributions of the following to research practice and research resources:

- *Medical literacy.* Medical professional knowledge competencies of general and specialized medical science and medical experiential knowledge applicable to medical practice; as distinct from health literacy (Bell 2010; Wentz 2011)
- *Health literacy.* Healthcare consumers' self-awareness about personal health and healthcare; the assumption is to treat the person and not just the health dysfunction (Berkman et al. 2011; Sørensen et al. 2012).

Other component literacies may be appropriate for specific research contexts and their research resources and practices.

RL makes sense as both a mindset and a skillset for self-critical systematic research planning and practice. One's ability to discern gaps between the state of the art of medical science and practice and the health status of populations gives rise to medical research opportunities in terms of specific research questions to be answered and problems to be resolved. Published medical research that is informed by functional, technological, information, medical, and critical literacies, and aligned with developing areas of personal and professional research interest, sharpens the focus and begins to define emergent research contexts. Knowledge advances through learning and self-awareness, self-critical review, and examination of the research context, defining methodological literacy needs and subsequent training and resources. Once the research context has been defined and a methodological research plan is developed, bioethics provides guidelines necessary for medical research practice. Conceiving, planning, and actualizing research necessitates a planning framework to communicate intentions and provide a roadmap for medical research practices. The research proposal serves as a turning point from critically reflective thought toward the practice of self-critical systematic inquiry.

47.3 Defining and Planning the Research Context: The Research Proposal

Research construed as a self-critical systematic form of inquiry invokes the need to evaluate research resources for planning to configure the interrelationships of content, process, and premise appropriately for the defined research context and to coordinate their effective and efficient use. The necessity of producing a tangible, communicable plan in order to actuate self-critical reflective research practice speaks to the need for research documentation. The research proposal is a research document that provides a template for thinking, planning, and communicating your intentions for research (Creswell 2003). It serves both as a journal for reflective research practice and the seed document of what will ultimately become the final research report.

The research proposal can either be a formal institutional or an informal personal document that serves to define your research context explicitly. As research planning proceeds and the research context begins to take form, specific resource needs begin to emerge from the answers to three initial questions:

- *What* do I intend to do?
- *How* do I intend to do it?
- *Why* does it matter that I do it?

The research proposal serves first as a locus for thinking in order to develop your ideas into a research context that is defined for systematic inquiry. As you progress toward specificity, the questions become more focused:

- *What* is the research question or problem?
  - What data are necessary to answer the question?
  - What data will resolve the problem?
- *How* should the data be collected?
  - Where?
  - From whom?
  - When?
  - How should the data be represented or organized?
  - How should the data be analyzed, interpreted, and explained?
- *Why* is the research necessary or significant?
  - What does it contribute to knowledge?
  - Who benefits from the research?
  - What risks does it entail?
  - Is the research feasible or practical?

The answers to these and subsequent questions will help to define your research context and the resources necessary for performing your research.

47.4 Evaluating Research Resources: Self-Reflective Research Practice

RL is a mindset and a skillset that is necessary for enabling the self-critical systematic inquiry that constitutes research practices. The process of systematizing inquiry into research begins with articulating questions, problems, observations, and ideas about reality to be organized and structured into an explicitly definitive research plan to guide the collection, organization, and transformation of data (measurements, observations, etc.) into information and knowledge that is meaningful. This is true both in the sense of the quantitative, objectified physical reality whose phenomena are measurable and the qualitative socially constructed realities that are experienced subjectively and uniquely, and shared to some extent as life experiences. Both represent constructs of what is real that must be conceptualized, either deductively as working hypotheses to be tested and validated though research before acceptance as the basis of theory to explain reality; or as theory built inductively and grounded in the data.

The relevance of research, its "meaningfulness," necessitates defining research contexts as sets of conditions where applicable assumptions about reality hold true or make sense. This is an important consideration for evaluation that is often demonstrated in the ways that researchers qualify their findings, limiting claims to defining contextual circumstances, i.e. "…under these conditions…" These conditions constitute a research context to qualify and limit the extent to which information or knowledge applies, is meaningful, or is able to be generalized. For example, medical research provides contextual evidence for medical decision-making in the system of evidence-based medicine (EBM) that serves as a framework for evaluating evidence for medical practice relative to configurations of circumstances (symptoms) that constitute manifestations of medical dysfunctions (Sackett et al. 1996). In a similar fashion, the contextual limitations of research as conditions point to the need for evaluating the decisions necessary for planning conceiving and research, including research resources.

It makes sense that if research is self-critical and systematic, one must evaluate research itself for its ability to produce knowledge claims that are meaningful in the sense of being credible and trustworthy. Researchers' accountability to be self-critical engenders the responsibility to evaluate research in multiple ways. These criteria for evaluation are themselves bound to contextual relevance when, for example, one uses research to answer questions, resolve problems, support claims, and provide evidence for research decision-making. Prior published research often provides the impetus for new research. Thus, medical students and residents who are trying to publish their research are both consumers of published research and potential producers of new research.

Researchers are obliged to evaluate research in light of the systems of inquiry that generate them, based on principles that impart significance and meaning to claims arising from the possibilities that research methodologies present through the application of specific research methods to generate data in order to account for the possibility of human error (Case 2007). The inherent need to evaluate research systematically argues for a system to decipher the quality and validity of your research resources.

The need for self-critical thinking to evaluate research and research resources suggests the use of frameworks established to demonstrate its evidence in discourse, including the discourse of critical self-reflection that provides the basis for changing our worldviews (Kember et al. 1999; Mezirow 1991). Using a linguistically based critical discourse approach enables us to functionally differentiate evaluation within the research system in terms of content (*What?*), process (*How?*), and premise (*Why?*). All three subsystems are bound to the others in the sense that the premise provides the rationale for process; process generates content; and content provides evidence for premise. Thus, *all three* must be continually evaluated in light of changes until your research reaches its conclusion. The interplay between content, process, and premise makes it difficult to discuss them separately.

Establishing and validating premise allows researchers to define the purpose(s) for research (Creswell 2003). Premise is crucial to establishing significance in terms of questions about whether findings are significant respective to choices of methodology and the research context that defines purpose. Different answers may provide an example of how research not only answers questions but also generates new lines of inquiry. Premise comes into play at the onset of considering research and recurs throughout when evaluating decisions about the research processes that will create content, including a self-critical dimension to examine and evaluate yourself as a research resource. This entails your ability to conceive and perform research that is significant and meets the needs and standards of the medical research knowledge community and precludes researcher bias (Pezalla et al. 2012). Evaluation of premise also includes weighing the significance of process and content. This includes the selection of methodologies and methods appropriate to the research context, the validity of findings and their power as logical evidence for claims, the conclusions and explanations that emerge from analysis of findings, and the limitations inherent to the meaning and application of the research.

Evaluation of content considers premise and often begins with a retrospective review of the published literature. Unless your research is completely novel to the point of being revolutionary, its origins will be found in the body of published knowledge that has been generated from prior research and validated by the community of medical researchers. Your research will contribute new links to the chain of knowledge that was forged by your predecessors. This prior research provides the foundation for your research, and the literature review section of your research proposal document establishes the connections between published research and new research. Research often answers questions, but new questions emerge to create opportunities for new research. These knowledge gaps or "holes" in the research literature frequently provide opportunities for research questions or problems to elaborate on or challenge content grounded in premise and process that has been established and accepted a priori as credible and trustworthy, i.e. truth.

Evaluation of content extends to new research as well. This includes the data generated from research methodologies and methods selected in accordance with the underlying premises that define the purposes of research. The circular chain of decisions from *what* to *how* to *why* and back necessitate evaluation of research findings, establishing strong connections between data and theory to explain it. Some methodologies are well structured for this purpose, e.g. statistical approaches that permit inference and predication from numerical data to the extent that system rules for data validity will allow. For qualitative methodologies, researchers themselves are the research instrument, suggesting further need for self-critical evaluation of research data and the outcomes of data analysis to ensure validity in accordance with the standards imposed by the chosen methodology and method.

Evaluation of process proceeds from premise and the need for data to answer a question or to resolve a problem. Self-critical research practice necessitates consideration of premise and purpose when deciding on the appropriate research methodology and its research methods to answer a research question or resolve a research problem. Evaluation of process also includes determining that the research methodology and methods you have chosen are applied appropriately in order to ensure the validity of the content (data) that is generated by process.

Evaluation of premise, process, and content provides a comprehensive scope of perspective for planning your research, and may help to make the differences between traditional criteria for evaluating quantitative and qualitative research and research resources more meaningful when planning and practicing self-critical systematic inquiry. Nevertheless, self-critical research practice requires that medical researchers create a construct of credibility and trustworthiness in their roles as both consumers and producers of research that is relevant to the research context. Self-critical research practice also assumes the applicability of a number of generic criteria such as currency, authority, consistency, relevance, and so forth. Evaluation and evaluation criteria change depending on the research system and the parameters of premise, process, and content. Overall, a major focus is on the validity of research, often differentiated as internal validity involving the elements intrinsic to the research context and research practices, and external validity, which addresses the limitations of inferences arising from researchers' interpretation or analysis of data, i.e. the scope of its meaning (Creswell 2003).

Researchers select quantitative methodologies and methods (processes) when the data that are appropriate for the research purpose or premise are numerical, typically used to measure or count empirically observable phenomena. Statistics refers to "a set of mathematical procedures for organizing, summarizing, and interpreting information" (Gravetter and Wallnau 2006, p. 4). This statistical approach to research is highly structured into tests for validating data that are appropriate to a range of research contexts defined by premise and purpose. For example, one can differentiate methods as parametric and nonparametric data based on its distribution curve. Here, content drives method. In addition, purpose or premise may drive choice, for example, when using data to describe for inference.

Consumers of quantitative research often cite statistical error, variance, standard deviation, probability values, confidence intervals, power coefficients, and other criteria as the basis for inferring validity and the extent to which one is able to predict events beyond the research context itself, often characterized as statistical validity. The variability of data is significant in the sense that data that fit a normal distribution curve or cluster in patterns are less random and more easily explained in terms of a statistical system. In medical research, this often involves experimental data to determine cause–effect relationships in medical interventions. The advantages are that numerical data are highly amenable to the mathematical operations that constitute the system of statistics, and the outcomes are unambiguous, e.g. one either proves or disproves a null hypothesis based on a statistical value. The problem comes when one considers the deeper meaning of data, arguing for alternative research methodologies and methods to complement quantitative approaches.

Qualitative research provides a complementary approach to quantitative research practices. Unlike the structured system of statistical methodologies, it is inherently ambiguous and complex, much like reality itself! Qualitative research is both descriptive and interpretive, invoking the role of researcher as research instrument. One can summarize the complexity of qualitative research based largely on premise or purpose. Qualitative research is committed to a naturalistic interpretation of its subject matter as part of a larger critique of the current state of affairs, a critical self-questioning approach that fits our definition of research (Denzin and Lincoln 2005). This often gives rise to the imposition of researchers' values in interpretations, making self-critical research practice essential.

Validity is a less-definitive concept in qualitative research than in quantitative research, and is often a construct pulled together specific to the research context. One example of such a construct involves the formulation of a research triangulation strategy, a form of self-critical research practice to address risks to credibility and trustworthiness arising from critique by peers. Triangulation is a self-critical research evaluation practice intended to show that independent measure supports, or at minimum, do not contradict findings and focuses on data source, data type, method, researcher, and theory (Miles and Huberman 1994). The construct may include method triangulation (process), participant checking by key informants (content), and researcher self-reflexivity (premise). The method triangulation attempts to bolster perceived deficiencies in one method with the strengths of another method, for example, using a focus group method in tandem with surveying. Checking allows participants to review and if necessary amend researchers' data to improve its accuracy, such as reviewing the transcript of an interview to elaborate or correct inconsistencies. Researcher self-reflexivity takes the form of self-critical reflection upon research practices intended to preclude researcher bias.

It is easy for medical researchers to become lost in the maze of research systems when attempting to decipher the validity and the quality of their research and research resources. This is why a systematic approach to evaluate the interplay of premise, process, and content makes sense as the point of origin for thinking about research, preliminary to planning and documentation. Understand how each of the three system elements may impact the others is essential to a holistic understanding of research. Self-critical research practice performed systematically provides a framework to ensure that medical research practices have been examined and found to be consistent with the standards of the medical research community of peers.

47.5 Meaning and Context: A Self-Critical Perspective

Meaning is contextual, and context is meaningful. Context and meaning are intertwined in the sense that each determines the other (Eggins 2004). Your ongoing experience as medical researchers will form your mindsets about research, specific research contexts, and increasingly specialized research resources to become realized as attitudes and preferences that risk becoming biases that filter reflective thinking and contribute to defining you as a researcher (Mishler 1979). RL provides balancing perspectives on the relation of your research context to other research contexts, and the cultural contexts in which they are situated. This demonstrates a form of higher-order reflective thinking necessary for self-critical systematic inquiry where the ability to shift focus between the tree and the forest will help to define and evaluate the relevant meaning and utility of your research practices and resources.

References

Bell, E.J. (2010). Climate change: what competencies and which medical education and training approaches? *BMC Medical Education* 10 (1): 31.

Berkman, N.D., Sheridan, S.L., Donahue, K.E. et al. (2011). Low health literacy and health outcomes: an updated systematic review. *Annals of Internal Medicine* 155 (2): 97–107.

Brody, J.L., Dalen, J., Annett, R.D. et al. (2012). Conceptualizing the role of research literacy in advancing societal health. *Journal of Health Psychology* 17 (5): 724–730.

Bruce, C. (2011). Information literacy programs and research: an international review. *The Australian Library Journal* 60 (4): 326–333.

Case, D.O. (ed.) (2007). *Looking for Information: A Survey of Research on Information Seeking, Needs and Behavior*. Elsevier, Ltd.

Compton, V., Compton, A., and Patterson, M. (2011). Exploring the transformational potential of technological literacy. *PATT 25 and CRIPT 8*.

Creswell, J.W. (2003). *Research Design: Qualitative, Quantitative, and Mixed Methods Approaches*. Sage.

Denzin, N.K. and Lincoln, Y.S. (eds.) (2005). *The Sage Handbook of Qualitative Research*. Sage.

Eggins, S. (2004). *Introduction to Systemic Functional Linguistics*. Bloomsbury Publishing.

Eisenberg, M.B. (2010). Information literacy: essential skills for the information age. *DESIDOC Journal of Library and Information Technology* 28 (2): 39–47.

Exner, N. (2014). Research information literacy: addressing original Researchers' needs. *The Journal of Academic Librarianship* .

Finch, P.M. (2007). The evidence funnel: highlighting the importance of research literacy in the delivery of evidence informed complementary health care. *Journal of Bodywork and Movement Therapies* 11 (1): 78–81.

Folinsbee, S. (2009). Workplace literacy: ethical issues through the lens of experience. *New Directions for Adult and Continuing Education* 2009 (123): 33–42.

Goodman, M.S., Dias, J.J., and Stafford, J.D. (2010). Increasing research literacy in minority communities: CARES fellows training program. *Journal of Empirical Research on Human Research Ethics: JERHRE* 5 (4): 33.

Gravetter, F. and Wallnau, L. (2006). *Statistics for the Behavioral Sciences*. Cengage Learning.

Ingerman, Å. and Collier-Reed, B. (2011). Technological literacy reconsidered: a model for enactment. *International Journal of Technology and Design Education* 21 (2): 137–148.

Kember, D. (1999). Determining the level of reflective thinking from students' written journals using a coding scheme based on the work of Mezirow. *International Journal of Lifelong Education* 18 (1): 18–30.

Kirsch, I. and Guthrie, J.T. (1977). The concept and measurement of functional literacy. *Reading Research Quarterly* 485–507.

Luke, A. (2012). Critical literacy: foundational notes. *Theory into Practice* 51 (1): 4–11.

McDaniel, C. (2004). Critical literacy: a questioning stance and the possibility for change. *The Reading Teacher* 472–481.

Mezirow, J. (1991). *Transformative Dimensions of Adult Learning*. San Francisco, CA: Jossey-Bass.

Miles, M.B. and Huberman, A.M. (1994). *Qualitative Data Analysis: An Expanded Sourcebook*. Sage.

Mishler, E.G. (1979). Meaning in context: is there any other kind? *Harvard Educational Review* 49 (1): 1–19.

Nolan, M. and Behi, R. (1995). From methodology to method: the building blocks of research literacy. *British Journal of Nursing (Mark Allen Publishing)* 5 (1): 54–57.

Pezalla, A.E., Pettigrew, J., and Miller-Day, M. (2012). Researching the researcher-as-instrument: an exercise in interviewer self-reflexivity. *Qualitative Research* 12 (2): 165–185.

Phillips, B., and Alston, O. (2013). An Organizing Framework for Literacy. Research-in-Progress Paper. Proceedings of the Nineteenth Americas Conference on Information Systems, Chicago, Illinois, August 15-17, 2013.

Poth, C. (2014). What constitutes effective learning experiences in a mixed methods research course? An examination from the student perspective. *International Journal of Multiple Research Approaches* 3453–3484.

des Anges Cruser, S.K.B., Ingram, J.R., Papa, F. et al. (2012). Practitioner research literacy skills in undergraduate medical education: thinking globally, acting locally. *Medical Science Educator* 22 (3): 162.

Sackett, D.L., Rosenberg, W., Gray, J.A. et al. (1996). Evidence based medicine: what it is and what it isn't. *BMJ* 312 (7023): 71–72.

Sørensen, K., Van den Broucke, S., Fullam, J. et al. (2012). Health literacy and public health: a systematic review and integration of definitions and models. *BMC Public Health* 12 (1): 80.

Steinerová, J. (2013). Methodological literacy of doctoral students–an emerging model. In: *Worldwide Commonalities and Challenges in Information Literacy Research and Practice*, 148–154. Springer International Publishing.

Stenhouse, L. (1981). What counts as research? *British Journal of Educational Studies* 29 (2): 103–114.

Wentz, D.K. (ed.) (2011). *Continuing Medical Education: Looking Back, Planning Ahead*. UPNE.

White, S. (2011). *Understanding Adult Functional Literacy: Connecting Text Features, Task Demands, and Respondent Skills*. Taylor and Francis Group.

Whitmarsh, J. (2009). Developing ethical literacy in postgraduate research. *International Journal of Learning* 16 (3): 207–217.

48

Critical Evaluation of the Clinical Literature

Jacopo Buti and Faizan Zaheer

School of Dentistry, The University of Manchester, Manchester, UK

48.1 Introduction

The clinician is faced with difficult diagnostic, prognostic, and therapeutic decisions on a daily basis. Good clinical decision-making involves finding the appropriate solution in a timely manner that is tailored for each unique challenge. This target would not be obtainable without assuming the knowledge of the scientific literature as the crux of modern day clinical practice.

Evidence-based medicine (EBM) is defined as "the conscientious, explicit, and judicious use of current best evidence in making decisions about the care of individual patients" and should be applied at every stage of delivery of patient care (Sackett et al. 1996). The clinician must be able to decipher and interpret scientific literature in light of the clinical problem at hand and readily gauge the quality and applicability of the evidence presented. Scientific evidence is then integrated with clinical knowledge/experience and patient expectations and values to make clinical decisions with the aim of improving patient outcomes and the overall quality of patient care (Figure 48.1).

In the past, the search for answers based on scientific literature was a very daunting process. It involved long hours of hunting through back issues of medical journals in the library. Now this process is made infinitely easier with computers and access to online medical bibliographic databases. Advances in scientific medical knowledge and an increasing number of publications on research and studies have led to a massive production of scientific data. The number of new medical research articles published each year continually increases, and more than 12 000 new articles, including reports of more than 300 randomized controlled trials (RCTs), are added to the MEDLINE database each week. It is crucial for clinicians to equip themselves with the necessary methodological, statistical, and computer skills in order to sieve through this vast body of available literature and identify the relevant scientific findings. Clinicians must be able to differentiate good-quality literature from the poor, critically judge the results of scientific studies, evaluate the effectiveness of one treatment approach compared to another, and verify the results of sound and controlled studies against information from uncontrolled studies or

A Guide to the Scientific Career: Virtues, Communication, Research, and Academic Writing, First Edition.
Edited by Mohammadali M. Shoja, Anastasia Arynchyna, Marios Loukas, Anthony V. D'Antoni, Sandra M. Buerger, Marion Karl and R. Shane Tubbs.
© 2020 John Wiley & Sons, Inc. Published 2020 by John Wiley & Sons, Inc.

Figure 48.1 Cornerstones of evidence-based medicine (EBM).

data published in non-peer-reviewed journals, in order to ensure sound decision making in patient care.

48.2   Critical Appraisal of a Scientific Article

Critical appraisal has been defined as the "…application of rules of evidence to a study to assess the validity of the data, completeness of reporting, methods and procedures, conclusions, compliance with ethical standards, etc. The rules of evidence vary with circumstances" (Last 2001). In other words, critical appraisal is a process that helps the reader to identify whether a piece of research has been done properly, if the reported information can be considered reliable and, for instance, how to interpret studies producing contradictory results on the same topic.

There is a multitude of tools available that can guide the clinician through the critical appraisal process. However, to date no "gold standard" approach has been established. The Critical Appraisal Skills Programme (CASP) provides a series of checklists that can be used to appraise a scientific paper based on the study design.[1] A checklist is available for each: systematic reviews (SRs), RCTs, cohort studies, case controlled studies, diagnostic test studies, qualitative research, and economic evaluations. A framework of critical appraisal aimed at evaluating the quality of clinical practice guidelines (CPGs) has also been developed by the Appraisal of Guidelines Research and Evaluation collaboration (AGREE).[2]

---

1 CASP tools and checklists are available at: www.casp-uk.net
2 AGREE Enterprise: www.agreetrust.org

CASP divides the critical appraisal process into three main steps:

1. Is the study valid? The first step is to evaluate the study's methodological quality and the impact of various biases on the study outcome. The type of study design also needs to be appropriately chosen to answer the research question at hand.
2. What are the results? If the study design is deemed valid, then a careful interpretation of the results is required in order to assess the statistical and clinical significance of the study.
3. Are the results useful? The final step is to assess if and how the findings of the study can be generalized to the population and be applied to clinical practice.

Although the methodological criteria and tools available to assess the validity of a study will vary according to its design, some general principles should always be considered in the evaluation of any research study. The proceeding sections of the chapter discuss key elements and concepts that are required to address each of these questions.

48.3 Is the Study Valid?

The first question requires addressing three elements: the research question, possible bias, and the grading system for study quality.

48.3.1 Research Question

48.3.1.1 Relevance
The research question embodies the idea behind and the purpose of the study. It provides a focused direction so that appropriate aims, objectives, and study design can be formulated. The validity and relevance of a scientific article is dependent on the validity and relevance of the research question. Overall, it is essential that a question be clinically relevant, achievable, clearly defined, and important for patient care.

The research question should also be pertinent to one's own field of work, as what might be of interest to some might be irrelevant to others. Contextual and prior expert knowledge in the area of the research will ensure a properly formulated research question.

48.3.1.2 Scientific Contribution
Investigating a well-designed research question can increase scientific knowledge by:

a. Providing a substantive new contribution on a specific topic
b. Increasing the knowledge on a previously researched topic confirming, extending (to a different target population), or withdrawing existing results
c. Summarizing the existing evidence on a topic

48.3.1.3 Pertinence to the Study Design
The validity of a scientific article is often based on whether the optimal study design has been chosen to address the focused research question. For example, an RCT would be inappropriate to investigate the effect of regular insulin administration on weight gain in patients with diabetes. Patients in this instance cannot be randomized to an insulin (test) group and a placebo (control) group as their allocation is determined by their clinical need for insulin administration.

48.3.1.4 Structure

A well-designed research question should include four parts based on the acronym PICO/PECO:

1. *Population*. The patient population to which this research question applies.
2. *Intervention/Exposure*. The intervention (in the case of experimental trials) or exposure (in the case of observational studies) that is being assessed.
3. *Control/Comparison*. The main alternative against which the intervention/exposure is tested/compared. Specific and limited comparisons permits a simpler research design and also facilitate an effective and targeted computerized search in reviewing the scientific literature. However, in some instances a control/comparison may not be needed/appropriate.
4. *Outcome*. The variable(s) that will be measured in order to assess the impact of intervention/exposure and (when appropriate) to compare it with the control/comparison. The outcome must include measurable variable(s) that are carefully chosen to demonstrate the desired effect (or lack thereof). For example, outcomes could include relief or elimination of specific symptoms, improvement or maintenance of function, enhancement of aesthetics, survival time, or prognosis of a disease/condition.

Let's consider a study aimed at investigating the effectiveness of a new hypoglycemic agent in managing type 2 diabetes mellitus; the following study characteristics are discernible:

a. *Population*. Patients suffering with type 2 diabetes
b. *Intervention*. New hypoglycemic agent
c. *Comparison*. Current standard medication (e.g. metformin)
d. *Outcome*. Glycemic control

Therefore, we can formulate this research question: What is the effect of "new hypoglycemic agent" on glycemic control in patients suffering with type 2 diabetes mellitus compared to a metformin?

48.3.2 Bias

Any scientific study may contain errors that reduce its validity. There are three types of errors:

1. Deliberate error (or fraud)
2. Accidental error
3. Systematic errors known as bias

Bias is "… any systematic error that results in inaccurate estimation of the effect of an exposure on an outcome" (Gail 2000). Bias can affect any phase of a study. This error is inherent within the design of the study and will likely repeat itself if the study was to be replicated. Biases are unwanted, as they have a tendency to influence the results of a trial and as a result impact the interpretation of the experimental intervention.

A common classification of bias includes the following.

48.3.2.1 Selection Bias

This occurs if the group of patients selected for the study is not sufficiently representative of the general population with the disease in question or if the two groups being compared have inherent differences in the baseline characteristics. This can happen because the sampling was not random, because an error occurred in the randomization process, or because sample size was too small to represent the entire spectrum of subjects in the target population. For example, if we were to recruit subjects for a study on patient compliance with the medical management of rheumatoid arthritis from a specialist rheumatology clinic, the study sample would miss out patients with rheumatoid arthritis being treated in primary care general practice setting and therefore not be representative of the general patient population with this disease.

A potential source of selection bias can also originate from the randomization process in a randomized controlled clinical trial. Random allocation of patients to the test and control groups must be concealed (allocation concealment) from the participants and researchers (both providers and outcome assessors) to prevent the risk of manipulation of a participant's inclusion into a specific group.

Failure to calculate the required sample size or incorrect calculation of it may result in a sample size that is not representative of the general population and reduce the validity of the study.

48.3.2.2 Performance Bias

Prior knowledge of the intervention or exposure and of the outcomes measured can lead to performance bias. For example, if a patient is given an analgesic, the psychological comfort gained from knowing that a painkiller has been administered may be enough to put the patient at ease and relieve symptoms regardless of the physiological effectiveness of the pain killer. Therefore blinding both patient and observer to the intervention is of utmost importance to ensure that a "true" response to the intervention itself is being measured rather than a response based on knowledge of which intervention was received.

In clinical trials, blinding of participants to the intervention is achieved through the use of a placebo. A placebo is administered to the control arm of the study and is indistinguishable from the intervention itself in terms of appearance and the manner in which it is administered. It, however, does not produce any direct therapeutic or physiologic effect. The subject is not aware of whether a placebo or the actual intervention has been given, which removes the risk of producing a response based on knowledge of the intervention. Sometimes, however, placebo administration can elicit a "nocebo" effect, which results in an adverse response by the participant to a harmless substance.

By virtue of participation in research, the subjects are at risk of giving a response that may not reflect their true behavior under natural circumstances. This concept is known as the Hawthorne effect.

48.3.2.3 Detection Bias

This can occur when there are differences in the conditions under which measurements are taken in each group. This could be due to a different mode of assessment applied to each or a lack of blinding of the observer. Detection bias can commonly occur if the observer is making a subjective assessment or if an invalidated assessment tool is being used. To avoid detection bias in such a scenario, it is essential that the person taking the measurements has adequate clinical expertise, whose accuracy and reliability have been evaluated beforehand and that he/she is blinded to the treatment administered to each participant.

48.3.2.4   Attrition Bias

This occurs when participants are lost to follow-up or withdraw from the study before the data collection is complete. If attrition is not accounted for, then a false overall outcome could ensue. For example, in a trial of a new drug, 40% of participants have withdrawn from the study due to lack of effect and a worsening of their condition. Data collected and analyzed from the remaining 60% (who may have responded well to the drug) are likely to show an overall positive effect to the drug. Since this does not include any data from the 40% on whom the drug did not work, the effectiveness of the drug will be grossly overestimated.

48.3.2.5   Reporting Bias

Reporting bias often occurs due to the tendency of researchers to only report significant findings of a study. Therefore, only positive correlations are more likely to be reported and negative or no correlations are underreported within a study.

48.3.2.6   Other Biases

Multiple other sources of biases are encountered. Brief descriptions of some of these are given below.

Foreign Language Bias   The authors of the study might make the error of not considering previous publications in less commonly spoken languages.

Learning Curve Bias   There usually is an improvement in results as the clinician's confidence with a procedure increases. Therefore, if a new procedure is being tested and if the groups are not randomized, or are incorrectly randomized, the results may be invalid as the procedures done at the beginning of the study are less likely to be successful.

Inappropriate Statistics Bias   This can occur as a result of inappropriately selected statistical unit or statistical test to support a particular hypothesis. One relatively frequent example occurs in oral implantology studies when a patient has more than one dental implant placed and the statistical unit analyzed is the implant rather than the patient. Some researchers may also change the level of significance of a statistical test from 5% to 10% if it favors their agenda.

Significance Bias   This can be found when the results have a statistical significance but have a negligible impact from the clinical standpoint (clinical significance).

Ecologic Bias   Ecological bias is when conclusions of studies proven true for populations are applied to individuals. In this regard, Nieri and colleagues have shown how conducting meta-analyses on aggregate data (means) rather on individual patient data can lead to error (Nieri et al. 2003).

Surrogate Variable Bias   A surrogate variable is a variable that is used in place of the actual variable, which is difficult or impractical to measure. The surrogate variable has a predictable dependent relationship with the actual variable. Distortion occurs if, for example, these variables are used to reach conclusions for long-term events. Studies on the effectiveness of drugs (i.e. flecainide) to treat patients with an history of heart attack that use suppression of ventricular arrhythmia as the surrogate variable instead of number of sudden cardiac deaths over time are an excellent example of this type of bias (Akhtar et al. 1990). Oftentimes patients may suffer sudden cardiac deaths without expressing the surrogate variable of ventricular arrhythmia.

"Fishing Expedition" Bias   Sometimes researchers collect a large number of data for a lot of variables, which they then analyze in the search of statistical significance: this is known as "fishing expeditions." To avoid this type of bias, it is essential to establish and record the study protocol beforehand.

Publication Bias   The unpublished, or gray, literature contains more frequent negative results of the tested treatments than the published literature. Many journals (and researchers) are much too keen to publish reports that give a positive result regarding efficacy of a new regimen, compared to the trials that did not find any difference or that report results that are negative with respect to the current trends in the literature.

Conflict of Interest Bias   Some corporate-sponsored studies may be withheld from publication if the results for the tested product are negative.

In conclusion, no scientific study or research can be totally error-free. For this reason, the strength of a study depends on the author's ability to reduce biases to a minimum.

48.3.3   The Grading System for Study Quality

Clinicians and academics often use a hierarchy of evidence to grade the reliability of a study. An alphanumerical gradation system is presented in Table 48.1 (adapted from Evans 2003). Studies that are higher up in position in this scale, such as systematic reviews and RCTs, are considered to be of better quality than those lower down such as case reports and expert opinions. Systematic reviews of RCTs provide the strongest evidence as they present pooled data from multiple studies. Randomized controlled trials are regarded as the gold standard in clinical trials as they aim to limit various biases that may be found in retrospective and noncontrolled studies. For example, blinding and randomization of participants to each group is not possible in a retrospective study. A research topic can also travel through an evolution of study design as it is further explored. Association between lung cancer and cigarette smoking was first identified by Bradford Hill by performing a retrospective case controlled study (Doll and Hill 1950). Once this was established, a more robust prospective cohort study was set up called "British Doctors Study" to further explore and quantify this association (Doll and Hill 1954).

It is important to note that this grading system is a ranking of study design and does not provide any information on the methodological quality or validity of a specific study therefore each study has to be assessed individually. Another limitation of this hierarchy system is that it is based on the effectiveness of an intervention and does not take into consideration the appropriateness or feasibility of implementing the intervention into practice. Often, it is not possible to utilize a robust study design due to limitations of the nature of the study. From the point of view of critical appraisal, we must consider the appropriateness of the study design in light of the research question.

Table 48.1 Grading of scientific evidence.

Grade	Type of study	Favorable aspects	Unfavorable aspects
1a	Systematic review (SR) of randomized controlled trials (RCTs)	SR uses data from several studies. SR makes it possible to aggregate different experimental conditions. If the considered studies are homogeneous, a meta-analysis can be performed, hence increasing the statistical power. Summarizes evidence from RCTs.	Publication bias. Ecological bias.
1b	RCT	The randomization process minimizes the presence of spurious variables.	Blinding of the operator to the allocated procedure is not always possible (i.e. in testing surgical interventions).
2a	Systematic review of cohort studies	SR uses data from several studies. SR makes it possible to aggregate different experimental conditions. If the considered studies are homogeneous, a meta-analysis can be performed, hence increasing the statistical power.	Publication bias. Ecological bias. Summarize evidence of cohort studies.
2b	Prospective cohort (controlled) studies	Comparison of different treated/exposed groups of population. The researcher does not know a priori the results of the therapies.	Lack of randomized allocation of the patients to different groups. Spurious variables can affect the treatment.
3	Retrospective cohort (controlled) studies	Comparison of different treated/exposed groups of population.	Lack of randomized allocation of the patients to different groups. Patient selection bias. The researcher knows a priori the results of the therapies. Spurious variables (i.e. smoke habits) can affect the treatment.
4	Case report/series	Illustration of a new therapeutic technique. Observation of a new phenomenon. Verification of a biological principle.	Conclusions cannot be generalized. Patient selection bias. Publication bias. Lack of a control group of clinical cases.
5	Expert opinion	Based on experience and knowledge. Useful when insufficient research evidence available.	Lack of data useful for statistical analysis.

48.4   What Are the Results? Assessing and Analyzing Results of a Study

The most basic step in appraising the results of a study involves understanding how the data was analyzed to produce these results. The application of statistics to medical research developed dramatically during the course of the twentieth century. In the present day, it is hard to find a scientific article, which does not include some statistical procedure. Clinicians have often been unable to keep up to date with the understanding and use of today's statistical language. In fact, "…Many people think that all you need to 'do' statistics is a computer and appropriate software. This view is wrong even for analysis, but it certainly ignores the essential consideration of study design, the foundations on which research is built. Doctors need not be experts in statistics, but they should understand the principles of sound methods of research" (Altman 1994).

Therefore, at least a basic understanding of statistics is needed for comprehending any scientific paper.

48.4.1   Variables

In statistics, a variable is a characteristic or a unit of description being observed in a study. It may assume a single value or a set of values to represent a statistical unit. Generally, in controlled trials, a single variable is chosen to compare the treatment arms. This variable, which has to be pertinent to the aims of the study, is also called the *outcome* variable or *dependent* variable or *endpoint*. The choice of the outcome variable determines the scale used for measurement.

48.4.2   Types of Data and Scales of Measurement

Data can be of two main types, each having further subcategories.

48.4.2.1   Categorical Data

Categorical data are values or observations that can be sorted into groups or categories with the difference among categories being qualitative.

- *Binary/dichotomous data*. Data can only have one of two values, e.g. gender (male/female).
- *Nominal data*. Categories have no ranking or numerical relationship to each other. For example blood group (A, B, AB, and O), hair color (black/brown/red/blonde) or ethnicity.
- *Ordinal data*. Categories have an intrinsic ranking or ordering. For example, smoking status (occasional/frequent/excessive).

48.4.2.2   Numerical Data

These are values or observations that can be measured:

- *Continuous numerical data*. Data can theoretically assume any numerical value, but usually fall within a certain range (e.g. blood pressure).
- *Discrete numerical data*. Data can only take certain values, usually integers (e.g. number of siblings). These are usually variables that are counted rather than measured.

Numerical data can be further expressed as interval or ratio data. Interval data are represented on a scale where the interval between each unit is of a fixed magnitude and does not have a natural zero point. Temperature in degrees Celsius is an example of interval data where the difference between 10° and 20° is the same as 90° and 100°. However, since 0.0° does not represent an absence of temperature, it is not a natural zero point. Due to the lack of this absolute zero point, 20° does not equate to being twice as hot as 10°. Ratio data has the same characteristics as interval data; however, it contains a natural zero point. For example, a variable like height and weight where a value of 0.0 represents absence of that variable.

48.4.3 Descriptive Statistics

Descriptive statistics provide simple summaries about the numerical information obtained by data collection. They can only be used to describe the *sample* that is being studied and the measurements that have been made. That is, the results cannot be generalized to any larger group (external validity). Demographic or clinical characteristics of the subjects such as the average age, the proportion of subjects of each sex, and the baseline measurements of clinical variables of interest are all examples of this and can usually be found summarized in tables. Categorical data are easily described by histograms or pie charts.

The information obtained about a variable from a series of measurements is usually described by the following values representative of the so-called *central tendency*:

- *Mean.* Arithmetic average, i.e. the sum of all the values divided by the total number of measurements.
- *Median.* Value such that half of the measurements are less than and half are greater than that value. The median is also called the 50th percentile or the $0:5$ quantile.
- *Mode.* The most commonly occurring measurement.

The *measurements taken for a* variable in a dataset are likely to be different. This variation can be described by:

- *Standard deviation and variance.* Measures of the variability (spread) of measurements across subjects. Standard deviation is the square root of the *variance*. The value derived for standard deviation can be used to gauge how large the variation in the data is. Notably, 95% of the measurements lie within two standard deviations either side of the mean.
- *Range.* The range gives the maximum and minimum values of the measurements.

48.4.4 Inferential Statistics

48.4.4.1 Sample and Population
When conducting research, the main aim is to find answers that are applicable to the general population and not just to the study sample. This is known as the *external validity* of a study. To assess the true impact of an intervention/exposure in a population, we would have to conduct a study in which every member of the population is included; however, since this is not feasible, we have to select a sample that is large enough and representative of the population. Inferential statistics can then be used to

make predictions or inferences about the whole population based on observations and analysis of the study sample.

48.4.4.2 Standard Error of the Mean (SEM)

If we replicated the same study an infinite number of times each time taking a different random sample from a population then each study will be described by its own unique sample mean and standard deviation. The mean value of the collective sample means will be representative of the true population mean. The standard deviation of the true population mean is known as standard error of the mean (SEM). A single study can be used to estimate SEM provided the study sample is large enough and representative of the population. Obviously, it is not possible to predict the actual value of the true mean of the population based on a single study, but SEM allows us to gauge the range of values within which the true population mean is likely to be.

48.4.4.3 Confidence Interval (CI)

Two SEMs either side of the true population mean give us a range of values representative of the 95% confidence interval (CI). Thus, the CI for a sample mean (x) will be: $x \pm 2*SEM$. This tells us that there is a 95% chance that the true mean of the population lies within this range. The larger we can make the study sample and the more representative it is of the actual population, the smaller we can expect the CI to be. As a result, we would be able to infer the results of our study and estimate the range of values within which the true population mean would lie more accurately.

Understanding the meaning of a CI is of primary relevance in the interpretation of study results. In systematic reviews, the effect size is often expressed in terms of mean difference between treatments, and their related CI is presented on graphs called *forest plots*.

48.4.4.4 Hypothesis Testing

In statistics, a hypothesis is a claim about something. Hypothesis testing is used when we are comparing two values such as mean difference between two treatment groups, to check if the difference between the values occurred by chance. Therefore, we first have to state a hypothesis and then apply the appropriate statistical test to assess what is the chance of our hypothesis being true. Hypothesis testing entails a stepwise approach as follows:

Step 1: State the hypothesis The *null hypothesis (H_0)* is stated first, and it assumes that there is no difference between the mean treatment effects of the tested therapies. Statistical tests are then used to calculate a p-value. This is representative of the probability of the null hypothesis being true.

The *alternative hypothesis (H_1)* is a statement that directly contradicts the null hypothesis. H_1 states that one therapy is superior to the other, and if the null hypothesis is not true, then the alternative hypothesis must be true.

Step 2: Set the criteria for a decision Since hypothesis testing is an inferential method, we have to understand that it works on the principles of probabilities and estimations rather than true values. We need to decide an arbitrary threshold below which we can say the probability of H_0 being true is too low and therefore must be rejected. This level of significance is typically set at 5%. Therefore, a p-value less than 0.05 means that there is less than 5% chance of H_0 being true. This value is below the 5% significance level and is therefore too low for us to accept. Thus, we would reject the null hypothesis. Conversely, this also means that there is a 5% chance that you will reject the null hypothesis when it is in fact true (also known as a type I error).

Step 3: Choice of the statistical test A statistical test provides a mechanism for making decisions about a research question using a mathematical formula. The choice of the correct statistical test for an experiment mainly depends on the nature of the outcome variable of interest and the sample characteristics. A description of individual tests and their application is beyond the scope of this chapter.

Step 4: Make a decision The value of the test statistic is used to make a decision about the null hypothesis. If the p-value is less than 0.05 (level of significance), then H_0 will be rejected, and as a result, H_1 must be true. This means that the difference between the two means is unlikely to have occurred by chance and is said to be statistically significant. Conversely, if the p-value is greater than 0.05, then we cannot reject the H_0 and the difference in the means is likely to have occurred by chance.

48.5 Are the Results Useful? From Research to Clinical Practice

Once we have the results, the next step is to determine whethere they are useful. This is where the study goes from research to clinical application.

48.5.1 Clinical versus Statistical Significance

The final step in evaluating a paper, and arguably the most crucial one, is to assess the following: how useful are the findings? What is the clinical relevance of the study? How will it impact my clinical practice? Often , studies (including RCTs) show statistically significant results, which do not have any clinical significance. It is important that in the process of appraisal we distinguish between statistical and clinical significance.

Minimal clinically important difference (MCID) is a statistical term, which describes the threshold above which the intervention is deemed to produce a clinically relevant outcome. If it falls below this threshold, regardless of its statistical significance, the effect produced is too small for the intervention to have an impact on patient care. For example, when assessing the results of an intervention on a 10-point pain scale, which is self-reported by the patient, a statistically significant reduction by one point may not necessarily be clinically significant. Due to the subjective nature of the scale, the overall difference between a score of 2 and 3 is more likely to reflect the variation in patient reporting rather than an actual reduction secondary to the intervention. In such an instance, it is important to predetermine the MCID value against which the results of the study can be compared.

Clinical significance, however, reaches far beyond results that are statistically significant. Multiple factors, such as the magnitude of therapeutic effect, the economic and biological cost, patient acceptance, operator skill level, and characteristics of the environment must all be considered. The cumulative balance of all of these factors must lie in favor of a significant patient benefit before the intervention can be considered for clinical application. For example, warfarin is more efficacious in reducing the risk of stroke, however, in patients with low stroke risk who require anticoagulant therapy often use aspirin as the drug of choice. Despite the pharmacological superiority of warfarin, the high risk of fatal hemorrhage outweighs its therapeutic effect.

48.5.2   External Validity

In simple terms, external validity describes the generalizability of the results of a study. For a study to have any impact on clinical practice, it must have high external validity. It is dependent on multiple factors with the main factors being the studied population and the experimental circumstances.

It is important that the sample population of a study is carefully selected so that it is representative of the population for whom the results are relevant. This point is discussed further in the section on selection bias above. The sample size can also impact external validity of a study. If a sample size is small, it may not compensate for the selection of outliers of the population, which may occur by chance, and hence give a completely skewed result. A well-designed study must include a sample-size calculation to ensure an adequately large sample is selected that can demonstrate the desired effect being studied.

Experimental circumstances can also affect the generalizability of a study. Often, the conditions of an experiment may not reflect real life scenarios and therefore the subjects may produce an altered response. This is known as the Hawthorne effect. It entails how the knowledge of an individual that he is being studied alters his behavior and response in a research setting.

It is challenging for researchers to undertake a study where perfect experimental conditions are created that prevent introduction of any biases while emulating real-life conditions simultaneously. Therefore, readers must take this account and critique literature with a realistic eye. Moreover, due to logistical or ethical issues, study designs with limitations may be the only source of evidence on a clinical issue and should therefore be treated with the appropriate merit.

48.6   The Research World

There are several auxiliary aspects of the research world, understanding of which can change the reader's perspective on scientific literature. Some aspects that the authors felt are important for this purpose are described below.

48.6.1   The Peer-Review Process

An author who wants to publish a study must firstly submit the article to the journal's editor. The article will then be passed on to reviewers who will assess the scope of the study, the appropriateness of the materials and methods and statistical tests used (internal validity), the congruity of the results as well as whether or not the article is

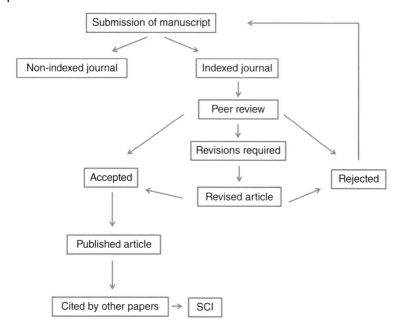

Figure 48.2 Peer-review process.

pertinent to the journal and its readership (external validity). This process is known as "peer review." Based on this, the article may be accepted for publication with no amendments required, or returned to the author(s) with requests for modifications prior to publication or it may be considered unsuitable for publication and rejected (Figure 48.2). This rigorous process ensures that only the highest quality of suitable research is published.

A committee of experts formed by the Institute for Scientific Information (ISI) is responsible for maintaining a database of journals that meet well-defined international standards and having a peer-review process prior to publication within the journal is an important prerequisite for its inclusion in the database.[3] Journals that do not implement this peer-review process are deemed to hold less scientific value.

48.6.2 Science Citation Index

Following publication, an article may be cited by authors in other scientific work. The Science Citation Index (SCI) has been developed to keep a record of these citations.[4] The impact of a paper on the scientific community can often be judged based on the number of citations it has received.

---

3 Institute for Science Information: http://isi-thomsonreuters.com
4 Thomson Reuters, Science Citation Index Expanded: http://thomsonreuters.com/science-citation-index-expanded

48.6.3 Impact Factor

The impact factor (IF) is a number assigned to a journal that indicates the frequency with which its articles are cited in a given period of time. It is often used as a marker for the quality of the journal.

IF is calculated using the formula:

$$IF = \frac{\text{Number of citations of the journal articles over the 2 previous years (year } n-1 \text{ and } n-2)}{\text{Number of articles published in the journal over the 2 years preceding year } n-2}$$

It is important to note that the IF should not be used as a criterion for evaluating a scientific paper, but rather, as an indicator of the standards of the journal in which the article is published.

48.6.4 Bibliographic Databases

Articles published in indexed journals are catalogued in online databases such as MEDLINE (produced by US National Library of Medicine) and EMBASE (produced by Elsevier). Publications go as far back as 1946 in MEDLINE and 1974 in EMBASE. MEDLINE can be accessed free of charge through PubMed. Articles can be searched using various search fields such as key words, titles, text word, subject headings, etc. There is a massive overlap in the two databases of the articles included, however, a little over 25% of articles are unique to each database. For this reason, it is recommended that both databases be searched for the topic of interest to ensure a comprehensive search.

The Cochrane Central Register of Controlled Trials (CENTRAL) provides the most comprehensive source of published controlled trials. It has been developed by the Cochrane Collaboration and includes records from MEDLINE, EMBASE, as well as other databases and reports identified through hand searching.

Apart from the main databases described above, other regional/national or subject databases can also be found. Gray literature (not published in journals or books) can be searched on databases such as INIST (produced by Institute for Scientific and Technical Information).

48.6.5 Conflict of Interest

The global economic crisis has led to a reduction in government spending on research. For this reason, researchers are increasingly forced to seek private funding in order to conduct studies and do their work. There is no law prohibiting this type of financial support, but it is important for the authors to maintain full disclosure about the source of their funding. The risk is that this type of cooperation might remain hidden and that papers might only present results that are favorable and beneficial to the sponsors of the project. Therefore, when reading a scientific article it is essential to consider whether the study was conducted with personal, government or corporate funding and only then reach the appropriate conclusions.

48.6.6 The Role of Clinical Practice Guidelines

Practitioners today can obtain additional benefits by applying CPG to their decision-making process when choosing the most appropriate treatment for their patients. In 2011, the US Institute of Medicine defined the CPG as "statements that include recommendations intended to optimize patient care that are informed by systematic reviews of evidence and an assessment of the benefits and harms of alternative care options" (Graham and Mancher 2011). The CPG are produced under the auspices of a medical specialty association, relevant professional society, public or private organization, government agency at the federal, state, or local level, or healthcare organization. Clinical guidelines must meet several criteria according to the US National Guidelines Clearinghouse. As a consequence, CPG are located at the highest level of the Grading System for therapies.

The above-described CPG differ from the generally called clinical guidelines in therapy. In fact, the latter often derive from consensus reports of expert opinions; they are not systematically based on the highest level of evidence, and they do not take social impact into account.

References

Akhtar, M., Breithardt, G., Camm, A.J. et al. (1990). CAST and beyond. Implications of the cardiac arrhythmia suppression trial. Task Force of the Working Group on Arrhythmias of the European Society of Cardiology. *Circulation* 81 (3): 1123–1127.

Altman, D.G. (1994). The scandal of poor medical research. *BMJ* 308 (6924): 283–284.

Doll, R. and Hill, A.B. (1950). Smoking and carcinoma of the lung. *BMJ* 2 (4682): 739–748.

Doll, R. and Hill, A.B. (1954). The mortality of doctors in relation to their smoking habits; a preliminary report. *British Medical Journal* 1 (4877): 1451–1455.

Evans, D. (2003). Hierarchy of evidence: a framework for ranking evidence evaluating healthcare interventions. *Journal of Clinical Nursing* 12 (1): 77–84.

Gail, M.H. (2000). *Encyclopedia of Epidemiologic Methods*. Wiley.

Graham, R.D.W. and Mancher, M. (2011). *Clinical guidelines we can trust*. Washington (DC): National Academies Press (US).

Last, J. (2001). *A Dictionary of Epidemilogy*, 4e. New York: Oxford University press.

Nieri, M. et al. (2003). Individual patient data. *Journal of Evidence Based Dental Practice* 3 (3): 122–126.

Sackett, D.L., Rosenberg, W.M.C., Gray, J.A.M. et al. (1996). Evidence based medicine: what it is and what it isn't. *BMJ* 312 (7023): 71–72.

Section VIII

Technical Writing

49

Writing for Your Audience
Maria J. Grant

Liverpool John Moores University, Liverpool, UK

49.1   Introduction

Every piece of writing you do should have a defined audience. While your enthusiasm for your topic area will carry you through the writing process, successful writing relies on your ability to define the audience who will ultimately read your work.

But who is your audience? Where can you reach them? And how does your audience inform your writing style?

49.2   Who Is Your Audience?

Good news! The audience for your writing is wider than you think. Every project has the potential for multiple audiences, although this does not mean that you can simply republish your writing in multiple places. Each piece of writing will, by necessity, be different with a unique and well-defined audience, or reader, in mind. To achieve this, you will be pleased to know that the form your writing takes can help clarify your intended audience and visa versa.

By virtue of the purpose of their reading, not all audiences will be interested in all aspects of your writing to the same level of detail. For example, a project report might be written for a management committee interested in a particular piece of your analysis and how it (and by association, your writing) can inform cost-benefit savings. On the other hand, a journal article arising from the same piece of work might seek to share an innovative methodology that others may wish to emulate, while a different journal paper or newsletter item on the project might focus strongly on your findings with the aim of influencing changes in local practice.

By defining an intended audience, you help set the parameters, the format, and the focus for writing for your audience.

49.3   Where Is Your Audience?

Having defined your audience, be they policy makers, senior management, colleagues or peers, you must now identify an appropriate outlet to reach them. Consider their

A Guide to the Scientific Career: Virtues, Communication, Research, and Academic Writing, First Edition.
Edited by Mohammadali M. Shoja, Anastasia Arynchyna, Marios Loukas, Anthony V. D'Antoni, Sandra M. Buerger, Marion Karl and R. Shane Tubbs.

perspective and where it is likely to be reported. In which publications will they typically be doing their continuing professional development (CPD) reading? Where do they themselves choose to publish? In answering those questions, you are creating a firm rationale for choosing your place and mode of publication. If you wish to publish a journal article, it can be helpful to look at your reference list and see from which journals you have cited papers. This can give a good indication that this journal has an interest in that area and be receptive to similar articles on the subject.

While writing for specific groups within your own sector is a good starting point, consider moving beyond the usual publications. Writing for other professional groups provides an opportunity to get your ideas out to a wider audience and can open up new opportunities and multidisciplinary collaborations. Consider who else might benefit from your work and discuss your ideas with colleagues. Is your work relevant or transferable to other sectors? Perhaps you worked with someone from a different discipline and could jointly write a piece for this audience? By considering different audiences, there is an opportunity to gain greater impact for your writing.

49.4   How Does Your Audience Inform Your Writing Style?

There is a duality to writing for publication in which you will need to define your audience in combination with deciding where you would like to publish, both of which should influence your writing style.

Always have your audience in mind throughout the writing process, not just at the start or end of your writing. Anticipate their needs. Answer questions your writing might stimulate, aiming to provide material that will inform their work, research or professional development. How can you anticipate their questions? Visualize a specific reader to keep you focused on what you should say and the questions they might ask by way of clarification, or ask a colleague to read your writing and highlight areas requiring further explanation. When writing for your audience, remember that, just like you, they are busy people, so make transparent the key messages they can expect to take away from your writing. Make it worthwhile for them to invest time in reading your text.

Whoever your chosen audience is, contextualize your writing in their literature. By using pertinent examples from their evidence base, you can demonstrate the significance of your writing within their discipline and body of knowledge and, mindful of your audience, draw out key messages and focus your discussion on addressing their concerns. By thinking about, and writing for, your specific audience, you can make a judgment about how much prior knowledge they have of the subject, how much can be implied and how much must be more clearly articulated.

A further consideration for your writing and a significant factor in whether your writing is accepted for publication is whether your writing meets the expectations of the editor of the journal or book in which you wish to publish. They will have a well-defined audience for their publications and will assess whether their audience will find your writing engaging and useful. Ensure that your writing conforms to a defined writing style for that audience by looking at past issues or sample chapters to ensure that you are writing in a similar style. Familiarize yourself with any author guidelines, and write with the publication's specific audience mind. It is not enough to simply identify where you want to publish to reach your audience; you will need to become acquainted with such publications so that your writing meets the expectations of the editor and, further down the line, those of the reader. In this way, your writing is likely to receive a more favorable reception.

Writing for your audience might seem like an obvious call to arms, and yet many writers forget this decisive factor, an element that will determine whether their writing can be accepted for publication and will ultimately be read.

50

Principles of Writing a Good Scholarly Paper

S. Jane Millward-Sadler

Newcastle University, Manchester, UK

50.1   What Is a Good Paper?

A good paper can be many things, but above all, it is a paper that the target audience finds readable and understandable. It is not as easy as it sounds to write such a paper. The following tips should help to demystify the writing process.

50.1.1   Simplicity

Use simple sentences. Try not to use one sentence where three will do. Overly long and complicated sentences are difficult to read or understand. The reader – and even the writer – can lose the thread before reaching the end. Break down the information into pertinent parts (e.g. sections and paragraphs) and use one sentence for each part. If the sentences seem too short and simple when the draft of your manuscript is complete, you can combine or extend them when you make revisions. This will also help to ensure that grammar and sentence structure are correct. Make sure that you use appropriate language. Do not be tempted to overcomplicate what you are trying to say (this is dealt with in more detail in Chapters 51–54).

50.1.2   Focus

It is very important to stay focused: Make all the points you wish to make, but make them clearly and concisely. One way of achieving this is to use bullet points to list the main pieces of information or the principal messages you intend to convey before you start to write the manuscript (see Chapter 53 for how to write the outline of the paper). You will be able to:

- Plan the content of your manuscript.
- Keep the information in the right order (i.e. structure the paper properly).
- Make sure you stick to the point.
- Avoid omitting anything important.
- Avoid repetition.

Do not try to cover all of your points at once. Deal with them individually or in small groups, as appropriate.

A Guide to the Scientific Career: Virtues, Communication, Research, and Academic Writing, First Edition.
Edited by Mohammadali M. Shoja, Anastasia Arynchyna, Marios Loukas, Anthony V. D'Antoni, Sandra M. Buerger, Marion Karl and R. Shane Tubbs.
© 2020 John Wiley & Sons, Inc. Published 2020 by John Wiley & Sons, Inc.

50.1.3 Punctuation and Grammar

Again, keep it simple. It is amazing how few people use punctuation correctly. Capitalize appropriately, sparingly (at the beginning of each sentence, as well as proper names), and consistently. Use colons before lists, and commas between the list components; commas also demarcate phrases and provide a breathing space if someone is reading aloud, but try not to overuse them. When a sentence has to begin with a number, write the number in letters rather than as digits.

50.1.4 Structure and Style

Start by looking at the editors' guidelines, and make sure that you know the answers to the following questions:

- What does the journal want?
- How do the editors want it presented?
- Do editors want separate or combined sections?
- How much detail does the journal require?
- How should the references, figures, and tables be set out?
- Is there a limit to the word count, or to the numbers of figures and tables that are permitted?

The answers to all such questions will help you to optimize the structure and style of the manuscript for your journal of choice.

50.2 Order of Writing

Write the Methods first. This will not be the first section in a paper, but it tends to be the easiest section to write and gets you into a good frame of mind for drafting the rest. Make sure that you include enough detail in the Methods, and do not use the figure legends for that purpose unless that is what the journal requires. Methodology should be sufficiently detailed for another investigator to understand what you have done and be able to repeat the experiment, but it should not be needlessly verbose.

After the Methods section is written, move on to the Results. Writing the Results before the Introduction makes it much easier to focus the Introduction on the parts of the literature that need to be covered to put your study into context. Be careful when you write the Results to describe your findings fully and clearly but do not discuss their meaning. The Results section should be crisp and concise, just stating the facts obtained, appropriately illustrated and supported by the figures and tables. Interpretation of the results and their relationship to the existing literature should be reserved for the Discussion section (unless the journal requires or permits a combined Results and Discussion section).

The Introduction is not a general summary of all of the research in your field. Review articles serve that purpose. It should cover the specific area explored in this particular manuscript and should do so adequately, but that is all. It should contain sufficient detail to set the scene and explain the purpose of the investigation, making it clear why this study is required to take the knowledge forward, and it should end with the question or hypothesis addressed in the research – i.e. the aim of your work.

You should write the Discussion section last. That is where you put the research into context, identify any drawbacks or problems you have encountered, speculate (briefly) on the implications of your results, propose any hypotheses you have formulated as a result of your findings, and outline the next steps to be taken in this area of study.

50.3 Draft, Draft, and Redraft

There are always parts of the manuscript that can be improved. There are always mistakes to be corrected. Nobody – even the most experienced author – writes a perfect manuscript on the first draft, , so everyone needs to make revisions. After writing your first draft, save it, then go away and come back to it later to read it with fresh eyes. It is also helpful to get a colleague to read through it and make comments. That will give you another perspective on your writing, which can be helpful and instructive.

50.4 Conclusions

Above all, get someone unconnected to the study to proofread your final manuscript. That will help to ensure that you have written what you intended to write rather than what you think you have written and will help to pick up any mistakes in spelling, punctuation, or grammar that have survived redrafting. The best person to proofread is someone who is sufficiently informed about the subject to understand and comment on the content, and is a native speaker of the language in which the manuscript is written.

After all of that has been completed, and you have prepared your figures and tables with their appropriate legends and compiled your reference list, you are ready to submit your manuscript to your chosen journal.

Further Reading

Gennaro, S. (2016). Mistakes to avoid in scientific writing. *J. Nurs. Scholarsh.* 48 (5): 435–436.

Grindstaff, T.L. and Saliba, S.A. (2012). Avoiding manuscript mistakes. *Int. J. Sports Phys. Ther.* 7 (5): 518–524.

Hoogenboom, B.J. and Manske, R.C. (2012). How to write a scientific article. *Int. J. Sports Phys. Ther.* 7 (5): 512–517.

Peat, J., Elliott, E., Baur, L. et al. (2002). *Scientific Writing: Easy When You Know How.* London: BMJ.

Porcino, A. and Moraska, A. (2015). Avoiding common writing mistakes that make your editors and reviewers cringe. *Int. J. Ther. Massage Bodywork* 8 (4): 1–3.

Swales, J. and Feak, C. (2004). *Academic Writing for Graduate Students: Essential Tasks and Skills*, 2e. Ann Arbor: University of Michigan Press.

51

Tips for Scientific Writing

Mark P. Henderson

Theoretical Medicine and Biology Group, Glossop, Derbyshire, UK

51.1   Professional Publications

Good scientific publications will:

- Add to the sum of knowledge.
- Help to advance the author's or authors' career(s).

  Therefore, when you decide to publish, you need to make sure that:

- Your addition to the sum of knowledge is significant.
- People who share your interest will want to read your article and will find it enlightening.

  That will give your work – and you, as author – recognition within the field.

  To present the work to the right audience, you need to select the right journal or other publication medium. To make it interesting and enlightening for that audience, you need to write in *good professional English*. Journal selection and article types are discussed elsewhere in this book. The focus of this chapter is "good professional English."

  English is the medical and academic lingua franca. Native speakers have, or can have, an advantage; native speakers of other languages are likely to need help. English-speaking colleagues or manuscript editing companies can improve a non-native author's writing, but quality is not guaranteed. After an article has been language-edited, it is best to seek a second opinion, preferably from an experienced English-speaking medical or scientific writer. Inexperienced writers, including native English speakers, are advised to seek help at the outset.

  "Professional" English denotes the language of an article intended for readers who share the author's knowledge base, skills, and interests. In medicine and the sciences, we use specialist vocabularies – i.e. words and phrases familiar to other doctors or researchers but unfamiliar to the rest of the world. Those words and phrases enable professionals to communicate *succinctly* and *unambiguously* with their colleagues. Authors of medical and scientific papers have no need to explain the vocabulary because their intended readers will understand it. Nonspecialists would not share that understanding without long explanations.

A Guide to the Scientific Career: Virtues, Communication, Research, and Academic Writing, First Edition.
Edited by Mohammadali M. Shoja, Anastasia Arynchyna, Marios Loukas, Anthony V. D'Antoni, Sandra M. Buerger, Marion Karl and R. Shane Tubbs.
© 2020 John Wiley & Sons, Inc. Published 2020 by John Wiley & Sons, Inc.

That does not justify obscurantism. Your readers will not be impressed by high-flown or deliberately difficult writing or unnecessary use of specialist vocabulary. They are busy, just as you are, so they will ignore your work unless your message is conveyed:

- As briefly as possible.
- As clearly as possible.
- As simply as possible.

Your professional vocabulary should help you to achieve brevity, clarity, and simplicity. It should never be used for the contrary purpose.

"Good" professional English involves:

- Choosing the right nonspecialist words to enhance brevity, clarity, simplicity, and precision
- Writing grammatically correct sentences that are constructed as simply and unambiguously as possible
- Punctuating correctly

51.2 Manuscripts Require Repeated Revision Before They Are Submitted for Publication

First drafts, even if written by experienced authors, are seldom, if ever, in "good professional English." They need to be revised, read critically by the authors, revised, and read and revised again. It is sensible to ask at least one experienced colleague who has not been involved in the case or the research to read a draft critically.

Almost every revision will lead to improvement. You will also receive criticisms of your paper from the referees after submission, so the revision process will continue until the article is finally acceptable to the journal editors.

That seems to be a long process, but it is worth taking time to ensure that the version of the article you submit is written in "good professional English." Acceptance of your work, and your own career advancement, could depend on it.

51.3 Establish the Right Mindset before You Write

Before you start to write your first draft, ask yourself:

- What significant new information do I have to share?
- Who is most likely to be interested in it?

Write short answers to those two questions, and keep them beside you while you draft your manuscript. They will help keep you focused. They will also help you to choose the most appropriate journal for submission of your article.

51.4   Remember Who Your Readers Are

Most of your readers already know about previous work in your field, probably as much as you do. So do not spell out the history (background) in full detail in your introduction. Focus on published work that is *directly* relevant to the new information you intend to share, particularly if it raises questions or controversies on which your findings might throw light.

Again, when you describe the methods you have used, be as brief as possible. For standard procedures, cite suitable references. Your readers will know them, so there is no need to write more than a sentence unless you have introduced modifications, in which case you need to be specific. Full details are needed only for entirely novel methods.

51.5   How to Convey Your Findings

New pieces of information or data that seem to support or refute your hypothesis should always be presented straightforwardly. When possible, use tables or graphs (or photographs, if appropriate). Some of your data might be inconsistent with received wisdom, or with your hypothesis, or cannot be explained. Make sure that those data are presented along with the rest of your information and are not marginalized. Contrary findings and controversies are often the engines that drive knowledge forward.

51.6   Keeping "Discussion" Sections under Control

In the Discussion section, do *not* summarize your findings. This will already be evident if you have described your results properly, so you would be repeating yourself. Find the most salient points (five or fewer) that emerge from your results, then use each point as a subheading, and discuss each in relation to the existing literature (and, if appropriate, to your hypothesis). Avoid speculation and conjecture, and keep any mention of future studies to no more than a sentence.

Some journals require a separate "Conclusions" section, in which the main inferences from your article are stated briefly. If there is no separate "Conclusions," use the final paragraph of your Discussion for the purpose.

Make sure your Conclusions do not reiterate what you write in the Abstract or Summary with which the paper begins. It is usually best not to write the Abstract until the rest of the manuscript is fully drafted and revised.

52

Perspectives of a Medical Editor

Marilyn Michael Yurk

Department of Neurological Surgery, Indiana University School of Medicine, Indianapolis, IN, USA

52.1   Everyone Needs an Editor

As an editor for more than 25 years, including 10 years as managing editor of an international medical journal, I have read thousands of articles before they appeared in journals in their (hopefully) much more polished forms. The other chapters in this book offer a wealth of advice about writing scientific articles, so I will focus, rather informally, on common problems and some techniques that can help young authors avoid those problems.

First, everybody needs an editor – even an editor needs an editor. All of us tend to overlook mistakes when we have read a manuscript over and over and yet again, so a second set of eyes is invaluable for detecting errors.

Even if you are lucky enough to have an editor at your institution who will check your manuscript, be sure that you and every one of your co-authors read your manuscript carefully several times before you submit it to a journal for peer review. And if you have an associate who is articulate and skilled at grammar and punctuation, ask that individual to check it, too. But please note that if your associate simply reads an article to correct grammar, spelling, and style, it does not mean that you should include that person as a co-author. Authorship should be reserved only for individuals who have made significant intellectual contributions to the research and writing of the manuscript. Many journals will ask for details about these contributions.

I have assembled here some quick tips that I hope will be helpful.

52.2   When Planning Your Foray into Academic Writing, Keep It Simple

Inexperienced authors sometimes choose to write a review on a broad topic for their first research manuscript because they know that articles about the topic are abundant, so they think the research will be easy. They then might find, to their dismay, that the article takes weeks or months of work. If articles about a review topic are plentiful, it might mean that you must devote many hours to reading those articles, sifting through conflicting statements, and determining which conclusions are the most reliable.

A Guide to the Scientific Career: Virtues, Communication, Research, and Academic Writing, First Edition.
Edited by Mohammadali M. Shoja, Anastasia Arynchyna, Marios Loukas, Anthony V. D'Antoni, Sandra M. Buerger, Marion Karl and R. Shane Tubbs.
© 2020 John Wiley & Sons, Inc. Published 2020 by John Wiley & Sons, Inc.

Even under the best circumstances, your first research article is sure to take far more time to write than subsequent articles, just because you are learning the style and format required for the articles. Choose your topic carefully. A case report on a very unusual condition or a small, narrowly focused study might be the best choice for your first article.

52.3 Learn to Use a Good Reference Software Program

In the hands of an experienced user, reference-citation software such as Endnote can be a godsend, saving the author many hours of tedious work. Most universities offer those programs free to students, staff, and faculty. Unfortunately, however, the learning curve for those programs is steep, and some hazards accompany their use – for example, a document can become corrupt if passed around among co-authors, especially if the co-authors use different software or types of computers and if several people add references.

I once spent an entire week editing a long chapter with four co-authors and 554 references. The document was corrupt because each co-author had used different software to enter references. The only way I could repair the document was to rebuild the text in a new document, transferring it paragraph by paragraph after deleting all of the fields that had been inserted by the reference programs, and re-entering the references using my own Endnote program. After all of the references were corrected, we found that many had been duplicates. There were actually about 500 references, so 10% of them dropped out.

A few specifics:

- If your university offers a course on a reference-citation program, take the course, and use the software that it most highly recommends. For more information on the available reference-management software, please see Chapter 65. You can learn the fundamentals from an effective teacher in two or three hours, but many of the programs are not easy to learn on your own.
- Appoint one person to enter all references into your document. If others want to add or change a reference, ask them to give the information to that individual.
- Learn to use PubMed unique identifier (PMID) numbers to pull in references automatically via your software and PubMed. Articles indexed in PubMed are each assigned a unique eight-digit number that can be used in both PubMed and Ovid to retrieve the article. For example, 23959415 is the PubMed number for "Analysis of single-staged resection of a fourth ventricular tumor via a combined infratentorial-supracerebellar and telovelar approach: Case report and review of the literature." It is much easier to type (or copy and paste) "23959415" into the PubMed search page than to enter the title of the article. Using Endnote, for example, you can use the PMID to read the reference into your Endnote library via its online search feature. That enables you to quickly and accurately insert most of your references into your document. You might still need to enter a few references manually, such as chapters, books, and very old or very recent articles.
- Learn how to change and edit styles with your reference software. That will allow you to format your references correctly for a specific journal, and if that journal rejects the

article, you can quickly reformat the references, do some minor revisions if necessary to comply with a different journal's guidelines, and submit the article to that journal. This is especially helpful if the first journal wanted references cited by appearance in the article, but the second journal wants them in alphabetical order.

- When moving text that includes reference citations within your document, be sure to *copy and paste* the text, and then go back and delete the original text. If you only *cut and paste* the text, you could delete references from your document library along with the cut, and you will then need to re-enter those references after moving the text.

52.4   Become Your Own Line Editor

Line editing is the process when a manuscript is carefully edited for tone, style, consistency, and precise choice of words. The purpose is to improve the text so that it is clear, concise, consistent, and grammatically correct (the four Cs). Slowly read through your text, and check spelling and grammar while you meticulously tighten and clarify the text. Repeat the process, and then repeat again. See Table 52.1 for some examples of line-edited text.

Although residents and fellows are highly educated, I have seen many of their articles marred by misspelled words, incorrect grammar, redundancy, and reference problems. If you submit an article in such a condition, and many do, it is far more likely to be rejected. Fortunately, however, a few hours of line editing can often transform the manuscript into an acceptable article.

If you edit your manuscript once or twice simply to eliminate unnecessary words, you can usually improve your text significantly. A few specifics:

- Spelling and grammar checkers highlight errors, but those aids are not enough by themselves. I often copy and paste words into Google or an electronic medical dictionary to check spelling and usage. I also make liberal use of my word processor's thesaurus to find more precise words.
- As a student or graduate, you probably can access the resources of your alma mater's medical library to check medical terms. For example, many university medical libraries offer online access to several medical dictionaries to faculty, students, and staff.
- Most journals prefer that you write the body of the article in the first person ("we did this") and the abstract in third person ("the authors did that"). That will help you avoid overuse of the passive voice. Conversion of sentences from the passive to the active voice whenever possible will, by itself, usually improve a manuscript immensely.

52.5   Minimize Acronyms

The Instructions to Authors for most academic journals ask authors to limit acronyms and define them at first use. The requirement is so basic and universal that it is hard to imagine that authors would overlook that request, but they usually do.

The purpose of an acronym is to make an article *easier to read*, but overuse of acronyms will make an article difficult to read. Use only standard acronyms commonly

Table 52.1 Examples of line editing. This table shows actual phrases and sentences from manuscripts written by young authors and a line-edited version.

Original text	Edited text	Comment
The majority of pineoblastoma tumors present in the pediatric population.	Most pineoblastomas occur in children.	Simplify.
A prospectively acquired aneurysm database maintained by the authors' institution was examined for information about patients.	We searched our institution's database for patients treated for intracranial aneurysms by clip ligation from 2000 through 2010.	Change to first person, active voice. Simplify. Add inclusion dates.
The patients consisted of 2 males and 10 females with the youngest patient being 24 and the oldest 81 at the time of diagnosis.	The patients included 2 men and 10 women (range 24–81 years old at diagnosis).	Simplify. Prefer woman, man, girl, or boy to male or female.
Imaging studies were available for all 12 of the patients in our series. Of the patients, (11) presented with hydrocephalus. The 12th patient had did not present with hydrocephalus.	Imaging studies were available for all 12 patients in our series. Eleven of the 12 presented with hydrocephalus.	The original is awkward, wordy, and grammatical incorrect. Delete the last redundant sentence.
The following information was obtained from patients: gender, age, presenting symptoms, presence or absence of hydrocephalus, history of previous resection, largest diameter of tumor, approach for resection, extent of resection, radiation, other adjuvant therapy.	We obtained the following information about each patient: sex, age, symptoms, presence of hydrocephalus, history of resection, size of tumor, surgical approach and extent of resection, and adjuvant therapy.	Change to first person, active voice, correct prepositions, simplify text.

used and only for long terms that you use repeatedly. Avoid acronyms for one- or two-word terms. It is *rarely* acceptable to create your own acronym but might be needed only for a long term used many times. In that case, check other articles on the same topic to determine whether there is a standard acronym for the term.

Of course, some acronyms are more readily recognized than the terms they represent – such as AIDS, DNA, CPR, MRSA. However, you must still define such acronyms at first use.

52.6   Beware of Predatory Journals

Many highly reputable academic journals charge publication fees. For example, most journals charge for color figures published in their print version, and many charge page fees. The American Heart Association journals (such as *Circulation* and *Hypertension*) have had author page charges for years, but they often waive the charges for authors who do not have publication funds.

On the other hand, a number of low-quality journals have recently emerged that exploit the author-pays model of academic articles. They have names that sound legitimate, but they publish low-quality articles of questionable scientific value (Beall 2014; Jones and McCullough 2014). Sometimes called *predatory journals,* those journals provide little or no legitimate peer review and might simply publish virtually every article submitted to them, usually only electronically, and only after the author pays a fee. Copyediting and production for those journals might be handled in India, Nigeria, or another developing country, and the staff members might not speak fluent English.

Be aware of the existence of predatory journals and avoid them. Do not submit your work to them, and do not agree to serve on their editorial boards. For names of some suspect organizations, see some of the work published by Jeffrey Beall (2014). Also, see Jeffrey Beall's chapter about open-access journals (Chapter 41).

52.7   Check Your Numbers – Then Check Them Again

I recently edited a manuscript by a resident who wrote: "Our final study group included *33 patients, 16 males and 16 females*" (italics added). Since 16 plus 16 equals 32, there were probably 32 patients. Of course, it followed that many of the statistics in the article were wrong.

This error appears surprisingly often, for two reasons. The first reason, obviously, is that the authors did not check their numbers carefully. The second is that the authors used a spreadsheet program – Excel in this case – to list the patients and characteristics, but didn't realize that the top line in the spreadsheet holds the column titles, not patient data. So even though information was entered on 33 lines, there were only 32 patients.

Wrong numbers and percentages may invalidate your results, and they are sure to negatively affect reviewers, so double-check your numbers and ask your co-authors to check them.

52.8   Fully Involve Your Co-authors

The scientific community has strict standards involving authorship. Authors must have made a significant intellectual contribution to the manuscript, and most journals expect that all authors approve the final manuscript before submission. Nevertheless, when there are multiple authors, some might be casual about their authorship duties.

If you are a young author and have taken the lead as first author, you will benefit most and produce better work if you fully involve each co-author in the creation of your manuscript. It is reasonable to ask more experienced co-authors to mentor you and validate your work.

Make sure that every co-author listed agrees with the methods, results, and conclusion. Ask if they have suggestions for the introduction and the discussion. And especially, make sure each author approves the manuscript before it is submitted and when it is revised after peer review.

52.9   Observe the Copyright Laws

A common misconception is that text and graphics on the internet are in the public domain and free for all to use. Sorry, the copyright laws cover online material just as they cover printed work. It is, however, permissible to quote internet material in part, referencing the authors and title, the date posted (if available), the uniform resource locator (url), and the date you accessed it.

52.10   Once Your Accepted Article Is in Production, Check Your Proof Carefully

I have yet to find a perfect human, and copyeditors, being humans, are not exceptions. They generally improve a manuscript significantly, even dramatically, but even copyeditors at the best journals make mistakes. They sometimes overlook misspelled words and incorrect punctuation, and, occasionally, they rework a sentence so that it is incorrect. Anyone can change the meaning of a sentence simply by adding a misplaced comma.

I have seen many significant errors created by copyeditors, tables edited so that data were altered, and incorrect figures included in articles. I once checked a proof in which the copyeditor had changed only one word – but that word was the name of a drug, and he changed it, not from a trade name to the generic name, which would have been correct, but to a different drug. I suspect that this editor might have used a spell-check program and clicked in the wrong spot.

Compare your proof with your original text, and verify authors' names and affiliations, tables, figures, and legends. The easiest way to do this is to compare your proof with the PDF (portable document format) file of submitted files that was generated by the journal when you submitted your last revision. Usually, PDF documents are available to download from the journal's website.

I hope that my suggestions are useful to readers. Scholarly research and publishing is an admirable pursuit that has the potential to improve the world. You should be commended for pursuing the challenge.

References

Beall J. List of predatory publishers 2014. Released January 2, 2014. Scholarly Open Access Website. Available from: http://scholarlyoa.com/2014/01/02/list-of-predatory-publishers-2014 [Accessed March 25, 2014].

Jones, J.W. and McCullough, L.B. (2014). Publishing corruption discussion: predatory journalism. *J. Vasc. Surg.* 59 (2): 536–537.

53

The Art of Organizing Your Research Content
Kristin N. Mount

College of Applied Health Sciences, University of Illinois at Chicago, Chicago, IL, USA

53.1   The Art of Organizing Your Research Content

The operative word for the development of a research paper is *write*. I am going to *write* a research paper. When will you be finished *writing* your paper? How much have you *written* already? This word is misleading. It conjures the idea of placing words on paper, one after the other, until the manuscript is complete. The *writing* of a research paper, however, encompasses a much grander process.

The realization of a formal research paper requires much more than simply writing. It involves the thoughtful collection of content and the organization and assembly of that content. Just as we would not build any complex structure without a plan, we also should not construct a paper without one. This plan involves the collection of materials, assessment of inventory (the content), and a well-designed blueprint, or in the context of a research paper, a thorough outline.

The plan begins with gathering your raw materials, or research content. That generally involves sorting through a collection of disorganized notes, including recorded data, methods, research findings, resources, etc. Those notes are your raw materials for the paper. They are necessary pieces of the structure, but they will not be useful until they are appropriately assessed and logically organized. The ultimate goal is to organize the content in a way that best articulates the research process and concept.

Like any other large undertaking, the planning and organization of a research paper are integral to its success. A research paper is traditionally divided into sections, which should inform your overall organization and the structure of your outline, and for which to sort your raw materials. The sections of a research paper minimally include an Introduction, Literature Review, Materials and Methods, Results, and Conclusions. In a scientific journal, the core sections vary slightly and include an Introduction, Methods, Results, and Discussion (Peat et al. 2002). The sections and their names might vary, depending on institutional standards and journal requirements.

A large task in organizing content is to understand the different sections of a research paper and allocate the content accordingly. After the scope of each section is understood, the appropriate content for the individual sections can then be evaluated and further organized in a more pointed context. Each individual section can be effectively treated as its own small paper.

A Guide to the Scientific Career: Virtues, Communication, Research, and Academic Writing, First Edition.
Edited by Mohammadali M. Shoja, Anastasia Arynchyna, Marios Loukas, Anthony V. D'Antoni, Sandra M. Buerger, Marion Karl and R. Shane Tubbs.
© 2020 John Wiley & Sons, Inc. Published 2020 by John Wiley & Sons, Inc.

While unadvised, a research paper can be written without any advanced planning. But this often results in a disorganized document that is difficult to follow. If you are submitting your paper to a journal, approaching your writing without a plan will likely increase the number of drafts you are required to revise (Peat et al. 2002). It is much more time-consuming – not to mention frustrating – to rearrange the content of a paper after it has been written. Not only is it more efficient to organize the research content before any writing is begun, it will also result in a more thoughtful and elegant piece of writing.

53.2   The Outline

An outline is an effective way to begin organizing the content of your paper. Start with the major sections of the paper and then elaborate further upon each. A well-developed outline gives the author a chance to view the overall arrangement of the content and edit the structure of each section, as necessary. It also allows the author to be sure all relevant content is included in the appropriate sections. Rearranging and editing the organization of your content after the paper is already written is much more time-consuming and will likely provide an unwelcome source of heartache.

A useful practice when developing an outline is to begin with the sections, assigning them each a Roman numeral in the order they are traditionally presented. Each section can then be divided into subsections, which correlate to the subheadings that might or might not be used in the final paper. Content can be further divided under the subsections, helping organize the paper down to the paragraph level. Word processing software, such as Microsoft Word, have built-in formatting tools that automatically suggest the appropriate outline structure as you create it. A common feature of such software also allows you to create an outline from an existing template.

While the outline is an important tool for organizing content, it is usually not included in the final paper or journal submission. Some make the mistake of attaching page numbers to their outline and using it as a table of contents, but that is not an appropriate practice, either. The outline can organize content all the way down to paragraph level, but a table of contents should only list the major sections and subsections of a paper.

53.3   Organization by Section

While this chapter focuses on the art of organizing content for research papers, any type of document intended for publication – be it a review article or a case report – can benefit from the techniques described herein. Specific sections of writing might vary; however, the approach to organizing and collecting content is applicable to all forms of scientific writing.

The pages and/or sections of a research paper typically appear in the order presented in Table 53.1. Keep in mind that the inclusion and formatting of individual sections depends on the requirements of the institution you are attending or the journal to which you are submitting. For example, a literature review is customarily required only in academic papers. The majority of journals, on the other hand, request that a focused review of literature be reserved for the discussion section of the paper. For case reports, the materials and methods section and the results section are often replaced by a case history section.

Table 53.1 Typical order of the sections of a research paper.

Title page
Table of Contents (sometimes not required)
Abstract
Introduction
Literature Review (sometime incorporated into discussion)
Materials and Methods
Results
Discussion
Conclusions (often incorporated as the last paragraph(s) of the discussion)
Acknowledgments
References/Citations
Appendices or supplemental materials

In an academic setting, you can request examples of research papers by those who attended the program before you, or ask your research advisor for specific formatting guidelines. Most journals or publications will provide instructions to authors, including basic formatting preferences. Time spent correctly formatting your paper can often improve the reception of your submitted manuscript by the journal's editors (Peat et al. 2002).

53.3.1 The Title Page or Cover Page

The layout and content requirements of the title page can differ among varying institutions and publications but follow the same basic guidelines. The title of the paper should be specific and offer an accurate description of the project without any extraneous information. The title should not include abbreviations of any kind; for example, "three-dimensional" should be used rather than "3-D" (for more on how to write an effective title, refer to Chapter 55). Besides the title, other information included on the title page is listed below:

1) The author's name.
2) The author's academic credentials and affiliations: This includes degree, department, and institution/university, the city, state, and country.
3) Running title or abbreviated title.
4) A statement of submission, for example: *Submitted in partial fulfillment of the requirements for the degree of Master of (degree name here) in the graduate college of (university name here).*
5) The year the research and paper were completed and defended, and grant information or source of support, if applicable.
6) A conflict-of-interest statement, for example: *The authors do not have any conflict of interest in connection with the submitted manuscript.*
7) Corresponding author's address and contact information.

53.3.2   The Table of Contents

A table of contents is not always required in a research paper. Its purpose, if included, is to help the reader conveniently find major sections and subsections of the paper. One of the most common mistakes in creating a table of contents is including too much information. As mentioned earlier, adapting your paper outline as the table of contents is a major *faux pas*. The outline is a tool by which students develop and organize content and should go into much more detail than the table of contents, but it should not be included in the paper.

The table of contents, on the other hand, should be succinct, including section titles and subheadings, if necessary. The treatment of subheadings should be consistent in form throughout. If you are required to add page numbers to the table of contents, be sure to double-check their accuracy *after* the paper is completed.

53.3.3   The Abstract

The abstract is a concise summary of the content of your paper. It briefly describes the scope of the project, its purpose and rationale, the methodology used, and the research conclusions. While it might give a small amount of background on the topic, the primary intent is to describe your research and findings using the least number of words. The abstract is often the last section to be written even though it is positioned early in the paper. The methodology and scope of a research paper commonly shift throughout the research process, so it makes sense to write the abstract after the rest of the paper is complete and the meaty sections have been finalized. For more on how to write an abstract, refer to Chapter 55.

53.3.4   The Introduction

The introduction section of your paper provides sufficient background information on your topic that leads clearly and logically to a formal research question. The research question is generally included at the end of the introduction or falls under its own subheading just after the introduction.

The introduction should be used to establish your research concept, the purpose of your research, and the research's relevance to your academic community. Your introduction should include any relevant background information on your research, an explanation of the significance of the research, and a thorough explanation of your research goal. Typically, the goal of the research is to determine an outcome; be careful not to introduce too many goals or questions. One is plenty to handle.

Although you will probably need to make reference to previous research in your area, save a detailed description of any articles or resources that you used for containment in the literature review section of your academic paper or the discussion section of your journal article.

53.3.4.1   Research Question

After the topic has been introduced and sufficient background information has been established, the next step is to articulate the research question. The question should be thorough but concise, arguable, and should not include any part that cannot be answered by the research. It is best to have a single, specific research question rather than multiple questions, and to be as clear as possible. The question should ideally allow for an expanded answer as opposed to a simple "yes" or "no." For example, instead of asking, "Does sun speed up the wound-healing process?" the more appropriate question is, "How does sun exposure affect healing times for human skin lacerations of one inch or less?"

53.3.5 Literature Review

The literature review is a thorough account of all of the sources you have used to collect information on your topic. It can include books, journal articles, pamphlets, videos, websites, and even live interviews. Its purpose is to provide a foundation of information that is valid to the new research you are conducting. The literature review is not a list. It should be written in paragraph form, organized by topic, and contain only information relevant to the research. Outlining this section alone will be a tremendous help before you begin to write.

The sources included in your literature review should be organized by topic and provide information on the different aspects of your research question. Consider the example of a project researching methods with which to create a 3D animation about cell mitosis for educational purposes. The topics of your literature review might include 3D techniques used to visualize to cellular events, the use of animation in science education, and the process of cell mitosis itself. If the same source was used for more than one of these topics, it may be cited each time, providing an appropriate context for the topic in question.

If interviews are referenced in the literature review, full names of those interviewed should be included, along with their credentials, position, title, and organization with which they are associated.

A separate literature review section is only customary in a dissertation or thesis. The majority of academic journals require that a focused literature review is included only in the discussion section of the paper. It should be focused on comparing the results from the new study to those from similar studies previously conducted.

53.3.6 The Materials and Methods

As its name implies, the materials and methods section (sometimes just called methods) describes the materials used in your research and the process by which your research was conducted. While information about both materials and methods is provided in this section, the two categories are not necessarily divided into respective subsections. This section is often organized around a chronological explanation of the procedure, and the materials are introduced throughout as part of the methodology.

If the description of materials warrants a separate subsection, the materials should be organized by type and should include anything that was necessary to complete the research. This should even include software, hardware, and lab equipment. Regardless of whether the materials are incorporated with the methodology or listed separately, relevant details should be included. Lab-equipment descriptions should specify types and sizes, along with the methods used to calibrate or standardize the equipment. Software should be listed using full names and version numbers (for example,

Microsoft Word v. 12.2, as opposed to just Word). Hardware should be listed using exact models and system specifications.

The methods portion of this section should describe in paragraph form the step-by-step process by which the research was conducted. A timeline can be used to establish the chronological events of your methodology but should be kept separate and is more appropriately included as an appendix (the appendices section is addressed later in the chapter). You can also describe changes of direction in the research here – as long as they are noted as such and rationalization is given.

A separate subsection titled "Participants" might be relevant if your research involves patients or human subjects. You must include appropriate, but sensitive, descriptors of the participants in this section. However, avoid using any information that can lead to the identification of a participant. Describe the method used to recruit subjects, collection of clinical information, and the use of the intervention, sham, or placebo in your study (Peat et al. 2002). A set of ethical principals should be applied when using humans in your research, which in some cases might be addressed specifically in the methodology under a separate subheading (Emanuel et al. 2014). For more on how to treat human participants in research, refer to Chapter 23 or National Institutes of Health (NIH) Bioethics Information and Resources web page.[1]

Statistical approaches used to analyze data should be included in the methods section, and are usually distinguished under a subsection titled "Statistical Methods." Specify here which statistical tests were used to analyze your data and why you used those tests. Discuss any variables affecting the results and define the p-value, or critical value, in order to determine the statistical significance of the research findings (Peat et al. 2002). More information on the different statistical approaches used in research papers can be found in Chapter 66.

53.3.7   The Results (or Findings)

The results section should be used to describe data that were gathered in your research without any interpretation or in-depth discussion. Explain in paragraph form what was determined using the specified methodology. This section is likely to incorporate a host of charts and graphs. While your results should be written in paragraph form, visualizing the data in charts and graphs will make trends easier to see. The informational graphics should be given figure numbers and captions and should be appropriately cited and placed within the text. Do not start the results section, or any section, of the paper with a figure. For more guidelines on how to effectively incorporate informative figures into your paper refer to Chapter 63.

The results section should be used to describe all results found in your research, whether or not they are ultimately relevant in answering your research question. Make sure to specify the results from your primary hypothesis of interest (*a priori* hypothesis) and those from hypotheses generated from post hoc/secondary analysis of the data (*a posteriori* hypothesis). The results should not be discussed or interpreted in this section. Highlighting specific data, analysis, and interpretation of the results should be saved for the discussion and conclusions sections. For additional information on how to write the results section, refer to Chapter 56.

---

1 http://www.fic.nih.gov/ResearchTopics/Pages/Bioethics.aspx.

53.3.8 The Discussion

The discussion section commonly replaces the conclusions section in a published journal article. The discussion section (much like the conclusions section) provides appropriate context for the research findings, summarizes the relevance of the research, and describes any limitations presented by the methodology or inconsistencies in the results (Peat et al. 2002). It is also appropriate to discuss any future direction or the potential impact of your research on the field.

Instead of a traditional literature review section, most academic journals expect that a focused review of literature is included in the discussion section of the paper. Specifically, you should compare your results to those published in previous studies similar in nature. That said, you should also present a conclusion that distinguishes your findings from previous research (Peat et al. 2002). Be careful to include only a discussion of the literature that is most relevant to your research. A tip to help you plan what literature to include is to make notes of the resources that specifically support your research and those at odds with it (Peat et al. 2002). For additional information on how to write the discussion, refer to Chapter 57.

53.3.9 The Conclusions

The conclusions section of your research paper should be used to analyze data from the results section and give the research meaning. In this section, you might write about what the results suggest or the trends that can be observed in them. Be sure to include a discussion about the limitations presented in your research process or methodology and any inconsistencies recorded in your research findings. A mention of study limitations is a sign of honesty and clarity, and will ultimately add value to your paper. In the final paragraph, you can discuss how your research can be applied and what it might lead to in the future. It is also appropriate to include some ideas about ways in which subsequent researchers can expand upon your research.

53.3.10 The References

The references section is used to list any and all resources used to conduct your research, including books, journal articles, and websites. Formatting for references depends on the style preferred by your particular institution or the journal to which you are planning to submit your work, and will likely be included in their guidelines.

53.3.11 Other Sections

There are other sections of the paper in which content can be organized. Those include the appendices, otherwise known as supplemental materials. Expanded informational graphics and comprehensive lists go here if they are too detailed to be included in the body of the manuscript. Appendices are typically labeled with letters (Appendix A, Appendix B, etc.) and are presented in the order in which they appeared in the paper. Each is given its own page.

In some cases, publications will distinguish appendices from supplements, the latter being generalized as material often existing in digital formats not appropriate for publication. That might include raw data files and source code, and should include a metadata document for each file or set of files describing the content.

If you need to include figures, they typically should be integrated throughout your research paper. Those might include charts and graphs, illustrations, screen captures, and photographs. Figures should be clear, readable, and relevant. They should be numbered and given appropriate captions. Be sure you've received the proper permissions and give credit where necessary. For more information on how to effectively incorporate artwork, refer to Chapter 63.

53.4   How to Approach Content Organization

While the prospect of organizing a complex collection of content for a research paper can seem daunting, like any other project, it can be made more manageable by beginning with larger pieces. The sections of the research paper defined in this chapter are those pieces.

When you first begin, you might have many pages of content that have no apparent order. It is helpful to look at each piece of information separately and decide where in each paper or section it belongs. There are many ways to organize the pieces of information in their appropriate sections. One simple method is to compose all content in a single document, then copy and paste the different pieces of information into new documents named for each section of the paper (Introduction, Materials and Methods, etc.). The result will be a separate document for each section, each with its own collection of content.

Each section can then be further rearranged and organized based on its subsections. Isolating the sections of your research into separate documents will allow you to focus on each individually. When each section has been sufficiently organized, all sections can be pasted back into a final, single document. At this point, the entire document can be read and reviewed as a whole, and any additional adjustments can be made.

Taking the time to organize your research content into a detailed outline before writing your paper will save you a great deal of time and energy in the long run. Often, the initial writing process is quite disorganized, simply because structure has yet to be conceived. The key to organizing your content is to have an initial plan, ideally in the form of carefully outlined sections and subsections, then break up the content and sort it within the outline. If you practice the organizational techniques laid out in this chapter, your final manuscript will be cohesive and organized, and this will increase the successful reception of your research concept.

References

Emanuel, E., Abdoler, E., and Stunkel, L. (2014). *Research Ethics: How to Treat People Who Participate in Research*. The National Institutes of Health (NIH) Clinical Center Department of Bioethics http://www.bioethics.nih.gov.

Peat, J., Elliott, E., Baur, L. et al. (2002). *Scientific Writing: Easy When You Know How*. London: BMJ Books.

54

Economy of Writing: How to Write Technical Content

Arthur C. Croft

Spine Research Institute of San Diego, San Diego, CA, USA

54.1 Planning

Unplanned excursions into technical writing usually become misadventures; like the ancient Chinese proverb concerning riding on the back of a tiger: once on board, one can never be sure of where he/she will disembark. You may have heard this admonition before, but because so many writers seem to ignore it, it bears repeating. Always plan your projects thoroughly before starting them. If you are conducting original research involving animals or humans, you may need Institutional Review Board (IRB) approval (see Chapter 22 for details). Most biomedical journals will require you to acknowledge that.

Make sure that your study design is perfectly sound by consulting with a methodologist or experimentalist well acquainted with the proper design of scientific studies. Many studies have gone unpublished because their authors failed to recognize a serious design flaw early on. Such fundamental errors are typically uncorrectable and can hopelessly doom the paper. On the other hand, the study might be published when the manuscript reviewers fail to spot the problem, and it become a lasting source of embarrassment to its authors when others recognize and report the faux pas.

As for the actual writing of the paper, it is a good idea to first target the top three journals in which you would like to publish. Always consider your target audience. In what aspects of your work will they be most interested? Being thoughtful of the audience means not only writing to their level of comprehension, but also articulating the information that is relevant to them while editing out the information that is not – which only clutters the report in the end. That can guide you in your writing of the paper. For example, if you are writing for an audience of mostly orthopedic surgeons, both the reviewers and the readers will be more interested in surgical procedures and outcomes than in the post-surgical management and rehabilitation. If you are writing for a journal that caters more to physiatrists, you might discuss the diagnostics and surgical indications (since their goal is to keep people out of surgery if possible), and then focus on the rehabilitation aspect. For how to choose a suitable journal for your manuscripts, see Chapter 42.

Another thing to consider is whether your project should be parsed into more than one paper. It is not uncommon to break up elements of a large research project into

A Guide to the Scientific Career: Virtues, Communication, Research, and Academic Writing, First Edition.
Edited by Mohammadali M. Shoja, Anastasia Arynchyna, Marios Loukas, Anthony V. D'Antoni, Sandra M. Buerger, Marion Karl and R. Shane Tubbs.

several reports. For example, if you had collected a large number of automobile crash injury cases, in one study, you might be looking at the most common types of injuries based on the type of collision. In another study, you might be considering what the outcomes looked like based on the type of injury. Or you might be looking at the diagnostic efficacy of obtaining certain types of radiographs or other imaging studies (magnetic resonance or computed tomography [CT] scans, etc.). Some authors, however, will submit only slightly altered manuscripts with different titles to two or more journals in order to increase their exposure and add to their list of published papers. In my view, the multiple publication of a single paper (sometimes referred to as salami slicing) should be discouraged. However, it is reasonable to publish multiple focused reports instead of one comprehensive and unfocused report from a large research project that has generated enough data and findings.

54.2 The Process

I mentioned earlier that I would focus on the production of a professional or scientific paper, but, for the most part, most of this chapter can easily be applied or adapted to any formal writing project. Let us start with the basic structure of a scientific paper and how an economical approach can benefit your writing. My background is biomedical and biomechanical, but the general structure for any scientific paper is the same. You will have a title, an abstract, and sections with names such as *Introduction*, a section called *Materials and Methods*, *Results*, *Discussion*, and perhaps also one named *Conclusions*. Understanding the different elements of technical writing and how to establish an organization that tells the "story" unambiguously and articulately is crucial.

54.2.1 Title

Your title should be concise and clear, and provide some insight into the subject of the study and its focus. An important consideration is the terminology you choose because it might very well affect the accessibility of your paper to people searching the various databases where it is indexed (e.g. PubMed, PsychINFO, OVID, CINAHL, Embase, and Cochrane Library). For example, if you are writing about "heart attack incidences," someone doing just a title search for the terms "myocardial infarction" or "MI" might miss your paper. If your subject was whiplash injuries and you titled your paper "neck injuries in motor vehicle accidents," a search for "whiplash" would likely miss your paper. Also, the term "accident" has been deprecated in favor of the terms "crash" or "collision." And in this case, it would also be helpful to know that in other countries the term "road traffic collision" is commonly used. Bear in mind that people often search the title, abstract, and keyword indexes simultaneously, making it less likely that they will miss your paper, but it is still a good idea to put ample thought into the title. For how to write an effective title, see Chapter 55.

54.2.2 Abstract

When I first started writing in the early 1980s, the abstract was generally unstructured. Some authors used the abstract to give the reader the condensed version of the paper, but as often as not, the abstracts provided little or no indication of what the authors actually found or reported. While the abstract serves as an overview of the research paper, it is the section that most frequently benefits from an economical writing style with focus, clarity, and succinctness. In the past couple of decades, most journals have transitioned to the structured abstract, which generally contains sections such as *Study Design*, *Objective*, *Summary of Background Data*, *Methods*, *Results*, and *Conclusions*. To this type of abstract is usually appended *Keywords*.

Almost every journal these days will have its own website, and somewhere on that website, you will find a page named "Information (or Instructions) for Authors." Before writing anything, you should go to the top-tier journals you decided upon and read their instructions for authors. Among other things, they will give you specific requirements concerning the content and size limitations of the abstract. Condensing papers into 300-word abstracts is often quite a trick, but it is terribly important because many potential readers will decide whether to order the paper after reading the abstract. And, although it is clearly not a good practice, some will cite the paper after just reading the abstract, so you want your abstract to be as complete and concise as possible. For how to write an effective abstract, see Chapter 55.

54.2.3 Keywords

The keywords are indexing terms that search engines associated with literature databases use, among other things, to find papers. The National Library of Medicine has a catalog of Medical Subject Heading (MeSH) terms. They even have a MeSH Browser, which can help writers find related terms.[1] This is another important area to pay attention to because the accessibility to your paper could hinge on your choice of MeSH terms. Looking at the terms used by other authors in your genre will also be helpful.

54.2.4 The Introduction

At this point, I should mention that technical writing should always be parsimonious. This is a technical paper, not a master's thesis. There are two primary purposes for the introduction: first, as a brief overview or summary of prior related research or knowledge in the area of your paper, and second, as a rationale for your research. It is not necessary to provide a running history of research, especially if our knowledge has changed several times through the years, but in some instances, the history might be interesting enough to allow some elaboration. Usually, it is sufficient to summarize our current knowledge. In some cases, when there may have been only one or a few prior related studies, they might have been limited in some way or there might have been conflicting results. In the cases of prior limitations, you may be overcoming those limitations with your study. In the case of prior conflicting studies, your justification for research might be to help shed more light on the subject and perhaps help to resolve the controversy. Perhaps one study had included only females and the second study, a mix of males and females. Or perhaps there was a problem with the representativeness or size of the sample population. Whatever issue you are trying to explore, resolve, or investigate in more detail, or in a novel way, this is typically the final statement you would make in the introduction.

---

1 Available from: https://www.nlm.nih.gov/mesh/MBrowser.html.

54.2.5   Materials and Methods

The details of how you conducted the study are provided here. It can be, and usually is, a dry section, but it is as important as any other because readers will usually want to know precisely what you did and how you did it. The Materials and Methods must be extremely specific in description but should be economical in terms of the explanation offered, which may be saved for the discussion section. Let us take an example of a randomized controlled trial first. You should mention informed consent, the name of your IRB, and that you did have IRB approval. You might start the section by explaining how you recruited the subjects and provide your inclusion and exclusion criteria. Those details need to be clearly articulated and will allow your readers to determine whether you accounted for certain criteria so that they can make the judgment as to how comparable your results might be to other published papers.

For example, we know that people who sustain whiplash injuries in rear-impact motor-vehicle collisions generally have more severe injuries than those who have whiplash injuries in front- and side-impact crashes. Most authors would consider a neck injury a "whiplash injury" regardless of the mechanism. However, technical-writing practices dictate that it is not enough to describe the type of injury but that the author must also define the mechanism. It is therefore very important for the authors to let the reader know whether their subjects were a mixed group or whether they were all victims of rear-impact crash injuries.

If your subjects were stratified as to sex and age, those groupings and their relative memberships should be provided. The randomization process itself should be explained because this is an area where bias can creep in. For example, subjects are sometimes reassigned *after* the randomization process. That might have occurred in the case of a surgical study because some subjects were taking blood thinners and their general practitioners did not want them to be taken off the medication in order to have surgery. There might have been a number of other reasons for group reassignment, but when that occurs, it no longer is a truly randomized study. After the study begins, patients are usually allowed to change groups if they want to, but there is a way to handle that from an analytical standpoint called *intention to treat*.

Inclusion criteria describes the characteristics that were required for the subjects to be included in the study, such as age over 18 years but under 65 years, general good health, and the ability to read and speak English (if they were going to be filling out forms or questionnaires), etc. *Exclusion criteria* would be those criteria that would prevent the person from participating in the study. Those might include certain kinds of pre-existing diseases, prior injuries, people on certain medications, pregnancy, or history of mental illness, and should be listed in your report. Those are also important features that your readers will want to know about in order to determine whether your subjects are comparable to those of other studies. For example, if one were studying the ability of exercise combined with medication to lower blood pressure, the readers would probably like to know how severe the blood pressure was in the study groups and whether it was comparable between the drug-plus-exercise and drug-only groups. It would also be important to know whether any of the people in the groups were exercising regularly before the trial.

It is possible to manipulate a study's design, whether intentionally or otherwise, to such an extent that the results are not likely to be reproducible in a sample of real-world patients where such manipulation does not usually occur. For example, a therapeutic modality such as ultrasound might be applied by a physical therapist to a group of patients with an injured extremity once a week for three weeks and the results compared to another group of people with similar injuries treated only with anti-inflammatory drugs for three weeks. Although the study design avoids the confounding that could result from other interventions by the therapist, we would say the study now lacks *external validity* because physical therapists would rarely apply ultrasound in the absence of other interventions.

A common problem in biomedical research is the failure to report compliance. I suspect the reason is that compliance is usually much lower than researchers care to admit, yet, without substantial compliance, the validity of the study will be called into question. For example, suppose we were investigating the effectiveness of taking a certain vitamin supplement upon rising and before bedtime during an exercise-based weight-loss regime. We randomly assign half of the volunteers to take the supplement in conjunction with the exercise program and the other half to simply complete the exercises. The exercises were supervised in the clinic so participation would be documented. A critical question must be asked of the supplement group: Did you actually take the supplement? Did you miss any doses? How many? If, say, 40% of the supplement group admitted that they did not take the supplement regularly, such an admission could really torpedo the study. That is why we rarely read that such questions were even asked.

Similar problems occur when using a placebo. In a study published several years ago, a little over half of the subjects in the placebo group stated that they knew they were taking a placebo, as opposed to the real drug. In that case, they can no longer genuinely be considered part of the placebo group because the assumption is that subjects in the placebo group do not know whether they are taking the real drug or a sugar pill.

This is the reason the *double dummy* method was developed. Suppose the real drug comes in a blue and white capsule and the placebo is a pill that looks like an aspirin tablet. We can create a double dummy by filling half of the colored capsules with the actual drug and half with powdered placebo. And we take the drug powder and form it into tablets. Then half of the placebo group gets their placebo in capsules and half in tablet form; similarly, for the drug group.

Measurement is another important consideration and should be concisely defined. If the basis of the study is the result of questionnaires, your readers would like to know whether those questionnaires have been validated in the past or by you. Has it been established that the demographic understands the questions on the questionnaire? It might be that there are technical terms that many might not understand. For example, it might be that some do not recognize that *influenza* is the same as the *flu*, or that *hypertension* is the same as *high blood pressure*. Likewise, use of terms such as "a lot" or "sometimes" might result in an unacceptable level of uncertainty. As it turns out, questionnaires should always be validated in a pilot study in order to be sure that the data will be meaningful and representative. In some cases, the entire questionnaire might be reproduced in an appendix. Measurement might also be by a laboratory test in which the normative ranges need to be defined, or by a device in which the measurement methods should be described.

Consider the reading of cervical radiographs (i.e. X-rays of the neck). It is most common to obtain them with the patient in a standing position, but some authors take them with the subject seated, and that can alter the curvature of the spine slightly. Some will be careful to position the patient's head while others will simply tell the subject to keep his/her head level. When describing imaging tests as part of your Materials and Methods, be sure to detail clearly in written terms how the participants were positioned for the test; account for the variance in methodology, which could affect the appearance of the cervical spine curve. If the goal is to measure the extent of range of motion by radiographing the subjects after fully flexing their necks forward and then fully extending them backward, some authors will have the subjects first go through a series of warm-up stretches. The author should mention whether warm-up stretching is used in the methods. And when radiographs are to be graded for, say, degenerative changes, authors often will either utilize an existing ordinal grading methodology or devise their own. Since that will introduce a degree of subjectivity, it is a good idea to make an effort to determine the degree of what is called the intra-rater and inter-rater reliability on a sample of the radiographs to be included in the report. The intra-rater reliability determines how researchers agree with themselves on different occasions. Inter-rater reliability is a determination of how well researcher A agrees with researcher B. That is usually statistically assessed using the intraclass correlation coefficient (ICC).

So, as you can imagine, with just about any device, machine, test, or assessment tool, one is likely to obtain slightly (or even markedly) different results unless rigorous standardization is adhered to. And, because absolute rigor is easier said than done, the next best thing is to describe the way you did your work in adequate detail. Furthermore, any method of standardizing the equipment should be clearly communicated in the report and should promote consistency in your methodology.

When using questionnaires as part of your methodology, operative terms need to be defined in the Materials and Methods section. In a classic study of a cohort of just over 2,000 whiplash claimants, which was taken from a Canadian insurance company database, the authors of the Quebec Task Force of Whiplash-Associated Disorders sent out a postal survey (questionnaire) more than a year after the claimed injury. Although the primary intent of the questionnaire was to inquire about the whiplash injury, it was disguised as a routine healthcare survey. They asked whether the claimants had returned to their *usual activities* (such as work or school). If they had, the authors deemed that they had recovered from their whiplash injury. Many argued that simply returning to work or school is not a suitable proxy for recovery from that kind of injury, and I would agree. Most patients and their physicians would define recovery as a return to health. In that case, the authors should have defined the use of "usual activities" in their questionnaire and their interpretation of the language to mean "recovered." As long as the authors provide the explanation, there would be no room for confusion based on unorthodox terminology. In epidemiology, we often refer to this as a problem of *surrogate outcome* (although in this case, *spurious* might be more descriptive).

Finally, your method of analysis should be described. That would include a clear statement of the statistical software used and the statistical methods used, along with any assumptions made for the selected statistical tests that are deserving of discussion or justification. If you used a one-tailed test rather than the usual two-tailed test, you should cite your rationale for doing that. If you used a more rigid P value for significance, such as $P = 0.01$ rather than the traditional and less stringent $P = 0.05$, you should explain your rationale. Likewise, discuss your application of a particular statistic that might violate some of the basic statistical assumptions. For example, an assumption of the most common statistical test, Student's t-test, is that the data are normally distributed and independent, and have equal variance. Because it is a fairly robust statistic, it is not terribly sensitive to some violations of those assumptions (most biomedical data, for example, is not normally distributed), but they should be mentioned. Even multiple t-tests, which violate the independence assumption, can be corrected using the Bonferroni corrected t-test.

Some journals now require that authors categorize their research study based on the hierarchy of evidence from the Centre for Evidence-Based Medicine in Oxford, England.[2] In the case of a meta-analysis, the American Psychological Association (APA) has promulgated Journal Article Reporting Standards (JARS), and Meta-analysis Reporting Standards (MARS) that can serve as a good guideline, although there are others as well.[3]

54.2.6 Results

When it comes to the reporting of randomized controlled clinical trials, it is generally recommended that authors rely on and follow the methodology recommended by CONSORT, which stands for Consolidated Standards of Reporting Trials.[4]

Ensuring an economical approach to reporting results, a hallmark of the system is the flow diagram, which looks like an algorithm. The flow diagram would start at the top with the total number of people eligible for the study (N = the total number) and then flow down to the number of people included; then down to the numbers randomized to each group, the number of people who changed groups, the number who dropped out, the number who had the first follow-up, second follow-up, etc.

In studies of a more simple design, at the very least, you will report descriptive statistics. Those would include measures of central tendency (mean, median, or mode), the frequency distribution, and measures of variability (range, standard deviation [SD], variance). Depending on the kind of study you have done, or research you have undertaken, your data can be represented in tables and/or diagrams. That is especially true if you are reporting a lot of numerical data. And keep the parsimony notion in mind here; if some of the data can be simplified in a few words rather than presented in a large table, it is best to do that. Some journals, especially engineering ones, allow you to have appendices, which are useful for data that are only likely to be of interest to a minority of readers. It is a good idea to check with the journal's guidelines for specific style preferences on reporting results. Also bear in mind that many journals will now allow you to provide ancillary files that can be placed on the journal's website. Those can include text or images and even video. Just remember that the internet is a movable feast. Somebody reading your paper five years from now might not be able to access those files, so don't make them an indispensable item to your report.

When reporting your results, resist the tendency to explain the data. The place to do that is in the next section, Discussion, or possibly the Conclusions section.

54.2.7 Discussion

To a large extent, this is one of the most important parts of your paper. This is where you summarize your findings and tie any loose ends together. The Discussion section

---

2 Available from: http://www.cebm.net/oxford-centre-evidence-based-medicine-levels-evidence-march-2009.

3 Available from: http://www.apastyle.org/manual/related/JARS-MARS.pdf.

4 http://www.consort-statement.org.

allows you put your research into perspective and try to make sense of the data and what the results really mean. It might be helpful to cite other literature here as part of that process. It is where you contrast your data with that reported by previous authors and explain any differences. In the process of doing that, it is acceptable to repeat some of your findings, but always bear in mind that exhortation concerning parsimony.

An important part of your discussion will be the limitations or potential shortcomings of your study. For example, there might have been some unavoidable degree of selection bias. There might have been some potential confounding variables for which you could not control, or you might have been unable to measure some parameters. Your follow-up might have been curtailed for some reason, or your sample size might not have been substantial enough, and that would have reduced the statistical power of the study. Or maybe your sample was skewed (more females than males, or more older than younger people).

This is also the section where you would describe some of the reasons why your data might have diverged from that of previous reports. Your patient sample might have been the result of a newer diagnostic test, which allowed you to intercede clinically at an earlier point in the natural history of a disease process. This could conceivably affect the survival rate through a process known as *lead time bias*. Or you may have relied on advanced imaging technology such as magnetic resonance imaging (MRI) that has higher sensitivity and specificity compared to the plain film radiology used by the authors of older studies. Perhaps your dropout rate was lower than expected. If your study design was different in an important way from other studies mentioned in your review, you should point out that the results of your study might not be entirely comparable to the other studies and explain why that is. When relying on historical controls, which we sometimes do with human skeletal research because we have large collections of skeletons from the nineteenth century, we need to be concerned about a potential *vintage effect* – a bias that might result from period differences in hygiene, healthcare practices, diet, etc.

While manuscript reviewers will always ask authors to cite potential limitations if they haven't already, those post-hoc admissions seem to have transformed into an automatic indemnity allowing authors to commit nearly any error. We manuscript reviewers prefer that the limitations are themselves limited.

54.2.8 Conclusion

In a single case report or in a case series, there might be nothing to say beyond the discussion. No conclusion would seem warranted in those instances. In just about any other kind of project, from a large literature review or meta-analysis to a cohort study or randomized controlled trial (RCT), you will draw some kind of conclusion.

One of the common errors made by authors is to draw conclusions that are not supported by their study design. That is a huge offense to the philosophy of writing economically. The conclusion should summarize the study results and benefits, but it should not offer any new information that hasn't already been presented. In particular, qualifying the results without recognition of the assumptions created by the methodology might lead to a misleading conclusion. A common error is known as the *ecological fallacy*. Suppose you conducted a cross-sectional survey of a small population that you felt was representative of the US population. Let us say you asked the subjects whether they suffered from chronic neck pain as a result of a car crash. And let us say 38% of the females claimed to have chronic neck pain compared with 19% of the males. It would then be perhaps tempting to conclude that females are twice as likely to develop chronic neck pain. It appears superficially that this is the case, but that requires us to make some assumptions, one of which is that females and males have the same rate of neck injury from car crashes. Do we know that? Does our study tell us anything about the acute injury risk? No. If you think about it, what we'd be doing here would be to assign longitudinal attributes to a cross-sectional study. Using the same example, my colleagues and I found, in fact, that females have a *risk ratio* (RR) of $2:1$ for acute injury risk, so they have a lower threshold for acute injury as compared to males. However, once injured that way, males share the same risk for developing chronic neck pain as females do. So any cross-sectional population survey will find about twice as many females with chronic neck pain from that kind of injury. Such an error does not always lead to erroneous conclusions, but authors should refrain from making assumptions.

Another example would be to look at the radiographs of peoples across the age range of 20–60 years. You'd find that degenerative changes in bones and joints increase with age. You would conclude from that that degenerative changes increase with age. In that case, your conclusion would be correct, even though your logic was flawed, since you again relied upon cross-sectional data to derive conclusions that really require a longitudinal design.

54.2.9 Bullet or Take-Home Points

Some journals will ask authors to append a set of bulleted or Take-Home points to their manuscript. Unlike that structured abstract, which serves as a highly condensed precis of the paper covering all of the main sections, the bulleted list would consist of just the five or six gems or clinical pearls you'd want your reader to remember. That is your chance to truly exercise your economical writing skills.

54.3 General Writing Issues

54.3.1 Referencing

Providing references in a technical paper is *de rigueur*, but there is a logic to it, and I will give you my take on it as a manuscript reviewer. Some things are so well understood or firmly established that is becomes unnecessary to provide foundational references. The human heart, like all mammalian hearts, has four chambers. We don't need to cite *Gray's Anatomy* to support that. Much of the remainder of medicine and engineering, on the other hand, seems to exist in a state of perpetual dispute. The more potentially controversial or the less well known the subject matter is, the more we like to see some reference material, preferably more than one, preferably of late vintage, and always of good pedigree (i.e. from papers in peer-reviewed, indexed journals). Book references should not be used in this way because books are not regarded as peer-reviewed.

Manuscript reviewers like to see balance. In just about every corner of science, there is likely to be differing opinions or contrarian theories, and one can usually find literature supporting both sides of any dispute. An unbalanced review of literature is one in which the authors provide only references to literature that they agree with or that supports their premises. I will usually call attention to that and ask them to discuss the literature on the other side of the question. In my own papers, I will provide the opposing theories and literature supporting them, but in my Discussion, I will outline the limitations, flaws, etc., of those studies to explain why their authors drew the conclusions they drew.

There are no clear guidelines as to the actual number of references required in any given discussion. In a master's thesis or dissertation, one can never use too many it would seem, but in technical writing, the more recent and of high quality one's references are, and the less polarized the literature happens to be on a given subject, the fewer the references that will suffice.

54.3.1.1 Don't Be a Reference Snob

It irks me when I cite a paper from the 1980s and someone criticizes me for using "outdated" references. Just because a study is 30 years old doesn't necessarily imply that it is outdated or no longer valid. While it is always preferable to cite the newest studies, the newest study on a topic is sometimes 20 or 30 years old. And sometimes, there are newer studies, but the larger or better-designed study was an older one. Cite it. Do not be a reference snob. We were not exactly doing research by candlelight in the 1980s.

54.3.1.2 Exceptions to the Rule

If you happen to be writing a book, rather than a technical paper, there really are no rules for citations. Personally, I have always felt it was just as important to provide citations in my textbooks and in chapters I have written, but some authors provide spotty references in textbooks, and some provide no references at all. I have a wonderful textbook called *Orthopaedic Biomechanics* (by Bartel, Davy, and Keaveny, published by Prentice Hall) that is a highly technical book, very nicely written, but is completely devoid of references. There is not a single one, other than the obligatory mention as source material under the captions of the handful of borrowed illustrations. Odd, I thought, as I was reading the book, but there may be a strategy to this after all. Say the authors were to write that the normal angle formed by the joint of the x and y bones in females is $32–38°$ and then cited other authors, Henderson and Jones, as a reference for that fact. Say that the reader of the book wanted to mention the same thing in a technical paper, who would he or she cite? Bartel, Davy, and Keaveny? No, because Henderson and Jones were the original source. But if the Henderson and Jones reference had never been provided, Bartel, Davy, and Keaveny would become the default citation.

54.3.1.3 The Reference Manager

How do you manage all of those references? When I was working on my first master's degree, I wrote my thesis in long hand and my (first) wife ever so kindly typed it on an IBM Selectric typewriter. Making changes and edits often meant retyping an entire page or more. And those hundreds of references had to be typed manually. I bought my first personal computer in 1983. It had 64K of RAM and no hard drive. It had *WordStar 1.0* – one of the first word processors. With only 64K of RAM (22K after DOS loaded itself), it did not even have enough RAM to run spell check, but life was suddenly much easier. But even then, I had to type the references manually. And it was still a major event when I needed to add a reference in the middle of a paper, which would then require me to manually renumber all of the following references.

But then came the reference-manager software, and my life became easier still (see Chapter 65 for more details). For those unfamiliar with the software, you need to pay attention here. I use a program called EndNote (Thomson Scientific, San Diego, CA). It installs in the ribbon of Microsoft Word. When I click on the EndNote button, my library opens, and I can find references with a click of another button, and they are automatically inserted into my document and simultaneously appended to the bottom of my report as a bibliography. If I later decide to add a new reference in the body of the document, EndNote will automatically renumber all of the references for me. If I decide to move a paragraph on page 3 to a new position on page 6, which would then make all of the reference numbers out of order, no problem: EndNote automatically renumbers all of them. I can even copy a paragraph containing references from one file into another file, and all of those references will be appended and renumbered in the new document.

It gets even better. Today, there are literally thousands of professional journals in medicine alone. And there are just about that many styles for references. For example, the APA style, which is used in all PhD dissertations and master's theses work, as well as in the social sciences and engineering literature, places the author names and years of publication in parentheses within the text block. Most medical journals use numbers instead, but some list their authors alphabetically so that if your first reference was to someone named Zimmerman, the actual reference number might be 41. Other journals list references sequentially so that the first reference is 1, and so on. So let us say your paper gets rejected by the first journal you send it to and now you want to submit it to a journal that uses a completely different reference style. More work, right? Not at all. You just change the style template in EndNote, and the references reformat in a couple of seconds.

How do you get the references into EndNote in the first place? You can take them directly out of PubMed and import them as a text file. No typing required. For example, let us say you are a neurologist. You go onto PubMed and do a wide-open search for carpal tunnel syndrome. You will find about 8,300 papers. Save them as a text file, import them to EndNote, and you now have 8,300 references in your library ready to cite. And that is free, of course, speaking of the economy of writing. I tell my students that they are absolutely crazy not to use a reference manager. I even give them a starter library.

54.3.2 Elements of Style

Like it or not, technical writing is dry, precise, and (we certainly hope) unambiguous. It is written to provide information, to stimulate thought. It is the antithesis of a novel, which is written to entertain and to stimulate the emotions and imagination. Therefore, we must resist any temptation to be humorous or sarcastic and avoid hyperbole or over-statement. Likewise, it is best to eschew slang, idiom, and common or lay terms when more technically correct or precise terminology is available.

Footnotes are a useful tool in writing because they allow the reader to quickly scan to the bottom of the page for additional information or explanation of a particular detail, but most technical journals avoid them.

In the past couple of decades, we have seen a great proliferation of technical journals, largely because of the advent of electronic publishing and the internet. The overall quality of the manuscripts, however, has diminished *pari passu*. Twenty years ago, if I happened to catch a typo, misspelling, or obvious grammatical error, I would correct it and return it to the editor along with my review. Finding little flaws was not my main purpose, of course. Usually, though, they were relatively uncommon because the authors would have caught most of them by the time they submitted the paper. These days, however, there are often so many glaring errors in spelling, punctuation, and grammar that I just pass on a general comment: "unpublishable in current form because of numerous errors in spelling, punctuation, and grammar." It is obvious that some of the papers are written by people who speak and write English as a second language, but I see no reason why the authors could not take the additional step of having a native English writer/speaker go through the manuscript and clean it up. In fact, if your audiences are international English speakers, you should be mindful also of national idioms. In England, when making a telephone call, one "rings someone up." In the United States, we "call" the other person. "Telephoning" would make this unambiguous. I remember my first trip to England. I asked a woman at a ticket counter where to catch a taxi, and she directed me to the "queue" just outside. I walked around for some time looking for a sign with a "Q" on it.

Ultimately, most of us will need to polish up our initial draft, and even when we think it is perfect, others are likely to read it and find parts that are confusing or ambiguous. The problem is simply that *we* know what *we* meant, but others might not. If a sentence or a paragraph can be interpreted in more than one way, it is definitely ambiguous and must be clarified or simplified.

Part of the second reading of the paper should once again concern the parsimony issue. Some of the things written might not need to have been written at all, allowing you to trim an entire paragraph or two, as painful as it might be. Thoughtfully review your own writing and ask yourself if the point is clearly articulated. If a sentence seems difficult to read all the way through, it might be beneficial to break the sentence into several, concise sentences. On the other hand, you might realize that most of a paragraph could actually be reduced to a single sentence. It might not be as poetic but know this: a four-page paper will be about twice as publishable as an eight-page paper, and getting into print as soon as possible is your primary goal. Even if the editor decides to publish your eight-page paper, it will probably take him twice as long to publish it in a journal that is still printed on paper. That will often mean the difference of another year or more.

54.3.3 Elements of Content

To augment the written word, it might be helpful to utilize tables, illustrations in the form of line drawings, photographs, or images depicting radiographs, MRIs, CT scans, etc. For tables, remember to check with Instructions for Authors for specific layout or structure for tables. Journals will have their own style and might have a specific layout you must adhere to. As mentioned, try to keep the number of tables to a minimum, and make sure that they are easy to understand and are fully labeled.

As a standard, all tables, illustrations, photographs, and images should be placed at the end of your file, each on its own separate page, and labeled appropriately. Illustrations taken from other published works will require permission for use in your paper or book. Such permission is not difficult to acquire, although it would be somewhat tedious if you were writing a textbook and planned on using dozens of illustrations. One way around acquiring permissions is to use the original illustration as a template or basis for a new drawing. The modified one can then be used without permission by simply adding a proper reference in the caption.

There are a couple of things to keep in mind when using illustrations or photographs or images. One is that the editor or publisher, in an ongoing struggle with space limitations, might reproduce them in a very small format. That can make it difficult for readers to see the details you hoped would be obvious. You should try to use simple line drawings that are not cluttered with a lot of detail. If reproduction in a certain size is important, you could certainly make that suggestion to the editor after the paper is accepted for publication.

One of the occasionally mortifying consequences of publishing radiographs, MRIs, and CT scans is that page designers in the publishing house might inadvertently place them upside down or on their side. In textbooks, a publisher will send you a *galley proof*, and you'd be able to catch those errors yourself. In a journal article, that practice is uncommon, so be sure to label the margins of the images with "top" and "bottom."

The same admonition made for tables should be made for graphs or charts. Use them as sparingly as you can because they greatly increase the size of the paper, and they can sometimes confuse the reader by providing more information than is really necessary. Believe me, I know how painful it is to cut three or four beautiful (in my mind anyway) graphs that I have spent hours making, but when my wife says she doesn't understand why I included them, I realize that I really did not need them. Snip, snip. Less is more.

Something I see quite often in the manuscripts I review are graphs that are not completely labeled. It might have "Time" along the abscissa, but it should list the units of measurement as in "Time (seconds)." You should also be consistent with the journal's style. If the journal uses SI (metric) units, be sure to submit your paper with metric units of measurement. Graphs should not have ordinals that do not include the entire scale. A graph that has as its y-axis, "Percent Improvement," for example, should not start at 40% and end at 70%. That will potentially distort the apparent magnitude of differences between graphed indexes. The graph should run from 0% to 100%.

Size is a final consideration. Digital cameras or other devices are capable of capturing resolution well beyond what is practically needed. And they usually save digital data in a 72 pixels-per-inch (ppi) format.

As another outgrowth of our digital transformation, I am seeing more and more color graphs. I like color as well as the next guy, but if the journal is still printed on paper and with black ink, those pretty color graphs and illustrations will necessarily be reduced to muddy shades of gray. And that often makes a graph difficult or even impossible to interpret. Amazingly, some publishers do not seem to notice. On the other hand, suppose the journals are electronic and publish in color, but then one of your students prints the article on a standard laser jet printer. We are back to an uninterpretable graph again. The solution is simple, although – based on the fact that I so rarely see it – not that obvious: Make your graphs chromatically independent, as shown in Figures 54.1 and 54.2.

54.3.4 Color Space

When printing in color on paper, we use the CMYK color space. It is an additive method, meaning that the colors or inks (cyan, magenta, yellow, and black) are added together to produce the desired colors. In the digital world, we use the RGB color space, which is subtractive and consists only of red, green, and blue. When maintaining the fidelity of

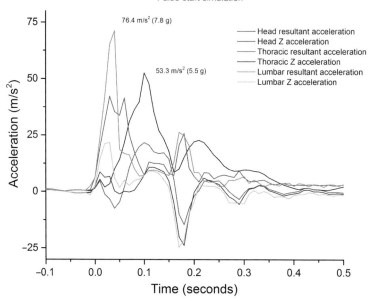

Figure 54.1 This was a color graph of an acceleration time history based on a multibody dynamic crash simulation. As you can see, when the red, blue, and green colors are reduced to shades of gray, it is difficult to interpret.

color is important, authors should realize that when they send an image produced with a digital device such as a camera to the publisher, they are probably sending an RGB image. The publishing house will need to convert this image into the CMYK color space in order to print it. And that conversion process, depending on the colors involved and the robustness of the software used to accomplish the task, can result in muddy colors that do not have the boldness of the original. For that reason, it is a good idea to do the conversion yourself, or have someone do it for you, using a high-end image-processing system such as Photoshop, so you can control the color conversion.

Radiographs, MRIs, and CT scans should be strictly grayscale images. That is also likely to be true for electrophoresis, DNA sequences, and other laboratory output. However, when people photograph such studies on view boxes with digital cameras, they are usually converted into the RGB color space. In other words, the camera assumes you want a color image so it adds a lot of color pixels that it gets from the yellowish, greenish, or blueish tones it picks up from the fluorescent tubes of the view box. As it turns out, this is a very easy problem to fix, which we will explain. Another common problem with those images is that they lack contrast, are too dark, or are too light. This is also easy to fix. The last problem is that no matter how well you try to align your digital camera squarely perpendicular to the film on the view box, it seems that you always end up with some degree of geometric distortion, which is obvious from looking at the distorted edges of the image. This is also easy to fix. Let us make all of the fixes in Photoshop as shown in Figures 54.3–54.6.

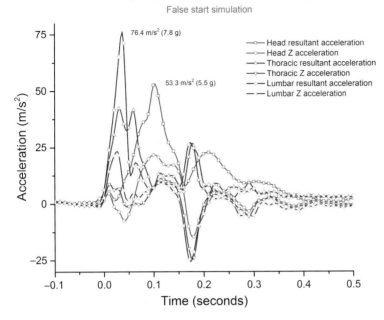

Figure 54.2 The same graph as in Figure 54.1 with the same vibrant colors. The only difference is that I added symbols to the colored lines. Now, it is no longer necessary to struggle with subtle shades of gray.

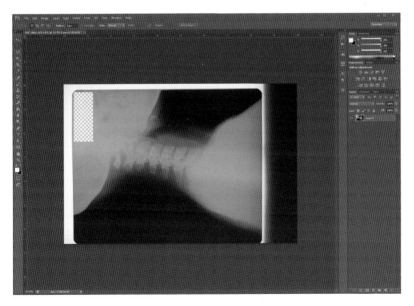

Figure 54.3 Although this color image is being reduced to grayscale for the book, you can see that there are a few problems with it. It is in the RGB color space and has a blueish tinge to it, it lacks contrast, it is on its side, it has extra junk around it, and it is geometrically a little distorted.

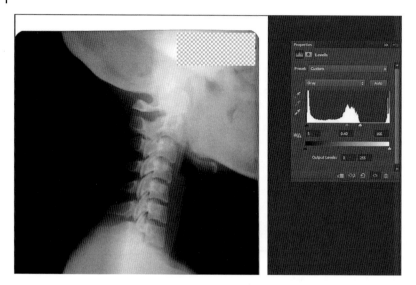

Figure 54.4 The first thing to do is go to the Image menu | Image rotation | 90 CW to rotate the image 90° counterclockwise. Then go to Image | Mode | Grayscale, and discard the color information. Then go to Window | Adjustments | Levels. Move the right side, light-colored triangle to the left so that it meets up with the right side of the white histogram. Move the left side, dark-colored triangle to the right to the beginning of the histogram. Then move the middle gray triangle so that it falls more under the main bulk of the histogram.

Save the file as a TIFF file, check the Copy box, and use LZW compression. This is a lossless compression and will not degrade the image, as opposed to JPG, which is a "lossy" compression algorithm.

You can label your images right in Photoshop, which is a pixel-based painting program, but vector-based programs are easier to use, more flexible, and will give crisp resolution at any magnification. Figure 54.7 shows a radiograph labeled with text on the right and a blowup of that text compared with a blowup of the text made using Adobe Illustrator below it.

54.3.5 Some Common Statistical Faux Pas

This is not a chapter on statistical methods, of course, but because I am writing it as a guide for writers based on my experience as a manuscript reviewer and editor, it seems worthwhile to mention some of the more common blunders made with statistics.

54.3.5.1 Extrapolating Beyond Your Data

Probably the most common mistake made by authors is to conduct a study, report their findings, and then draw conclusions that extend beyond what they actually obtained from the data. It is quite acceptable in your Conclusions section to make a rational or intuitive guess that extends beyond the data, as long as you acknowledge that you are doing that. Only in the Conclusion section is it acceptable to offer a rational or intuitive extrapolation beyond your data, and you must acknowledge that you are doing that.

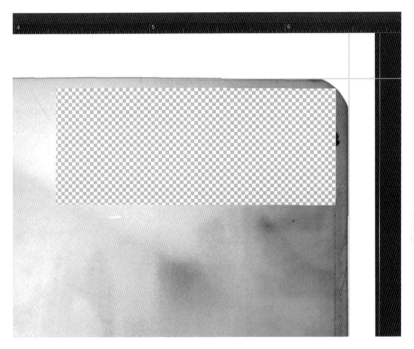

Figure 54.5 Now select the main image layer again. In order to square up the image back to its original 2D geometric proportion, bring some guides down from the rulers to serve as true vertical and true horizontal. Then go to Edit | Transform | Skew. Now you just drag the corners until the image is square.

54.3.5.2 The Absence of the Statistical Process

A fairly prevalent problem I see in the engineering literature is a failure to use statistics at all. For some reason, not all engineers see the importance of statistical procedures, and the most we get from many engineering journals are raw proportions and counts.

54.3.5.3 Selecting the Wrong Statistic

This could be a chapter unto itself, but suffice it to say that having a statistics program is no substitute for understanding the statistics you plan to use. If you have any doubts at all, I would strongly suggest that you have your study design vetted early on by a competent statistician.

54.3.5.4 Casting the Net Wide and Far

When you collect a lot of data and subject it to multiple t-tests or correlations, it is likely that by chance alone, you will find some relationships that reach statistical significance even though, in reality, the relationships are not always truly important. This process is sometimes referred to as data dredging or, more colorfully, whaling in the village pond. And remember that, like many statistics, independence is an important assumption. Take a study on the effectiveness of a new drug designed to reduce the destruction of joints in rheumatoid arthritis patients. It was a double-blind study in which half of the subjects received the drug and the other half the placebo. After 12 weeks, there was a statistically significant difference in terms of joint changes between the groups. Looking at the numbers however, one discovered that although there were only 38 people

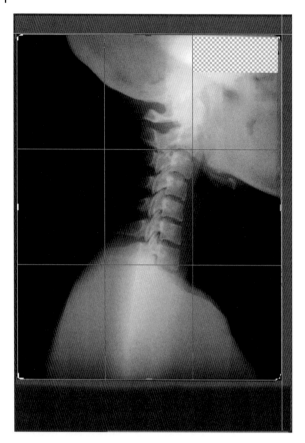

Figure 54.6 The last two things will be to crop the image and to resize it. Select the crop tool and crop it vertically and horizontally. Then go to Image | Image size, and set the resolution to 300 ppi, and set the height to 5 in. or so. If the resolution changes as you do this, you might have to check or uncheck the resize image box.

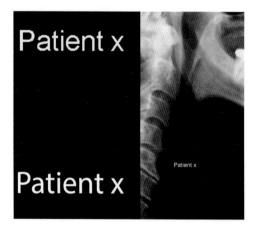

Figure 54.7 The radiograph on the right was labeled with "Patient x" in Photoshop. When that region is enlarged, however, it is pixelated (above left). When that radiograph is imported into Adobe Illustrator and labeled, it can be enlarged without the pixilation effect because it is a vector-based program (below left).

in the study, there were 380 joints (10 from each participant). Certainly, we would not assume that a woman's knee is really independent from her elbow in terms of the drug's effectiveness. There were only 38 independent samples – not 380.

54.3.5.5 The Mathematical Coupling Error

One of the easiest ways to reach statistical significance is using correlational statistics. Here again, however, correlations do not always imply cause-and-effect relationships, as the old hack about the positive correlation between the number of bluebirds in the sky and the incidence of motorcycle crashes in Minnesota in the spring reminds us. A slippery variant is when one of the variables contains an element of the other.

Consider the correlation of two parameters, X and Y. Also assume that part of Y is directly related to X. For example, consider body mass index (BMI). It is a ratio of weight in kilograms divided by height in meters squared. Suppose we correlated BMI with height. Although the problem with this example would be obvious to just about anyone, there are subtle examples of that kind of error that sometimes are not so obvious.

This is referred to as the *mathematical coupling* error. To see how it works, we can take two sets of 100 completely random numbers using a program like MATLAB, place them in two columns, and run a correlation on them in any statistics program or even Excel. You will get an R-value indicating that there is no correlation, as we would certainly expect. If you plot the values, you will get an essentially flat histogram. Now subtract the X column from the Y column and label the results Z. Now correlate the X column with the Z column (or the Y with the Z), and you will obtain an R-value of about 0.77 and an R^2 of about 0.6. The numbers will always come up no matter how many times you try this, yet from a statistical standpoint, they would be considered highly significant in spite of the fact that we started with completely random numbers.

54.3.5.6 Misinterpreting the Meaning of *p*-Values

What does the "p" stand for? Some sardonic statisticians might tell you it stands for *passport to publish*. Many seem to get confused over the actual meaning of p. The p-value relates the *probability* of the trial outcome, not the probability of the hypothesis. $p < 0.05$ means only that there is less than a 5% chance that your results were simply due to chance – not that there is a more than a 95% chance that your hypothesis is correct. A comparable problem is the conditional probability of a child with a fever having appendicitis. That would be represented $[P(A|F)]$ and is read "the probability of appendicitis given a fever." The inverse would be the probability of a child having a fever, given that he has appendicitis $[P(F|A)]$. The probability of the first one would be quite low because most children with fevers do not have appendicitis. The probability of the second, however, would be very high because appendicitis is almost always associated with fever. In general, confidence intervals (CIs) are preferable to p-values, and some journals require them.

54.3.5.7 Using SEM Instead of SD

The SD is part of the descriptive statistics you use to qualify your data. It is a measure of the variance of your sample. In most cases, we want to see as little variance as possible, so we like to see small numbers. The standard error of the mean (SEM) describes the variation of sample means around the true, unknown population mean. To arrive at the SEM, we divide the SD by the number of observations (n) in our sample. Accordingly, SEM will always be a smaller number than the SD, so many authors mistakenly choose to report the SEM. With regard to your sample data, the meaningful value is SD, not SEM.

54.3.6   Gail's Fallacy

Here is an example of how an overlooked detail during the planning stage of an otherwise well executed study torpedoed it. In a large, multi-center study investigating methods for managing chronic low-back pain, subjects for the study were recruited from specialty clinics across the country. They were then randomized into two groups: one received nonsurgical and noninvasive care such as physical therapy, acupuncture, massage, and recommendations for home exercise, weight loss, etc.; the other group underwent lumbar spine surgery. At follow-up, many months later, using a number of outcome assessment methods, the authors determined that the surgical group had the best overall outcome. The study was published in one of the most prestigious spine journals in the world. But there is a significant design flaw that essentially invalidates the study's findings, and it is called *Gail's fallacy*. It works this way:

   All of the subjects for the study suffered from long-term, chronic, low-back pain. They were recruited from specialty-spine clinics. It was known upfront that half of them would be randomly assigned to spine surgery. That means that all of these volunteers had already failed to respond to noninvasive forms of treatment for back pain such as physical therapy, home exercise, weight loss, etc. None of them, however, had ever had back surgery. By randomly assigning half of them to receive more noninvasive treatment, the outcome of that subgroup essentially took the form of a self-fulfilling prophecy: It had already been shown to be ineffective. The surgery group was exposed to a completely new intervention, so it was more likely that at least some of them would respond favorably – *Gail's fallacy*. And there were other design flaws as well: The authors excluded subjects who had spinal stenosis – a condition in which the vertebral canal diameter is reduced. While they did not give a rationale for that exclusion, it is noteworthy that other studies have shown that the outcomes of surgery for that condition are not as sanguine as they are for the treatment of disc herniations. One might suspect a little cherry picking (i.e. selection bias) was at play. In the real world of spine surgery, people with spinal stenosis do have spine surgery. Thus, the study might also be said to be lacking in *external validity*.

54.3.7   Authorship Contribution

Some journals will ask for a brief explanation of the specific contributions of each of the authors of your paper.

54.4   Epilogue

I certainly hope that this chapter adds something to your understanding of the goals, standards, and basic methodology of expository writing. As I write this, of course, I am not privy to any of the other chapters in this textbook, but, judging from the tentative table of contents I was given, I am sure that this chapter is only one of many that might initially discourage the budding science writer. But do not be intimidated. Writing papers is actually fun and rewarding on a couple of levels. This book gives you a set of blueprints that can only help you in avoiding disappointment by minimizing outright rejections, by reducing the number of revisions you'll have to write, and by making your submissions of higher quality in terms overall content, illustrations and tables included. And when your papers do get rejected, remember that even the best writers get rejections once in a while. Remember that the prestigious journal *Nature* turned down the paper by Hans Adolf Krebs (1900–1981) on the citric acid cycle for which he was later awarded the Nobel Prize, in 1953. Besides, there is likely to be another journal that will publish your paper with enthusiasm. And like any undertaking, the more writing you do, the more accomplished and polished you will become, and the easier it will be for you to publish the next paper (or textbook).

55

Writing an Effective Title and Abstract

Sandra Buerger

College of General Studies, Boston University, Boston, MA, USA

55.1   An Effective Title

Whether you are writing a book or a journal article, or preparing a presentation, the title represents the first interaction your audience will have with your work. Quite often, your prospective audience will decide whether to read your paper based on the title alone. That makes the title a vital part of your work and one that should be crafted with careful consideration. The title has two main goals: to accurately and briefly convey the contents of the paper and to draw the reader in.

The title is by necessity brief. Readers will be dissuaded by long titles, and most publishers prefer short titles. Many journal editors also view a too-wordy title as a red flag for a submission. Some journals might even have a word limit on titles. Beyond those basic requirements, readers also tend to be drawn toward shorter titles. Long titles will seem tedious to a potential reader. The reader might, in fact, even skip over a long title and look for a paper on the same subject with a shorter, catchier title. As the goal is to draw the reader in and have him look on to your abstract, you want to have something that will entice the reader. Thus, you want to craft a title that is concise, informative, and clear. The use of jargon or abbreviations should be eliminated or reduced as much as possible, and only minimal punctuation should be used.

Of course, in addition to drawing in the reader, the title must accurately reflect the content of the paper. You must be cautious not to overstep the scope of the paper. Sweeping statements or words – e.g. the description of a phenomena as "ubiquitous" – will leave the reader disappointed and annoyed if your paper does not live up to the promise of the title. The title should be a specific summary of your introduction and results. It should quickly convey to readers whether the paper is relevant to their interests.

One recommendation is to begin the title with the main subject of your paper. That will ensure that your readers are immediately aware of the topic of your paper, and they can decide whether it is relevant to what they are looking for. Another possible approach is to use a question as a title. The risks in this approach include that the question can mislead the reader on the topic being covered, plus question titles are unpopular with reviewers.

The final consideration for your title is the use of keywords. As most papers are located via search engines, it is important to think of words that might be used to search for your title and include them so that your title will appear high on the list of search results.

A Guide to the Scientific Career: Virtues, Communication, Research, and Academic Writing, First Edition.
Edited by Mohammadali M. Shoja, Anastasia Arynchyna, Marios Loukas, Anthony V. D'Antoni, Sandra M. Buerger, Marion Karl and R. Shane Tubbs.

In summary, your title should be concise, accurate and draw the reader in. Writing a list of potential titles and having a trusted mentor or colleague review them is often a useful way of determining the best title for your paper.

55.2 Preparation of the Abstract

A well-written abstract is an essential part of any scientific paper or presentation. Although there are different kinds of abstracts, they all have the same centrality to scientific communication both as stand-alone material and as a gateway to pique the interest of other scientists. The abstract of a journal article is often available on databases such as PubMed and Medline. Abstracts for conferences might be available on the conference website. The abstracts are generally the first part of a manuscript to be read by peer reviewers and to be viewed by the potential readers; it will influence whether the entire paper is of sufficient merit or interest to be read in its entirety. At a conference, attendees will scan abstracts to decide which talks or poster sessions to attend. Thus, having a clear, concise, and interesting abstract that accurately portrays the content of your material is vital for effective scientific communication. This chapter will discuss the features of a quality abstract and approaches to take in the preparation of the abstract.

55.2.1 When to Prepare the Abstract/Starting to Write

The abstract should ideally be prepared after the rest of the writing is completed or the presentation has been prepared. After you have a completed manuscript, you should begin outlining the salient points for each subheading (background, purpose, materials, and methods, results, and discussion). The important point here is to include only the main ideas of each section. This will give you a basic outline from which to prepare your abstract.

Following the preparation of those points, you can begin to write the abstract. The first draft of the abstract can be written without regard to word count. Be sure to include all the information that you think is vital. Following the initial preparation, you can begin to shorten and pare down in order to get to the appropriate word count.

55.3 Features of a Quality Abstract

55.3.1 General Features

The foremost feature of a quality abstract is that it presents, in a clear and concise manner, the major points of the study. The publisher often prescribes the conciseness of the abstract in terms of a word count, and also at times in a specified structure. For any abstract, the first step should be to review the requirements of the publisher to ensure that you are constructing an abstract that will meet all the requirements. Although exact required word counts do vary, a typical word count is not more than 250 words. That leaves no room for any superfluous words or phrases.

You should also avoid the use of jargon or subfield-specific abbreviations. Using such wording will limit your audience to those who are experts in your specific subfield. In order to obtain a wider readership, it is vital to express your work in a fashion that is clear and not overly reliant on background knowledge that might not be available to all readers.

55.3.2 The Structured Abstract

Some publishers will require a structured abstract, usually in the form of subheadings. In that case, it is important to follow the instructions of the publisher exactly. Deviation from the instructions might result in rejection or at the very least a requirement to reformat the abstract. In either case, it will not make a favorable impression on the editors or peer reviewers. In the section below, the typical parts of the abstract are described. If no subheading requirements are provided, the following parts should be included as a single paragraph.

55.3.3 Parts of the Abstract (for Original Research Articles)

The abstract is made up of five essential parts that mirror the sections of a scientific publication. Those are (i) background, (ii) purpose, (iii) materials and methods, (iv) results, and (v) discussion and implications. Each section is discussed below.

Background. The background section of the abstract should consist of a single sentence or, at most, two sentences. In the sentence, describe the most relevant (direct) background information. Along with the purpose section, you should be including the main points of the introduction in the background section.

Purpose. The next statement should describe the significance of the study – in other words, why the study was undertaken. The sentence should be very similar to, if not exactly like, the final sentence of your introduction. Your introduction can, however, provide more detail than can easily be fit into the single sentence that should be provided for this part in the abstract. In that case, you must condense the information you have provided in the introduction.

It is also important in this section to be sure not to overstate the purpose. You must be accurate in detailing what sort of questions your research will answer. Overstatement and exaggeration will turn readers off and make them skeptical of your work.

Materials and methods. This section should include a statement of the main method that was used. Details on methods need to be confined to the methods section within the manuscript and should not be included in the abstract. Simple statements that clearly articulate what was done should appear here. Additionally, if you are including statistical analysis, a short description of the methods used should be supplied in this section. This section should not be omitted to save space. Remember, many readers will not get beyond the abstract and as such, it must stand alone. What was done needs to be clearly articulated.

Results. Here, you should state your main result: the answer to the main question that was stated in the section on the purpose of the study. It is important to include only the main result in this section and not secondary observations that might have come up during the course of the study. If you have included statistical methods in your study, the essential elements of the computed statistics should be displayed here. This is usually expressed as a value in parenthesis along with the main result.

Discussion and implications. The final section of the abstract should mirror the discussion section. As in the purpose section, it is very important not to overstate the significance and implications of your results. The statement must be confined to the question that was answered and not extrapolated to include conclusions on unrelated studies. After you have completed the writing of all these sections, you should continue on to editing the abstract.

55.4 Editing of the Abstract

55.4.1 How to Edit for Brevity

Editing the abstract to abide by the required word count can be challenging. It is important to strike the correct balance between brevity and effective communication. During this editing process, you do not want to eliminate any important communication, but you must eliminate any nonessential words. To avoid frustration, it is often recommended to take time between editing sessions. This will help you view the abstract in a clear light and avoid missing mistakes and potential areas to shorten.

The next step is to have your co-authors examine the abstract. All co-authors should have a chance to comment and make appropriate changes to the abstract before submission.

55.4.2 Review by Outside Authors

After you have drafted your abstract, it is important to have others read over your work. Although this might be uncomfortable and expose you to criticism, it is a necessary part of the process. You should choose people you respect, both from your subfield and those who might represent a wider audience for your paper. That will allow editing and comments both from experts within your subfield, who can make comments on the accuracy of statements, and comments from a wider audience who are likely to provide feedback that will allow you to see where you could improve clarity. Finally, the added benefit to having others read your abstract is the correction of minor errors (e.g. spelling and grammar) that you might miss through your own editing.

55.4.3 Submission

Following the completion of editing and review by trusted colleagues, you should submit the abstract either as a part of journal article or as an individual conference abstract. After it is submitted, you should no longer worry. If you have followed all of the above steps, it is likely that your abstract is prepared correctly and contains only minor (if any) errors. Minor errors, while not ideal, are understandable. If the merits of the publication and the overall quality of the abstract warrant publication, such minor errors will not be a major impediment and can be corrected during the final editing process.

In summary, writing an abstract is deceptively challenging. Although short, much essential information needs to be contained within the abstract. That means that you must condense your ideas and information into a very small space and at the same time retain a clear message. Adding to the importance of this preparation is the fact that the abstract is often the first and sometimes the only exposure to your work for the reader. As such, you should be willing to put a significant amount of time and effort into both the preparation and editing of the document.

Further Reading

Boullata, J.I. and Mancuso, C.E. (2007). A "How-To" guide in preparing abstracts and poster presentations. *Nutrition in Clinical Practice* 22: 641.

Cole, F.L. and Koziol-McLain, J. (1997). Writing a research abstract. *Journal of Emergency Nursing* 23 (5): 487–490.

Peat, J., Elliott, E., Baur, L. et al. (2002). *Scientific Writing: Easy When You Know How*. London: BMJ.

Pechenik, J. (1997). *A Short Guide to Writing About Biology*. New York, NY: Longman.

Pierson, D.J. (2004). How to write and abstract that will be accepted for presentation at a national meeting. *Respiratory Care* 49 (10): 1206.

Trimble, J.R. (2000). *Writing with Style*. Upper Saddle River, NJ: Prentice Hall.

Weinert, C. (2010). Are all abstracts created equal?? *Applied Nursing Research* 23: 106–109.

56

Writing the Results Section

Bulent Yalcin

Department of Anatomy, Gulhane Medical Faculty, University of Health Sciences, Ankara, Turkey

56.1 Introduction

The Results section is the soul of an article (Shidham et al. 2012). Many authors find it useful to begin writing the paper with this section and to build the rest of the paper around it (Cetin and Hackam 2005). The Results section contains all the data needed to support, or refute, the hypothesis proposed in the introduction (Cetin and Hackam 2005). It is therefore vital to use clear language to objectively present key findings in an orderly and logical sequence while referring to illustrative materials (figures, graphics, tables, etc.) that represent the data (Kallestinova 2011). This section is also a mirror of the methods section. For every method (what you did), there should be a corresponding result (what you found), and vice versa. Ultimately, findings in the Results section should match and answer the research questions posed in the introduction. Keeping this goal in mind will help you to be more concise in writing the Results section, and to decide which findings to present and which to leave out (Kotz and Cals 2013).

56.2 Content

Authors should selectively include only those data and experimental details that are related to the research question of interest. Findings that do not relate to the research objectives should not be mentioned (Fathalla and Fathalla 2004). Various international guidelines exist for reporting observational and experimental studies. These guidelines not only address the desired constituents of the Results section (Tables 56.1 and 56.2), but also provide tips for constructing other sections of an original research article or systematic review. In general, the Results section should weave a coherent story and must communicate the findings to the reader in a logical, transparent manner. Therefore, it is appropriate to describe the results in a manner that makes logical sense, as opposed to describing the experiments in chronological order of execution (Cetin and Hackam 2005). A common sequence of contents is: (i) recruitment/response, (ii) characteristics of the sample, (iii) findings from the primary analyses, (iv) secondary analyses, and (v) additional or unexpected findings that relate to the research (Kotz and Cals 2013).

A Guide to the Scientific Career: Virtues, Communication, Research, and Academic Writing, First Edition.
Edited by Mohammadali M. Shoja, Anastasia Arynchyna, Marios Loukas, Anthony V. D'Antoni, Sandra M. Buerger, Marion Karl and R. Shane Tubbs.

Table 56.1 2007 STROBE statement checklist for reporting observational studies.

	Item no	Recommendation
Title and abstract		
	1	(a) Indicate the study's design with a commonly used term in the title or the abstract. (b) Provide in the abstract an informative and balanced summary of what was done and what was found.
Introduction		
Background	2	Explain the scientific background and rationale for the investigation being reported.
Objectives	3	State specific objectives, including any prespecified hypotheses.
Methods		
Study design	4	Present key elements of study design early in the paper.
Setting	5	Describe the setting, locations, and relevant dates, including periods of recruitment, exposure, follow-up, and data collection.
Participants	6	(a) *Cohort study* – Give the eligibility criteria, and the sources and methods of selection of participants. Describe methods of follow-up. *Case-control study* – Give the eligibility criteria, and the sources and methods of case ascertainment and control selection. Give the rationale for the choice of cases and controls. *Cross sectional study* – Give the eligibility criteria, and the sources and methods of selection of participants. (b) *Cohort study* – For matched studies, give matching criteria and number of exposed and unexposed. *Case-control study* – For matched studies, give matching criteria and the number of controls per case.
Variables	7	Clearly define all outcomes, exposures, predictors, potential confounders, and effect modifiers. Give diagnostic criteria, if applicable.
Data sources/ measurement	8[a)]	For each variable of interest, give sources of data and details of methods of assessment (measurement). Describe comparability of assessment methods if there is more than one group.
Bias	9	Describe any efforts to address potential sources of bias.
Study size	10	Explain how the study size was arrived at.
Quantitative variables	11	Explain how quantitative variables were handled in the analyses. If applicable, describe which groupings were chosen, and why.

Table 56.1 (Continued)

	Item no	Recommendation
Statistical methods	12	(a) Describe all statistical methods, including those used to control for confounding. (b) Describe any methods used to examine subgroups and interactions. (c) Explain how missing data were addressed. (d) *Cohort study* – If applicable, explain how loss to follow-up was addressed. *Case-control study* – If applicable, explain how matching of cases and controls was addressed. *Cross-sectional study* – If applicable, describe analytical methods taking account of sampling strategy. (e) Describe any sensitivity analyses.

Results

	Item no	Recommendation
Participants	13[a)]	(a) Report numbers of individuals at each stage of study, e.g. numbers potentially eligible, examined for eligibility, confirmed eligible, included in the study, completing follow-up, and analyzed. (b) Give reasons for nonparticipation at each stage. (c) Consider use of a flow diagram.
Descriptive data	14[a)]	(a) Give characteristics of study participants (e.g. demographic, clinical, social) and information on exposures and potential confounders. (b) Indicate number of participants with missing data for each variable of interest. (c) *Cohort study* – Summarize follow-up time (e.g. average and total amount).
Outcome data	15[a)]	*Cohort study* – Report numbers of outcome events or summary measures over time. *Case-control study* – Report numbers in each exposure category, or summary measures of exposure. *Cross sectional study* – Report numbers of outcome events or summary measures.
Main results	16	(a) Report the numbers of individuals at each stage of the study, e.g. numbers potentially eligible, examined for eligibility, confirmed eligible, included in the study, completing follow-up, and analyzed. (b) Give reasons for non-participation at each stage. (c) Consider use of a flow diagram.
Other analyses	17	Report other analyses done, e.g. analyses of subgroups and interactions, and sensitivity analyses.

Table 56.1 (Continued)

	Item no	Recommendation
Discussion		
Key results	18	Summarize key results with reference to study objectives.
Limitations	19	Discuss limitations of the study, taking into account sources of potential bias or imprecision. Discuss both direction and magnitude of any potential bias.
Interpretation	20	Give a cautious overall interpretation of results considering objectives, limitations, multiplicity of analyses, results from similar studies, and other relevant evidence.
Generalizability	21	Discuss the generalizability (external validity) of the study results.
Other information		
Funding	22	Give the source of funding and the role of the funders for the present study and, if applicable, for the original study on which the present article is based.

a) Give information separately for cases and controls in case-control studies and, if applicable, for exposed and unexposed groups in cohort and cross-sectional studies.

Source: Reproduced from von Elm et al. (2007) with permission from BMJ Publishing Group Ltd.

Visual representations such as tables and illustrations make findings easier for the reader to understand. Tables should be used to show the exact values of more data than can be summarized in a few sentences of text; or when the objective of presenting the data is to illustrate specific interrelationships (Fathalla and Fathalla 2004). If exact values are important, a table is preferable to a graph; when trends and relationships are more important, a graph is more efficient. Illustrations should be used only for a specific purpose in the Results section: as evidence to support the argument, since "seeing is believing." Illustrations can be an effective way to present data; however, using them for emphasis or just to stress a point is not normally appropriate in this section. This use of illustrations is better for oral presentation than a written paper (Fathalla and Fathalla 2004).

56.3   Language, Style, and Organization

The Results section should present a clear, concise, and objective description of the findings. Long and confusing sentences, interpretation, and references to other work should be avoided. The Results section can encompass research conducted over a period of time, and the authors should provide their readers with meaningful background into their research to establish the necessary context. Results should be written in the past tense using an active voice, meaning the subject performed the action. Wordiness convolutes sentences and conceals ideas. One common source of wordiness

Table 56.2 International guidelines for reporting results of animal studies, diagnostic accuracy studies, parallel group randomized trials, systematic reviews, and meta-analysis.

The ARRIVE 2010 guideline: Animal Research – Reporting *In Vivo* Experiments (NC3Rs Reporting Guidelines Working Group 2010)	Baseline data	1) For each experimental group, report relevant characteristics and health status of animals prior to experiment.
	Numbers analyzed	2) Report the number of animals in each group included in each analysis. Report absolute numbers (e.g. 10/20, not 50%).
		3) If any animals or data were excluded in the analysis, explain why.
	Outcomes and estimation	4) Report the results for each analysis performed, with a measure of precision (e.g. standard error or confidence interval).
	Adverse events	5) Give details of all important adverse events in each experimental group.
		6) Describe any modifications to the experimental protocols made to reduce adverse events.
STARD 2015: The Standards for Reporting Diagnostic Accuracy Studies (Bossuyt et al. 2015)	Participants	1) Provide flow of participants, using a diagram.
		2) Provide baseline demographic and clinical characteristics of participants.
		3) Present distribution of severity of disease in those with the target condition.
		4) Present distribution of alternative diagnoses in those without the target condition.
		5) Report time interval and any clinical interventions between index test and reference standard.
	Test results	6) Provide cross tabulation of the index test results (or their distribution) by the results of the reference standard.
		7) Present estimates of diagnostic accuracy and their precision (such as 95% confidence intervals).
		8) Report any adverse events from performing the index test or the reference standard.
CONSORT 2010 Statement: guidelines for reporting parallel group randomized trials (Schulz et al. 2010)	Participant flow (a diagram is strongly recommended)	1) For each group, report the numbers of participants who were randomly assigned, received intended treatment, and were analyzed for the primary outcome.
		2) For each group, report losses and exclusions after randomization, together with reasons.

Table 56.2 (Continued)

	Recruitment	3) Provide dates defining the periods of recruitment and follow-up.
		4) Elucidate why the trial ended or was stopped.
	Baseline data	5) Provide a table showing baseline demographic and clinical characteristics for each group.
	Numbers analyzed	6) Present, for each group, the number of participants (denominator) included in each analysis and whether the analysis was by original assigned groups.
	Outcomes and estimation	7) For each primary and secondary outcome, report the results for each group and the estimated effect size and its precision (such as 95% confidence interval).
		8) For binary outcomes, present both absolute and relative effect sizes.
	Ancillary analyses	9) Give results of any other analyses performed, including subgroup analyses and adjusted analyses, distinguishing pre-specified from exploratory.
	Harms	10) Report important harms or unintended effects in each group.
The PRISMA Statement: Preferred Reporting Items for Systematic Reviews and Meta-Analyses (Moher et al. 2009)	Study selection	1) Give numbers of studies screened, assessed for eligibility, and included in the review, with reasons for exclusions at each stage, ideally with a flow diagram.
	Study characteristics	2) For each study, present characteristics for which data were extracted (e.g. study size, PICOS, follow-up period) and provide the citations.
	Risk of bias within studies	3) Present data on risk of bias of each study and, if available, any outcome level assessment.
	Results of individual studies	4) For all outcomes considered (benefits or harms), present, for each study: (a) simple summary data for each intervention group, (b) effect estimates and confidence intervals, ideally with a forest plot.
	Synthesis of results	5) Present results of each meta-analysis done, including confidence intervals and measures of consistency.
	Risk of bias across studies	6) Present results of any assessment of risk of bias across studies.
	Additional analysis	7) Give results of additional analyses, if done, such as sensitivity or subgroup analyses and meta-regression.

Table 56.2 (Continued)

MOOSE 2010 Guideline: Meta-analysis of Observational Studies in Epidemiology (Stroup et al. 2000)	Presentation	1) Present a graph summarizing individual study estimates and the overall estimate.
		2) Present a table with descriptive information for each included study.
		3) Present results of sensitivity testing such as subgroup analysis.
		4) Present indication of statistical uncertainty of findings.

is unnecessary intensifiers (Kallestinova 2011). Adverbial intensifiers such as "clearly," "essentially," "quite," "basically," "rather," "fairly," "really," and "virtually," not only add verbosity to sentences, but also lower the credibility of the report. For example, "The outcomes in the present study *clearly* show that ...." Adverbial intensifiers appeal to the reader's emotions but lower objectivity. Another source of wordiness is the use of nominalizations – when nouns are derived from verbs and adjectives – paired with weak verbs, including "be," "have," "do," "make," "cause," "provide," and "get," or following constructions such as "there is/are" (Kallestinova 2011). For example, "The outcomes in the present study *will make surgeons be aware of* ...." Words such as "remarkably" or "strikingly," which imply interpretation of the findings, should be avoided. Consistent language should be used in describing similar results; the use of synonyms is discouraged as they can confuse the reader (Kotz and Cals 2013).

The findings of primary data analyses (i.e. those obtained from testing a priori hypotheses) should be mentioned early in the Results section in order to emphasize their importance. Next come the secondary analyses. If you have additional (unexpected) findings, they should be stated as such, and it should be made clear if any findings are from secondary (post-hoc) analyses, which are aimed at generating new hypotheses from the existing data (Kotz and Cals 2013). Bonferroni correction is usually used to adjust probability (p) values when multiple statistical comparisons are performed for the same analysis (Armstrong 2014). The purpose of this adjustment is to circumvent the problem of type I error in multiple comparisons: it is said that if 10 independent tests are reported for the same analysis, each conducted at the 0.05 significance level, there is a 40% chance that at least one significant difference will be found even if there are actually no differences (Elliott and Woodward 2007).

The results should be organized into subsections, each presenting the purpose of the experiment, your experimental approach, data (including tables, figures, schematics, algorithms, and formulas), and data commentary (Kallestinova 2011). The Results section should be presented in logical sequence, beginning with the general findings and flowing smoothly toward more specific ones. Do not repeat any information in the text that has been presented in tables and figures, except to emphasize or highlight a few important findings (Setiati and Harimurti 2007).

56.4 Accuracy of Findings and Presentation of Data

The importance of accuracy cannot be overstated. Check and recheck data and numbers and be certain they add up correctly (Johnson 2008). Sufficient detail should be given to allow other scientists to assess the validity and accuracy of the results (Fathalla and Fathalla 2004). Because many journals have a strict word limit for papers, avoid repetition or redundant language and focus on communicating a deliberate message. Remember that splitting the content into separate small papers is not an ideal practice (Alexandrov 2004).

Data in a study generally include statistics, which should be adequately described but not take over the paper (Fathalla and Fathalla 2004). Always use the same order when presenting data; for example, report findings from the experimental group before those from the control group. Provide effect sizes, such as odds ratios or relative risks (whenever applicable), together with their 95% confidence intervals in addition to *p*-values (Kotz and Cals 2013). Make consistent use of meaningful decimals for reported figures; unless you have a very large sample size, present numerical values to only one decimal place. Furthermore, present measures of central tendency together with their appropriate measures of variability: mean (standard deviation) or median (interquartile range). Always present the absolute number of cases in addition to relative measures; for example, "The percentage was 22% (33/150) in the intervention group compared with 15% (23/150) in the control group" (Kotz and Cals 2013). The word *significant* is often used in everyday language to stress something that is important or substantial, but in a scholarly article it is better to use the phrase "statistically significant" to report a difference proven by a statistical test (Kotz and Cals 2013).

56.5 Use of Tables and Illustrative Materials

Tables should be prepared before the text is written, as they contain the bulk of the evidence on which the paper is based and are particularly useful when a large body of data is to be presented (Kotz and Cals 2013). For studies with fewer than 30 subjects or studies addressing a rare medical condition, tabular presentation of the details of each case is recommended. The table encompasses all findings and makes them understandable to readers, who can easily follow the observations for each case.

Typically, at least two or three tables will be placed after the reference section of the manuscript, each occupying a single page. The first table often details the demographics and clinicopathological data. Tables also allow you to cross-tabulate, for example, different types of relapse from different types of treatment (Veness 2010). The data presented in the tables should not be repeated in the text; a table should be readily understood without reference to the text. Each table should have a title, footnotes with abbreviations, and comments, if necessary. Tables should be sequentially numbered and cited within the text (Veness 2010).

The title should describe the content of the table exactly. It should not include any unnecessary words or repeat the column and row headings. Short or abbreviated headings for each column and row should be used and, if necessary, a footnote defining nonstandard abbreviations can be provided. Footnotes can also be used to specify the statistical tests used to obtain the findings (Fathalla and Fathalla 2004).

There should be no ambiguity about the purpose of the columns and rows. When column headings are grouped, a straddle line should be used to eliminate any confusion about the column groupings. Row headings that are grouped can be indented to indicate each grouping (Fathalla and Fathalla 2004).

Line drawings and photographs are the two common types of illustration and should be prepared according to journal guidelines. The figures are numbered in the order in which they are cited in the text. Letters, numbers, and symbols (including those on axes) should be legible when the figure is submitted. Usually, line drawings and graphs are submitted as black and white images, but make sure that different plots are distinguishable (e.g. a solid line versus a dashed line) (Gaafar 2005). Care should be taken to avoid repeating the findings in the figure and the body of the text (Cetin and Hackam 2005). If an image is copyrighted or created by someone else, it is necessary to obtain permission to reproduce it. Captions or legends should be written in detail, allowing the reader to identify the relevant structures identified in an image.

References

Alexandrov, A.V. (2004). How to write a research paper. *Cerebrovasc. Dis.* 18 (2): 135–138.

Armstrong, R.A. (2014). When to use the Bonferroni correction. *Ophthalmic Physiol. Opt.* 34 (5): 502–508.

Bossuyt, P.M. et al. (2015). STARD 2015: an updated list of essential items for reporting diagnostic accuracy studies. *BMJ* 351: h5527.

Cetin, S. and Hackam, D.J. (2005). An approach to the writing of a scientific manuscript. *J. Surg. Res.* 128 (2): 165–167.

Elliott, A. and Woodward, W. (2007). *Statistical Analysis Quick Reference Guidebook*. SAGE.

von Elm, E., Altman, D.G., Egger, M. et al. (2007). Strengthening the reporting of observational studies in epidemiology (STROBE) statement: guidelines for reporting observational studies. *BMJ* 335: 806–808.

Fathalla, M.F. and Fathalla, M.M.F. (2004). *A Practical Guide for Health Researchers*, 134–136. Eastern Mediterranean Series, Egypt: WHO Regional Publications.

Gaafar, R. (2005). How to write an oncology manuscript. *J. Egypt. Natl. Cancer Inst.* 17 (3): 132–138.

Johnson, T.M. (2008). Tips on how to write a paper. *J. Am. Acad. Dermatol.* 59 (6): 1064–1069.

Kallestinova, E.D. (2011). How to write your first research paper. *Yale J. Biol. Med.* 84 (3): 181–190.

Kotz, D.I. and Cals, J.W. (2013). Effective writing and publishing scientific papers, part V: results. *J. Clin. Epidemiol.* 66 (9): 945.

Moher, D. et al. (2009). Preferred reporting items for systematic reviews and meta-analyses: the PRISMA statement. *PLoS Med.* 6: e1000097.

NC3Rs Reporting Guidelines Working Group (2010). Animal research: reporting *in vivo* experiments: the ARRIVE guidelines. *J. Physiol.* 588: 2519–2521.

Schulz, K.F., Altman, D.G., Moher, D. et al. (2010). CONSORT 2010 statement: updated guidelines for reporting parallel group randomized trials. *Ann. Intern. Med.* 152: 726–732.

Setiati, S.I. and Harimurti, K. (2007). Writing for scientific medical manuscript: a guide for preparing manuscript submitted to biomedical journals. *Acta Med. Indones.* 39 (1): 50–55.

Shidham, V.B., Pitman, M.B., and Demay, R.M. (2012). How to write an article: preparing a publishable manuscript! *Cytojournal* 9: 1.

Stroup, D.F. et al. (2000). Meta-analysis of observational studies in epidemiology: a proposal for reporting. Meta-analysis Of Observational Studies in Epidemiology (MOOSE) group. *JAMA* 283: 2008–2012.

Veness, M. (2010). Strategies to successfully publish your first manuscript. *J. Med. Imaging Radiat. Oncol.* 54 (4): 395–400.

57

Writing the Discussion Section for Original Research Articles

Ayhan Cömert[1] and Eyyub S. M. Al-Beyati[2]

[1] School of Medicine, Department of Anatomy, Ankara University, Ankara, Turkey
[2] School of Medicine, Department of Neurosurgery, Ankara University, Ankara, Turkey

57.1 Introduction

The practice of writing a scientific paper is a required skill for every researcher and often the manifest objective. Searching for evidence-based answers to a set question or hypothesis and articulating the findings are the main motivations for a research study.

An oft long and laborious journey, a scientific study starts with an idea that is developed into a formal hypothesis, which is ultimately tested using various methods. The obtained results are then thoroughly analyzed and the study is concluded with a thoughtful interpretation of the collected data. Reporting the study in the form of a research paper is as vital as carrying out the study itself. Until it is published in a proper medium, it is as if the study was never performed. Scientific publications constitute the primary communication tool between researchers and scholars all around the world.

A scientific paper that follows a fairly rigid structure includes various sections that are important for new researchers to understand when attempting to shape their first paper. These sections generally include an Introduction, Materials and Methods section, Results section, Discussion, and Conclusion. It is essential to maintain a uniform structure in order for the audience to readily comprehend the content. Organizing the data in such a manner helps the reader find information where it is expected (Gopen and Swan 1990). However, writing an effective research paper is not easy, and it consists of a creative process that requires a disciplined approach.

Scientific writing requires more than just the data to be exciting. There are many factors that impact successful writing. These are not natural skills that accompany a lucky researcher, but learned abilities acquired with reading and writing, frequently in the light of patience.

The art of science writing consists of presenting complex data in basic form and relating the exact message to the reader avoiding any misunderstanding. To reach this goal, managing information in an objective way is the core requirement for every researcher. Nurturing young writers to defend quality rather than quantity is a major concern regarding scientific writing (Looi 2009).

A Guide to the Scientific Career: Virtues, Communication, Research, and Academic Writing, First Edition.
Edited by Mohammadali M. Shoja, Anastasia Arynchyna, Marios Loukas, Anthony V. D'Antoni, Sandra M. Buerger, Marion Karl and R. Shane Tubbs.
© 2020 John Wiley & Sons, Inc. Published 2020 by John Wiley & Sons, Inc.

57.2   Purpose of a Discussion Section

The Discussion section, typically reserved as a follow-up to the results section, is considered to be the most important and often the most difficult section to write. The writer aims to explain what the findings mean and how they inform the research question posed in the introduction. An interpretation of the findings should be presented and discussed objectively, evaluating the importance and relevance of the research by comparing it with previous studies (Hess 2004; Sterk and Rabe 2008).

The content included in the Discussion section is organized to resemble an inverted pyramid, whereas the Introduction section presents information in a traditional pyramid construction. Together, they form bookends to the report in the shape of an hourglass asking a question and describing the results of the research in order to answer it (Cals and Kotz 2013). In fact, the Introduction and Discussion sections are complementary: what is presented in the Discussion section represents the impetus of the researcher to propose the hypothesis and generate the study itself.

57.3   What and How to Discuss

The soul of the Discussion section is in providing real objectivity in its approach to the study and its findings, self and nonself. The content to be written in the Discussion section is what makes the researcher interested in pursuing the study in the first place. Taking notes during the research process and reevaluating them over time may be helpful when finally writing the paper, especially the Discussion section. There are some important rules to consider when organizing this section. By following these rules, an outline of thoughts can be formed, which allows the writer to put findings into a context while maintain fluency, logical queue, and simplicity in syntax.

Long paragraphs make it difficult to command the attention or concentration of the reader; short paragraphs can be more effective by being to-the-point and delivering a precise message. Throughout the entire paper, including the Discussion section, use an active voice where possible, even though it is traditional to use a passive voice (Guyatt and Haynes 2006). In addition, it is better to use the same verb tense and point-of-view that were used in stating the research question.

57.3.1   Main Study Findings

The Discussion section is based on working out the possibility of the obtained results in answering the primary question(s). Begin with putting major findings forward as an effective first step. It is important not to re-write the results, but rather discuss them in the context of the research problem put forth in the introduction. When it is necessary to specify the data, it may be more appropriate to use the summary estimates or make reference to the tables or figures included in the results section.

Explaining why the findings are important and how they contribute to the answer are important, but care should be taken to avoid inflating the significance of the results. Inconvenient results are a part of almost all studies; trying to hide or reconstruct them should be strictly avoided. Always maintain honesty and academic integrity. A scientific study is made with the aim of posing and separating truths from falsehoods, and compelling results from those deemed fruitless. Therefore, any unexpected – even if negative – findings may strengthen the study rather than impairing it.

57.3.2   Expectations and Literature

The scientific study is based on a hypothesis that explains or answers a question; there are always one or more expectations that accompany a study. Comparing the findings with these expectations, and clarifying their compatibilities or incompatibilities, should be thoughtfully made in the Discussion section.

Linking the findings with each other and providing logical explanation may help the reader to grasp the relationship more easily. It is critical not to defend a singular view that seems to be more compatible with the results; approaching all possible explanations objectively is more appropriate and makes the paper more valuable.

During comparison of the data and results with previous studies, start with self-results. Differences and similarities should be reported meticulously and with extreme objectivity. Try to substantiate the reasons of why the results are compatible or incompatible with those in the literature, defending the hypothesis where necessary, in order to clarify any additional uncertainty. Attention should be paid not to intellectually attack studies that report contrasting results. Refrain from name-calling.

Stating the contribution of findings to the field of study is important. For example, in a medical paper, underlining how the obtained data can provide better patient care, diagnostic evaluation, or a new treatment approach will resolutely strengthen the article. Explaining why the study is important and how the data obtained can contribute to existing literature may launch the study forward.

57.3.3   Strengths and Limitations

Every study, even the most thorough one, will have some limitations in their nature (design and methodology limitations affecting the generalizability of the findings). Clearly identifying the study strengths and limitations may provide a more transparent view that could pave the way for better understanding. There may be more than one method to approach a study. Selection of the subjects, measurements methods, and analytical approach adopted in the study should be critically discussed and evaluated. It is advisable to point out alternative methodologies that can be used to answer the research question of interest.

Trying to evaluate the paper as an independent reader is essential. Examine all possible weak points of the study before any reader or editor does. In their deliberation, editors and reviewers are nearly always right (Looi 2009). So, be careful not to leave much room for them to debate.

57.3.4   Further Studies

There is no single study that answers all questions related to one issue. Scientific approach prefers working step-by-step toward the reality and better understanding of the topic at hand. From this viewpoint, it is routine for the researcher to arrive at new questions during the course of research. This may be valuable in leading the scientific community to design future studies to clarify, for example, aspects of the initial findings and answer new questions raised. The discussion section may end with suggestions for further studies related to the original research.

57.3.5 Take-Home Message

Including a final take-home message serves as brief summary of the entire study into one or two sentences; this is the main piece of information that the reader should remember regarding the study. Some journals require a summary statement under the conclusion section; otherwise, ending the discussion section with the take-home message is also appropriate.

A scientific paper usually reflects what a team of scientists have endeavored in pursuit of one or more questions. It is not just a few pages of text but an in-depth analysis and explanation of the research proposed and conducted. Presenting such a profound process simply and clearly is a task of both science and art at the same time. Reading published papers by established science writers, and effectively thinking and writing in this manner, is the best way to advance scientific writing practices.

References

Cals, J.W. and Kotz, D. (2013). Effective writing and publishing scientific papers, part VI: discussion. *J. Clin. Epidemiol.* 66 (10): 1064.

Gopen, G.D. and Swan, J.A. (1990). The science of scientific writing. *Am. Sci.* 78 (6): 550–558.

Guyatt, G.H. and Haynes, R.B. (2006). Preparing reports for publication and responding to reviewers' comments. *J. Clin. Epidemiol.* 59: 900–906.

Hess, D.R. (2004). How to write an effective discussion? *Respir. Care* 49 (10): 1238–1241.

Looi, L.M. (2009). Nurturing young writers: sustaining quality, not quantity. *Singapore Med. J.* 50 (11): 1044–1047.

Sterk, P.J. and Rabe, K.F. (2008). The joy of writing a paper. *Breathe* 4: 224–232.

58

Reporting a Clinical Trial

Dirk T. Ubbink

Department of Surgery, Amsterdam University's Academic Medical Center, Amsterdam, The Netherlands

58.1 Evidence-Based Medicine

Evidence-based medicine (EBM) is the cornerstone of current medicine. It synthesizes clinical expertise, best available evidence from medical literature, and the patient's situation and preferences (Figure 58.1). This may sometimes imply adjusting, or even correcting, the caregiver's personal expertise based on the evidence and patient's preference in order to ensure the highest quality of care. This "life-long learning" is essential because the medical profession is a dynamic one, changing constantly due to technical and scientific progress and a broadening disease spectrum. Hence, high-quality research is mandatory to underpin our clinical practice.

A randomized clinical trial (RCT) is considered as the highest level of evidence among individual studies. It typically involves inviting eligible patients to be enrolled in a scientific study in which they are randomly allocated to two (or more) treatment arms. Then the patients are followed for a certain period of time after which the predefined outcomes are assessed and compared between the treatment arms. Admittedly, uptake of (even the positive) findings from high-quality trials and systematic review in daily practice is often slow and sometimes negligible. For example, decades passed before β-blockers were prescribed at discharge for patients after an acute myocardial infarction. It even took centuries to introduce citrus fruits as routine nourishment on trading ships sailing around the world to prevent scurvy!

Dissemination of research findings is likely to be a biased process (Song et al. 2010). Reasons for this are that clinicians do not become aware of the results, perceive the results as either invalid or not relevant to their patients, or simply do not remember to use the treatment (Glasziou et al. 2008). An additional barrier is the clinicians' inability to carry out the treatment on the basis of the information provided in the published reports (Glasziou et al. 2008). Some trial descriptions may leave you wondering exactly how to carry out treatments such as a "behavioral intervention," "wound care," or "weight reduction program." Thus, the better the description of a clinical trial, the higher the odds its message will be adopted.

A Guide to the Scientific Career: Virtues, Communication, Research, and Academic Writing, First Edition.
Edited by Mohammadali M. Shoja, Anastasia Arynchyna, Marios Loukas, Anthony V. D'Antoni, Sandra M. Buerger, Marion Karl and R. Shane Tubbs.
© 2020 John Wiley & Sons, Inc. Published 2020 by John Wiley & Sons, Inc.

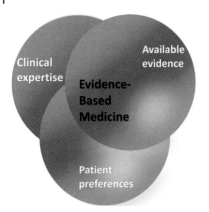

Figure 58.1 The synthesis of clinical expertise, available evidence and patient preferences into evidence-based practice.

58.2 Quality of Clinical Trials

A clinically irrelevant research question, an improper study design, a poorly conducted study, and floppy reporting lead to a waste of effort and money (Chalmers and Glasziou 2009) or even public criticism in the media (Sheldon 2010). Even today, the quality of reporting of RCTs leaves room for improvement (Péron et al. 2012; Balasubramanian et al. 2006). Inadequate reporting of trials is associated with biased estimates of treatment effects. First, positive study results tend to be published more often than indifferent or negative study results, also known as publication bias (Crawford et al. 2010; Dwan et al. 2013). Second, adverse outcomes of an intervention may be neglected or reported selectively (also known as reporting bias, caused by the researchers), which overly stresses the beneficial outcome of the intervention (Dwan et al. 2013). This outcome reporting bias is underrecognized (Kirkham et al. 2010). Third, the definition of essential terms or presentation of the results may differ among publications (Kirkham et al. 2010; Chalmers and Glasziou 2009; Brölmann et al. 2013). For example, in wound care trials, "wound healing" may be defined as complete epithelialization, >95% closure or reduction in wound surface area, while quality of life results may be presented as changes from baseline or as compared with a healthy population.

These sources of bias emphasize the need for full and transparent reporting of clinical trials, which will allow readers to assess the validity, strengths, and limitations of the research performed, and may also protect clinicians from integrating biased results in their clinical decision-making. Fortunately, the reporting quality can be improved with the help of the updated international Consolidated Standards of Reporting Trials (CONSORT 2010) statement (Moher et al. 2010; Schulz et al. 2010).[1] Many scientific journals have endorsed this statement and have incorporated it in their instructions for the authors (Turner et al. 2012; Hopewell et al. 2008). Similarly, the PRISMA (Preferred Reporting Items for Systematic Reviews and Met-Analyses) statement offers useful guidance when conducting systematic reviews or meta-analyses of available RCTs.[2]

---

1 Available from http://www.consort-statement.org.
2 Available from http://www.prisma-statement.org/statement.htm.

58.3   Writing Your Manuscript

The journal you choose to submit your manuscript will certainly influence the way you should plan writing your trial report. The journal's online "instructions for authors" will guide you through this process. In addition, the checklist belonging to the CONSORT statement is helpful for a complete report (see Table 58.1). The following sections will give more detailed instructions on how to write the common sections of a manuscript on a clinical trial. The standard sections are: Abstract, Introduction, Patients and Methods, Results, and Discussion. Requirements as to these sections may differ slightly among journals. A carefully chosen manuscript title, a well-structured abstract, and keywords (if required) will help index your publication for easy retrieval as a clinical trial (Hopewell et al. 2012).

58.3.1   Title

Avoid using titles that are vague, too long, and mind-boggling. The title should be judiciously composed to point out the major aim, result, or finding of your study and should avoid abbreviations. A deterrent example is "Nanotheranostics in the LAPS syndrome: a multicenter double-blinded randomized controlled trial." Preferably, make the title catchy or timely to help gain the editors', reviewers' and readers' attention. The CONSORT 2010 statement requires identification of the study design in the title (Moher et al. 2010). A "randomized controlled trial" seems somewhat like a pleonasm, as any randomized trial has a control group. Rather, "randomized clinical trial" is more informative, discerning it from laboratory or animal experimental studies. "Double-blinded" is an unclear phrase, as it does not indicate who actually was blinded; the physician, the patient, or the outcome assessor?

58.3.2   Abstract

Structured abstracts of a RCT should include the objectives of the trial; its design (preferably including method of allocation and masking); participants (i.e. description, numbers randomized, and number analyzed); interventions intended for each randomized group and their impact on primary efficacy outcomes and harms; conclusions; and source of funding. Stating the level of evidence, i.e. mentioning RCT as the study design, in the abstract should not be omitted as it further improves the immediate assessment of the type of study described, particularly when other scientists search for your RCT as part of a systematic review. At the end of the abstract, the trial registration name and number should be stated, as well as the database in which it was registered. Common and equally valid primary registries all have specific databases (Table 58.2).

58.3.3   Introduction Section

A good introduction should resemble a funnel; broad at its beginning, narrowing down toward the end. The first paragraph should cover the current size and severity of the problem addressed and its importance in general, including the overall impact on health, costs, and social gain. A convincing identification of the key issues in current practice and literature is required to set the scene for your trial. Summarize, refer, and elaborate

Table 58.1 The CONSORT 2010 checklist of information (to be included when reporting a randomized trial).[a]

Section/topic	Item no.	Checklist item	Reported on page no.
Title and abstract			
	1a	Identification as a randomized trial in the title	
	1b	Structured summary of trial design, methods, results, and conclusions (for specific guidance, see CONSORT for abstracts)	
Introduction			
Background and objectives	2a	Scientific background and explanation of rationale	
	2b	Specific objectives or hypotheses	
Methods			
Trial design	3a	Description of trial design (such as parallel, factorial) including allocation ratio	
	3b	Important changes to methods after trial commencement (such as eligibility criteria), with reasons	
Participants	4a	Eligibility criteria for participants	
	4b	Settings and locations where the data were collected	
Interventions	5	The interventions for each group with sufficient details to allow replication, including how and when they were actually administered	
Outcomes	6a	Completely defined prespecified primary and secondary outcome measures, including how and when they were assessed	
	6b	Any changes to trial outcomes after the trial commenced, with reasons	
Sample size	7a	How sample size was determined	
	7b	When applicable, explanation of any interim analyses and stopping guidelines	
Randomization:			
Sequence generation	8a	Method used to generate the random allocation sequence	
	8b	Type of randomization; details of any restriction (such as blocking and block size)	
Allocation concealment mechanism	9	Mechanism used to implement the random allocation sequence (such as sequentially numbered containers), describing any steps taken to conceal the sequence until interventions were assigned	
Implementation	10	Who generated the random allocation sequence, who enrolled participants, and who assigned participants to interventions	

(Continued)

Table 58.1 (Continued)

Section/topic	Item no.	Checklist item	Reported on page no.
Blinding	11a	If done, who was blinded after assignment to interventions (e.g. participants, care providers, those assessing outcomes) and how	
	11b	If relevant, description of the similarity of interventions	
Statistical methods	12a	Statistical methods used to compare groups for primary and secondary outcomes	
	12b	Methods for additional analyses, such as subgroup analyses and adjusted analyses	
Results			
Participant flow (a diagram is strongly recommended)	13a	For each group, the numbers of participants who were randomly assigned, received intended treatment, and were analyzed for the primary outcome	
	13b	For each group, losses, and exclusions after randomization, together with reasons	
Recruitment	14a	Dates defining the periods of recruitment and follow-up	
	14b	Why the trial ended or was stopped	
Baseline data	15	A table showing baseline demographic and clinical characteristics for each group	
Numbers analyzed	16	For each group, number of participants (denominator) included in each analysis and whether the analysis was by original assigned groups	
Outcomes and estimation	17a	For each primary and secondary outcome, results for each group and the estimated effect size and its precision (such as 95% confidence interval)	
	17b	For binary outcomes, presentation of both absolute and relative effect sizes is recommended	
Ancillary analyses	18	Results of any other analyses performed, including subgroup analyses and adjusted analyses, distinguishing prespecified from exploratory	
Harms	19	All-important harms or unintended effects in each group (for specific guidance, see CONSORT for harms)	
Discussion			
Limitations	20	Trial limitations, addressing sources of potential bias, imprecision, and, if relevant, multiplicity of analyses	
Generalisability	21	Generalisability (external validity, applicability) of the trial findings	
Interpretation	22	Interpretation consistent with results, balancing benefits and harms, and considering other relevant evidence	

(Continued)

Table 58.1 (Continued)

Section/topic	Item no.	Checklist item	Reported on page no.
Other information			
Registration	23	Registration number and name of trial registry	
Protocol	24	Where the full trial protocol can be accessed, if available	
Funding	25	Sources of funding and other support (such as supply of drugs), role of funders	

a) We strongly recommend reading this statement in conjunction with the CONSORT 2010 Explanation and Elaboration for important clarifications on all the items. If relevant, we also recommend reading CONSORT extensions for cluster randomized trials, noninferiority and equivalence trials, nonpharmacological treatments, herbal interventions, and pragmatic trials. Additional extensions are forthcoming: for those and for up-to-date references relevant to this checklist, see http://www.consort-statement.org.

Table 58.2 Clinical trial registries.

Registry	Host	Website
ClinicalTrials.gov	US National Library of Medicine	http://www.clinicaltrials.gov
ISRCTN (International Standard Randomized Controlled Trial Number) registry	Current Controlled Trials Ltd., a sister company of BioMed Central	http://www.controlled-trials.com
Nederlands Trial Register	Dutch Cochrane Center	http://www.trialregister.nl

on the scientific background related to your research question. Strive to summarize the best available and up-to-date evidence, preferably by referring to a pertinent systematic review. This will help identify what is known about the topic in the literature and what is yet unknown. The unknown should then be the topic of the trial.

At the end of this section, you can narrowly focus on your research question. The last paragraph of the Introduction should specifically describe the research question(s) your trial is going to address. The best way to formulate this is by using the PICO-format, describing the *P*atient's problem, the *I*ntervention under study, the *C*omparator or standard intervention, and the (primary) *O*utcome parameter(s) of interest (Da Costa Santos et al. 2007). You might also want to describe the future perspective of "what (will happen) if we have the results."

58.3.4   Patients and Methods Section

Not too long ago, some journals printed the Methods section of their papers in a smaller font size. In the current EBM-era, this section receives the most attention when the paper is critically appraised, as it tells the reader how well the study has been conducted and, hence, how valid the results of the study are. The CONSORT checklist also pays special attention to this section of a RCT report (see Table 58.1).

In general, the description of your patients, materials, methods, and procedures should be in such detail that any research-oriented reader can evaluate and repeat your work if he would. Again, the PICO-system may be helpful to structure this description. Help your reader interpret the quality of the trial (did the authors do the best they could?) and the possible sources of risk of bias (can the reader believe the results?), as research methods are jeopardized by important risks of bias that may invalidate the findings. A previously published trial protocol is useful to refer to in the paper on the trial results as it saves space when describing the methods used.

Specific items and tips are (in the order of how you would describe them):

- State your adherence to the CONSORT – a statement that an increasing number of journals require – as well as the submission of a completed checklist.
- State your definitions clearly, especially when confusing terminology exists. For example, in trials on wound care it is important to characterize "wound etiology," the method of "wound debridement," measurement units and techniques of "wound healing" assessment, and local and systemic types of "wound care" applied.
- For most, but in particular pharmacological, treatments the description would need to include the dose, route, timing, and duration of administration, and any monitoring employed.
- For some complex interventions, graphical display can further clarify the methods (Perera et al. 2007; Hooper et al. 2013).
- State the approval of the medical ethics institutional review board of the contributing center(s) for conducting the trial. Also, state that the trial has been conducted according to good clinical practice (GCP) guidelines and the principles of the Declaration of Helsinki on the ethical principles for medical research involving human subjects (WMA 2013).

Next, address how you handled potential sources of bias, as these are regularly underreported or even overlooked, which may cause justified suspicion of bias:

- Define the criteria used for patient inclusion and exclusion, including the stage and duration of the disease studied, which will determine the applicability of the trial outcomes.
- Describe an a priori sample size calculation of the study. Without a power analysis, the reader cannot assess whether observed differences are meaningful and clinically relevant or at risk of falsely positive (type-I error) or falsely negative (type-II error) outcomes.
- Specify the method of sequence generation (randomization), such as a random-number table or a computerized random number generator, which provides information to assess the likelihood of selection bias in the group assignment. Note that using date of birth, day of the week, or patient identification numbers are no longer acceptable means of *randomization.*
- Describe any advanced methods of randomization, such as block-randomization or stratification. If stratification was applied, report why the chosen stratification variables are likely or known to influence the outcome in a substantial way.

- Describe how patient allocation was concealed, e.g. by sealed envelopes. If allocation is not (reported as) concealed, this is one of the major sources of bias (Odgaard-Jensen et al. 2011). For multicenter trials, a central randomization office is commonly applied to overcome selection bias.
- Report who generated the allocation sequence, who enrolled patients, and who assigned patients to the trial groups. Even a perfect randomization schedule can lead to bias if implemented incorrectly.
- Describe who (i.e. physicians, patients, outcome assessors) were blinded to the treatment, given rather than stating "single-" or "double-blinded trial." Especially in surgical trials, blinding of surgeons and patients may be virtually impossible. Blinding of the outcome assessor, i.e. choosing an independent person to judge the (main) study outcomes, is vital to avoid assessment bias.
- Describe the standard – and possible additional – treatment procedures, including any incidental co-interventions, which may interfere with the effect of the intervention under study. For example, in a trial comparing preoperative antiseptic techniques on surgical site infections, giving postoperative antibiotics when deemed necessary will likely influence the wound infection rate, and possibly with a disparate effect between both study arms.
- Define your primary and secondary study outcomes. Explain why these are relevant (to the patient), and how these will be measured. Primary outcomes are those considered most clinically relevant and represent the measures of greatest therapeutic benefit. The required sample size is calculated based on differences expected regarding these primary outcomes. Secondary outcome measures may provide information on therapeutic effects of secondary importance, side effects, or tolerability. The usage of surrogate outcome measures, i.e. indirect measures of a clinically relevant outcomes, should be avoided.
- Describe whether the outcome measures chosen are evidence-based and validated (e.g. Short Form-36 for generic quality of life, Numeric Rating Scale for pain, or Amsterdam Linear Disability Score for general ability assessment) whenever possible, in particular regarding patient-reported outcomes (Reeve et al. 2013).
- Report how and at which time-point(s) your outcome measures were assessed and why these time points are considered as patient-relevant.
- The duration of follow-up should be specified. This is especially important for the reader to interpret outcomes like survival, infection rates, quality of life, or tumor recurrence. Also, authors of systematic reviews can only perform meta-analyses of trial data when the follow-up periods in trials are known and similar.
- Mention if an intention-to-treat analysis was intended. This is the analysis of choice, as it best mimics clinical practice. Intention to treat gives a pragmatic estimate of the actual benefit of a treatment policy rather than of the potential benefit in patients who receive treatment exactly as planned, which is called a "per-protocol" analysis.
- Per-protocol analysis may also be valuable to assess the actual effect of an intervention per se. However, this means that patients who actually received an intervention are selected for the analysis, irrespective of their allocation at randomization. Thus, the advantages of the randomization, being the equal distribution of possible confounders over the study arms, are lost. These possible influences should be described.

- Finally, do not disclose only who sponsored the study but also their contribution to the data analysis, interpretation of the results, or their publication. This is an underrecognized reason for selective reporting of trials (Melander et al. 2003). An unrestricted grant gives the best guarantee that the sponsor has no involvement as to the interpretation and description of the study results.

58.3.5 Statistical Analysis

When describing the statistics used, the reader should be able to judge the validity of the analyses and subsequent findings. Statistics should scientifically corroborate your clinical outcome rather than a means of finding a statistically significant result to publish.

The universally accepted definition of a p-value <0.05 as significant implies that when doing 20 statistical tests as part of the data analysis, one type-I error (falsely positive result) is likely to occur. Hence, predefine and limit the number of statistical analyses and always report any planned (subgroup) analyses to avoid post-hoc analyses that may have the appearance of data-dredging. Mention the analysis software and version used.

p-values are not more than arbitrary measures of the precision of the outcome measure and depend heavily on the study sample size. Thus, these p-values are not equal to clinical relevance. For example, a large, multinational clinical trial in patients with intermittent claudication may demonstrate that a certain vasodilating drug significantly ($p < 0.001$) extends walking distance with 20 m, but this improvement can hardly be considered clinically relevant. In contrast, a small single-center trial, powered to find a difference of 200 m, may not show an significant treatment effect ($p = 0.07$) when the mean improvement found was "only" 180 m, but this would still be a relevant improvement for the patient.

When outcomes are dichotomous (yes/no, dead/alive) risk differences, relative risks, or odds ratios should be calculated and presented with their 95% confidence intervals. The numbers needed to treat (NNT) or needed to harm (NNH), which can easily be derived from the risk differences, much better indicate the clinical impact of the results found (Tramèr and Walder 2005). Also for continuous data, a mean difference can be provided with a 95% confidence interval. If trial data are skewed, present not only the median values (with interquartile ranges) per study arm but also the mean values (with standard deviations) to enable future meta-analyses.

58.3.6 Results Section

This section should start with a summary of the study setting, number of patients included, protocol violations (if any), and any peculiarities about the patients' baseline characteristics and disease details in each study arm. Descriptive statistics should give information about adherence to treatments, cross-overs, adverse events, concurrent treatments, eventual follow-up duration, and loss to follow-up, including the reasons for this.

Incomplete outcome data in trials with a long follow-up duration are virtually inevitable. Attrition bias occurs when predominantly patients with either the best or the worst outcome remain. The first situation (e.g. when the patients not responding to the therapy have died or tried their luck at another hospital) will lead to overestimation of the trial results. The opposite (e.g. when the best patients do not attend their outpatient check-ups anymore) happens when only the worst cases remain. An additional problem arises when this loss to follow-up differs between the study arms. This should be examined and described by means of a missing data-sensitivity analysis, assuming best- and worst-case scenarios.

The Results section should include the following components:

- The first figure in the RCT report should, according to CONSORT, depict a flowchart of the numbers of patients available and suitable for inclusion, follow-up and analysis (see Figure 58.2).
- The first table in the RCT report invariably shows the detailed baseline characteristics of patients in each study arm. As these study arms have been generated by randomization, any differences between the two (or more) patient groups are purely by chance. Stating p-values with these differences is therefore not meaningful. Any obvious baseline differences should be addressed and corrected for in the analysis.
- Not only the beneficial but also the adverse effects of the studied interventions should be presented to enable the reader to weigh all aspects of the intervention.
- Report the outcomes as predefined in the Methods section to prevent selective reporting, showing the benefits and harms.
- Present these outcomes whenever possible as absolute numbers so that the numerator and the denominator are clear.
- Time-dependent parameters (e.g. time to wound healing) should be analyzed by means of survival analysis (Kaplan–Meier and log-rank analysis) rather than by presenting means or rates (e.g. number of wounds healed after 12 weeks).

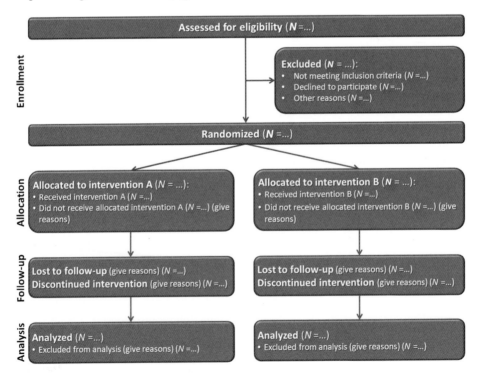

Figure 58.2 Flow chart template of patient inclusion, follow-up, and analysis (from Schulz et al. 2010).

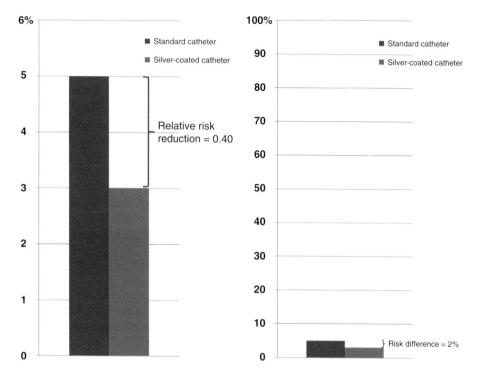

Figure 58.3 Example of the different framing of the same results. The right-sided graph shows the relative risk reduction of using silver-coated urinary catheters on urinary tract infection rates. The left-sided graph shows the absolute risk reduction of the same intervention. Note that asymptomatic bacteriuria was used here as a surrogate outcome.

In general, the data should be presented in an unambiguous, unprejudiced manner. Framing of the results in solely relative outcome measures (relative risks) in the text or graphs may be tempting, but is misleading (Gigerenzer et al. 2010). Consider this example: silver-impregnated urinary catheters may reduce the risk of urinary tract infections with about 40%, which corresponds with a relative risk of about 0.60 (Schumm and Lam 2008). This effect sounds impressive, but it actually reduces the infection rate from 5% to 3% of all patients requiring a urinary catheter. This means a NNT of 50. This may not be worth the substantial extra costs involved if 100 patients are given a silver-coated catheter to avoid a urinary tract infection in two of them. A graphical representation of the same reduction is shown in Figure 58.3.

An RCT may stop earlier than expected, e.g. when a planned interim analysis shows larger than expected benefit or harm for the experimental intervention, the funding is exhausted, or researchers cannot recruit sufficient eligible patients. This underpins the value of installing and supporting an independent data safety and monitoring committee. With this board a priori stopping rules may be defined and interim analyses planned and done. In case of early stopping, a full reporting of why a trial ended is important for evidence-based decision-making. Otherwise, no (unplanned) interim results should be reported.

Finally, some medical disciplines already have agreed-on certain specific guidelines or reporting standards to adhere to, e.g. regarding the reporting of adverse effects of radiation therapy in oncology (Trotti et al. 2007), trials on wound care (Brölmann et al. 2013), thrombolysis for acute stroke (Higashida et al. 2003), or endovascular revascularization for intracranial atherosclerosis (Schumacher et al. 2010).

58.3.7 Discussion Section

This is usually considered to be a challenging section to write. The most common pitfall here is that results are merely being reiterated rather than being interpreted. The Discussion section should help readers understand the consequences of the findings in terms of their impact on practice, policy, education, or future research.

The following items are essential parts of the Discussion section:

- An answer to the research question: The first sentence may start with "This trial shows that ...," summarizing the main findings in general terms.
- An interpretation of the results and comparison with the findings from other studies previously published and possible explanations of any differences: This section gives the opportunity to give explanations, comparisons, and the extent to which the results are generalizable, i.e. might be extrapolated to other patient groups.
- A self-criticism, admitting the limitations of the study and how these may or may not seriously affect the outcomes. Also the strengths of the study may be addressed. Too often, however, a positive message about the effectiveness of the intervention is sought to warrant the effort put into the trial and to increase the chance of acceptance by the journal. The benefits of the intervention should be fairly weighed against its possible harms, required learning curve, applicability in other settings or patient groups, and inherent costs.
- A discussion of the implications of the trial results for current clinical practice (does it really benefit my patients?) and future research should be provided.
- Depending on the journal style, you may conclude the Discussion with a take-home message about what the reader should do "as of tomorrow" in their clinical practice with the results of this trial in a final subheading "Conclusions."

References

Balasubramanian, S.P., Wiener, M., Alshameeri, Z. et al. (2006). Standards of reporting of randomized controlled trials in general surgery: can we do better? *Ann. Surg.* 244 (5): 663–667.

Brölmann, F.E., Eskes, A.M., Sumpio, B.E. et al. (2013). Fundamentals of randomized clinical trials in wound care: reporting standards. *Wound Repair Regen.* 21 (5): 641–647.

Chalmers, I. and Glasziou, P. (2009). Avoidable waste in the production and reporting of research evidence. *Lancet* 374: 86–89.

Crawford, J.M., Briggs, C.L., and Engeland, C.G. (2010). Publication bias and its implications for evidence-based clinical decision making. *J. Dent. Educ.* 74 (6): 593–600.

da Costa Santos, C.M., de Mattos Pimenta, C.A., and Nobre, M.R. (2007). The PICO strategy for the research question construction and evidence search. *Rev. Lat. Am. Enfermagem.* 15: 508–511.

Dwan, K., Gamble, C., Williamson, P.R. et al., Reporting Bias Group. (2013). Systematic review of the empirical evidence of study publication bias and outcome reporting bias – an updated review. *PLoS One* 8 (7): e66844.

Gigerenzer, G., Wegwarth, O., and Feufel, M. (2010). Misleading communication of risk. *BMJ.* 341: c4830.

Glasziou, P., Meats, E., Heneghan, C., and Shepperd, S. (2008). What is missing from descriptions of treatment in trials and reviews? *BMJ* 336 (7659): 1472–1474.

Higashida, R.T., Furlan, A.J., Roberts, H. et al. (2003). Trial design and reporting standards for intra-arterial cerebral thrombolysis for acute ischemic stroke. *Stroke* 34 (8): e109–e137.

Hooper, R., Froud, R.J., Bremner, S.A. et al. (2013). Cascade diagrams for depicting complex interventions in randomised trials. *BMJ* 347: f6681.

Hopewell, S., Altman, D.G., Moher, D. et al. (2008). Endorsement of the CONSORT Statement by high impact factor medical journals: a survey of journal editors and journal "instructions to authors". *Trials* 9: 20.

Hopewell, S., Ravaud, P., Baron, G. et al. (2012). Effect of editors' implementation of CONSORT guidelines on the reporting of abstracts in high impact medical journals: interrupted time series analysis. *BMJ* 344: e4178.

Kirkham, J.J., Dwan, K.M., Altman, D.G. et al. (2010). The impact of outcome reporting bias in randomised controlled trials on a cohort of systematic reviews. *BMJ* 340: c365.

Melander, H., Ahlqvist-Rastad, J., Meijer, G. et al. (2003). Evidence b(i)ased medicine–selective reporting from studies sponsored by pharmaceutical industry: review of studies in new drug applications. *BMJ* 326: 1171–1173.

Moher, D., Hopewell, S., Schulz, K.F. et al. (2010). CONSORT 2010 explanation and elaboration: updated guidelines for reporting parallel group randomized trials. *BMJ* 340: c869.

Odgaard-Jensen, J., Vist, G.E., Timmer, A. et al. (2011). Randomisation to protect against selection bias in healthcare trials. *Cochrane Database Syst. Rev.* 4: MR000012.

Perera, R., Heneghan, C., and Yudkin, P. (2007). Graphical method for depicting randomised trials of complex interventions. *BMJ* 334 (7585): 127–129.

Péron, J., Pond, G.R., Gan, H.K. et al. (2012). Quality of reporting of modern randomized controlled trials in medical oncology: a systematic review. *J. Natl. Cancer Inst.* 104 (13): 982–989.

Reeve, B.B., Wyrwich, K.W., Wu, A.W. et al. (2013). ISOQOL recommends minimum standards for patient-reported outcome measures used in patient-centered outcomes and comparative effectiveness research. *Qual. Life Res.* 22 (8): 1889–1905.

Schumacher, H.C., Meyers, P.M. et al. (2010). Reporting standards for angioplasty and stent-assisted angioplasty for intracranial atherosclerosis. *J. Neurointerv. Surg.* 2 (4): 324–340.

Schulz, K.F., Altman, D.G., Moher, D., and CONSORT Group (2010). CONSORT 2010 statement: updated guidelines for reporting parallel group randomised trials. *BMJ* 340: c332.

Schumm, K. and Lam, T. (2008). Types of urethral catheters for management of short-term voiding problems in hospitalized adults. *Cochrane Database Syst. Rev.* 2: CD004013.

Sheldon, T. (2010). Dutch probiotics study is criticised for its "design, approval, and conduct". *BMJ* 340: c77.

Song, F., Parekh, S., Hooper, L. et al. (2010). Dissemination and publication of research findings: an updated review of related biases. *Health Technol. Assess.* 14 (8), iii, ix–xi, 1–193.

Tramèr, M.R. and Walder, B. (2005). Number needed to treat (or harm). *World J. Surg.* 29 (5): 576–581.

Trotti, A., Pajak, T.F., Gwede, C.K. et al. (2007). TAME: development of a new method for summarising adverse events of cancer treatment by the Radiation Therapy Oncology Group. *Lancet Oncol.* 8 (7): 613–624.

Turner, L., Shamseer, L., Altman, D.G. et al. (2012). Consolidated standards of reporting trials (CONSORT) and the completeness of reporting of randomised controlled trials (RCTs) published in medical journals. *Cochrane Database Syst. Rev.* 11: MR000030.

World Medical Association. (2013) WMA Declaration of Helsinki – Ethical Principles for Medical Research Involving Human Subjects. http://www.wma.net/en/30publications/10policies/b3. Last accessed: June 4, 2014.

59

Publishing a Case Report

S. Jane Millward-Sadler[1] and Mohammadali M. Shoja[2]

[1] Newcastle University, Manchester, United Kingdom
[2] Division of General Surgery, University of Illinois at Chicago Metropolitan Group Hospitals, Chicago, IL, USA

59.1   Introduction

Case reports are the most basic type of descriptive clinical study, and they document an unusual clinical scenario often representing the first clue in identification of a new disease or a novel treatment or formulation of a new hypothesis concerning potential disease risk factors (Hennekens and Buring 1987; Martins et al. 2005). The collection of related case reports from a short span of time is referred to as case series (Martins et al. 2005). The usefulness of case reports in clinical practice is best illustrated by the story of discovery of acquired immune deficiency syndrome (AIDS). In 1981 and 1982, a series of cases on Kaposi's sarcoma, pneumocystis pneumonia, mucosal candidiasis, and other opportunistic infections among homosexual men was reported by clinicians (Gottlieb et al. 1981; Durack 1981). Centers for Disease Control and Prevention (CDC) issued warning to the medical community on the unexpected outbreak of these uncommon diseases in young homosexual men who had no apparent reason to acquire them, which up to that time were thought to occur only in patients with severely compromised immune systems (CDC 1981, 1982; Durack 1981); this prompted rigorous investigations that ultimately led to the identification of a new blood-borne disease (now referred to as AIDS), which not only occurred among homosexual men but also among intravenous drug users and recipients of blood products (Hennekens and Buring 1987). Michael Gottlieb, who reported one of the first cases of this new disease in the *New England Journal of Medicine* in 1981, was a young assistant professor of immunology at the University of California Los Angeles (UCLA) Medical Center (Fee and Brown 2006). It is argued that writing a case report fosters reflective inquiry and advances the practitioner's clinical reasoning skills (Higgs et al. 2008).

59.2   Clinicians Are Rich Sources for Unique Medical and Surgical Cases

In medical practice, beginning during medical school training, clinicians quickly learn that no two patients are alike. Most conditions affecting patients are commonplace and

A Guide to the Scientific Career: Virtues, Communication, Research, and Academic Writing, First Edition.
Edited by Mohammadali M. Shoja, Anastasia Arynchyna, Marios Loukas, Anthony V. D'Antoni, Sandra M. Buerger, Marion Karl and R. Shane Tubbs.

are amenable to routine treatment. However, sometimes a patient has an unusual disease that requires something other than routine surgery or therapy; the patient might have a relatively common condition with a distinctive or unique presentation, or something that proves resistant to standard management. In order to determine how best to treat and manage such a patient, it is necessary to (i) discuss the case with colleagues, e.g. within the hospital (be willing to seek collaboration of your colleagues in management of unusual cases), and (ii) search the literature for similar cases.

59.2.1   Collaboration

Clinicians are unique in having both direct contact with the patient *and* full access to all the relevant history and medical notes. Jointly and severally, they and their colleagues have the knowledge needed to make sense of a medical condition and decide on an appropriate course of action.

In the most challenging and interesting cases, collaboration among colleagues with different specialities and varied professional experiences can prove invaluable. Discussion among colleagues with different viewpoints and approaches can, and often does, help to identify the best way forward. This is not only useful for managing the individual case; it can also enlighten every practitioner involved, enriching their knowledge and their future practice.

59.2.2   Using the Literature to Help with Individual Patients

When the collective knowledge and wisdom of a group of practitioners proves insufficient, it becomes necessary to search the literature for guidance to help the patient. Someone, somewhere, is likely to have encountered a similar case and to have published an account of the history, presentation, treatment, and outcome. These publications, normally in the form of case reports, constitute an essential resource when you are faced with uncertainty and new dilemmas. Even when a collaborative effort among colleagues identifies an approach to treatment, a search of the medical literature is necessary, as it can help to confirm, modify, or expand this approach.

59.3   The Importance of Publishing Case Reports

Just as you and your colleagues can obtain valuable information from the literature, so other clinicians around the world can benefit from reading about your experience. When you have managed a genuinely challenging case – an unusual condition requiring nonstandard or innovative treatment, a more common condition with a distinctive presentation, or one that has proved resistant to the usual therapeutic approaches – you are likely to help your peers if you write it up for publication.

The most useful case reports describe new treatments, unusual presentations, newly identified problems such as emerging diseases, or the side effects of current treatments. If, during a case you have managed, you have identified something novel and significant in one or more of these areas, you should consider sharing your experience with the medical community. Case reports of this nature are not only a valuable resource for informing the treatment of other patients both nationally and globally, but can also provide an inspiration for larger-scale and more general studies.

59.4   When Should a Case Report Be Written?

As a practicing physician or surgeon, you might, together with one or more colleagues, treat a case that seems sufficiently interesting to merit communication to the international medical community. A case can be "sufficiently interesting" for several reasons. For example:

- The presentation is distinctive: The case is very rare in terms of symptoms, history, or the demographic characteristics of the patient. If there are no reports of similar presentations in the literature, it could be unique.
- The treatment given is novel or controversial, e.g. it is a novel surgical technique.
- The progression of the patient following treatment is in some way unusual, or the outcome is unexpected.

Publishing interesting and informative case reports can help one's career. However, personal need or wish to publish in the interests of career enhancement (yours or a colleague's), or competition with others, is not sufficient in itself to make a case worthy of publication.

59.5   If You Publish Your Case Report, Who Will Be Interested in Reading It?

If you decide that the case is "sufficiently interesting" to publish, you need to identify for whom it is interesting and for what reasons. This also helps you to determine which of the many available journals would be most suitable. If the case is extremely rare or unique, then, by definition, your colleagues around the world are unlikely to meet a similar one. This could make it academically interesting, and should not deter you from publishing if you honestly believe that other surgeons or physicians will be intrigued, but it should give you pause for thought. Would a letter to the editor be more appropriate than a full case report? This said, many diseases and presentations that have become familiar to the medical community began somewhere and were noticed by someone first. This is a decision that you will need to consider carefully, probably in consultation with your immediate colleagues.

However, if the case is of a kind likely to be familiar to others, but the progression or outcome is unusual, or if the treatment is novel or controversial, then you could have better grounds for publishing a fully-fledged case report than you would if the case were a real rarity. Depending on how the patient recovered, or failed to recover, you can encourage your peers to follow your example, or identify pitfalls and drawbacks that they could avoid with other patients.

Never be afraid to report a failure if you believe it to be instructive. No medical practitioner can be 100% successful throughout his or her career. If your patient failed to recover, or died, you will probably be reluctant to report the case to the international community. Nevertheless, your experience could offer vital information about treatment that could help to prevent similar instances of morbidity or mortality, so you might consider it right to publish your findings.

59.6   Do You Have Sufficient Material for a Case Report?

Having considered the foregoing points, a good criterion on which to base your answer to the question "Do I have enough material for a case report?" is to search the literature for related cases and for studies (clinical or scientific) relevant to the pathogenesis and/or treatment of the patient's condition. Does your case fit comfortably into a discussion of this literature? Does it throw any light on unresolved controversies or unanswered questions? Does it challenge received wisdom in a constructive way? Remember that in the Discussion section of your report you will be expected to relate your case to previous publications and clinical and practical issues relevant to it.

If your answer is yes, and your colleagues and potential co-authors agree, then you have good reason for writing a case report.

If the answer is no, or if you and your colleagues are unsure, it might be better to write a letter to the editor of an appropriate journal rather than submit a fully-fledged case report. Should the editor think you have been too cautious, the editor might suggest that you write a full report after all.

Table 59.1 is a self-evaluation checklist that authors can consider in drafting a case report or in planning to do so.

59.7   Choosing the Right Journal

59.7.1   Traditional or Open Access?

Irrespective of whether you have decided to submit a full case report or merely write a letter to the editor, you need to choose a suitable journal. Many traditional (print-copy) journals have case study sections and have fairly high impact factors, which make them attractive. There are also print-copy journals that specialize only in case studies. The drawback with most traditional outlets is that the time that elapses between submission and publication can be many months. If you consider it important to report your case quickly, this could be significant. Another consideration is that, although print journals now have online versions as well, these tend to require a subscription for the content to be viewed, so they are not freely accessible to everyone.

The alternative to a traditional journal is the range of open-access (which are often purely online) journals, which tend to have lower impact factors but promise much more rapid publication. These are also freely accessible, so they reach a much wider audience. If you think it urgent to share your case with the rest of the world, you might prefer to consider this option. However, open-access journals often require the authors to pay a fee for publication, known as the article processing charge, although letters to the editor are often published free (for more information on open-access journals, see Chapter 41). Most traditional journals levy no fee for publication, since their funding comes from readership subscriptions, but there are exceptions.

59.7.2   Journal Scope and Readership

Whichever type of journal you choose – traditional or open-access – it is obviously sensible to choose a journal aimed at readers who are likely to be interested in your

Table 59.1 Case report self-evaluation checklist.

Section	Description	Present
Authorship	All authors meet the International Committee of Medical Journal Editor (ICMJE) criteria for authorship.	
	Authors are listed in the order of contribution to the paper.	
	A reasonable number of authors are listed.	
Patient privacy	All identifying information has been removed from case report materials.	
	Consent from the patient to publish the case and/or approval from a privacy officer/Institutional Review Board (IRB) has been obtained.	
Title	The title is an accurate, succinct description of the case.	
Abstract	The abstract is ≤250 words in length.	
	The abstract is written in a structured format.	
	The objective/purpose is clearly stated.	
	The most important parts of patient management are highlighted in the case report (methods) subsection.	
	Key outcomes are presented in the results subsection.	
	The discussion summarizes what the case contributes to the literature and states the overall conclusion learned from the study.	
	Key indexing terms from PubMed medical subheadings are provided.	
Introduction	The purpose is clearly stated.	
	The health problem and its significance are clearly stated (e.g, prevalence, incidence, morbidity, financial, and social costs).	
	Definitions for pertinent terms or concepts are provided.	
	Literature on this problem was reviewed in relation to diagnosis and treatment.	
	The importance of the study or how it contributes to the literature is related.	
Case report	The case is described in a concise and clear manner.	
	The case is presented in chronological order.	
	Pertinent patient characteristics are described.	
	Salient aspects of the patient's health history are clearly described.	
	Positive results and significant negative results pertinent to the examination are concisely presented.	
	Appropriate outcome measures were utilized for clinical measurement.	
	Novel diagnostic or assessment strategies are fully described.	
	References to support the validity/reliability of novel diagnostic tests are present.	
	All unusual terms and patient variables are operationally defined.	

(Continued)

Table 59.1 (Continued)

Section	Description	Present
	A diagnosis is presented.	
	Treatment procedures are clearly and concisely presented.	
	Important outcome measures have corresponding data reported before/after care.	
Discussion	The case is compared to what is known in the literature.	
	Differential diagnoses are discussed.	
	A rationale for the management of the patient is provided.	
	Interpretations of the results are offered by the author(s).	
	The author(s) proposes a mechanism for the observed changes.	
	Limitations of the study are offered.	
	Suggestions for future research are made.	
Conclusion	The conclusion relates to the purpose of the paper.	
	New information learned from the case is summarized.	
	The conclusion is approximately one paragraph in length.	
Acknowledgments	Written consent from those acknowledged is obtained.	
References	The author(s) provides appropriate and adequate references.	
	References are prepared as per the journal instructions for authors.	
Tables	Tables present data using inter-relating horizontal rows and vertical columns.	
	Tables have a corresponding title.	
	Tables are self-contained, needing no text to support them.	
	Permission to reprint a previously published table is obtained from the publisher.	
Figures	Figures are self-contained, needing no text to support them.	
	Permission to reprint a previously published figure is obtained from the publisher.	
	Written permission to publish photos of models or identifiable people is obtained.	
	Figures are prepared according to the journal's instructions to authors.	
General	The case is objective and devoid of unsubstantiated claims.	
	The case is clearly presented.	
	The case is prepared in accordance with the journal's instructions for authors.	
	The length of the case report is 1000–2500 words or less than the maximum.	

Source: From Green and Johnson, 2006; reproduced with written permission.

case. Some journals focus on a narrow range of topics, while others cover much wider scopes. As a general rule, the wider-scope journals attract more readers and therefore more contributors, so there is more competition for space and the acceptance rates are correspondingly low. If such a journal publishes your report, more people will see it, but there is a higher chance that your submission will be rejected. The editors will accept it only if they consider it to be particularly interesting and important.

If you choose to submit your report to a journal with a narrower focus, you will have a greater chance of acceptance. However, the total readership will be lower, and the impact factor of the journal is also likely to be lower than wider-scope journals, although the percentage of readers who are specifically interested in the type of case you describe is likely to be higher.

If you are unsure about the suitability of your case for a particular journal, whether online or print, the areas of interest are usually stated in the authors guidelines, which will also give guidance on the type of articles likely to be accepted. This tends to be a good place to start when you are trying to decide where to publish your article and what type of article to write.

In summary, you need to:

- Decide if your case is of interest to others, and if so who.
- Identify which type of journal is likely to be most interested in your type of case.
- Determine whether you have enough material for a full case report or would be better with a letter to the editor.
- Investigate the literature to determine what has been published about your particular type of case, and work out how your case fits into and adds to current knowledge.
- Write and submit your case report or letter to the editor.

59.8 Ethical Considerations: Patient and Institution

Before dealing with specific ethical issues concerning a case report, we shall provide a quick review of the medical ethics in general and its application in one of most popular clinical research methods (i.e. clinical trials). Then, we will build on ethical considerations as they pertain to publishing case reports.

59.8.1 Principles

A simple definition of *ethics* is "a set of rules or principles underpinning moral behavior." *Medical ethics* denotes a set of rules determining right and wrong behavior in respect of patients, colleagues, the institution in which you work, and the wider community. There are three main principles:

1) You should always display respect for others – patients, colleagues, people you encounter outside work, and (if your research involves them) experimental animals.
2) You should always seek to do good (or the best that you can) and to avoid causing harm. This "principle of beneficence" has been fundamental to medical practice since the time of Hippocrates, and it applies to medical research, too.
3) You should try to ensure that both the burden and the benefits of research are distributed fairly, in the interests of justice.

59.8.2   Why Do We Need Ethics?

The need for ethics in medicine is obvious, but it is worth making the following major reasons explicit:

- We need ethics to protect the patient's dignity and ensure it is maintained at all times during treatment and throughout any clinical trial in which a patient participates. The patient's rights, safety, and well-being must be protected. This means that risks, fear, pain, and distress must always be minimized.
- We must prevent exploitation of patients and clinicians by companies, institutions, or individuals.

59.8.3   Ethical Considerations in Clinical Research

59.8.3.1   Patients
- Patient anonymity must be maintained at all stages, particularly when publication of findings is considered.
- Written informed consent has to be obtained from the patient whenever possible. If it is not possible, e.g. if the patient is dead, too mentally incapacitated, or too young to give informed consent, then consent must be obtained from the next-of-kin.
- Assent is also required from children old enough to form their own opinions but not yet legally of age to give informed consent.
- The patient should be made aware of his or her right to withdraw consent at any time, including during the trial or after a manuscript is written and about to be submitted for publication.
- To ensure that these rules are followed, the patient must be enabled to understand the objectives and methods of the trial and the material to be included in a publication.

59.8.3.2   The Institution
- The trial must receive local ethical approval.
- The design of the trial must be considered and approved by the institution's ethics board.
- The trial must conform to the law of the country in which it is to be conducted.

59.8.3.3   Experimental Animals
- All animals used in experiments must be treated humanely.
- No animals should be subjected to experiments without sound medical and scientific reasons.
- Licenses for animal experimentation must be obtained by the institution and by the researchers carrying out the experiments.
- Written approval for animal experiments must be obtained from all relevant bodies.
  With respect to patients, institutions, and experimental animals, the researchers must be prepared to provide written proof of approvals and consent if their work is audited, or if the publishing journal requests such evidence. It is essential that careful records are maintained throughout research and for a predetermined period after the study has ended to ensure that they are available should they be required.

59.8.4 Ethical Issues in Publishing Case Reports

When it comes to managing a new or unusual case and publishing a case report, all the above principles and rules apply. From ethical standpoint, one can look at a case report as a clinical trial with the sample size of one! In most cases, if not all, a written, informed consent should be obtained from the patient in order to publish a case report. The patient should be fully informed of what portion of his health information, including any pictures or radiological images, will be included in the report and of the methods of de-identification.

59.9 From Case Studies to Large-Scale Clinical Studies

Although individual case studies are valuable, they have obvious limits. To establish the prevalence of a condition, the success of a particular treatment regimen, and the incidence of side-effects of a treatment, etc., larger-scale studies are needed. Such studies fall into two major categories, *observational* and *treatment* (Table 59.2).

Table 59.2 The main types of observational and treatment studies.

Observational studies	Cohort studies	Either prospective or retrospective. A *prospective* cohort study follows a group of otherwise similar individuals (the cohort) who differ in regard to the factor or factors being studied, and determines the relationship of these factors to a stated outcome. In a *retrospective* cohort study, the medical records or histories of two (or more) cohorts differing in a certain characteristic are compared with respect to a particular outcome.
	Case-control studies	Used by epidemiologists to identify factors that might contribute to a disease: two groups that differ in outcome are compared on the basis of one or more possible causal factor(s), thus providing odds ratios for the factors studied.
	Cross-sectional studies	Examine the incidence (prevalence) of a condition in a whole population at a particular time point. Such studies can be used to support inferences about cause and effect, but they determine absolute and relative risks, not odds ratios (as case-control studies do).
	Longitudinal studies	Differ from cross-sectional studies because they examine the population at several time points instead of just one particular time, and therefore provide evidence of changes in the prevalence of a condition.
Treatment studies	Randomized controlled trials	Considered the gold standard for treatment study design.
	Adaptive clinical trials	Evaluate patients' reactions to a drug from the beginning of and through the trial, and modify the trial accordingly.
	Non-randomized (or quasi-experimental) trials	Examine the effect of a treatment on the target population without introducing the random-selection element characteristic of randomized controlled trials.

References

Centers for Disease Control (CDC) (1982). A cluster of Kaposi's sarcoma and *Pneumocystis carinii* pneumonia among homosexual male residents of Los Angeles and Orange Counties, California. *Morbidity and Mortality Weekly Report* 31 (23): 305–307.

Centers for Disease Control (CDC) (1981). Kaposi's sarcoma and pneumocystis pneumonia among homosexual men – New York City and California. *Morbidity and Mortality Weekly Report* 30 (25): 305–308.

Durack, D.T. (1981). Opportunistic infections and Kaposi's sarcoma in homosexual men. *New England Journal of Medicine* 305 (24): 1465–1467.

Fee, E., Brown, T.M., and Michael, S. (2006). Gottlieb and the identification of AIDS. *American Journal of Public Health* 96 (6): 982–983.

Gottlieb, M.S., Schroff, R., Schanker, H.M. et al. (1981). *Pneumocystis carinii* pneumonia and mucosal candidiasis in previously healthy homosexual men: evidence of a new acquired cellular immunodeficiency. *New England Journal of Medicine* 305 (24): 1425–1431.

Green, B.N. and Johnson, C.D. (2006). How to write a case report for publication. *Journal of Chiropractic Medicine* 5 (2): 72–82.

Hennekens, C.H. and Buring, J.E. (1987). *Epidemiology in Medicine*. Philadelphia: Lippincott Williams & Wilkins.

Higgs, J., Jones, M.A., Loftus, S. et al. (2008). *Clinical Reasoning in the Health Professions*, 3e. Amsterdam: Elsevier Health Sciences.

Martins, S., Zin, A., and Zin, W. (2005). Study design. In: *Anaesthesia, Pain, Intensive Care and Emergency Medicine* (ed. A. Gullo), 719–748. Trieste: Springer.

60

Writing Editorials

Naomi Andall, Bharti Bhusnurmath, Shivayogi Bhusnurmath and Marios Loukas

Department of Anatomical Sciences, St. George's University School of Medicine, Grenada, West Indies

60.1   What Is an Editorial?

An editorial is a critical piece that analyzes and reviews an article in a journal. It methodically examines the important issues discussed in that journal article with varied perspectives to appraise the validity of the article and its clinical relevance. It is important that it remains objective by representing both substantiating and opposing points equally. Furthermore, an editorial should not be written to reflect the author's own biases but be constructed with reputable supporting evidence (Day and Gastel 2012; Fontanarosa 2014; Morgan 1985a,b; Peh and Ng 2010; Pond 1974).

60.2   Who Can Write an Editorial?

The author of an editorial is the editor or editor in chief of the publication. Editorials can also be written by invited academics who are content experts. They use their expertise in the field to critically assess a journal article or paper. Some journals select viewers of a journal to contribute to an editorial piece (Day and Gastel 2012; Fontanarosa 2014; Morgan 1985b; Peh and Ng 2010). Independent academics also submit their editorial manuscripts to journals. Their editorials are put through a review process before it is accepted. Once it is determined that the writer has followed all the guidelines outlined for submitting an editorial, the editor appoints three reviewers. These are experts in the field who make comments and suggestions to improve the manuscript. The publisher can then provide the writer with these revisions to allow for resubmission. However, this can also be the stage where the manuscript is rejected (Day and Gastel 2012; Peh and Ng 2010; Hames 2008).

60.3   Contents of a Typical Editorial

The structure of an editorial is more concise compared to that of a research paper. The word limit can range between 450 and 1000 words with the final manuscript being no more than two pages (Fontanarosa 2014; Morgan 1985a,b; Peh and Ng 2010). The

A Guide to the Scientific Career: Virtues, Communication, Research, and Academic Writing, First Edition.
Edited by Mohammadali M. Shoja, Anastasia Arynchyna, Marios Loukas, Anthony V. D'Antoni, Sandra M. Buerger, Marion Karl and R. Shane Tubbs.

writer has to keep in mind that the editorial should be informative and present relevant, evidenced-based perspectives on a significant issue.

The following structure can serve as a guide for the contents of an editorial.

60.3.1 Title and Introduction

The title should be short, indicative of the topic, and have an immediate effect on the reader. It is important that the title has words and phrases that will be commonly searched for by other academics (Fontanarosa 2014; Peh and Ng 2010).

The opening paragraph is the official statement of the problem, question, or argument (Fontanarosa 2014; Peh and Ng 2010). It should give a brief background on the topics that will be discussed later on in the editorial (Fontanarosa 2014; Morgan 1985b; Peh and Ng 2010).

60.3.2 Discussion and Analysis

The Discussion section of an editorial should include a brief summary of the journal article. This summary should include the title, aim, author, methodology, results, discussion, and conclusion sections of the journal article being reviewed so that the reader does not necessarily have to refer to the journal article before reading the editorial. This spares the reader valuable time and allows for a good understanding of the study or article being addressed (Fontanarosa 2014).

The Discussion and Analysis sections should reinforce and develop the views presented in the editorial by providing an analysis of the supporting and contradicting evidence of the article's findings (Fontanarosa 2014; Morgan 1985b; Peh and Ng 2010). In this section, the writer can also consider the strengths and limitations of the research design and methodology, statistical significance, interpretation, and clinical relevance of the journal article that is being reviewed by comparing the article to other similar studies that had been conducted (Day and Gastel 2012; Fontanarosa 2014).

60.3.3 The Final Message

The conclusion of an editorial should suggest answers to the problem raised in the beginning of the editorial or provide future directions (Day and Gastel 2012; Fontanarosa 2014; Morgan 1985b; Peh and Ng 2010). An editorial is also expected to include all contributing authors' names, the contact information of the principal author, and organizational and institutional affiliations (Peh and Ng 2010). Any conflicts of interest should also be mentioned if present (Fontanarosa 2014; Peh and Ng 2010).

60.4 Key Point for Writing Editorials

As previously stated, an editorial should be fair and balanced. The evidence presented needs to include material that represents all relevant perspectives on the topic discussed, even in cases where the author might have his own standpoint on the issue. The editorial must also be appropriately cited, as with any other publication, and written in a way that is easily read and understood by its readers (Fontanarosa 2014; Peh and Ng 2010; Hames 2008).

Writers should refer to the journal's Instructions to Authors to get an understanding of the rules pertaining to writing an editorial in that publication (Day and Gastel 2012; Peh and Ng 2010; Hames 2008). It contains important guidance and information about the word limit, the number of images or graphs that are allowed, the amount of references required, fees, and other specific submission details (Day and Gastel 2012). After the editorial has been edited and reviewed, it will then be ready for publication. Editorials should be submitted as soon as possible after the article has been released because there could be particular rigid deadlines held by the journal publisher (Day and Gastel 2012; Fontanarosa 2014; Morgan 1985a).

Academics tend to shy away from writing editorials under the impression that these works are not regarded as highly as original research articles. In reality, editorial writing should be encouraged because it enhances the experience of the readers by providing a critical analysis of the current perspectives in clinical and academic medicine. Editorials also provide a meaningful platform for important issues to be discussed (Fontanarosa 2014; Hoey et al. 2002; Morgan 1985b; Peh and Ng 2010).

References

Day, R. and Gastel, B. (2012). *How to Write and Publish a Scientific Paper*, 154. Philadelphia: ISI Press.

Fontanarosa, P. (2014). Editorial matters. Guidelines for writing effective editorials. *JAMA* 311: 2179–2180.

Hames, I. (2008). *Peer Review and Manuscript Management in Scientific Journals: Guidelines for Good Practice*, 25–36. Wiley.

Hoey, J. and Todkill, A. (2002). An editorial on editorials. *CMAJ: Can. Med. Assoc. J.* 167: 1006–1007.

Morgan, P. (1985a). How to write a letter to the editor that the editor will want to publish. *Can. Med. Assoc. J.* 132: 1344.

Morgan, P. (1985b). Scientific editorials- a precious and scarce element in medical journals. *Can. Med. Assoc. J.* 132: 315.

Peh, W.C.G. and Ng, K.H. (2010). Writing an editorial. *Singapore Med. J.* 51: 612. Retrieved 04 June 2014 from http://smj.sma.org.sg/5108/5108emw1.pdf.

Pond, M. (1974). Editorials: a new editorial philosophy emerges. *Am. J. Public Health* 64: 97.

61

Writing a Letter to the Editor

Haley J. Moon and Joel A. Vilensky

Department of Anatomy and Cell Biology, Indiana University School of Medicine, Fort Wayne, IN, USA

61.1 What Is a Letter to the Editor?

A letter to the editor is a document that serves as a dynamic communication device inviting authors and readers to provide commentary on a previously published article. It is generally a short, written communication that is subjective in nature, raising concern or articulating your specific point of view on part or all of a published piece. In academic journals, letters to the editor are sometimes considered open, post-publication reviews of an original paper. Some journals publish the short report of a new study or report of a unique case(s) as a letter to the editor. That allows for the efficient use of the journal space and speedy publication of the article. The Nobel Prize-winning article published in *Nature* in April 1953, in which James Watson and Francis Crick reported the double-helical model for the DNA structure, was a letter to the editor conveying a groundbreaking discovery. The significance of the letter to the editor, thus, should not be underestimated.

In this chapter, we will only examine the class of letters that serve as commentary on a previously published article. There are two types of commentary letters to the editor: those delivered in a supportive tone or those criticizing the work in question. Besides offering a medium for communication, a letter to the editor often is viewed by an invested audience as an invaluable resource because it contributes to their understanding of the published article. Supportive letters often help to strengthen the original article. A well-written letter to the editor might even benefit an otherwise obscure article, garnering attention for future research. Composing a letter to the editor is a valuable exercise for inexperienced authors. It provides them confidence and precious practice in the publishing process before tackling an original article for a peer-reviewed journal. Letters to the editor also help create visibility and can open up new opportunities within one's own professional network.

61.2 How to Approach Writing a Letter to the Editor

When writing a letter to the editor, it is important to establish credibility by providing valid references in support of your observations. A letter to the editor should offer

A Guide to the Scientific Career: Virtues, Communication, Research, and Academic Writing, First Edition.
Edited by Mohammadali M. Shoja, Anastasia Arynchyna, Marios Loukas, Anthony V. D'Antoni, Sandra M. Buerger, Marion Karl and R. Shane Tubbs.
© 2020 John Wiley & Sons, Inc. Published 2020 by John Wiley & Sons, Inc.

helpful suggestions, referring readers to other research and publications that defend your point of view or that shed new light on the topic.

Each journal has its own specific set of instructions for submitting letters to the editor, which should be followed in order to be considered for publication. Beyond the journal's guidelines, your chance of publication is greatly increased by adhering to a few key points that ensure a successful letter:

- Commit to the tone of the letter; decide whether you are criticizing the content of the article or to offer support or praise.
- Articulate your message clearly and to the point. Most journals have a word limit, and digressions might hurt your chances at publication.
- Be interesting and witty. While you want your message to be focused, your letter should also be memorable and aim to increase the pleasure of the reader.
- Maintain a professional approach in your writing, even when providing critique. While letters that are critical of an article – or offer a controversial perspective – are often more interesting than those that praise it, remember that derogatory comments likely will result in rejection. You can disagree and give constructive criticism to the article's author, but your letter should always remain courteous and tasteful.
- Ask your peers, colleagues, teachers, and mentors to read over your letter before submission. Editing your own work is difficult. It is advisable to step away from the letter for a day or two upon completion so that you can look at it with fresh eyes. This can be valuable before sharing it with others to proofread.
- Take care to comply with the journal's instruction on correspondence practices and timelines. Most journals will only accept letters for consideration within three to six months after the publication of the article in question.

61.3 What Happens When Your Letter Has Been Accepted?

After your letter to the editor has been accepted by the editor of the publication, your responsibility as a content contributor is not quite over. The editor will most likely ask the author of the original article to draft a reply to your letter. Many publications like to print both the letter to the editor along with the response at the same time in order to create a dialogue. Because of that, your letter might not necessarily be published in the subsequent issue.

Typically, accepted letters to the editor and their responses will be reviewed for grammar, style, and length before publication. Depending on the journal, the editing process might be minimal for such writing. Published letters to the editor should preserve the message and tone delivered in the original submission.

Following acceptance and a short round of editing, your letter will be sent to the publisher. A proof of your letter (how it will look in print) will be sent to you so that you may review it for any errors and answer any questions put forth by the publishers. During the proofreading stage, it is your responsibility to clear up any typos and respond to the inquiries. Most journals require a timely response, so it is critical to reply by the requested deadline or the letter might be published "as is." Remember that after the letter has been published, it cannot be revised or altered.

Writing letters to the editor is a way to expand your professional network as well as it can increase exposure. Good letter writing is an important skill that will enhance your career, providing you with valuable experience in the publication process.

Further Reading

Cappell, M.S. (2010). Is lumping peer-reviewed case reports together with non-peer-reviewed comments for publication as letters to the editor appropriate? *American Journal of Gastroenterology* 105: 1901–1902.

Dotson, B. (2013). New practitioners forum: writing a letter to the editor. *American Journal of Health-System Pharmacy* 70: 96–97.

Hacker, D. (2000). *A Writer's Reference*, 3e. Boston, MA: Bedford Books of St. Marten's Press.

Lemery, L. (1993). Develop your career through writing. Professional Perspective Column. *Laboratory Medicine* 24: 703–704.

Lemery, L. (1997). *How to Write for A Specific Publication*, vol. 7. ASCP-AMS Leadership Line.

Peh, W.C. and Ng, K.H. (2010). Writing a letter to the editor. *Singapore Medical Journal* 51: 532–535.

Watson, J.D. and Crick, F.H. (1953). Molecular structure of nucleic acids; a structure for deoxyribose nucleic acid. *Nature* 171 (4356): 737–738.

62

Writing a Book Review

James Hartley

School of Psychology, Keele University, Staffordshire, UK

62.1   Introduction

Suppose you have been asked by the editor of a journal to write a review of a new book on a topic of your concern. What do you do?

Assuming that you agree to carry out the task, there are some practical details to consider. First, you need to consider the journal's style and its handling of book reviews – how long are the reviews, in general? Are the reviews meant to be read by the general public or experts in the field? Book reviews generally contain:

- A statement of what the book is about.
- An indication of how the book in question fits in with a larger, broader context.
- An evaluation of the strengths and weaknesses of the arguments.
- Possibly some comments on the quality of the printing, the use of illustrations, the price, and a recommendation (should people buy this book, or would it be better as a library copy?).

62.2   Strategy for Crafting Book Reviews

There certainly are several published guidelines discussing the approaches and strategies for writing book reviews. One concern is the nature of the text to be reviewed. Is it a student textbook, a practical manual, a set of readings, or someone's thesis? Each type of book will require a different approach. But there will be commonalities. Table 62.1 shows the strategy recommended by Lee et al. (2010) for crafting book reviews.

The process of writing a book review can be divided into three stages (Hartley 2010):

Stage 1. Writing preliminary notes – jotting down key points and ideas that come to mind while reading the book.

Stage 2. Composing and sequencing the preliminary notes into an initial rough draft.

Stage 3. Re-sequencing, editing and polishing the text several times in order to produce a final version. This stage involves multiple revisions of the review, re-sequencing the text and changing the order of arguments, and finally polishing particular words and phrases.

A Guide to the Scientific Career: Virtues, Communication, Research, and Academic Writing, First Edition.
Edited by Mohammadali M. Shoja, Anastasia Arynchyna, Marios Loukas, Anthony V. D'Antoni, Sandra M. Buerger, Marion Karl and R. Shane Tubbs.

Table 62.1 Strategy for crafting book reviews (based on Lee et al. 2010).

Initial segment	Capture the reader's attention.
	Outline the aims and scope.
Main segment	Describe the central ideas of the book.
	Explore key arguments of the review.
	Discuss strengths and weaknesses of the book.
	Highlight the book's uniqueness.
	Note the book's contribution to the field.
Concluding segment	Balance the book's achievements and weaknesses in order to support your final assessment.

62.3   Negative and Positive Book Reviews

Most book reviewers claim that the ways in which they write book reviews vary according to the book in question (Hartley 2010). In one study, 51 scientists ranked the following items in a book review in terms of their importance (cf. Hartley 2006) (Table 62.2). Almost two-thirds of the respondents recalled reading dreadful reviews. Some of the things that they said about such reviews were that they were pointless, uninformative and boring, a mere listing of the contents, pretentious, unkind, and careless, personally abusive about the author's credentials or written to cherish the reviewer's ego.

Some book reviews provide devastating critiques of the original text. Negative reviews are entertaining for the reader, but display possibly more about the reviewer than the quality of the book in question. Some of the phrases that are jotted down when reviewing a text actually have hidden messages (Hartley 2010). The reviewers often use phrases

Table 62.2 Items important in book review (1 is the most important; 2 and 3 are of intermediate and low importance, respectively).

Rank	Item
1	A straightforward overview of what the book is about
2	A critique of the argument of the book
	An evaluation of the book's academic credibility
	A comparison with other books in the field
	An assessment of the book's usefulness for its intended audience
	Information about the intended audience
	Information about the number of pages
	Information about the price
3	A substantial, as opposed to a brief, discussion
	An attempt to position the book in its historical context
	A well-known person as the author of the review
	A chapter-by-chapter account

Table 62.3 The statements and phrases with hidden connotations.

Statement or phrase	Connotation
This blockbuster of a text ...	*This is an enormous book.*
I hoped that I would learn from this text.	*I did not learn this from this text.*
One might expect to find	*One did not find . . . .*
And therein lies the rub.	*This is not what it seems.*
A grand recycling is going on here.	*This has all been published before.*
This is a surprising book.	*This is much better than I expected.*
This is a mixed bag.	*Not much in this but one or two are chapters worth thinking about.*
This is a useful book for the library.	*Not worth having at home or reading.*
For the most part, this is a thorough, lucid, and well-argued book, but a few weaknesses can be noted.	*That's completed the praise bit; now let's get down to the criticisms.*
The discussion is somewhat abstruse.	*I could not understand much of discussion.*
In my view, more scholarly references would be better for the readers of this text than the par-boiled information referred to on websites.	*This is a lightweight text and/or my scholarship is superior to the author's.*
The author has presented opposing views fairly, although instances of bias are detectable by the omission of some critical references.	*He has left out a key paper on . . . .*

and sentences that mean something else (Hartley 2010). Some examples are presented in Table 62.3.

It is more entertaining to report such hidden negative comments than any positive ones. Most book reviews tend to be positive rather than negative (Hartley 2006). In the 1970s, there was a tendency in psychology for men to review books by men more favorably than books by women (Hartley 2006). Praise and criticism were offered equally in the reviews of books in the arts and social sciences, but book reviews in the sciences contained twice as much praise as criticism (Hartley 2006).

62.4   Editorial Guidance

Different journals and editors provide different, but helpful, advice on what should or should not be written in book reviews. Some journals provide lengthy notes, for instance, providing potential reviewers with information on the aims and scope of the journal, together with a paragraph on what a review might contain (Table 62.4). Some points are specific to the journals concerned, some more general. Some of them provide style requirements. Other journals go further to make special recommendations of what sort of arguments and subjects should be included in the review. There is a possibility that a reviewer, having examined a book, might not wish to review it. Those books should, therefore, be returned. Some journals comment on the reviewer's ethical obligations.

Table 62.4 Perspectives of select journals on book reviews.

Example journal	Truncated guideline
Journal of the Medical Library Association	"Reviews should contain a brief overview of the scope and content (of the book being reviewed) so that readers can determine the book's interest to them. Reviewing each chapter of a book is not necessary. For a research or historical work, please comment on its significance in relation to the focus area as well as to the field as a whole. For an applied or descriptive work, be sure to comment on its usefulness. In both cases, compare the book with similar publications in its area and indicate its potential audiences, where relevant."
Law and Politics Book Review	"The editor encourages reviewers to devote special attention to the political assumptions and discussions in the book under review .... Professional ethics require that you do not review a book when an overriding sense of personal obligation, competition or enmity exists."
PsyCritiques	"We are seeking reviews that are incisive... integrative... balanced... and provocative."
American Journal of Physics	"It is not required that every review contain at least one negative remark. Selective detail is refreshing. Encyclopedic detail – as in a chapter-by-chapter outline – is rarely called for."

Others also require the book reviewers to sign certain disclaimers (e.g. that they have not been in dispute with the book's author) before their review can be published.

62.4.1 Unsolicited Book Reviews

While some editors do not consider unsolicited book reviews, there are quite a few who accept or even strongly encourage unsolicited book reviews, provided that they meet the required standards. A few journals have developed a database of potential book reviewers where you can register to be informed and invited to review future books in the areas of your interest and expertise.

62.5 Checklist for Writing a Book Review

As noted above, a range of recommendations and several checklists are available for reviewers to consider when writing their reviews. Table 62.5 is a checklist that can be used when writing a book review or in planning to do so.

Table 62.5 Checklist for writing a book review.

Item
An early paragraph saying what the book is about and putting it in context
Information about the intended audiences, and how well the book meets their needs
A critique of the argument of the book
Any supporting references (if allowed)
Remarks on the strengths and limitations of the book
A note on how well the text is supported by tables/diagrams/illustrations/web references
A note on the format, length and price (or value for money)
Whether an e-version of the text is available
Book information (accurate details of the authors' or editors' names, title of the publication, edition, date of publication, publisher and place of publication, ISBN number, format (hardback, paperback, e-book), number of pages, and price (these details may be provided by the publisher)

References

Hartley, J. (2006). Reading and writing book reviews across the disciplines. *Journal of the American Society for Information Science and Technology* 57 (9): 1194–1207.

Hartley, J. (2010). The anatomy of a book review. *Journal of Technical Writing and Communication* 40 (4): 473–487.

Lee, A.D., Green, B.N., Johnson, C.D. et al. (2010). How to write a scholarly book review for publication in a peer-reviewed journal: a review of the literature. *The Journal of Chiropractic Education* 24 (1): 57–69.

63

Use of Illustrations and Figures to Enhance Scientific Presentations and Publications

Marion Karl

Lure Animations, Reno, NV, USA

63.1 Introduction

Medicine and biology are dynamic and beautiful sciences that require visualization to fully understand the relationship between form and function. Thoughtful integration of illustrations and figures in your paper can elegantly communicate your research while concomitantly articulating and elevating your message. This chapter discusses how to evaluate medical illustrations and effectively incorporate artwork into your research papers.

63.2 What Is Medical Illustration?

The term illustration is defined as a genre of image-based representations that describe the scientific subject. Illustrations are commonly 2D drawings, which can be delicately rendered acutely generalized. They can stand alone or be diagrammatic in nature, meaning that the representations are used to outline information in steps or chronicle changes in structure or function.

The practical use of illustrations in medical publishing is a necessary and useful device. It is a means of exploring structures that are difficult to distinguish in gross specimens and can be used to describe biologic mechanisms invisible to the eye. The practice of medical illustration not only captures the aesthetic character of biologic form, but also showcases physical features that significantly enhance medical instruction. Orientation, spatial relationships, and unique structural and functional characteristics are best documented visually; illustrations provide the editorial capacity to enlighten readers about the subject in a clear and instructive context.

Plates from an anatomy atlas or images of a gross specimen do not always supply the reader with the entire essence of the dissection, but a medical illustration that communicates depth, structural relationships, and specific anatomical details might be published side-by-side with the corresponding research to help the audience reach fuller comprehension of the subject. Medical illustrations can provide context and illuminate alternative views or cross-sections. When looking to incorporate illustrations into your

A Guide to the Scientific Career: Virtues, Communication, Research, and Academic Writing, First Edition.
Edited by Mohammadali M. Shoja, Anastasia Arynchyna, Marios Loukas, Anthony V. D'Antoni, Sandra M. Buerger, Marion Karl and R. Shane Tubbs.

paper, seek artwork created by a certified medical illustrator. People qualified in the field of medical illustration should demonstrate a professional knowledge of anatomy and medicine, and have a portfolio of examples to substantiate their experience.

63.3   Types of Illustrations

Let us discuss the relevant types of illustrations to consider for publication. Detailed illustrations can be used to meticulously describe complex anatomy or visualize subtle anomalies that are elusive in medical imaging or photographs. Simple line drawings are often appropriate to focus attention and help the audience understand basic relationships. Medical illustrations are furthermore useful to demonstrate comparisons of differentiating structures, pathology, and classification systems (e.g. tumor grading or fracture types).

If your research involves surgical techniques, diagrammatic illustrations can explain the steps necessary to understand the procedure, detail anatomical landmarks, and orient readers to the surgical field. And, when appropriate, medical imaging (radiographs, magnetic resonance images (MRIs), computed tomography scans, and angiograms) can help readers visualize patient data as it supports the surgical approach; however, such imaging must be used with great sensitivity and should always protect the patient's anonymity.

Research of cellular mechanisms can make use of schematic figures or drawings to illustrate the mode of action, molecular structure, or chemical pathway. In some cases, scanning electron micrographs, histologic slides, or molecular illustrations can be used to acquaint your audience with cellular form or tissue structure. Graphic representations, such as tables and charts, are useful to describe changes over time or to compare specific analytes. Infographics are used to help readers visualize quantitative data and are furthermore effective in describing relationships between various biologic factors (e.g. the reciprocal activities of insulin production and blood sugar homeostasis).

Your research will dictate what types of illustrations are appropriate. Understanding the different types of illustrations, from line drawings to infographics, and how they can benefit your paper is the first step when organizing your content.

63.4   Illustrations and Research Design

The integration of appropriate images should ideally be considered while the concept of your paper is being developed and not simply as appended material to the written content. A good paper is economical in terms of description needed to communicate the thesis. That does not mean that it should be overly simplified or abbreviated, but rather that the topic should be deliberate and clearly articulated.

An image can help conserve an otherwise complex description and improve comprehension. For example, if you are describing a surgical procedure used to correct a rare condition in which anomalous structures must be navigated, an illustration can orient the audience and provide valuable information to guide the surgical approach.

Whether visualizing a complicated disease or biologic process, it can be appropriate to use a variety of images to tell the story of your subject. How do you demonstrate the relationship between the cellular mechanism of a condition and its clinical manifestations?

Cancer, for example, alters not only the function and structure of the cell, but that of the organ and body system. Schematic illustrations are appropriate to represent genetic changes and demonstrate the resulting dysfunction of cellular constituents; relate these illustrations to secondary imagery showing how the pathologic cells grow and collectively harm adjacent structures or parts of the body.

Visualizing the relationship between cellular dysfunction and physical symptoms can help the viewer more fully comprehend the disease epidemiology, which might be necessary to understand the therapies used to manage the condition.

The relative success of your paper might depend on the use of medical illustration not only to supplement anatomic descriptions but also to provide context and visualization of the consequence or result of the pathologic state. Finding the right artwork early on can inform the structure of your paper or presentation, and even inspire the written content.

63.5 Sourcing Illustrations

When sourcing illustrations, remember that the quality of the artwork depends not only on the ability to effectively communicate information, but also on scientific accuracy and validity. Having reviewed the different types of illustrations, let us consider the quality of the artwork, how to distinguish a good illustration from a bad illustration, copyright requirements, and artwork format for publication.

Most importantly, you want to make sure that the anatomy is described accurately. While there are many resources available, inaccurate anatomic illustrations abound. Look for artwork by illustrators who have been educated in biomedical visualization or whose experience demonstrates familiarity with medical subjects. Use your own medical expertise and those of your colleagues or advisors to discern the validity of the artwork.

If using stock imagery, be selective, as many popular websites do not require that their medical or health-based images meet standards of anatomic accuracy. It is a good idea to expand your search beyond standard stock sites. A growing number of sources for medical and scientific images continues to emerge as the demand for quality illustrations increases.

The difference between good artwork and bad artwork can affect the reception of your paper. Artwork is considered bad if the anatomy is represented inaccurately, if the view is difficult to understand or disorienting, or if relevant details are lacking. A good illustration drives focus on the subject, which is neatly and accurately represented. If the view is illusive, then obvious landmarks or directional cues (e.g. anterior, posterior, superior, inferior) should be used to help orient the viewer. If labels are included, make sure that they are well integrated and appropriate for your topic.

Artwork selected for your paper should be thoughtful of the audience. Is there enough detail to communicate the subject? Or, on the other hand, will too much detail be lost on the reader? Thoroughly consider how the illustrations and figures will affect the audience – do they help clarify the research or just create confusion?

If you intend to publish your paper, buying appropriate stock images or contacting contributors who are willing to license or sell their artwork are safeguards against possible copyright infringement. Publishing illustrations without licensing or getting the artist's permission can result in a lawsuit; however, a lawsuit is simple to avoid if you contact the artist or buy the artwork directly.

Use care not only in selecting specific artwork, but in its presentation. Print standards for images require appropriate formatting and high-resolution files that are normally a minimum of 300 dpi; screen resolution required for a PowerPoint presentation or online publication is 72 dpi. When preparing your manuscript, request specific artwork guidelines from the journal to which you are submitting your research. Most publications will provide instruction on preferred file formats and size requirements.

The ability to distinguish and source illustrations can help elevate your research and findings. If you require a more specific visual handling of the subject, consider forming a collaboration with a medically trained artist.

63.6 Collaborations

The field of medical illustration yields a rich collaboration between artists and scientists; through such collaborations, high-quality images congruous with your research design can be conceived.

Available sources include the Association of Medical Illustrators (www.ami.org), the *Medical Illustration Source Book,* a network of medical illustration and animation studios, and students in a biomedical visualization program. Many artists in the field would welcome an opportunity to contribute to your publication. Consider an illustrator whose work complements your research; for example, if your research requires you to describe an uncommon procedure, find an artist whose portfolio demonstrates familiarity with surgical techniques and instrumentation. On the other hand, if you need to communicate mode of action, look for an artist whose strength is in visualizing cellular and molecular subjects.

Certified medical illustrators have been educated or have experience in creating illustrations for the medical and biological sciences. Discuss with the artist the type of illustration that would most appropriately support the research. Sharing your research with the artist will help him or her conceptualize and better understand the subject. Provide the artist with specific details of what view you need illustrated, whether or not to include titles or labels, and which format is required for your publication.

63.7 Conclusion

Whether you source existing artwork that fits with your subject or collaborate on unique graphics, incorporating illustrations into your paper or presentation will make your research more easily and more elegantly understood. The appropriate use of artwork in a research paper enables you to better clarify and articulate your topic. The better your readers are able to comprehend your research concept, the more effective and successful your paper will ultimately be.

Further Reading

Boersema, T., Zwaga, H., Hoonhout, H. et al. (eds.) (1999). *Visual Information for Everyday Use* (digital ed.). London: Taylor & Francis, Ltd.

Tufte, E.R. (1990). *Envisioning Information*. Chesire, CT: Graphics Press LLC.

64

How to Prepare Supplemental Materials for Scientific Publications

Barbara J. Hoogenboom

Department of Physical Therapy, Grand Valley State University, Grand Rapids, MI, USA

64.1 Introduction

The dissemination of scientific writing and research is important to the advancement of science and to foster evidence-based practices in all disciplines. Multiple opportunities exist for publication of the results of bench research, clinical trials, systematic reviews/meta analyses, case reports, and clinical commentary. Diverse possibilities for medical and scientific publication exist, and most publications expect that the submission be visually and graphically appealing in order to present data and results, describe interventions, and enhance reader interest. Because paper-based (print) publications have been the norm in the past, dissemination of video, audio, or large data files has been limited. However, with the growing popularity of online journals and websites that accompany printed journals and other publications, utilization of a variety of supplemental materials has become reality. Authors and audiences can benefit from advances that are rapidly occurring in digital publishing. Many contemporary publishers encourage the use of supplemental materials in order to enhance the readability, completeness, and graphic appeal of a publication. Thus, authors should consider the use of such resources. The purpose of this chapter is to describe the range of supplemental materials, how to decide what to include as supplemental materials, and how to prepare them. Medical and healthcare professionals can benefit from the creation and access to interactive, three-dimensional presentation of concepts, data, and treatments.

64.2 Definition of Supplemental Materials

Typically, the body of a scientific paper contains graphic displays of data, results, and pictures or drawings of procedures in what are referred to as *figures*. These include graphs, charts, photographs, and line drawings. Figures serve to visually display information in two dimensions (two-dimensional), clarify ideas and procedures, and to allow readers to "see for themselves" what is being described in writing. Such inclusions enhance readability, keep the interest of a visual or graphic learner, and give access to information in broader ways than just the written word.

A Guide to the Scientific Career: Virtues, Communication, Research, and Academic Writing, First Edition.
Edited by Mohammadali M. Shoja, Anastasia Arynchyna, Marios Loukas, Anthony V. D'Antoni, Sandra M. Buerger, Marion Karl and R. Shane Tubbs.

However, it is common for publications to limit the number of figures, tables, or graphs that can be contained within a manuscript. That could be due to page-count limitations, the expense of producing graphics or color figures, or policy/procedural limitations of a given journal. Despite great advances in technology, medical publications are typically restricted to the presentation of ideas and results in two dimensions.

Supplemental materials are those materials typically not included in the text of a paper or textbook chapter, but that would still be of interest to the reader. They include, but are not limited to, multimedia files such as audio files, video files, three-dimensional interactive models, and large equations or gene sequences. Examples include an audio description of patient presentation or history and physical that can accompany a figure, an interactive three-dimensional display of a gene sequence, or a video of a complex examination technique or exercise. If a picture is worth a thousand words, how much learning could occur from the viewing of interactive models, or audio and video files?

A typical solution for the utilization of such supplemental materials is to make them available for download, via the website of either the publisher (journal or textbook) or the author. Many medical and allied health journals have adopted the model, and supplemental files are accessible for the readership of the journal by visiting the website. Additionally, several medical and rehabilitation textbook publishers have utilized large data-storage websites, such as Access Physical Therapy, to house digital media associated with paper textbooks. However, as electronic publication becomes more accepted, the possibility of embedding such supplementary files directly within the paper or chapter is exciting. Works are beginning to emerge in the context of enhanced e-textbooks that allow for the reader to click on a figure and have a video play an example of a technique or exercise. An example is *Musculoskeletal Interventions: Techniques for Therapeutic Exercise, 3rd Edition,* published by McGraw-Hill Medical publishers. This comprehensive physical therapy textbook has a paper format and an enhanced e-book that has links to 300 video clips that are accessible by simply clicking on a still figure and the video opens and plays.

The current standard for the final product of a published journal article is the portable document format (PDF). PDFs are cataloged and accessible via many databases. With a PDF, it is possible to embed audio, video, and interactive three-dimensional models, accessible by a mouse click on a static picture in the online paper (Ziegler 2011; Kumar 2010; Ruthensteiner and Hess 2008). New programs allow the PDF to be "readable," similar to a journal, where the pages turn. Such presentation of information is not only user-friendly and graphically appealing, but can become interactive with embedded audio, video, or image files (Table 64.1). Clearly, the use of embedded multimedia content is most appropriate for use in electronic publishing (Ziegler 2011). As technology continues to advance, so, too, should the use of supplemental materials, both separate from and embedded within journal articles and textbook chapters.

64.3 How to Decide What Is Supplementary

What should be considered supplementary in the context of a journal article is often described by submission criteria or Instructions for Authors provided by the publisher of the journal. Please note that the requirements vary substantially between journals and publishers, so investigate upfront and comply with the requirements for format and what is to be considered supplemental material. Often, online data supplements or other supplemental materials are encouraged as an enhancement to the print article. Some journals only accept supplementary materials that are "essential to the scientific integrity and excellence of the paper" (Science, Instructions to Authors, accessed, March 2019) and "cannot be presented in the main text because of its nature." (Journal of Cell Science, instructions to authors, accessed March 2019).

64.4 Types of Supplemental Materials and How to Prepare Them

Again, it should be noted that publishers of journals and textbooks will dictate the exact preparation instructions for supplemental materials (Table 64.1). However, some general information might be of assistance to the reader of this chapter:

1) Audio files are usually submitted as MP3 files (.mp3).
2) Video files most typically should be submitted in QuickTime (.mov) or as mpeg files (.mp4, .m4v). Some publishers accept audio/video interface (.avi), flash video (.flv, .f4v), Real Media (.rm) or .wav files. Many publishers have a maximum file size ranging between 5 and 10 MB. Please note, it is very important that the video files be created with minimal clutter or distractions in the background, only audio that goes with the purpose of the video, and in high definition (HD) and as high of a resolution as possible. Videos should be taken using a tripod to minimize unnecessary vibration and movement of the video device while shooting.

Table 64.1 List of formats that can be used in direct placement into portable document format (PDF) documents using Adobe Acrobat.

Multimedia type	File format or type	File extension
Audio	MP3 audio	mp3
Three-dimensional image	Autodesk 3D Studio	3ds
	Product Representation Compact (PRC)	prc
	Stereolithography	stl
	Universal 3D	u3d
	Virtual Reality Modeling Language	vrml, wrl
	Wavefront Object	obj
Video	Flash video	flv, f4v
	MPEG	mp4, m4v
	QuickTime Movie	mov
	Shockwave Flash	swf

Used with open access permission from Ziegler (2011).

3) Three-dimensional images often depend on which format was used when the image was created. Examples include Autodesk 3D Studio (.3ds), Product Representation Compact (.prc), Universal 3D (.u3d), and Virtual Reality Modeling (.vrml, .wrl).
4) Animations are typically submitted as .gif or .jpeg files.
5) Others include oversized tables, expanded methods or results, lengthy or technically complex appendices, large data sets or databases, or calculations, equations or genetic sequences. Those are commonly submitted as Microsoft Word tables (.docx), an Excel spreadsheet (.xlsx), or as a PDF (.pdf).

For most journals, supplementary material should be presented as self-contained, appropriately labeled documents, using terminology such as:

> Supplementary material to (author name), followed by the title of the paper, author name, and author affiliation, followed by the name of the file in text. For example, Video 1, 3D Image 2.

Additionally, a detailed file for legends of each of the supplemental materials is of utmost importance. Highlight in each legend the key points of the video, three-dimensional image, or audio file. Direct the reader/learner to what to glean or discern from the supplemental file.

64.5   Online Storage and Access

Whether stored on a journal's website or embedded directly into a publication, the access to supplemental materials is important. Ideally, supplemental materials could be accessed from handheld devices and tablets, as well as from computers. Such material could significantly affect the way in which students, clinicians, and researchers access and use data and graphic presentations of techniques.

64.6   Conclusions

The future of the creation, utilization, and storage of supplemental files is bright and could enhance the dissemination of ideas, data, and outcomes. Didactic possibilities for teaching with interactive and three-dimensional figures provided in journal articles and textbooks, and in online repositories, are being discovered continually. The use and storage of digital files could revolutionize journal and textbook publication, creating more interactive materials. These types of materials could provide enhanced comprehension for a wide variety of learners with distinct and different learning preferences. Smartphone, tablet, and computerized access to materials will only continue to improve.

References

Kumar, P., Ziegler, A., Grahn, A. et al. (2010). Leaving the structural ivory tower, assisted by 3D PDF. *Trends in Biochemical Science* 35: 419–422.

Ruthensteiner, B. and Hess, M. (2008). Embedding 3D models of biological specimens in PDF publications. *Microscopic Research Techniques* 71: 778–786.

Ziegler, A., Mietchen, D, Faber, C et al. (2011). Effectively incorporating selected multimedia content into medical publications. *BMC Medicine* 9: 17–22.

65

Reference-Management Software

Paul Tremblay[1] and Thomas P. Walker[2]

[1] *Sidney Druskin Memorial Library, New York College of Podiatric Medicine, New York, NY, USA*
[2] *Holy Spirit Library, Radnor, PA, USA*

65.1   Introduction

Researchers seeking to contribute to their fields by investigating material and resources toward the completion of a manuscript come across a tsunami of sources and information. They are faced with an increasing number of published or unpublished studies, articles, reviews, presentations, proceedings, etc., originating from many sources in all possible media. Some chapters in this book explore the art of searching, other the techniques of writing, but how does one organize, store and insert countless citations or build a bibliography? In the past, style manuals such as the American Psychological Association (APA) or the American Medical Association (AMA) handbooks were referenced in order to properly format and cite resources, but it could take weeks to compile a consistent, working list of sources used within a paper, especially when working in collaboration. In order to comply with the rules of academia and to cope with the increasing numbers of citation styles, software producers have created various programs to assist researchers in acknowledging, managing, and formatting their different resources.

For researchers, the workflow typically involves searching and finding sources in an online database, exporting the citations to the proper software, and then inserting the citations in a text document and bibliography. Historically, and at their most basic level, reference-management software (RMS), also called citation managers, have been designed to format and store citations of resources found online or offline that could be later used and cited within a document; for the most part, researchers use an RMS program to configure a bibliography or list of works cited according to a given official format such as Modern Language Association (MLA), Harvard, Chicago, or Turabian. However, many RMSs have evolved to cover the formatting of documents (in-text citation and works cited, etc.) with the integration of said software within word processor applications such as Microsoft Word; this integration allows for a whole document to be modified from, say, MLA to Harvard with a few clicks. The most advanced programs can format a document according to a specific journal's own requirements. Other RMSs go further by incorporating style beyond citations in comprising paragraph alignments, spacing, etc. Indeed, some RMSs offer native search modules within the RMS's own interface where one may search other proprietary

A Guide to the Scientific Career: Virtues, Communication, Research, and Academic Writing, First Edition.
Edited by Mohammadali M. Shoja, Anastasia Arynchyna, Marios Loukas, Anthony V. D'Antoni, Sandra M. Buerger, Marion Karl and R. Shane Tubbs.

or free databases like PubMed, as well as an institution's own Online Public Access Catalog, resulting in an even more seamless process of searching, adding, organizing, storing, and incorporating citations. Links to full-text PDFs are now stored as well as Uniform Resource Locators (URLs). It is no wonder that some of those RMSs are now marketed at "research assistants."

A very quick and unscientific survey of medical literature, for instance, shows that contemporary journal articles of approximately seven to eight pages are accompanied by a bibliography of over 50 items. Book chapters will typically display hundreds of citations. A point could be made that, in these technological times, RMSs are a necessity in order to successfully complete a publishing project, the same way researchers need a word processor to write it, or a spreadsheet and statistical software package to calculate statistical findings.

This chapter will cover the basic functioning of RMS, from the free plug-ins to subscription-based packages (individual and institutional). Word processors have successfully implemented their own referencing modules: since Word 2007, for instance, Microsoft includes a reference ribbon enabling users to insert in-text citations and bibliographies according to mainstream formats. These added features may not search external proprietary databases directly from Word, though, and they may not format an entire paper (margins, spacing, etc.) according to a specific style.

It goes without saying that RMSs exist within the context of academic integrity and plagiarism dealt in much more detail in other chapters of this book; however, full-fledged RMSs greatly assist researchers in managing and citing their sources and therefore reduce chances of plagiarism, accidental or not.

There are many different programs offered on the market, and readers, from new to experienced researchers, should assess which is most compatible for their needs. A recent survey found no less than 30 citations managers in the market, for free or for a fee. Some disappear or merge (ProCite used to be a researchers' favorite until it was bought and merged last year by Thomson Reuters, who also owns EndNote). A great number of websites compare RMSs, and we strongly encourage researchers to look up these ever-changing resources (refer to the Bibliography at the end of this chapter for a select number of documents and their URLs).

65.2 Situational Example

Two researchers are collaborating on a project on "current treatments for diabetic foot ulcers in young children." Their literature search takes them to PubMed, ScienceDirect, the Cumulative Index of Nursing and Allied Health Literature (CINAHL), and other proprietary medical databases. They have to explore Cochrane Library for systematic reviews and end up with hundreds of citations, not including some institutional videos, streaming media, books (offline and electronic, whole and chapters), interviews, and others. Our two researchers, working in different institutions and cities, one with an Apple computer (Mac) and the other with a personal computer (PC), come to the conclusion that, although their citations are stored on a computer somewhere in a text document, it would be difficult to manage and cite them manually. It would, in fact, be reminiscent of writing the whole project on a typewriter with a style manual on the side. They use a citation-manager software package, for free at first, which will at least store the items and format them in the desired style (say, in this case, AMA). One of them will have access and control of the RMS.

The process is relatively seamless: for instance, a resource is found in an online database, and the citation of this resource may then be exported in a given format to the user's own RMS account. She or he then opens a Word document (after downloading and installing the word processor plug-in), synchronize to the RMS account to add the in-text citation and build a bibliography in a desired format, in this case AMA.

They decide to email citations and attachments to each other, which seems to be a reasonable plan. But how does one suggest, correct, and review a text without changing the in-text citations? Having to add an item to the bibliography? New citations are pouring in every day, making it tricky to change the text and the bibliography at any given moment.

The researchers then take a deep breath and share the RMS account, which means that now both researchers have to download and install the word processor plug-in. They start importing citations from different databases to their RMS account, directly in their text files; an in-text citation can be added automatically and the item is added likewise to the bibliography. They have a choice to import citations from the RMS's own interface or to export citations from a database to their RMS account. However, had our researchers worked separately with different RMS software, they could still potentially (although not intuitively) export/import files containing citations.

Most software allows for file sharing with export/import modules. Users may save files in their EndNote libraries, for instance, export them, and email them as email attachments. Users of RefWorks, Zotero, or other RMSs will import, convert, and store the new files and citations. Institutional RefWorks users may share attachments of their own texts within their RefWorks interface. Our researchers may find sources in proprietary databases or the general web, export them to their RMS account with no problems (most databases have a built-in feature allowing users to export directly to mainstream RMSs). However, one of our writers finds an excellent source in Google Scholar. Again, no sweat, Google Scholar offers a "cite" option that permits users to export sources to their preferred RMSs. Should the researchers use Zotero, for instance, all they would have to do is use the drop-in feature or type a URL, International Standard Book Number, a Digital Object Identifier, or PubMed ID in the appropriate box and the records automatically fill in. We have found issues saving and exporting directly from PubMed to Zotero Standalone, for instance; however, there are easy ways to save in RIS or XML and export/import (Ahmed and Al Dhubaib 2011). Some RMSs allow for an entire folder of data, or discrete parts of datum, to be shared through a URL emailed to a nonsubscriber (e.g. RefWorks with RefShare).

The fact that our Mac-based researcher and her PC-based counterpart may share the same interface is a step in the right direction. However, collaboration does not have to stop at sharing between our two researchers. As mentioned previously, some RMSs have evolved by morphing, so to speak, as "research social media." They allow users to look up what research items are most used by other researchers, meaning that our two writers' own list is also partially transparent to outsiders.

After some time, the two researchers decide that their project is finished and submit it to a peer-reviewed journal. Things do not go very well and they decide to resubmit to another journal, who prefers another citation style, let's say American Psychology Association. Not a headache – our writers save a copy of the previous file and create ("save as") a new document to be formatted entirely in APA. Should the journal advocate its own style, mainstream RMSs usually have this option, or at least may add the journal's style upon request.

The preceding scenario is typical and involves choices based on needs, systems, compatibility, etc. Researchers have many options, and we will attempt to cover all major software, for free or for fee.

65.3   What Is on the Market?

How did the two researchers search and find an appropriate RMS? Many RMSs are available as open sources and others are proprietary, but researchers can expect some basic features in RMS, notwithstanding the cost. Mainstream RMSs include EndNote, JabRef, Mendeley, Zotero, and RefWorks. We do not endorse any of them but simply acknowledge the different products available (Table 65.1). We have attempted to omit RMSs that are platform-specific; to provide only a few examples, Bookends works with Mac products only, whereas Biblioscape and Citavi function exclusively in a Windows environment; therefore, we have overlooked them. Most RMSs have free options up to a specific amount of data (300 MB, 2 GB, etc.), and offer different plans for more inclusive services, individualized or institutionalized. Some of the open source RMSs includes:

- *JabRef*[1]. Fully web-based; offers less features (for instance, only 19 document types to cite from); no for-fee service as of publication.
- *Mendeley*[2]. More recognized as a citation social media, offers less features than mainstream RMSs. Easy to use and more than adequate for students. It also offers a for-fee service for over 2 GB of data. Mendeley is web-based but Mendeley Desktop stores data on the users' workstations (Figure 65.1).
- *Zotero*[3]. Versatile and has been a favorite among academics since its inception in 2006. Available as a web-based application or an installed program on selected workstations (Zotero Standalone). The word processor add-in installs automatically for both; Zotero is relatively intuitive and comes with a legion of help, instruction, video, users forum, etc. There is no phone assistance. The service is free to up to 300 MB of data at time of publication.

Some of the proprietary RMSs are:

- *EndNote*[4]. One of the workhorses of the trade, EN offers "templates" to format entire documents according to styles and journal formats. EndNote Basic is free but lacks many features. Produced by Thomson Reuters, Inc., it is often used at the institutional level. Users install EndNote on their workstations and in their word processor program; data are stored in "libraries" in their computers (Figure 65.2). The support is very good, as it includes round-the-clock phone, email, and chat assistance, as well as a YouTube channel (EndNote Training).

---

1 Jabref: http://jabref.sourceforge.net.
2 Mendeley: www.mendeley.com.
3 Zotero: www.zotero.org.
4 EndNote: www.endnote.com.

Table 65.1 Some individual reference management software and their features.

Characteristics and features	EndNote	JabRef	RMS		
			Mendeley	RefWorks	Zotero
License	Proprietary commercial	Open source	Open source	Proprietary commercial	Open source
Cost	Individual and institutional accounts based on EFT; free on a trial basis. Basic EndNote is free.	Free	Free (Paid subscription available for over 2 GB of storage)	Individual and institutional accounts; no free option (except on a trial basis)	Free (Paid subscriptions available for 2 GB and 25 GB of storage.)
Installation feature	Download application, install on workstation.	Web-based	Download application; however, it is web-based.	Web-based	Web-based; Option to download application (Zotero Standalone)
Operating systems platforms	Win, OSx	Win, OSx, Linux, Unix	Win, OSx, Linux	Win, OSx, Linux, Unix	Win, OSx, Linux, Unix
Mobile (Phones and Tablets)	iOS but no Android or Windows apps are available at time of publication.	Can be downloaded for iOS devices	Yes	Yes, but different URL	Yes

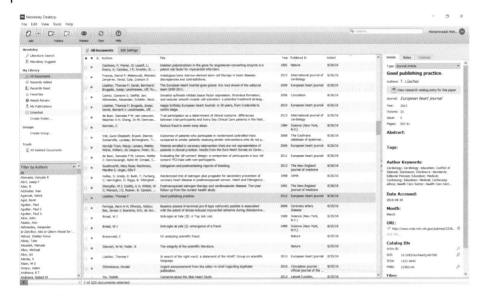

Figure 65.1 Mendeley desktop user interface.

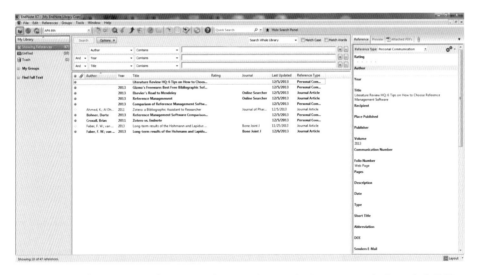

Figure 65.2 Endnote user interface. Researchers may import citations automatically to their RMS or create them manually (as shown).

- *RefWorks*[5]. Another star of the industry, RefWorks is web-based and like EN it is quite intuitive and also commonly used at the institutional level. One of the differences between EndNote and RefWorks is that RefWorks stores all your citations online, meaning you can access your folders and files from anywhere; it also installs a word processor plug in. It comes with an excellent support and training component, quite similar to EndNote, but chat is not available.

---

5 RefWorks: www.refworks.com.

65.4   What to Expect from an RMS

RMSs should not be confused with some useful albeit very simple citation generators. The free version of NoodleTools,[6] for instance, does generate citations according to a desired format once fields are filled, but it is then up to the researcher to copy and paste the final citation within a document. This is sufficient for many students, however, for seasoned researchers working in collaboration on multiple publishing projects, this can be wearisome and somewhat unmanageable. Mainstream RMSs are designed to store citations in a searchable database (Figure 65.3), itself connected to users' word processors, in order to easily insert in-text citations and build bibliographies; researchers may create folders and subfolders to store their citations and search them by authors, titles, keywords, etc., remotely or not (in effect, cataloging and indexing the records), and export them to other RMSs or computer drives. The evolution of

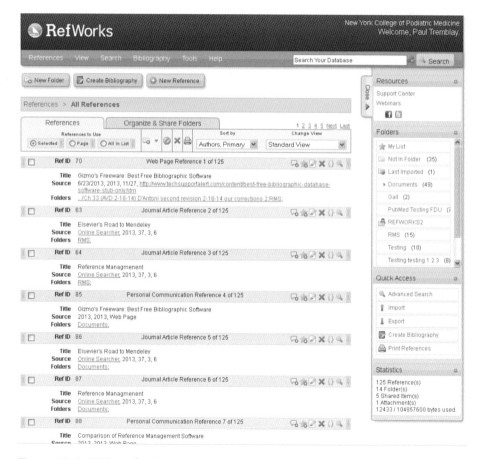

Figure 65.3 An RMS interface becomes a database on its own, as researchers can search their own citation folders and external databases. They can elect to import/export citations to other RMSs off the same page.

---

6 NoddleTools: www.noodletools.com.

RMSs from simple citation formatting tools to more robust software tend to reflect a trend described as a desire and a need by the research community to streamline the research process. "Ultimately, today's researcher is interconnected and searching for value beyond the citation" (Hensley 2011). Even a casual observer could detect the blurring of borders between databases, RMSs, and word processors. The movement is obviously toward cloud-based collaboration and sharing, as RefWorks Flow, Zotero, and Mendeley demonstrate (Farkas 2012; Hensley 2011). The convergence of social media, searching, annotating, collecting data, writing, citing, sharing and editing within the same interface points in the direction that RMSs are now heading. In other words, modern researchers desire tools that not only generate citations but also organize and share their sources and, ultimately, their finished projects.

An eventual cloud-based collaboration example is already being practiced in the form of shared accounts in which several users add and edit citations, PDF files, and even their own texts. It entails that all users have to download the word processor plug-in, but once done, researchers may partner and share information from different workstations (and locations) but within the same RMS interface.

Some of the concerns and questions researchers should ask before choosing a RMS follow:

1. Ease of use of the software, from download, to installation, to implementation. Does the new program slow your workstation or network to a crawl?
2. Support and training provided by the software company. Most RMSs have videos available for just about any kind of topic; as mentioned, mainstream RMSs even host YouTube channels. Is there a phone number that you can call in case of emergency? Can you easily find a support email address or chatting module? Does the documentation speak to seasoned researchers only, or does it also address the novices' concerns?
3. Vast choice of citation styles (EndNote offers over 5000 and Zotero claims 6400, perhaps an overkill for some writers, especially students).
4. The ability to cite nontraditional sources and media; most of the citations will be for monographs or periodical articles, but more sources from web pages, printable document format (PDF), images, television shows, paintings, conferences, proceedings, musical scores, course syllabi, streaming audio or video, etc. are used. Studies have been published on the evolving landscape of research and publishing and the decline of text-only material (Harris 2006).
5. Integration with word processors and to some email-based word processors (Google Docs, for instance). Historically, word processor integration has changed RMSs from just copy-and-paste citation generators to automated style and format software. It involves installing a plug-in to your word processor on your workstation (Figure 65.4); users work on their RMS's interface, synchronize on their word processor, and add citations and bibliographies once the citations have been imported (or exported, depending on your workflow) from a database to a word processor document. The caveat is that you have to install the plug-ins (or add-ins) on all workstations you are using. Some do it automatically (Zotero), others have to be installed (EndNote and RefWorks). A glitch has been observed by the authors when it comes to slightly older word processor versions. Also, some Zotero files converted to RefWorks did not read well once we were using the "review" feature of Word. So

Figure 65.4 RMSs integrate with word processors. Here, you can see "Add-Ins" (containing a version of Zotero) and a special tab for RefWorks, both in the ribbon.

far, a seamless email-based word processor integration is still a thing to come, as Google Doc, for one, does not communicate very well with mainstream RMSs, with the exception of Zotero. Amazingly enough, an RMS like JabRef does not integrate seamlessly with Word and is more comfortable with open-source and email-based word processors.

6. Multiple users (sharing or working in collaboration); extremely important and a time-saving device for co-authoring. A single account or a selection of folders may be shared. We deal with this issue elsewhere in this chapter.

7. "Platforms" (i.e. the operating systems) and the browser being used. Most RMSs are platform independent, be it Unix, PC, or Mac, and increasingly so with mobile devices from tablets to smart phones, although some issues remain. Users have to look up software compatibility in this respect. Zotero used to need third-party apps (Zandy, for instance) to function on mobiles, but it is relatively seamless now with plug-ins. Operating systems are usually not a problem unless users are relying on an older version or an extremely recent one (Windows 8.1 had reportedly some issues, but they have been dealt with). Another potential hurdle would be with browsers, from Safari to Google Chrome. Again, past glitches are being ironed out by software designers. Zotero is a Firefox extension (Figure 65.5) and has been known to hit road bumps with Internet Explorer, among others.

8. Ability to migrate data from one RMS to another or at least allow users to save data before moving to another institution. For instance, is it possible to migrate data collected on EndNote to RefWorks, or vice versa? We will address these issues later in this chapter.

Here are seven features that might be desirable, depending on users' needs:

1. Most RMSs offer journal-specific styles; some publications may be added following users' suggestions.

2. Native database search capability (such as PubMed) off the RMS's own page is quite a useful feature (Figure 65.6).

3. Whether the software is web-based (RefWorks) or workstation-based (EndNote) is of course dependent on a writer's needs, but this feature should be considered closely before subscribing to an RMS: data on a hard drive will remain there, whereas data on a remote server may be lost once the subscription ends.

4. Export/import capabilities. Most if not all proprietary databases allow users to export citations to RMSs, and it might be a good idea to verify if your RMS of choice is included in an "export" option, but not *all* databases are connected to *all* RMSs,

Figure 65.5 Zotero is a Firefox extension. Once downloaded, a few RMSs remain within eyesight and can be opened and used at any time.

Figure 65.6 Example of an RMS's native search box where you may search and export citations from different databases.

sometimes for coding incompatibility such as Z39.50 (which, ironically, is a protocol used between different proprietary platforms, institutions, and especially library catalogs in order to enable users to "read" other databases). Elsevier favors Mendeley, which it owns. RefWorks belongs to ProQuest, so it is a given that there is a commercial connection between software in some cases; however, there are ways to export in RTF or RIS format, and your RMS of choice should be able to import this way.

5. An interesting conundrum exists at this junction, and some RMSs are struggling with it: it involves extracting metadata from a PDF file already saved in one of your workstations. Check if your RMS of choice can successfully (and easily) perform this function.
6. Entire paper style formatting (paragraph, spacing, alignment, etc.) is an interesting addition, but not a vital feature for most users.
7. To be redundant about it: the ability to export data from one RMS to another. Does one software use a file format unrecognizable by another? Is it convertible? Verify the "export/import file format" of the RMS of choice, as it may have compatibility issues (or would be a source of trouble) to import from another specific RMS.

65.5   Working in Collaboration and Sharing

It is possible, even feasible and encouraged, to share resources. Co-authorship is one of the motives, of course, but some RMSs act as citation social media, in a manner of speaking. Mendeley allows easy sharing between two or more authors but also the possibility to look up what is being used, which resources are most or least popular at a given time, and more. RefWorks with Share and RefWorks Flow are also moving in this direction, and Zotero has been a proponent of this practice from the outset. For the most part, the data being shared consist of folders of citations, of course, but researchers may also trade their own texts (RefWorks allows attachments between subscribers within its own interface).

It goes without saying that PDF files of full-text articles may also be traded; however, the slippery slope of copyright infringement is in sight. It is up to users to abide by copyright laws and regulations and not disseminate copyrighted material without publisher and/or authors' consent.

Researchers may also share with other persons without giving them permission to add or edit the records. Subscribers just have to tag the folders or files as "read-only" for specific users.

65.6   In-Computer versus Cloud

It is more than just an academic decision: EndNote stores your citations in a "library" located in your workstation. RefWorks, Zotero, among others, save theirs in a remote server. Should you subscribe to a RMS and either drop the subscription or leave the institution that is paying your membership, you may lose all your data, unless you migrate it to a personal account. Zotero Standalone downloads as a separate application on your computer. This being said, a migration is far from intuitive and involves some "exporting" on one end (and saving in a specific format, be it txt or XML), and "importing" at the other end, but is feasible and may even be performed in email attachments. We recommend periodically exporting folders (in bulk or selectively) to an email account to save citations to a computer drive. The file may display in unformatted style, but at least the records are still there.

65.7 All Is Well? What Does the Future Hold?

The scenario as played out earlier in this chapter obviously reflects an ideal situation: the two researchers had to adopt an RMS compatible with both Apple and PC, and mobile devices of any kind. However, most RMSs do not accommodate every platform or provide every formatting style known to humankind. All in all, as in all things digital and electronic, the problem that remains is not only compatibility but consistency. Although a convergence of tools, from database searching to word processor integration, is already de rigueur with mainstream RMSs, we foresee a much more seamless experience to benefit users, comprising cloud-based, platform-independent (including mobile devices) referencing software integrated with a full-fledged office suite (from word processor to spreadsheet and presentation software), capable of mining nearly all proprietary and open-source databases and providing researchers with the ability to intuitively save folders and files both online and offline. We are not there yet, however, as harmonizing all components, from browsers to databases to platforms and between commercial entities continues to be a difficult and frustrating concern.

Just a few of the hurdles preventing a full convergence, as already mentioned, are copyright infringement related to sharing resources, as well as its close kin, proprietary software understandably reluctant to seamlessly communicating with another software due to different incompatible formats, but these issues and facts of academic life exist outside the realm of RMSs in any event.

Most software producers are still struggling with several key issues, including the ease (or lack thereof) of exporting, importing, migrating, interfacing between different RMSs, or between one RMS and one database, integrating to a word processor, or even between an RMS and some browser (Zotero versus a non-Firefox Mozella, for instance). Researchers must contend with these issues in terms of selecting an RMS. Many web pages and much ink have been used to discuss difficulties of directly importing websites as well as PDF metadata into one's RMS account. Some RMSs are better suited than others at accomplishing this task, but the landscape is constantly changing. What about converting a preexisting document (already loaded with citations and a bibliography) to either EndNote, Zotero, or RefWorks? Researchers may have to tweak and redo their documents in order to adapt it to a RMS.

Other predicaments include RefWorks which, for instance, cannot host ScienceDirect on its own native search interface, as ScienceDirect is XML-based and RefWorks is Z39.50. This is eased by the fact that ScienceDirect offers a way to export directly to RefWorks from ScienceDirect's own page. The authors of this chapter worked on EndNote and RefWorks concurrently and purposefully to demonstrate some issues and unearth some solutions: for instance, opening a Word document, itself formatted in EndNote, into another workstation connected to RefWorks was not a seamless process. It involved the EndNote user saving his "library" into a .txt format, exporting it, emailing it, and then the RefWorks user had to open it, import it, and voilà, we have a conversion. Full integration with Google Docs will eventually happen, but the process is awkward with EndNote and RefWorks: you may create a Google Docs, even post some in-text citations, but users need to export the document to Word, add a bibliography, and re-export it to Google Docs. Zotero is slightly more seamless, consistent with its ease of drag-and-drop operation.

Although the preceding paragraphs may seem gloomy, we (and researchers in general) acknowledge the overwhelming need for RMSs. Compatibility issues progressively fade away with new technology, itself a source of integration problems, although the web is full of tricks and solutions and we strongly suggest reading software producers' literature and other users' forums. In other words, absolutely everything is possible, but in some instances, such as Zotero, it may take some programming background.

However, there is no turning back: Any serious researcher, in either a solo or co-authorship endeavor, is opting for one RMS or another, or two at a time (Farkas 2012). For the instructor training students on a RMS, this becomes a plagiarism teaching moment as well. RMSs do not prevent plagiarism but assist researchers into avoiding pitfalls.

Publishing or grant request research projects involve endeavors that typically take months or years to complete. RMSs provide researchers with the ability to store, list, format, and share all references to be used and modified them for multiple papers at an instant notice, leaving researchers time to investigate and write. Although some word processors do offer some built-in reference features, disciplined researchers will benefit from RMSs, as these programs will continue to play an important part in archiving and retrieving references.

References

Ahmed, K. and Al Dhubaib, B. (2011). Zotero: a bibliographic assistant to researcher. *Journal of Pharmacology and Pharmacotherapics* 24 (4): 303.

Farkas, M. (2012). Tools for optimal flow: technology-enable research workflows. *American Libraries* 43 (7–8): 23.

Harris, B. (2006). Visual literacy via visual means: three heuristics. *Reference Services Review* 34 (2): 213.

Hensley, M.K. (2011). Citation management software: features and futures. *Reference and User Services Quarterly* 50 (3): 204.

Further Reading

Gizmo's Freeware (2013) *Best Free Bibliographic Software* [WWW]. Available from: http://www.techsupportalert.com/content/best-free-bibliographic-database-software-stub-only.htm. [Accessed 04/11/2017].

Pahi (2017) *Comparison of Reference Management Software*. [Wikipedia] 5th April. Available from: http://en.wikipedia.org [Accessed 04/11/2017].

Bohner, D. (2013) *Reference Management Software Comparison – 3rd Update – June 2013* [WWW]. Available from: http://mediatum.ub.tum.de/doc/1127557/1127557.pdf. [Accessed 11/27/2013].

Croxall, B. (2011) *Zotero* vs. *Endnote* [WWW]. Available from: http://chronicle.com/blogs/profhacker/zotero-vs-endnote [10/03/2013].

Steeleworthy, M. and Dewan, P. (2013). Web-based citation management systems: which one is best? *Partnership* 8 (1): https://journal.lib.uoguelph.ca/index.php/perj/article/view/2220/2781#.U35C7dJdWSp.

Section IX

Biostatistics

66

Basic Statistical Analysis for Original Studies

Ganesh N. Dakhale, Sachin K. Hiware, Avinash V. Turankar, Mohini S. Mahatme and Sonali A. Pimpalkhute

Department of Pharmacology, Government Medical College, Nagpur, Maharashtra, India

66.1   Introduction

Statistics is an art and science that primarily deals with ways to gather and analyze data. It can be defined as the treatment of numerical data obtained from a group of subjects. It is possible with the aid of statistics to achieve greater precision at minimum cost by using available resources effectively. Statistical methods are important for drawing valid conclusions from research data. This chapter deals with the application of basic biostatistical concepts for designing and undertaking studies and for analyzing and interpreting the study data. Statistics implies both data and statistical methods. The branch of statistics applied in biology, medicine, and public health is known as biostatistics.

66.2   Measures of Central Tendencies

Mean, median, and mode are the three measures of central tendency. The mean is the most common and widely used of the three. It is calculated by adding up the individual values ($\sum x$) and dividing the sum by number of items (n). Suppose the fasting blood sugar levels of seven diabetic patients are 130, 135, 140, 140, 145, 150, and 155 mg/dL. The sum of blood sugar from these seven patients is 995 mg/dL. The mean value is 995/7, or 142.14 mg/dL.

When all the values of a variable are sorted from the lowest to the highest value (or vice versa), the middle value is known as median; half of the sample has values below the median and the remaining half has values above. The median in the case of even numbers of observations is taken arbitrarily as the average of the two middle values. The mode is the most frequent value or the point of maximum concentration. In example above, the mode is 140. The mode is rarely used in medical studies.

A Guide to the Scientific Career: Virtues, Communication, Research, and Academic Writing, First Edition.
Edited by Mohammadali M. Shoja, Anastasia Arynchyna, Marios Loukas, Anthony V. D'Antoni, Sandra M. Buerger, Marion Karl and R. Shane Tubbs.

66.3 Types of Data

There are two main types of variable, namely, qualitative and quantitative. A qualitative variable is either nominal or ordinal and quantitative variables are classified as discrete or continuous. It is very important to understand the type of variable as the choice of statistical method for analysis largely depends on it.

1. *Categorical (or nominal) variable.* Data are simply assigned names or categories based on the presence or absence of certain characteristics without any ranking between categories. For example, ethnicity is a categorical variable. This type also includes binominal (binary or dichotomous) data with two distinct values or levels such as sex (male or female), outcome after cancer surgery (recurrence or remission), or response to a treatment (improvement or no improvement).
2. *Ordinal variable.* Generally, these types of data are expressed as scores or ranks. There is a natural order among categories, so they can be ranked or arranged in that order. Pain levels may be classified as mild, moderate, and severe according to severity. Since there is an order between the three grades of pain, these types of data are called ordinal. There is no such order in nominal data.
3. *Discrete variable.* The data have integral but not fractional values, e.g. the number of cigarettes smoked per day by a person.
4. *Continuous variable.* A continuous variable can take any value within a given range. Continuous data are either interval or ratio data. Interval data lie on a scale that shows the range or distance between two adjacent units of measurement, but the zero point is arbitrary, e.g. temperature and year. Ratio data lie on a continuous, ordered, and constant scale with a natural zero, e.g. blood pressure, weight, and height.

66.4 Distribution of Data

Although the universe is full of uncertainty and variability, a large set of biological observations always tends toward a normal distribution, and this is called regression toward the mean. This unique behavior of data is the key to the whole of inferential statistics. Overall, data distributions are divided into two types, Gaussian (normal) and non-Gaussian.

66.4.1 Gaussian (Normal) Distribution

A normal distribution is also known as a Gaussian distribution (named after the German mathematician Carl Friedrich Gauss). If the data are symmetrically distributed on both sides of the mean and form a bell-shaped curve in the frequency distribution plot, the distribution of data is called normal or Gaussian. The normal, bell-shaped curve reflects the ideal distribution of continuous data. The mean, median, and mode are equal in this distribution. Computer software assists in checking whether the distribution of the raw data is normal. In an ideal Gaussian distribution, 68% of the data fall within the range ± 1 standard deviations (SD) of the mean, 95% within the range ± 2 SD, and 99.7% within the range ± 3 SD. Even if the distribution in the original population is far from normal, the distribution of the sample tends toward normal as the sample size increases. In a

Positive Skew

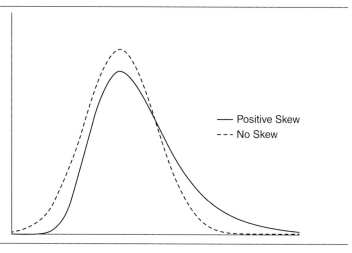

Negative Skew

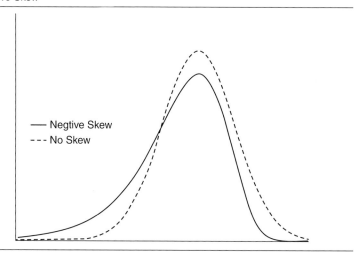

Figure 66.1 Diagram showing normal distribution curve with negative and positive skews. μ = Mean, σ = Standard deviation. Source: Wiley Study Guide for 2018 CIMA Exam. Hoboken, NJ (2018).

normal distribution the skew is zero. If the mean is greater than the median, the curve is positively skewed, and if it is less the curve is negatively skewed (Figure 66.1). The measures of central tendency in a data set are not necessarily equal.

66.4.2 Non-Gaussian Distribution

If the data are skewed to one side then the distribution is considered non-normal. It may be a bimodal or Poisson distribution among other possibilities. When the distribution is non-Gaussian, different tests such as the Wilcoxon, Mann-Whitney, Kruskal-Wallis, and Friedman tests can be applied depending on the study design and research question.

66.5 Transformation (Normalization) of Data

For the purpose of precision and ease of analysis, certain data may have to be transformed. Drug doses are converted to their log values and plotted on dose-response curves to generate a straight line so that analysis becomes straightforward. Similarly, asymmetrically distributed data can sometimes be transformed so the degree of skewness is reduced. This will make the analysis easier. It is possible to convert interval data into ordinal or nominal data. For example, consider Hb level data from 15 patients. These are interval data. If the patients are classified into normal (Hb 12–14 g/dL), mildly anemic (Hb 9–11 g/dL), moderately anemic (Hb < 7 g/dL) and severely anemic (Hb < 5 g/dL), the interval data are converted to ordinal data. If classified into normal (>11 g/dL) and anemic (<11 g/dL), then they become nominal data. The final judgment will depend on the purpose of the study and in such situations the clinician's decision is crucial.

66.6 Measures of Variability

Measures of central tendency provide useful summary information about the data but are not sufficient to establish how the measurements of the variable are dispersed or scattered about the mean. Therefore, the following indices of variability are important in biostatistics. *Range* is an index of variability amongst observations. It takes into account the highest and lowest values. It is not preferred over other variability parameters as it does not consider intermediate values and has no further use in mathematical calculations, as the standard deviation and standard error of the mean (SEM) have.

66.6.1 Standard Deviation (SD)

SD is an improvement over mean deviation as a measure of dispersion and is used most commonly in statistical analysis. SD describes the variability of the observations about the mean. To describe the scatter of a sample, SD is the most useful measure of variability. To calculate it we need to determine its square, called the *variance*. Variance is the average square deviation around the mean ($\bar{x}$), and standard deviation is the square root of the variance:

$$\text{Variance} = \sum (x - \bar{x})^2 / n \text{ OR } \sum (x - \bar{x})^2 / (n - 1)$$

$$\text{SD} = \sqrt{\text{Variance}}$$

SD helps us to predict how far the given value is from the mean so we can predict the coverage of values. SD is appropriate only if data are normally distributed. If individual observations are clustered around the sample mean (M) and are scattered evenly around it, the SD helps to calculate a range that will include a given percentage of observations. For example, if $N \geq 30$, the range $M \pm 2\,\text{SD}$ will include 95% of the observations and the range $M \pm 3\,\text{SD}$ will include 99%. It is better to report mean and SD by "mean (SD)" rather than "mean ± SD" to minimize confusion with confidence interval (CI) (see below). SD is very useful for sample size calculations.

66.6.2 Standard Error of Mean (SEM)

The sample estimates of statistics differ from population parameters because of chance or biological variability. A measure of this variability is provided by the SEM, which is calculated as (SEM = $SD/\sqrt{n}$). Thus SEM is a measure of chance variation; it does not mean error or mistake. SEM is always less than SD. SD applies to the sample and SEM to the population mean. It is helpful to work out the limits of confidence within which the population mean should lie. This can be illustrated by taking the example of the diastolic blood pressures (DBP) of 100 doctors. Suppose the mean is 90 mmHg and SD = 7; SEM = $SD/\sqrt{n} = 7/\sqrt{100} = 7/10 = 0.7$. With 95% confidence limits, the DBP of the doctors would be mean DBP ± 2 SEM = $90 \pm (2 \times 0.7) = 90 \pm 1.14 = 88.86$ and 91.14 mmHg. So the confidence limit of the DBP of the doctor population is 88.86–91.14 mmHg. If the DBP of another doctor is 84, it can be said with confidence that this doctor does not belong to the same population.

66.6.3 Confidence Interval (CI)

The confidence interval is the two extremes between which 95% of observations would lie. It quantifies the precision of the mean; the limits within which 95% of the mean values, if determined in similar experiments, are likely to fall. The confidence interval indicates the variation of the data. Statistical calculations combine sample size and variability (SD) to generate a confidence interval for a population mean. A wide interval implies larger variation. The value of t corresponding to a probability of 0.05 for the appropriate degree of freedom is obtained from the table of distribution. The 95% confidence limit for the mean is obtained by multiplying this value by the standard error (SE). This value depends on the degree of confidence; this is traditionally 95%, but it is possible to calculate intervals for any degree of confidence. Nowadays, computer software packages directly calculate the confidence interval. The important difference between the p-value and confidence interval is that the confidence interval represents clinical significance whereas the p-value indicates statistical significance. Therefore, in many clinical studies, the confidence interval is preferred to the p-value and some journals specifically insist on it. For example, a change of blood pressure of 4 mmHg in a trial could be statistically significant but not clinically significant. In conclusion, range and SD are descriptive in nature and not sensitive to sample size (N), whereas SE and CI are inferential in nature and are sensitive to N.

66.7 P-Value and its Importance

The p-value is the probability (range zero to one) that the relative frequency of occurrence of an observed/estimated difference or association is due to chance. If the p-value is small, it is concluded that the difference between sample means is unlikely to be a coincidence. In medical research, the p-value most commonly considered the threshold of significance is less than 0.05, or 5%. However, on justifiable grounds, a different standard such as $p < 0.01$ or 1% can be adopted. Whenever possible it is better to give the actual p-value instead of $p < 0.05$. If the population value is placed within 95% confidence limits, only in 5% of cases may it be wrong. "Significant or insignificant" indicates whether a value is likely or unlikely to occur by chance. When the *p*-value is <0.05, the probability of obtaining the difference between groups purely by chance is less than 5%. If the *p*-value is >0.05, the difference is considered statistically nonsignificant.

66.7.1 Errors

No experimentation or observation can be totally free from errors. But errors must be recognized and identified if they are to be eliminated as far as possible. There are two types of errors, type I and type II.

66.7.1.1 Type I Error (False Positive)

A type I error is also known as an α error. It is the probability of finding a difference when no such difference actually exists. This, for example, can result in the acceptance of an ineffective treatment as an effective one. Such an error, although not unusual, can reveal itself in subsequent studies and thus finally be rejected. Assume that in a trial a new drug A is proved to have an antihypertensive effect. If a type I error is committed in this experiment, then subsequent trials on this drug will be likely to refute the claim that drug A has an antihypertensive effect; this can lead to drug A being taken off the market. A type I error is actually addressed in advance of the study as the investigators set the level of significance (α criterion).

66.7.1.2 Type II Error (False Negative)

This is also known as a β error. It is the probability of failure to detect a difference that actually exists. This, for example, could result in the rejection of an effective treatment as an ineffective one. An error of this type can be more serious than a type I error because once a treatment is labeled ineffective, there is a chance nobody will try or test it again. Researchers should be very careful about reporting type II errors. This type of error can be minimized by taking bigger samples and by employing sufficient doses of the tested drug in the trial.

66.8 Outliers

When data are analyzed, it is sometimes found that one value is very different from the others. Such a value is called an outlier. If this value is obtained due to chance, then it should be kept in the final analysis as it belongs to the same distribution as the population of interest. If it is due to an error, then it should be deleted to avoid invalid results.

66.9 Various Statistical Tests and their Uses

The statistical tests used for analyzing data have to be chosen carefully. Tests that are not appropriate for a given data set and study design will lead to invalid conclusions. In general, the two types of statistical test are parametric and nonparametric. The parametric tests include the *t*-test, analysis of variance (ANOVA), linear regression and Pearson correlation. The nonparametric tests include Wilcoxon, Mann-Whitney U test, Kruskal-Wallis test, Friedman ANOVA, and Spearman rank correlation. Variables that follow normal distributions can be subjected to parametric tests, and those that do not are suitable for nonparametric tests.

66.9.1 Parametric Tests

Parametric tests make certain assumptions about the population from which the sample is obtained. One of these assumptions is that the study variable is distributed normally in the population.

66.9.2 Student's *t*-Test

This is still the most popular test used in original research involving the use of small samples and is the most sensitive to interval data. The *t*-test is always used when the sample size is 30 or less. If the sample size is more than 30, a *Z*-test is applied. There are two types of *t*-tests, paired and unpaired.

When two measurements on the same subjects after two consecutive treatments are to be compared, a paired *t*-test is used; for example, if the effect of drug A on blood sugar is to be compared between start of treatment (baseline) and after one month of treatment.

When two measurements on two different groups are to be compared, an unpaired *t*-test is used; for example, if the effects of drugs A and B on mean change in blood sugar after one month from baseline are to be compared between the two groups.

When there are more than two groups, a novel test such as ANOVA is desirable to examine differences in the data before various combinations of means are tested to determine individual group differences. If ANOVA is not performed, multiple tests between different pairs of means will alter the alpha (α) level for the experimentation as a whole. Therefore, ANOVA protects the researcher against error inflation if there are differences among the means of the groups.

66.9.3 One-Way ANOVA

This statistic compares three or more unmatched groups when the data are categorized in one way; for example, to compare a control group with three different doses of drug *X* in rats. In this case, there are four unmatched groups of rats, so it is necessary to apply one-way ANOVA. We should choose a repeated measures ANOVA test when the trial uses matched subjects; for example, the effect of atenolol on each subject before, during, and after treatment. Matching should not be based on the variable to be compared but on the basis of age or other variables. The term *repeated measures* applies strictly when the trial involves repeated treatments of one subject. ANOVA works well even if the distribution is only approximately Gaussian. Therefore, these tests are used routinely in many fields of science. The *p*-value is calculated from the ANOVA table.

66.9.3.1 Post-hoc Tests

Post-hoc tests are modifications of the *t*-test. They account for multiple comparisons and for the fact that the comparisons are interrelated. ANOVA only shows whether there is significant difference among the various groups. If the results are significant, ANOVA does not tell at what point the difference between the various groups subsists.

But a post-hoc test is capable of pinpointing the exact difference among the groups being compared, so it is statistically useful. There are five main types of post-hoc test, namely: Dunnett's, Tukey, Newman-Keuls, Bonferroni, and the test for a linear trend between mean and column number.

66.9.3.2 Selection of the Appropriate Post-hoc Test

1. *Dunnett's post-hoc test.* This test is used when one column represents the control group and all the other columns are to be compared with it but not with each other.
2. *Bonferroni, Tukey's or Newman's test.* These post-hoc tests are used when all pairs of columns are to be compared with each other. If there are more than five groups, Tukey's or Newman's test is used instead of Bonferroni.

66.9.4 Two-Way ANOVA

This test is also called two-factors ANOVA. It determines how a response is affected by two factors; for example, if a response to three different drugs in both men and women is to be measured. We will limit our attention to one-way ANOVA here as we are dealing with basic biostatistics.

66.10 Nonparametric Tests

Nonparametric tests are less powerful than parametric tests. Generally, *p*-values tend to be higher, making it harder to detect real differences. Therefore, it is best to try to transform the data first. Sometimes a simple transformation will convert non-Gaussian data to a Gaussian distribution. A nonparametric test is considered only if Tukey's or Newman's test is used and the outcome variable is on a *rank or scale* with only a few categories. In this case, the population is far from Gaussian or one or a few values are off scale, too high or too low to measure.

66.10.1 Chi-square Goodness-of-Fit Test

The chi-square test is a nonparametric test of proportions. It is not based on any assumption or the distribution of any variable. Although different, it follows a specific distribution known as the chi-square distribution, which is very useful in research. It is most commonly used when the data are frequencies such as the numbers of responses in two or more categories. The major limitation of this test is that it tests the *association between two events or characters* but does not measure the strength of association. It only indicates the probability (*p*) that the association is attributable to chance. It involves the calculation of a quantity called chi-square (χ^2), from the Greek letter "chi" (χ), pronounced "Kye." It was developed by Karl Pearson.

66.10.1.1 Applications

1. *Test of proportion.* This test is used to find the significance of a difference in two or more than two proportions.

2. *Test of association.* The test of association between two events in binomial or multinomial samples is the most important application in statistical methods. It measures the probability of association between two discrete attributes. Two events can often be studied for their association such as smoking and cancer, treatment and outcome of disease, level of cholesterol and coronary heart disease. In these cases there are two possibilities: either they influence or affect each other or they do not. In other words, it can be said that they are dependent on or independent of each other. Thus, the test measures the probability (p) or relative frequency of association due to chance and also whether two events are associated or dependent on each other. Variables used are generally dichotomous, e.g. improved/not improved. If the data are not in the format, the investigator can transform them into dichotomous data by specifying upper and lower limits. A multinomial sample is also useful for determining the association between two discrete attributes, such as the association between numbers of cigarettes smoked per day (≤ 20) and the incidence of lung cancer. Since the table presents the joint occurrence of two sets of events, the treatment and outcome of a disease, it is called a contingency table (Con – together; tangle – to touch). Since it has two rows and two columns it is also called a 2×2 contingency table.

66.10.1.2   Preparation of a 2×2 Table

When there are only two samples, each divided into two classes, it is called a four cells or 2×2 contingency table. In a contingency table, we need to enter the actual number of subjects in each category. We cannot enter fractions or percentages or means. The top row usually represents exposure to a risk factor or treatment and the bottom row is mainly for controls. The outcome is entered as the columns on the right side with the positive outcome as the first column and the negative outcome as the second. A particular subject can be in only one column, not both. Table 66.1 illustrates this.

Even if sample size is small (<30), this test is still applied using the Yates correction, but the frequency in each cell should not be less than five. Yates's correction is not applicable to tables larger than 2×2. When the total number of items in a 2×2 table is less than 40 or the number in any cell is less than five, Fischer's test is more reliable than the chi-square test.

66.10.2   Wilcoxon Matched Pairs Signed Ranks Test

This test is mostly used when the data are in the form of ranks or scores. For large numbers it is almost as sensitive as Student's t-test. It is used when the data are not normally distributed in a paired design. It is also called the Wilcoxon Matched Pair test. It analyzes only the difference between the paired measurements for each subject.

Table 66.1 A 2×2 contingency table.

Exposure (novel treatment)	Outcome (clinical improvement)	
	Yes	No
Yes	20	8
No	6	22

Table 66.2 Common statistical tests used for analysis of data depending on hypothesis or question addressed.

Goal	Parametric test (If Gaussian distribution)	Nonparametric test (If non-Gaussian distribution)
Compare two unpaired groups	Unpaired *t*-test	Mann-Whitney test
Compare two paired groups	Paired *t*-test	Wilcoxon-Signed Ranks test
Compare three or more unmatched groups	One-way ANOVA	Kruskal-Wallis test
Compare three or more matched groups	Repeated measures ANOVA	Friedman test
Association between two variables	Pearson correlation	Spearman correlation
Only two possible outcomes		Chi-Square test/Fisher's test
Predict value from another measured variable	Simple linear or Nonlinear regression	Nonparametric regression
Predict value from several measured variables	Multiple linear regression	

ANOVA = Analysis of variance.

66.10.3 Mann-Whitney Test

This test is the same as the Wilcoxon Matched Pairs Signed Ranks test except it is used when two unpaired groups are to be compared.

66.10.4 Friedman Test

This test is similar to repeated measures ANOVA. It is used when the data are not normally distributed and compares three or more paired groups. The values are ranked in each row from low to high.

66.10.5 Kruskal-Wallis Test

This test is similar to one-way ANOVA. It is used if the data are not normally distributed and compares three or more unpaired groups.

Whenever possible, when selecting a statistical test, ask yourself what kind of data you have collected. Then refer to Table 66.2 for selection of the appropriate test.

66.11 Statistical Power

Any study should have at least 80% power to be scientifically sound. If the power of a study is less than 80% and the difference between groups is not significant, then it can only be inferred that no difference between the groups could be detected. It cannot be inferred that no difference exists; the power of the study is too low to identify any real difference. If we increase the power of the study, then the sample size also increases. A more powerful study is required to have a greater chance of identifying existing differences. Power is calculated by subtracting the beta error from 1. Hence power is (1 − beta). The power can be calculated after completion of the study; this is called a posteriori power calculation. It is always better to decide the power of study at the outset of the research. Nowadays, the power of the study can easily be calculated by using appropriate computer software and entering the desired values.

66.12 Determination of Sample Size

Adequate sample size is very important in biomedical research. A small sample will give invalid results, whereas a large sample requires more cost, manpower, and time. The following points are to be considered carefully whenever sample size is calculated for estimating means. First, the minimum expected difference between the groups is assessed. Sample size cannot be calculated without an estimate of variability of the items. Therefore, we have to determine the SD of the variables. Standard deviation can be calculated either from a pilot study or from previous/similar studies. Set the level of significance called the alpha level at $p < 0.05$ and the power of the study at 80%, then feed these values into a computer with suitable software to obtain the required sample size. Various software programs are available free of cost for this purpose. Appropriate allowances should be given for noncompliance and dropouts. Then the final sample size for each group in study is inferred. For example, a new drug for hypertension is to be compared with placebo. The question is how many patients will be required at 90% power and 5% significance to detect an average blood pressure difference of 5 mmHg between the new drug and placebo groups assuming a SD of 10 mmHg. In this case, the minimum expected difference is 5.0, SD is 10, alpha level is 0.05, and power of study is 90%. After putting all these values into the computer, the sample size is calculated at 85. If we add five to allow for noncompliance and dropout, then the final sample size for each group will be 90.

66.13 Establishing a Statistical Relationship

66.13.1 Correlation

Correlation measures the closeness of linear association between two continuous variables. It is expressed in terms of a coefficient. Correlation coefficient values represented by r are always between +1 and −1. The maximum value of 1 is obtained if all points on the scatterplot lie on a straight line. The value will be zero if the variables are not correlated. The association is positive (Figure 66.2) when the values of the x and y axes tend to be high or low together (positive relationship). Conversely, the association is negative (Figure 66.2) when a high y-axis value tends to go with a low value on the x-axis (inverse relationship). The larger the correlation coefficient, the stronger the association. It should be remembered that correlation between two variables does not necessarily suggest a cause–effect relationship. This test is normally used for forming a hypothesis or suggesting areas for future research. It indicates the strength of association for any data in comparable terms (e.g. correlations between height and weight, parity and birth weight, socioeconomic status and hemoglobin).

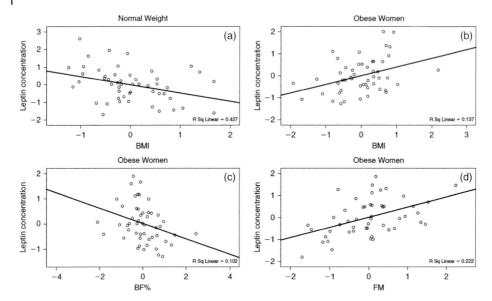

Figure 66.2 Correlation plots showing positive or direct (b and d) and negative or inverse (a and c) relationships: (a) serum leptin concentration versus body mass index (BMI) in normal weight women (a), and (b) serum leptin concentration versus BMI, body fat percent (BF%), and (d) fat mass (FM) in overweight/obese women. Source: Ostadrahimi et al. (2008).

66.13.2 Regression

Regression analysis is a process of estimating the relationship between a dependent variable and one or more independent variables. If the dependent variable is continuous, the analysis of interest is linear regression. If the dependent variable is categorical, it is logistic regression. The variable showing change is called the dependent variable (or outcome) and those not changing and influencing the dependent variable are called independent variables (or predictors). Apart from estimating the association, regression analysis also provides the following information:

1. It predicts dependent variables from corresponding changes in the independent variable.
2. In a restricted sense, linear regression indicates causal connections between two variables. However, caution should be exercised in drawing conclusions about causality based on regression or correlation. Sometimes such conclusions are misleading.
3. The equation obtained after regression analysis can be used to predict the value of the dependent variable if the independent variable values are available.

66.14   Interpretation of Interactions

It is easy to find the interaction between the dependent variable and a single independent variable using either qualitative or quantitative tests of significance. However, the scenario becomes more complicated if another independent variable is suspected of influencing the change in the dependent variable; for example, whether music on/off (independent variable) influences memory recall (dependent variable). Here one will apply a *t*-test to the data. But suppose it is suspected that being introvert or extrovert affects recall with the music on or off (i.e. the precise question that would be framed if you want to know the interaction between them). In this case, one has to make a 2×2 contingency table as in Fischer's exact test and apply a test of significance.

If this indicates nonsignificance, then the other independent factor is not interacting. But if it turns out to be significant, then one has to examine the actual mean (SD) values, distinguish the influence of the extra independent factor separately and conclude accordingly.

66.15 Determination of Statistical Causality

To establish causality between exposure and outcome is important in biostatistics. It has implications for prevention and many other remedial aspects of disease or outcome. Association can be estimated by two kinds of analysis, i.e. correlation and association. Correlation measures the strength of association between two parametric variables. The larger the correlation coefficient, the greater the strength of association. However, a strong correlation does not necessarily mean causation. The correlation can be arbitrary and can occur by chance. The number of sunny days and runs scored by a cricket player may be correlated, but there need be no causal link. Another kind of analysis is by chi-square, which is done when outcome and exposure both are discrete and presentable in a contingency table. Austin Bradtor Hill has suggested some criteria to establish the causation aspect of association. These include:

1. *Strength of association.* The greater the relative risk or odds ratio, the higher the chances of association and causation; e.g. an odds ratio of magnitude five is more likely to indicate association and causation than a ratio of two.
2. *Temporality of causation.* The exposure should precede the outcome. This is established in a cohort design.
3. *Consistency of association.* The more studies suggest association, the greater the likelihood of causation.
4. *Availability of theory or mechanism.* If there is some plausible explanation or mechanism for the outcome, the chances of causation are higher.
5. If a dose and effect relationship is evident from studies, this favors causation.
6. Availability of experimental evidence to show cause and effect indicates causality.

66.16 Adjustment of Covariates in Clinical Trials

The purpose of this adjustment process is to increase the credibility of the data generated in clinical trials by showing that a treatment effect is not due to imbalance in the patient characteristics (covariants). The adjustment procedures can be employed before or after the trial. They are termed pre-adjustment and post-adjustment procedures. A pre-adjustment procedure includes such methods as stratification and minimization. Stratification of patients holds good if the strata are limited to three. Minimization is another way to reduce imbalance; it involves dynamic allocation of patients. Some statisticians also recommend a combined approach of stratification and minimization. For post-adjustment, a stratified test is preferred over an unstratified test. It is always better to decide which covariates should be included in the analysis and the method for handling missing values in the protocol itself in order to reduce the scope for data manipulation. The choice of test to be applied in a post-adjustment procedure depends on the usual factors such as type of data.

66.17   Statistical Software Packages

Statistical computations are nowadays more feasible, owing to the availability of computers and suitable software programs. Commonly used software includes MS Office Excel, Graph Pad Prism, SPSS, Instant, Dataplot, Graph Pad Instat, STATA, SAS, and Sigma Graph Pad.

Acknowledgments

This chapter is an updated version of the authors' article published in *Indian Journal of Pharmacology* (Dakhale et al. 2012). The authors are grateful to the editors and publisher of the journal for grating the permission to reproduce this update.

References

Dakhale, G.N., Hiware, S., Shinde, K. et al. (2012). Basic biostatistic for postgraduate students. *Indian J. Pharmacol.* 44: 435–442.

Ostadrahimi, A., Moradi, T., Zarghami, N. et al. (2008). Correlates of serum leptin and insulin-like growth factor-I concentrations in normal weight and overweight/obese Iranian women. *J. Women's Health (Larchmt)* 17 (8): 1389–1397.

Further Reading

Bland, J.M. and Altman, D.G. (1996). Transforming data. *BMJ* 312: 770.

Charan, J. (2010). Mean ± SEM or mean(SD)? *Indian J. Pharmacol.* 4: 329.

Curran-Everett, D. and Benos, D.J. (2004). Guidelines for reporting statistics in journals published by the American Physiological Society. *Am J. Physiol. Regul. Integr. Comp. Physiol.* 287: R247–R249.

Dukes, K.A. and Sullivan, L.M. (2007). A review of basic statistics. In: *Evaluation Techniques in |Biochemical Research* (ed. D. Zuk), 50–56. Cambridge, MA: Cell Press.

Gerstman, B.B. (2008). *Basic Biostatistics, Statistic for Public Health Practice.* Sudbury, MA: Jones and Bartlett Publishers.

Ghosh, M.N. (2005). *Statistical Methods*, 3e. Kolkata: Bose Printing House.

Kim, J.S., Kim, D.K., and Hong, S.J. (2011). Assessment of errors and misused statistics in dental research. *Int. Dent. J.* 61 (3): 163–167.

Lang, T. (2004). Twenty statistical errors even you can find in biomedical research article. *Croat. Med. J.* 45: 361–370.

Lang, T. and Altman, D. (2013). Basic statistical reporting for articles published in clinical medical journals: the SAMPL guidelines. In: *Science Editors' Handbook* (ed. P. Smart, H. Maisonneuve and A. Polderman), 5–9. European Association of Science Editors.

Lew, M.J. (2012). Bad statistical practice in pharmacology and other biomedical disciplines: you probably don't know P. *Br. J. Pharmacol.* 166: 1559–1567.

Ludbrook, J. (2008a). Analysis of 2×2 tables of frequencies: matching test to experimental design. *Int. J. Epidemiol.* 37: 1430–1435.

Ludbrook, J. (2008b). CEPP Guidelines for the use and presentation of Statistics. Available from http://www.blackwellpublishing.com/pdf/CEPP_guidelines_pres_stat.pdf.

Ludbrook, J. (2008c). Letter to the editor. *Clin. Exp. Pharmacol. Physiol.* 35: 1271–1274.

Ludbrook, J. (2008d). Statistics in biomedical laboratory and clinical science: applications, issues and pitfalls. *Med. Princ. Pract.* 17: 1–13.

Mahajan, B.K. (2010). *Methods in Biostatistics*, 7e. New Delhi: Jaypee Brothers Medical Publisher (P) Ltd.

Medhi, B. and Prakash, A. (2010). *Practical Manual of Experimental and Clinical Pharmacology*, 1e, 123–133. New Delhi: Jaypee Brothers Medical Publisher (P) Ltd.

Motulsky, H.J. (1999). *Amazing Data with GraphPad Prism*, 65–89. San Diego, CA: GraphPad Software Inc. www.graphpad.com.

Nanivadekar, A.S. and Kannappan, A.R. (1990a). Statistics for clinicians: nominal data (I). *J. Assoc. Physicians India* 38: 931–935.

Nanivadekar, A.S. and Kannappan, A.R. (1990b). Statistics for clinicians: introduction. *J. Assoc. Physicians India* 38: 853–856.

Nanivadekar, A.S. and Kannappan, A.R. (1990c). Statistics for clinicians: interval data (I). *J. Assoc. Physicians India* 39: 403–407.

Prabhakar, G.N. (2006). *Biostatistics*, 180–188. New Delhi: Jaypee Brothers Medical Publisher (P) Ltd.

Rao, K.V. (2007). *Biostatistics: A Manual of Statistical Methods for Use in Health, Nutrition and Anthropology*, 2e. New Delhi: Jaypee Brothers Medical Publisher (P) Ltd.

Saracio, G., Jennings, L.W., and Hasse, J.M. (2013). Basic statistical concepts in nutrition research. *Nutrition in Clinical Practice.* 28 (2): 182–193.

Sathian, B. and Sreedharan, J. (2007). Meaning of p-value in medical research. *WebmedCentral Biostatistics* 3 (5): WMC003338.

Sathian, B. and Sreedharan, J. (2012). Importance of biostatistics to improve the quality of medical journals. *WebmedCentral Biostatistics* 3 (5): WMC003332.

Sokal, R.R. and Rohlf, F.J. (2009). *Introduction to Biostatistics*, 2e. Dover Publications.

Wayne, D. (2005). *Biostatistics: A Foundation for Analysis in the Health Science*, 8e. Wiley.

67

An Overview of Systematic Review and Meta-Analysis

Anthony V. D'Antoni and Loretta Cacace

Division of Anatomy, Department of Radiology, Weill Cornell Medicine, New York, NY, USA

67.1 Systematic Reviews and Meta-Analyses

Evidence-based medicine (EBM) requires physicians to stay abreast of the current biomedical literature related to risk factors, diagnosis, treatment, and prevention of disease in order to provide the best care for their patients. This is a difficult task because the EBM paradigm now exists in a world of information overload. The number of published studies is voluminous, and the heterogeneity that exists among them makes comparing their results arduous. This fact demonstrates why evidence-based review articles are so important for EBM. Review articles concisely summarize the entire body of evidence on a specific topic, making them an integral part of the EBM paradigm (Straus et al. 2011).

Review articles are usually systematic reviews (SRs) conducted by a group of experts, some of whom may be epidemiologists. These reviews use a highly refined scientific research method that minimizes bias and other errors inherent in traditional narrative reviews. An SR must have explicit inclusion and exclusion criteria for the studies it retains. The procedure by which the authors found the individual studies is outlined in a flowchart so that readers have a visual representation of how the final sample size is reached. An SR also includes the designs with related commentaries for the individual studies retained in the review (Margaliot and Chung 2007).

Two kinds of SRs exist: qualitative and quantitative. This chapter focuses on the quantitative SR, which is also called a meta-analysis. A meta-analysis is a type of SR that contains a series of quantitative methods for systematically combining (pooling) data from many individual studies (typically from randomized controlled trials [RCTs], but also from observational studies) that allow authors to draw conclusions from these data (Chung et al. 2006). In an RCT, researchers are able to control (to a certain extent) the intervention of interest and other potential confounding variables that would normally plague an observational study. In the field of epidemiology where observational studies are the norm, the experimental RCT is therefore considered the gold standard.

The meta-analysis is a review article containing a compilation of results of numerous experimental studies, making these review articles, essentially, the "best of the best" evidence to establish the efficacy of a treatment for a specific disease. The statistical advantage of pooling data from numerous RCTs also contributes to the strength of the

A Guide to the Scientific Career: Virtues, Communication, Research, and Academic Writing, First Edition.
Edited by Mohammadali M. Shoja, Anastasia Arynchyna, Marios Loukas, Anthony V. D'Antoni, Sandra M. Buerger, Marion Karl and R. Shane Tubbs.
© 2020 John Wiley & Sons, Inc. Published 2020 by John Wiley & Sons, Inc.

evidence provided by meta-analyses. Pooling the data increases the sample size, increasing the statistical power of the analysis, which allows authors to estimate a more precise treatment effect. To summarize, the meta-analysis of RCT's is the best form of evidence clinicians can use when they need a quick assessment of the current literature concerning a treatment of interest (Chung et al. 2006).

Despite being the gold standard of evidence, meta-analyses are somewhat underutilized. This may be due to the fact that for authors, meta-analyses are difficult to conduct and for physicians these studies are notoriously difficult to critically appraise. The purpose of this chapter is to help physicians understand the structure of meta-analyses so that they can critically appraise them and incorporate their results in clinical practice.

67.2 Structure of a Meta-Analysis

Meta-analyses published in the peer-reviewed biomedical literature usually contain the following sections: title, structured abstract, introduction, methods, results, and discussion (Chung et al. 2006).

67.2.1 Title

Like other research papers, the title of a meta-analysis should be clear and to the point. This is not often the case, however, and readers are sometimes left wondering if a paper is a meta-analysis or qualitative systematic review. The title should explicitly include the word "meta-analysis" to avoid confusion and clearly describe the topic of the meta-analysis.

67.2.2 Structured Abstract

Most biomedical journals that publish meta-analyses require authors to provide a structured abstract (details are usually found in the instructions for authors) when they submit their manuscript to the journal for publication consideration. A structured abstract usually contains Introduction, Methods, Results, and Discussion sections, although some journals differ in the names of the sections (e.g. using "background" instead of "introduction"). The advantages of a structured abstract are that it is easier to read and standardizes information so that meta-analyses on the same topic published in different journals can be more easily compared.

67.2.3 Introduction

The introduction section should contain information that allows readers to quickly understand the value of the meta-analysis and its clinical relevance. Authors of the meta-analysis must demonstrate that an exhaustive literature search was conducted by referencing the peer-reviewed literature (Chung et al. 2006). When appropriate, epidemiologic and economic information (e.g. annual cost to treat the condition) about the disease should be included. The introduction is viewed as the opportunity for the authors to entice the reader to read further and should explicitly state the purpose of the meta-analysis. Operational definitions of the disease and the treatments (both conventional and experimental) must be included in this section.

The literature review should enable readers to clearly visualize the gaps in the literature, i.e. is this meta-analysis the first to assess this treatment? Have there been other meta-analyses of its kind, but not focused on certain groups of subjects? Is this analysis focused on different outcomes than other previous meta-analyses? Essentially, the authors should justify why conducting a meta-analysis was necessary, and they should address how their analysis is going to contribute to the literature. Any controversies related to the treatments should also be mentioned and any treatment guidelines that may exist should be referenced. The final paragraph of the introduction should clearly frame the research question using the PICO (patients, intervention, comparator, and outcomes) strategy, with emphasis on the comparative treatments (see Chapter 48) (Chung et al. 2006).

67.2.4 Methods

The methods section is the heart of the meta-analysis because here is where any methodologic biases are revealed. In the past, traditional narrative review articles were written by experts in their fields and published in peer-reviewed journals. A major problem with these reviews was author bias and lack of a systematic and reproducible methodology. Authors of these reviews often wrote them using published papers that supported their views while omitting others, which could ultimately led to biased reviews (Margaliot and Chung 2007). Bias is any aspect of the research study design that could potentially alter the results to wrongfully support or reject the hypothesis of the interest.

In a systematic review, the process by which the authors find the individual studies must be exhaustively explained so that it can be replicated. The authors should list all of the databases (e.g. PubMed, CINAHL, etc.) that were used to find individual studies and include the medical subject headings (MeSH) and non-MeSH terms that were used to search the literature, including any Boolean operators. Some meta-analyses, like those published in the Cochrane Reviews, include appendices with this information. The dates that were used for each search should be included, and the termination date should be "present," which indicates that the search was as up-to-date as possible at the time. All languages (both domestic and foreign) that were used to search for individual studies and any unpublished clinical trials like those in the Cochrane Central Register for Controlled Trials should be disclosed in the methods.

The inclusion and exclusion criteria used to screen studies should be listed. The individual studies found by the literature search should have been critically appraised, and a valid and reliable rubric should have been used for this endeavor. The Jadad scoring system is commonly used to assess the methodologic quality of RCTs (Margaliot and Chung 2007). These scores (on a scale from 1 to 5) consider important aspects of the RCT study design, such as blinding and randomization. Some experts would rather retain only studies with high Jadad scores (4 or 5) for the meta-analysis, while others have discouraged use of such scales on the ground that they are not validated. The process by which the data were extracted from the retained studies should be described and any difficulties in the extraction process explained (Margaliot and Chung 2007). The statistical models (fixed versus random effects) used to analyze the data should be described and justification given for their use. The statistical software packages used to analyze the data should also be mentioned.

By providing the details of this methodology, the authors are able to explain how they were able to minimize bias. For example, by explaining the selection criteria, authors are able to convey to the reader how they minimized selection bias that used to be an issue in earlier review articles. Further, at the level of the individual study included in the analysis, bias can be further eliminated by only including those that blinded and randomized throughout their study design. When an experiment is blinded, it means that the subject, the researcher, and/or the statistician do not know who was assigned which treatment, ensuring that analysis and measurement of outcome variables remain unbiased (Gordis 2009). Randomization is the random allocation of the treatment options among subjects in order to avoid the bias that certain types of people get the treatment as opposed to others (i.e. more women than men had the treatment) (Gordis 2009).

Inclusion of all of these details within the methods of the meta-analysis is crucial to their success in providing accurate and detailed data to researchers and clinicians (Margaliot and Chung 2007).

67.2.5 Results

The results section should be clear and concise. A flowchart that demonstrates how studies were retained in the meta-analysis should be included in the results. Such a chart includes the number of abstracts and other resources screened, number of full-text studies retained, and number of studies excluded and the reasons for their exclusion. The flowchart should conform to the guidelines set forth by the Consolidated Standards of Reporting Trials (CONSORT) (Chung et al. 2006). Recently, however, the guidelines of Preferred Reporting Items for Systematic Reviews and Meta-Analyses (PRISMA[1]) to replace CONSORT have been adopted by a number of high-quality biomedical journals. Information on the quality of the individual studies should be provided in this section and may be presented in a table. If heterogeneity existed among the individual studies, then an estimate of whether the differences were significant using a chi-square test should be reported. Results of descriptive and inferential statistics are tabulated so readers can easily identify the statistical tests performed and if they reached significance.

Pooling of the data from homogeneous studies results in calculation of a summary estimate in the form of a forest plot. Forest plots are usually created using mean differences between groups, relative risk, or odds ratio (Figure 67.1). The plot includes a list of the individual studies (by author name and publication year), number of subjects in each group (control and treatment), and the point estimate (95% confidence interval) of each study. The point estimates are combined into a summary estimate at the bottom of the plot (typically a diamond shape) that may favor the control or treatment. In the case of forest plots demonstrating mean difference between groups, a vertical line through 0 corresponds to no effect of treatment. Therefore, a 95% confidence interval that includes 0 means that the result is not significant ($p > 0.05$). If the forest plot includes relative risk or odds ratio, then a vertical line through 1.0 indicates that there is no measurable association between the treatment and disease of interest. In this case, a 95% confidence interval that includes 1.0 means that the result is not significant ($p > 0.05$) (Gordis 2009).

---

1 The Prisma Statement (2015): Transparent Reporting of Systematic Reviews and Meta-analyses. http://www.prisma-statement.org

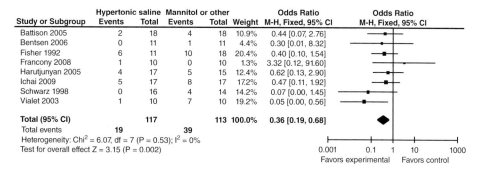

Study or Subgroup	Hypertonic saline Events	Total	Mannitol or other Events	Total	Weight	Odds Ratio M-H, Fixed, 95% CI
Battison 2005	2	18	4	18	10.9%	0.44 [0.07, 2.76]
Bentsen 2006	0	11	1	11	4.4%	0.30 [0.01, 8.32]
Fisher 1992	6	11	10	18	20.4%	0.40 [0.10, 1.54]
Francony 2008	1	10	0	10	1.3%	3.32 [0.12, 91.60]
Harutjunyan 2005	4	17	5	15	12.4%	0.62 [0.13, 2.90]
Ichai 2009	5	17	8	17	17.2%	0.47 [0.11, 1.92]
Schwarz 1998	0	16	4	14	14.2%	0.07 [0.00, 1.45]
Vialet 2003	1	10	7	10	19.2%	0.05 [0.00, 0.56]
Total (95% CI)		**117**		**113**	**100.0%**	**0.36 [0.19, 0.68]**
Total events	**19**		**39**			

Heterogeneity: Chi² = 6.07, df = 7 (P = 0.53); I² = 0%
Test for overall effect Z = 3.15 (P = 0.002)

Figure 67.1 Example of a forest plot from a quantitative systematic review comparing the rates of treatment failure or insufficiency with hypertonic saline versus mannitol or normal saline for intracranial hypertension. M-H = Mantel-Haenszel. Source: Mortazavi MM, Romeo AK, Deep A, et al. (2012). Hypertonic saline for treating raised intracranial pressure: literature review with meta-analysis. *Journal of Neurosurgery* 116 (1): 210–221. Reproduced with permission.

A funnel plot may be used to demonstrate if publication bias exists in the meta-analysis. A funnel plot is essentially a scatterplot of treatment effect and sample size of the individual studies. Readers should look for an inverted funnel that is symmetrical because this suggests no publication bias exists in the meta-analysis. In contrast, an inverted funnel that is asymmetrical suggests a relationship between treatment effect and sample size, and this should alert readers that publication bias exists in the meta-analysis (Lau et al. 2006).

67.2.6 Discussion

Similar to other study designs, the Discussion section of a meta-analysis includes commentary by the authors in relation to their original hypotheses (i.e. did the meta-analysis accept or reject the null hypothesis at the 95% confidence level?). Here is where the authors discuss the implications of their findings and how they fit into data from previous studies and current standards of care. Study limitations such as heterogeneity of individual studies, low internal validity, and low external validity (generalizability) are addressed in this section. The impact of the meta-analysis on clinical practice and the usefulness of conducting future meta-analyses should also be mentioned by the authors in this section.

67.3 Role of Meta-Analysis in Evidence-Based Medicine

EBM is the conscientious use of current best evidence when making decisions about the care of individual patients (Straus et al. 2011). Physicians use research evidence, clinical experience, and patient values in the EBM paradigm in order to tailor treatments to patients and provide high-quality care. While this principle seems intuitive and straightforward, the fact is that EBM is challenging to practice because it requires that physicians remain current with the tsunami of published biomedical information. This is precisely why meta-analyses are powerful tools for the EBM physician. In the field of EBM, a hierarchy exists among the levels of evidence[2], with meta-analyses existing at the highest level possible. The fact that meta-analyses pool data from numerous RCTs allows physicians to quickly incorporate powerful and summative data into the clinical decision-making process, which results in the best evidence for the care of patients.

The important role that meta-analyses have in the era of EBM cannot be overstated. Medicine is in a state of perpetual evolution as a result of novel research that is continually published and disseminated. Meta-analyses, therefore, can allow physicians to keep up-to-date with the volume and pace of these published studies provided they possess the ability to critically appraise meta-analyses and infuse their results in clinical practice.

67.4   Critical Appraisal Sheet for Systematic Reviews

The Centre for Evidence-Based Medicine (CEBM, www.cebm.net) of the University of Oxford, England, is an important resource for physicians interested in EBM. CEBM contains many freely downloadable critical appraisal sheets, including one for SRs. We encourage readers to use the SR critical appraisal sheet when reading meta-analyses and to incorporate meta-analytic results in the clinical setting whenever possible, even when presenting cases for grand rounds.

References

Chung, K.C., Burns, P.B., and Myra Kim, H. (2006). A practical guide to meta-analysis. *Journal of Hand Surgery* 31 (10): 1671–1678.

Gordis, L. (2009). *Epidemiology*, 4e. Philadelphia, PA: Saunders Elsevier.

Lau, J., Ioannidis, J.P.A., and Schmid, C.H. (1997). Quantitative synthesis in systematic reviews. *Annals of Internal Medicine* 127: 820–826.

Margaliot, Z. and Chung, K.C. (2007). Systematic reviews: a primer for plastic surgery research. *Plastic Reconstructive Surgery* 120 (7): 1834–1841.

Straus, S.E., S. Richardson, W., P., G. et al. (2011). *Evidence-Based Medicine: How to Practice and Teach it*, 4e. Edinburg: Churchill Livingstone Elsevier.

---

2 Centre for Evidence Based Medicine (2009) Levels of Evidence. Available from http://www.cebm.net.

68

An Introduction to Meta-Analysis

Arindam Basu

School of Health Sciences, University of Canterbury, Christchurch, New Zealand

68.1   Introduction

Health information has grown exponentially in recent years, and evidence-based practice is the new paradigm of healthcare. As a result, health professionals in all disciplines are expected to be up to date with the most relevant and best current evidence. However, it is impossible to keep up with the sheer volume of information (Wyatt 2000; Zipkin et al. 2012). This calls for careful curation of available information. In the absence of robust information synthesis, nonsystematic, narrative information or reviews that are based only on the reviewers' prior experience or expert opinions cannot be reproduced, and as a result, these reviews cannot be updated. With the rapid pace of innovation and new studies that supersede old information, such reviews are at risk of getting outdated. In addition to this, the well-established dictum of best evidence positions meta-analysis of randomized controlled trials (RCTs) or interventions at the peak of evidence on which one can base one's practice of healthcare (Evans 2003). Consequently, carefully conducted, reproducible reviews that employ framing of relevant research questions, well-specified algorithms to search literature databases and research databases of primary studies, and synthesis of information are very important to conduct knowledge synthesis and develop practice guidelines. Such explicit reviews of the literature are known as systematic reviews. In meta-analysis, findings from similar but unique studies are statistically synthesized. In this chapter, the principles and practices of meta-analysis will be presented.

In general, health professionals conduct three different types of reviews, depending on the aim or goal of the review. Where the goal of the review is to collate all available evidence or information on a particular topic of interest (omnibus review), a comprehensive narrative review of the topic from different perspectives is usually conducted. For example, Sonia Ancoli-Israel et al. (2003) conducted a comprehensive review of the literature to identify studies on the role of actigraphy in the study of sleep and circadian rhythms that addressed management of ambulatory care sensitive conditions. The aim of this review was to provide a large, comprehensive compendium of the role of actigraphy.

Another purpose of conducting a review may be to identify the best evidence to manage a particular disease or health condition (best practice review). In these reviews, the

A Guide to the Scientific Career: Virtues, Communication, Research, and Academic Writing, First Edition.
Edited by Mohammadali M. Shoja, Anastasia Arynchyna, Marios Loukas, Anthony V. D'Antoni, Sandra M. Buerger, Marion Karl and R. Shane Tubbs.

reviewers not only compare benefits of the different treatments or assess the association between different exposure conditions and health outcomes but also compare and contrast risks. The reviewers thus compare the relative effectiveness and harms and their costs and effectiveness associated with those treatments or approaches.

Third, the most popular type of review in medicine and public health is where investigators aim to summarize the association between exposures and health outcomes or effectiveness of different treatments for specific conditions either in contrast with no treatments, placebos, or alternative treatments, but such associations are studied using a well-defined approach. The authors, adhering to evidence-based principles, stepwise frame answerable questions, identify relevant studies, critically appraise the studies, and, on the basis of that critical appraisal, abstract key information from the studies and summarize the findings from individual studies to arrive at an evidence-based answer to the question posed. The comparisons can be conducted among three or more diagnostic or treatment approaches. In terms of methodology, narrative synthesis of information or narrative reviews are conducted to narratively summarize key information contained in the articles considered for review.

Meta-analysis thus refers to a specific type of systematic review that has the following three characteristics:

1. *Numerical or statistical pooling of the study results.* In meta-analysis, data from the studies are weighted and the results are pooled to form a series of summary estimates to estimate an overall effect size
2. *Only two comparisons.* At any time, only two interventions or alternative conditions are compared. Another distinguishing feature of meta-analysis is that, here at any point, only two alternative treatments are compared. These treatments can be a novel intervention versus a placebo, or a novel intervention versus another intervention (treatment as usual), or two alternative interventions or two alternative conditions. This is true of parallel arm meta-analysis that is the subject of this chapter. In contrast, in network meta-analysis (a growing field whose discussion is beyond the scope of this chapter), direct and indirect comparisons are made among networks of studies, allowing for multiple comparisons of the studies.
3. *Included studies should be similar.* This is an important consideration in meta-analysis, and formal approaches exist to indicate the extent to which studies are homogeneous or heterogeneous.

In addition to these distinguishing characteristics, the steps of meta-analysis follow those of any systematic review: in selecting explicit research question, and methodology that is reproducible and synthesis of information. The steps of meta-analysis are explained below.

As illustrated in Figure 68.1, meta-analysis can be conceptualized as a more or less linear process consisting of the following sequential steps:

1. Frame the research question.
2. Search the literature and conduct initial screening of studies based on their titles and abstracts.
3. Critically assess studies based on a close reading of their full texts to identify key data elements and appraise their risk of bias.
4. Abstract data from individual studies.

Figure 68.1 Steps of conducting meta-analysis.

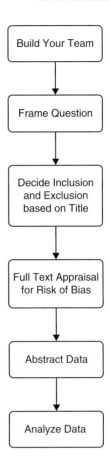

5. Assess the homogeneity of the included studies from statistical and other perspectives (methodological and clonal homogeneity).
6. Pool statistical results abstracted from each of the identified studies to arrive at a summary statistic, and examine the different subsets of studies by performing sensitivity analyses (also referred to as subset analyses or meta-regression).

These steps of meta-analysis indicate that a range of skills are needed, and meta-analysis therefore is a multidisciplinary team-based activity. Consider, for example, that you are planning to conduct a meta-analysis for available evidence on the effectiveness of meditation plus medications, as opposed to medications alone for the control of hypertension. In order to conduct such a meta-analysis, you will need the skills of a physician who is knowledgeable about management of hypertension, but also of an expert who can advise on mindfulness meditation, an expert who can conduct literature search on the topic, and a statistician who can statistically combine data for arriving at a summary estimate. In addition, you will need researchers who can maintain a database full of articles and abstract data from the articles for further processing. In summary, meta-analysis is an interdisciplinary team work. Framing a research question includes domain knowledge and skills, while at the same time, critical appraisal of the studies to identify risks of biases will need skills where the researcher should not allow domain knowledge to be biased at conclusions of specific studies. Therefore, a study team should include both experts and nonexperts. Given the "explosion" of information in literature and research studies, the study team will need a person with skills in literature search to narrow down the exact quanta of literature needed for a successful meta-analysis (a health information specialist can bring on board such skills). Hence, before a meta-analysis can be conducted, it is important for the analyst to put together a team consisting of individuals, with at least the following set of skills:

1. One or more domain experts who can identify and scope the health problem of interest
2. A health information specialist who will bring on board skills of literature search and retrieval of studies
3. A statistician who will analyze complex data from the studies as studies can present data in different ways and often sophistical statistical skills are required in identifying or estimating key data elements for meta-analysis
4. A database expert who can store and curate records or bibliographies

Also, since meta-analysis is based on previously conducted studies and studies whose results have already been obtained, this is necessarily retrospective. Besides, in a primary study, the trialist or the investigator works with individuals. The meta-analyst on the other hand, accesses the primary studies, and these primary studies, rather than individuals, are the source of information. Therefore, quality of meta-analysis depends on the quality of the primary studies: if the original studies are of poor quality, then either the meta-analysis cannot be conducted or if conducted, that meta-analysis is open to the same biases and subsequently leads to propagation of errors in the original studies. The individual steps are explained as follows.

68.2   Step One: Frame a Study Question

68.2.1   How to Select Questions

Framing a research question is the first step in meta-analysis. In general, health problems where answers are not always clear or where increased precisions are needed are good candidates for meta-analysis. For example, Peck et al. (2013) conducted a meta-analysis to assess the magnitude and direction of the difference in blood pressure response to ACE inhibitors between black and white populations. In another example, Babu et al. (2013) addressed whether a job stress factor was associated with hypertension, as previous studies were too diverse and therefore pooling of studies would enable framing a response. These examples show the power of meta-analysis — it is possible to pool together results from small studies that themselves might be underpowered or inconclusive, yet when combined with each other in a meta-analysis, point to overall conclusions or figures that provide stronger estimates of the association between interventions or exposure and outcomes. Thus, questions that are either "not settled," or are based on small but inconclusive studies are good candidates for meta-analysis. The questions can be best framed using the participant-intervention-comparator-outcomes (PICO) format, as follows.

68.2.2   PICO Format

A well-formatted question directs the course of action and specific steps taken in a meta-analysis. The research question is formatted using participants (P), intervention (I) or exposure (E) depending on whether the meta-analysis is about interventions to be tested against each other or is about association of a specific exposure (against another) for a particular outcome; comparator (C) – who or what is being compared with the intervention or the exposure under study, and finally outcome (O) – the specific health outcome of interest, in that order.

The role of participants is important in meta-analysis, as the same topic can result in different research questions, depending on the participant profiles. For example, if your interest is in studying the risks of hormone replacement therapy for breast cancer, the studies can be very different, depending on whether you are only going to be interested in pre-menopausal or post-menopausal women (age as participant character becomes important). Similarly, if you are interested in studying roles of antihypertensive therapy, gender can be an important variable. The intervention or exposure will need to be specified as the scope and relevance of a meta-analysis, and often, whether a meta-analysis can at all be attempted, depends on how broadly or how tightly the intervention is defined. For example, imagine that you are planning to conduct a meta-analysis comparing a combination of mindfulness-based stress reduction (MBSR) and drugs with drugs alone for the control of stress-related symptoms among breast cancer survivors. While this is a well-defined intervention in itself, the number of studies that you can identify may be limited.

On the other hand, you could retrieve a larger number of primary studies if you were to relax the intervention to include "any form" of meditation rather than MBSR. Then again, you would have to sacrifice homogeneity of studies (loosely, similarity of studies, explained later) and in turn, this consideration alone might lead to a different form of summarization rather than conducting a meta-analysis; for instance, you could shape up the review not as a meta-analysis but as an omnibus review or an overview of the effectiveness of any form of meditation for the control of hypertension. You might have to abandon a meta-analysis and end up doing a systematic or narrative review. Therefore, precision in the definition of intervention or exposure criteria can be a major decider for the meta-analysis. Likewise, comparators and outcomes to be studied for a research are crucial to its success.

Table 68.1 shows the PICO criteria for a meta-analysis of the effectiveness of MBSR for control of hypertension among elderly (64+ years old) hypertensives.

Table 68.1 Explanation of PICO.

Condition	Definition or explanation
Participant	Mention the participants, e.g. all hypertensive adults both sexes age 64 years and above.
Intervention	Prescribe mindfulness-based stress reduction plus medications.
Comparator	Prescribe medication alone.
Outcomes	Assess blood pressure control (systolic and diastolic).

68.2.3   Too Narrow Versus Too Broad Meta-Analysis

Framing of the research question sets the tone for a meta-analysis. A meta-analysis can be very narrow in scope; a too narrowly defined meta-analysis can result in retrieval of too few studies. Scoping a meta-analysis is not necessarily an easy task, as this involves taking into consideration several factors: availability of supporting information, background data on the problem being studied, composition of the team and resources, and also the potential impact of the meta-analysis on the problem being studied. If the meta-analysis is too narrow, then the results cannot be generalized to larger population. For instance, generalizability of the results is a problem with randomized controlled trials, as they tend to be very specific, and using few randomized controlled trials in a meta-analysis may not be very helpful for generalization of the study findings. This is almost akin to conducting a subgroup analysis in a meta-analysis where only a select subgroup of available studies are considered; additionally, too narrowly focused meta-analysis often results in selection bias of studies, particularly with experts who are prone to exclude studies that do not meet their inclusion and exclusion criteria, as these are very tightly defined.

On the other hand, if the meta-analysis is too broadly defined, it becomes a very time-consuming exercise, as the number of search results is very large, and it takes time to analyze the volume of studies that are retrieved. Additionally, as studies would be very diverse, risks of studies being heterogeneous is high, and in turn leads to problems of interpretation of data. Large number of studies that are dissimilar to each other because of the diversity and over inclusive nature of the search results in what is often referred to as "mixing of apples and oranges." In a tutorial on conducting Cochrane Reviews, Higgins and Green (2006) have stated that such mixing is fine when the object of the study is to know about "fruits" but not when finer characterization about either apples or oranges is the objective of the study. Given this dilemma, a possible middle path for a large or broad-based question might be to start with a series of smaller meta analyses that compare only two conditions or two interventions at a time and subsequently adding the studies to a thematic whole so that the overall large topic can be addressed from a robust methodological perspective.

68.3   Step Two: Search the Literature and Conduct Initial Screening

After a study question is framed, the meta analyst then proceeds to conduct a search of the literature databases. The exact phrases and combinations of words used to search the databases depend on the criteria already set up in the scoping of the meta-analysis. In general, more than one database will be searched with different techniques and combination of keywords, and use of Boolean logic. For biomedical literature, PubMed (www .pubmed.org) is a large and widely used database. Many academic and research institutions and universities have libraries that, in turn, host their own selection of databases in which they buy licenses. In addition, meta search engines such as Google Scholar provide excellent starting point for exploration of the studies.

68.3.1   Use Controlled Vocabulary

Meta-analysis depends on retrieval of primary studies. Authors can express titles and abstracts in many different styles and use different types of headlines to express the core messages. Although how to write titles and structure abstracts are now quite standardized, authors are free to use expressions and statements that they best know. As a result, there is a need for specific keywords or expressions when articles and journals are deposited to electronic databases so that publications can be easily retrieved. These keywords together make up what is known as controlled vocabulary. Controlled vocabulary therefore lists specific keywords under which primary studies are curated by database curators.

For example, in PubMed, MeSH (medical subject headings) constitutes such a vocabulary where different medical terms are organized in hierarchical order. Use of controlled vocabulary is an extremely useful strategy to identify studies. The most widely used controlled vocabulary for biomedical studies is used by the National Library of Medicine, the PubMed, or the Medline Database. The controlled vocabulary is known as MeSH. The curators or maintainers of the PubMed/Medline Database say that the MeSH vocabulary "imposes uniformity and consistency to the indexing of biomedical literature. MeSH terms are arranged in a hierarchical categorized manner called MeSH Tree Structures and are updated annually."[1]

68.3.2   Use Specialized Databases

For meta-analysis of intervention trials, randomized controlled trials and clinical trials are included, and data from these trials are abstracted and analyzed. For observational epidemiological studies such as cohort and case-control studies, these studies are sought in literature databases and data are abstracted from these studies and synthesized. Therefore, curated databases that contain specifically randomized or nonrandomized trials are useful for location of trials for conducting meta analyses. While PubMed/Medline and EBSCO (Europe focused) are two major sources of both observational studies and randomized controlled trials for all conditions, two major sources of clinical trials are http://clinicaltrials.gov and controlled clinical trials registry database. These databases not only contain information about completed trials but also additional information about ongoing trials and trials that are currently recruiting participants. These make searching for studies easy and also provide opportunities to easily search for gray literature or studies or trials that may have been completed but whose results are not yet available in published format.

68.3.3   Use Boolean Logic-Based Searching of Literature

While availability of the specialized databases and generic databases have made it easy to access articles and original data for analyses, these would still need strategic and careful searching using a range of techniques. Generally, words and phrases in the

---

1 For more information, see http://www.nlm.nih.gov/bsd/disted/pubmedtutorial/015_010.html.

Table 68.2 Use of Boolean operators for searching mindfulness meditation-related studies.

Boolean operator	Example
AND	Mindfulness and Meditation will retrieve all citations that only have both "mindfulness" and "meditation" in it.
OR	Mindfulness OR Meditation will retrieve all citations that have either Mindfulness OR Meditation or both mindfulness and meditation in it.
NOT	Mindfulness NOT Meditation will retrieve all citations that have Mindfulness in it but not meditation.

text, title, abstract, and words/expressions in the controlled vocabulary are used to effectively search these databases. Use of Boolean expressions AND (narrows down the searches to only specific terms), OR (expands the searches to include all the terms or phrases used), and NOT (excludes the searches and narrows down to specific terms) are used, along with wildcard entries (Table 68.2). A common strategy for searching of articles is to start with terms describing "outcomes" first, then add related terms that describe the "intervention" or "exposure" terms, and finally, add terms that define or describe the "study design" related terms. For example, if you were to search for all studies on mindfulness-based meditation and control of hypertension, you might start with "hypertension" or "high blood pressure;" then follow up the search with "mindfulness-based meditation," and then terms descriptive of "randomized controlled trials" or "controlled clinical trials." Finally, use of years, and languages often limit the searches. In addition to manually constructing search terms, the researchers also frequently make use of specialized search terms and combinations of search terms that are made available for specifically conducting searches. Specifically, researchers who conduct Cochrane Reviews can avail themselves of the services offered by the Cochrane Trialists or Cochrane Coordinators who maintain and curate databases of search terms that are validated for specific types of studies to be retrieved and these are used.[2]

Table 68.2 shows an example of mindfulness meditation for the control of hypertension and use of search operators.

68.3.4 Use Validated Search Filters

Search algorithms can become quite complicated, as different databases have different types of controlled vocabulary, but also there are issues around usage of wildcards and other notations that help researchers to search individual databases. Increasingly, specialized search algorithms to search specific types of articles for data analysis are becoming available. The ISSG Search Filter resource is a large repository where search filters are made available for different types of study designs and different databases for researchers to use. For more information, check out the ISSG Database Search Filter site.

---

2 See https://www.cochranelibrary.com/advanced-search/search-manager.

68.3.5   Understand Gray Literature and Hand Search

In searching for information, relying only on published articles in peer reviewed journals often cause a problem in that there is a bias where articles or publications that have
positive findings are over represented (cite Cochrane and mention page). As a result, if
only those studies that were published in peer reviewed journals or published in public
domain are included in the meta-analysis, and publications that were either not published because they failed to present positive findings and therefore not included in the
meta-analysis, the meta-analysis would itself be biased and would result in potentially
erroneous estimates. Therefore, the meta-analyst must conduct an active search for all
publications and data that might not have been published or otherwise in archives or in
authors' personal collection that, although the research was completed, still remained
unpublished. Such publications are known as gray literature and must be included in any
meta-analysis. While it is not easy to retrieve such publications, usually contact with
known experts, and active searching of conference abstracts, contacting first authors,
or searching of trial registers are warranted to identify these sources of information.
Omission of gray literature leads to *publication bias* (also referred to as file drawer problem), meaning that articles or publications that are otherwise publishable (they are well
conducted trials of good quality of evidence) but they never see the light of day (and consequently, are likely to be confined to the file drawers of the individual investigators for
a variety of reasons. Perhaps the articles submitted to the journal are rejected because
the editors of the journals do not have any interest in publishing negative findings, or
the investigator decides not to send such articles to any journal as the journal is perceived not to publish them. Irrespective of the reason, publication bias must be formally
examined during meta-analysis. At present, rather than formal testing of the extent of
publication bias, visual methods such as construction of *funnel plots* are available and
should be reported in any meta-analysis.

Besides reporting of publication bias, the reference lists of all the retrieved articles
must be read and then the titles and abstracts of each reference article on that list should
be read and an attempt must be made to identify additional studies. This is known as
hand searching and involves searching where necessary library archives or paper copies
of journal articles. Thus, in summary, searching for publications involves teamwork and
careful construction of search filters and algorithms, and it is a recursive process where
the articles are searched exhaustively until a number of studies are finally retrieved that
can answer the research question.

68.3.6   Need for Inclusion and Exclusion Criteria

A review of the titles and abstracts of the articles or publications retrieved and collected
in the first pass are then reviewed based on their titles and abstracts. The flowchart
(Figure 68.2) starts with all the retrieved publications in the first pass, and then moves
downward progressively to show the reason for exclusion of the publications, both at the
stage of only reviewing the titles and abstracts but also on review of the full text of the
articles. A rule at this stage is to follow the inclusion and exclusion criteria strictly, and in
cases where the analyst is in doubt, whether an article or publication can be included or
excluded, the advice is to include the article, as it can always be excluded on close reading in the second pass. The list of excluded articles are then kept separately at another

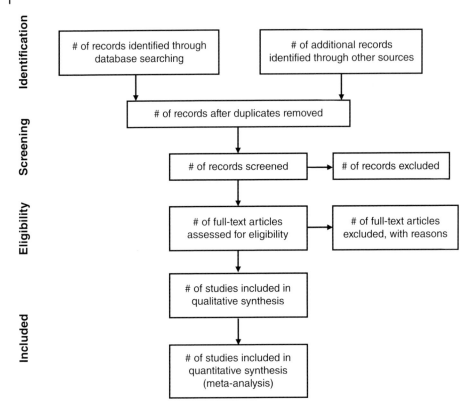

Figure 68.2 PRISMA Chart: the preferred reporting items for systematic reviews and meta-analysis chart (see Moher et al. (2015) for details).

database and the analyst proceeds with the included studies. For each included study at this stage, the full text is obtained and data are abstracted from the studies and the studies are also assessed for their risk of bias. Additionally, in systematic reviews, not all article are included for meta-analysis, and therefore, a separate listing of the number of articles that are kept for narrative synthesis versus articles kept for more detailed quantitative syntheses are indicated at this stage.

68.4   Step Three: Conduct Risk of Bias Appraisal of Full Texts

In meta-analysis, data from primary studies are combined to arrive at a summary estimate of the association between two variables. Where the objective is to estimate the overall effectiveness of a particular treatment, this process involves pooling results from primary studies to arrive at a summary estimate to assess whether compared with alternative treatments or placebos or compared with no treatment at all, the treatment under review was effective in achieving the outcomes set out in the studies. For observational studies, the aim is to assess whether the pooled estimates of the odds ratios or relative risk estimates indicate a valid association between the two entities. It follows that in meta-analysis, the analyst is not only interested in the presence of the evidence but also whether the evidence is comparable across the studies included in the meta-analysis. As a result, if the studies themselves are of poor quality, then the overall quality of the meta-analysis will not only be poor; it may also end up propagating the errors that compromise internal validity in primary studies.

In order to establish internal validity of a trial or an observational study, the investigators of studies should address three related entities – play of chance (the study should have sufficient number of participants to rule out random association), biases that can arise in the course of conducting the study, and controlling for confounding variables.

As an example, consider an investigator who is interested in studying the association between excessive coffee drinking and the risk of pancreatic cancer and decides to conduct a case-control study of individuals with and without pancreatic cancer and will measure their coffee consumption. With this measurement, the investigator will research whether excessive coffee consumption as defined under conditions of the study is associated with pancreatic cancer. In order to establish internal validity in such a study, the investigator must specify ahead of conducting the study how many participants should be included in each arm (cases and controls). She should also consider alcohol drinking or other variables that are associated with both pancreatic cancer and alcohol intake, as potential confounding variables. Additionally, such observational epidemiological studies are open to selection and response biases. In case of randomized controlled trials, the process of randomization ensures controlling of known and unknown confounding variables. However, for both randomized controlled clinical trials and observational studies, several different types of biases are possible:

1. *Selection bias.* Indicates systematic differences in comparison groups that occur as a result of how the groups being compared were selected for the study. A similar bias results if the respondents of a study differ in which they respond to survey questions (response bias). Selection bias is a particularly important problem in meta-analysis and in appraisal of risk of bias, it is important to keep in mind two forms of biases:
    A. In case of RCTs and other forms of clinical trials, it is important to ensure that true randomization was done by way of using random numbers table rather than leaving chances for systematic allotment of treatments and alternative conditions.
    B. A very important point is to ensure that the allocations to the intervention and control arms were truly concealed (that is neither the investigator nor the patient was aware of where the intervention and control arms were allocated. This principle is known as double blinding). Allocation concealment is particularly important in clinical trials to ensure chances of selection and performance bias.
2. *Performance bias.* This results from systematic differences in the care provided between the intervention and the comparison group in case of RCTs.
3. *Attrition bias.* This results from systematic differences in how participants have withdrawn from the trials.
4. *Detection bias.* Systematic differences in which the outcomes are assessed. In review of RCTs, it is essential to critically examine how participants were allocated to treatment and alternative conditions and how such allocations were concealed from not only the participants in the study but also for investigators. Such concealment is referred to "blinding" or "allocation concealment." Studies that fail to demonstrate robust processes of allocation concealment are likely to report significant selection

biases or reporting biases and therefore these studies are at significant risk of studies with inaccurate estimations of the extent of associate between the treatment and outcomes.

68.5   Step Four: Abstract Data from Individual Studies

Abstraction of data from individual study is critical for meta-analysis, and the process is now fairly standardized when specific summary of findings forms (SoF forms) are used. These forms have been developed by the Grading of Recommendations Assessment, Development and Evaluation (GRADE) Working Groups and provide detailed instructions as to how to use the data abstraction forms.[3] These forms are also standard components of software such as RevMan used to conduct meta-analysis published by the Cochrane Collaboration. In general, the abstraction of information from primary studies includes the first (or corresponding) author of the study, the year the study was reported, the population on which the study was conducted, the intervention or the exposure that was studied, and comparison groups, the outcomes, the effect estimates, and elements of information that indicate quality of the study. In general, the *Cochrane Handbook* (Higgins and Green 2006) recommends the following elements of data to be abstracted:

1. Title of the review and name of the coder
2. A key or identifier for every primary study included in the review
3. A field where you indicate that the study is eligible or not (this is somewhat redundant as you can sense that all studies included in the review are included here, but also there may be misses and this is where this extra field is helpful particularly if you have more than one coder for your project
4. Type of trial or study design (RCT, others, before-after study, crossover)
5. Whether allocation concealment was done (adequate, unclear, inadequate, not done, not relevant)

- Participant characteristics
- Depend on the study or review itself, if it is reasonable to believe that participant characteristics might influence the outcome or research in some ways then that characteristic must be included. Example: In your study on the efficacy of mindfulness meditation for hypertension, you may want to include ethnicity of the participants on whom studies were done (unless you restricted) as an element of the data:
  1. Age
  2. Gender
  3. Settings (hospitals, emergency rooms, offices, nursing homes, prisons, others, community setting)
  4. Diagnostic criteria for the outcome of interest. This is important for a number of reasons, particularly for hypertension or others as to what or how did the investigators ascertain the outcome? What criteria were used?

---

3 GRADE Handbook is available at: https://gdt.gradepro.org/app/handbook/handbook.html.

5. Interventions. For drugs used route of administration, dosage, timing; for other kinds of interventions, who administered the intervention, how often,
6. Comparator condition. As in above.
7. Outcomes. Follow GRADE guidelines (see Guyatt et al. 2013a, b).

68.5.1   Outcomes

In both intervention research (RCTs, clinical trials) and in observational studies (case-control studies, cohort studies, others), health effects or health-related phenomena that depend or arise out of an intervention or an exposure are referred to as outcomes. Examples of outcomes include recovery from an illness (whether patients recovered from an infection), death or survival (whether the patients survived after five years from the detection of breast cancer), or length of stay at a hospital following an intervention.

Outcomes can be measured or expressed in terms of grades of responses from patients or participants in a trial. That is, on a scale of 1−5, where is a patient in terms of pain following a procedure? The number that the patient expresses is the outcome of the procedure on a scale of 1−5 and is expressed in an ordinal measure and an outcome of the procedure).

In general, four common types of outcomes are described in the literature:

1. *Dichotomous or binary outcomes.* Dichotomous outcomes indicate one of the two states of existence. For example, "survival" in the form of reporting of either dead or survived a procedure or a disease. Another example might be if the patients or participants in a trial or a study report whether they were diagnosed with a disease or not diagnosed with disease; that outcome would be an example of a binary outcome, or dichotomous outcome. These outcomes are measured in terms of proportion of the participants with desired outcomes in a sample. For example, Brewer et al. (2011) conducted a randomized controlled trial on 88 nicotine-dependent adults to test the efficacy of MBSR and compared this with American Lung Association's Freedom from Smoking treatment. They measured the effectiveness of MBSR using self-report of participants on their smoking behavior and calculated rates of smoking cessation following either treatment. This is an example of how dichotomous or binary outcomes are reported Brewer et al. (2011).
2. *Continuous outcomes.* These include outcomes that are measured on a scale where the boundary of the levels of the measurement of outcomes overlap. Such examples include measures of systolic and diastolic blood pressure measured in mmHg, measures of blood sugar control measured by mmols/L or measured by HbA1c, etc. The measures are reported in terms of either mean differences or standardized mean differences (explained below).
3. *Ordinal data.* These data are measured on rank-ordered scales. Examples include quality of life on a scale between, say, 1 and 5, where 1 indicate very low values and 5 indicates very high values. These are again measured in terms of percentages of participants who are in each of the scores on the scales.
4. *Time to event.* A time-to-event outcome denotes a length of time between initiation of the intervention (for clinical trials or intervention trials) or initiation of the observation (for observational epidemiological studies). For example, if in a study

the objective of the investigators is to compare the length of time from admission to discharge between intervention and control arm participants, then that outcome is a time-to-event outcome. Similarly, in cohort studies, often investigators are interested in studying the length of time before the first case of disease appears following exposure (and those who were not exposed to specific exposure variables). Here as well, the length of time to the emergence of disease in participants is considered as a "time-to-event" outcome. In this chapter, we shall discuss binary and continuous outcomes but skip other two outcomes as beyond the scope for this chapter.

68.5.2 How to Abstract Data for Binary Outcomes and Continuous Outcomes

For abstraction of data from binary outcome variables, it is helpful to construct a 2×2 table and fill in the cells as follows. In Table 68.3, A is the number of participants who received the intervention or who were in the exposure arm of the study and ended up with the outcome of interest. Likewise B, C, and D refer to the number of participants in the trial or study where corresponding relationships between exposure or intervention and outcomes were recorded. Accordingly, the risk ratios are defined as follows:

Rate ratio or relative risk estimates $= [A/(A + B)]/[C/(C + D)]$ (Or, rate of outcome among the exposed or intervention group versus rate of outcomes among the control group.)

Odds ratio $= A \times D/B \times C$

For RCTs, clinicians are also interested in studying risk differences and associated numbers needed to treat, expressed in the form of inverse of risk difference: (1/Risk difference).

The risk difference is given in the following formula: $[A/(A + B)] - [C/(C + D)]$.

However, the interpretation of risk difference requires that the baseline risk be kept in perspective. For example, Nissen et al. (2004) conducted a randomized controlled trial to compare the effect of intensive versus moderate lipid-lowering therapy on progression of coronary atherosclerosis. In order to do this, they conducted a double-blind, randomized active control multi-center trial (Reversal of Atherosclerosis with Aggressive Lipid Lowering [REVERSAL]) at 34 community and tertiary care centers in the United States comparing the effects of two different statins administered for 18 months. Intravascular ultrasound was used to measure progression of atherosclerosis. Between June 1999 and September 2001, 654 patients were randomized and received study drug; patients were randomly assigned to receive a moderate lipid-lowering regimen consisting of 40 mg of pravastatin or an intensive lipid-lowering regimen consisting of 80 mg of atorvastatin. They found that C-reactive protein decreased 5.2% with pravastatin and 36.4% with atorvastatin. Based on this information alone, the risk difference for reduction of

Table 68.3 A crosstab showing abstraction of data elements for studies with binary outcome variables.

Exposure or intervention	Outcomes occurred	Outcome did not occur
Exposure/intervention occurred.	A	B
Exposure/intervention did not occur.	C	D

C-reactive protein was $36.4 - 5.2\% = 31.2\%$ and favored atorvastatin over pravastatin. When translated to numbers needed to treat, this would indicate about three patients would be needed to be treated with atorvastatin (compared with pravastatin) to register one additional person's benefits for C-reactive protein Nissen et al. (2004).

In cases where outcomes variable is continuous, two measures are mean difference and standardized mean difference. The formulae for these two measures are as follows:

MD = Mean of the outcome variable for participants in the treatment arm − Mean for the outcome variable for participants in the control arm.

SMD = Mean for those in the treatment arm − Mean for those in the control arm/Pooled SD.

The SMD is used for those outcome variables where the units of measurement for the same outcome variable are different for different studies. Consider a meta-analysis of control of blood sugar among diabetics where the outcome variable is concentration of HbA1C. This variable can be reported as measured in National Glycohemoglobin Standardization Program (NGSP) percentage values; alternatively, this can be measured in terms of International Federation for Clinical Chemistry (IFCC) reference method in mmol mol^{-1} units. As these are different units of the same outcome, a common comparable measurement strategy of the outcomes is in the form of SMD in the measure.

In general, different software packages are used for meta-analysis, and these software packages often specify the level of outcome variable and entities required for analysis. For example, in the popular software package STATA, if you conduct meta-analysis using the routine referred to as "metan," you will need to provide either three variables, four variables, or six variables, depending on the type of information you can provide the software. If you decide to use odds ratios, hazard ratios, or rate ratios or relative risks along with their 95% confidence intervals, then you will need to provide the software three measures of effect estimates: the odds ratio (or the relative risk or rate ratio or hazard ratio; the upper, and the lower boundary estimates of the confidence interval). Usually these values are provided after transforming them to their logarithmic values, and the software, in turn, provides the estimates after converting the figures. On the other hand, if you can construct the 2×2 tables, then the four data points as outcome variables can also be provided; finally, if the effect estimate of interest is MD or SMD, then the software needs to be provided with the sample size, the mean difference between treatment and control group, and the corresponding SD.

A common problem is that, often study authors present data in the form of numbers of participants, relative risk estimates, and p-values alone, rather than providing the readers data enough to abstract the numbers of participants in each arm to enable construction of 2×2 tables for the concerned study. In these cases, depending on the sample size presented, either z-values or t-values are estimated, and then from these estimates standard deviations are worked out (these are the situations where statistical advices become essential to proceed). In other instances, you can easily construct a 2×2 table for some studies, while for others, authors may present only risk ratio estimates with 95% confidence intervals indicating lower and upper limits of the confidence intervals and the number of participants in the study (in each arm). In these cases, the analyst needs to estimate the point estimates of effect and associated standard deviation (SD), but all studies for the meta-analysis should use the same measures. If some studies report relative risks and other studies report numbers that would enable estimation of

2×2 tables, then for all studies relative risks are reported along with standard deviations to estimate the summary effect measure.

68.6 Step Five: Assess Whether the Studies Are Homogeneous

In a meta-analysis, the researchers answer the following three questions:

1. What is the overall (summary) relationship between the treatment/intervention/ exposure and the health outcomes? Or, put another way, is there evidence that the intervention is associated with the outcomes under study?
2. Is this association consistent across the studies that constitute the meta-analysis?
3. Are all studies that would have been captured contribute to the pooling of results? Is there a bias introduced in the way studies were collated to introduce selection bias because of the way studies were selected for publication (publication bias)?

The first two questions are addressed in a meta-analysis by pooling of data from studies and by statistically testing for the presence of heterogeneity in the study findings. In general, when a meta-analysis is conducted, a number of different studies with diverse populations and different measures are included. These studies are based on different population and as such are different from each other in several respects (they do share commonalities such as study design, study aims, and objectives). As a result, although studies are so selected to have very similar interventions (as a matter of fact, identical or same interventions or same exposures) and outcomes (health outcomes), there can still be differences in profiles of participants or differences in the quality of the studies, or the methods used in the studies. The differences in the participant profiles are known as *clinical heterogeneity*, and differences in the method of execution of the studies themselves are known as *methodological heterogeneity*.

Beyond these two sources of heterogeneity, variability is also observed in the magnitude and direction of the effect size between the intervention or exposure and health outcomes. This diversity is referred to as statistical heterogeneity and refers to the extent that the results differ from each across the different studies included in the meta-analysis. As long as these differences are so small that they do not statistically significantly differ from a centrally pooled estimate, these studies are known as *statistically homogenous*. Such homogeneity can be tested commonly in two ways:

1. Simple chi-square test of homogeneity. Here, the number of studies are tested in a framework of hypothesis testing. The null hypothesis states that the effect sizes of individual studies are same, while the alternative hypothesis states that the effect sizes are different from each other (heterogeneity). In the chi-square test (the measure is also known as "Cochran's Q") with $N - 1$ degrees of freedom (where N = number of studies included in the meta-analysis) provides a measure of heterogeneity. If the p-value is less than 0.05 (or less than 0.10 or a prespecified cut-off), then the studies are considered to be statistically heterogeneous. However, measurement of heterogeneity in this manner has the problem that if the meta-analysis includes a large number of studies, then the chance of statistical heterogeneity also proportionately increases even if the studies are similar to each other to a large extent. To address these concerns, another test, I-square test of heterogeneity is used.

2. I-square test of homogeneity. The I-squared test of heterogeneity is expressed by the following formula, I-squared $= [(Q - df)/Q] \times 100$ and expressed as a percentage, where $Q =$ Cochran's Q (see above), and $df =$ degrees of freedom where given by $N - 1$ where $N =$ number of studies. According to Higgins and Thompson (2002), I-squared is interpreted as thresholds as follows:

3. 0–40%: the heterogeneity might be important
4. 30–60%: may represent moderate heterogeneity
5. 50–90% may represent substantial heterogeneity, also 75% and above: considerable heterogeneity

Interpretation of tests of homogeneity. Use the *p*-values carefully and if the *p*-value is greater than 0.05 or 0.10, this indicates that the studies cannot reject the null hypothesis. The null hypothesis in this case indicates that the studies are not heterogeneous or that the studies are homogeneous. This interpretation needs caution as if there are large number of studies, then the *p*-value can be less than the cut off yet the studies may be homogeneous (Higgins and Green 2006).

68.6.1 What Happens When the Test of Homogeneity Fails

In case of statistically significant heterogeneity, the analyst must explore features of studies that explain such heterogeneity. This can be done through careful review of the methods and results of individual studies or studies in groups; also, analysis of subgroups of studies based on predefined criteria (also quality assessments) are important. A common statistical way to conduct subgroup analyses is to conduct meta-regression, where the pooled effect estimate is regressed on identified features of the studies and searched for regression lines that explain how individual study features might be associated with the variability in the effect estimates.

In case of significant heterogeneity where there are also significant differences in the directions of the effect estimate, an average estimate of the effectiveness may be misleading (e.g. imagine conducting a meta-analysis on the effectiveness of an intervention on survival for post-operative patients; now, in that meta-analysis, you found statistical heterogeneity; over and above, some studies indicate that interventions have protective effects on the survival, i.e. they increase survival] while other studies indicate that the interventions actually increase the risk of death or reduce the survival probability). In this situation, a meta-analysis is best not conducted by statistically pooling data from the studies, but a narrative review can be attempted to summarize the key features of the studies and summarize study findings; also, a systematic review need not contain meta-analysis. It is sufficient to conduct a narrative synthesis of data and explore the causes of such heterogeneity. Another option might be to perform a random-effects meta-analysis – the studies are assumed to be part of a "universe of similar studies," and therefore this meta-analysis can accommodate the fact that the studies can be so dissimilar with each other that their effect sizes may vary significantly. In summary, in presence of statistical heterogeneity, in all cases, an exploration of the causes of heterogeneity should be attempted, using subgroup analysis; additionally, if there is also additional variability in the direction of the effect measures, then it is best not to pool the study results but conduct a thorough exploration of the causes of such heterogeneity. In cases where the direction of effect measures do not vary, in addition to exploration of the causes of such heterogeneity, a random effects meta-analysis may be conducted.

68.7 Step Six: Conduct Fixed Effects or Random Effects Meta-analysis and Perform Sensitivity Analyses

In those situations where formal statistical pooling of the study results are warranted, the goal of meta-analysis is to arrive at a summary measure of the overall effect estimate based on individual study effect sizes. These individual studies are obtained based on specified research questions and active search of the literature databases and indeed other sources of information, such as trial registries and often studies are obtained in consultation with individual authors and investigators. In establishment of the association between an intervention and an outcome (alternatively, an exposure and an outcome), two questions aim to characterize the nature of this association:

1. Is this association summarized here in the meta-analysis a definitive based on the studies that have been studied?
2. Is the association summarized an indicator of the direction and magnitude of the association in a general sense that is expected in such an association?

These questions may seem pedantic. Consider the first question: In a meta-analysis this question is based on an assumption that the studies included in the meta-analysis by themselves sufficiently summarize the relationship between the intervention (or exposure) and the outcome. Such a model of meta-analysis to characterize the relationship is known as fixed effects model meta-analysis. The second question indicates that the studies included in the meta-analysis actually constitute a sample of studies out of all studies possible in a universe of similar studies. According to Diana Petitti (1999), these two models of meta-analysis answer two different questions. While fixed effects model answer a question for instance, whether intervention X was associated with the outcome O in the studies analyzed, a random effects model answers quite another question: whether intervention X is likely to be associated with the outcome O. In turn, random effects meta-analysis results in summary estimate that is closer to the null estimate and that, the associated confidence interval is wider than what is obtained in a fixed effects meta-analysis.

68.7.1 Example: Do White Responds Better than Blacks to ACE Inhibitors for the Treatment of Hypertension?

Peck et al. (2013) conducted a meta-analysis of RCTs to test the relative responses to ACE inhibitors of Whites compared with Blacks, and identified 16 studies where the authors investigated differences in systolic and diastolic blood pressures of Whites and Blacks. For this particular illustrative example of pooling data for meta-analysis, we pool only results of diastolic blood pressure differences between Blacks and Whites and conduct a meta-analysis. We presume that the initial steps of meta-analysis – framing of research question, searching of the literature, identification of the studied, and critical appraisal of the studies themselves and identification of the risk of bias – were all completed, so these are not repeated (Peck et al. 2013). Here, we are interested in three aspects of meta-analysis:

1. What evidence or effect size exists to suggest that there is a difference in response for Blacks and White respondents in their diastolic blood pressure to ACE inhibitors?

2. Is this response or effect size consistent among the studies considered for this meta-analysis?
3. Is there a significant publication bias in the studies?

We first test whether the studies were sufficiently homogeneous or whether the studies were grossly heterogeneous. The readers are encouraged to read the main meta-analysis (see reference section), but in this case, our interest is only on statistical tests of heterogeneity. We note that otherwise in terms of recruitment of the participants, the methodology of research, the studies were similar (i.e. the studies were clinically and methodologically similar). Let's review the statistical tests of heterogeneity:

- Heterogeneity chi-squared $= 36.75$ (df $= 15$) $p = 0.001$
- I-squared (variation in WMD attributable to heterogeneity) $= 59.2\%$

These figures suggest statistically significant heterogeneity. In both chi-square tests and I-squared tests (see above), we also note that the studies are moderately heterogeneous. However, as the studies were otherwise found to be similar and this question (whether there is difference in response to ACE-inhibitors for different races is a generic question), therefore, a random effects meta-analysis may be appropriate for this question, rather than attempting either a fixed effects meta-analysis or discarding meta-analysis altogether. This leads us to answer the second question, "What evidence exists about the pooled response of the White and Black races in response to ACE inhibitors as measured by diastolic pressure?"

The graph in Figure 68.3 is known as a "forest plot." Here, individual studies are plotted on the y-axis. The pooled estimate or the summary effect size is presented in the form of a diamond-shaped figure – the forest plot. The sample size of each study is indicated by a shade around their point estimate. The dark straight line around which the studies are arranged indicates the null. In this case, the mark is 0, indicating no difference in diastolic blood pressure changes recorded between Blacks and Whites; while positive direction indicates that Whites have recorded larger changes and negative values indicate Blacks have recorded larger changes. The dotted line with a diamond-shaped figure at the bottom indicates the pooled estimate, in this case measured using the DerSimonian-Laird method of random effects meta analyses. This figure shows that the pooled estimates indicate that compared with Blacks, Whites registered a larger change in the diastolic pressure (2.18 mmHg, 95%CI:1.28–3.09) and thus, this figure suggests that Whites, compared with Blacks, are more likely to respond to ACE inhibitors in their diastolic blood pressure measurements. In this meta-analysis, studies that have a larger sample size were given larger weights compared with smaller studies.

Is this conclusion is based on considering all relevant studies?

Figure 68.4 shows a funnel plot of the association between mean differences in diastolic blood pressure in response to ACE inhibitors among Blacks and Whites. Data taken from Peck et al. (2013) study. Note that the left lower part of the graph is empty indicating publication bias.

To answer whether the meta-analysis was based on all relevant studies, evidence of publication bias is tested using a funnel plot (the boundaries of such a plot takes the shape of an upturned funnel, hence the name). A more formal statistical test is not available for further analysis of this graph. As can be seen in this funnel plot, weighted mean differences in the diastolic pressures of the two groups are plotted on the x-axis and standard error of the mean differences are plotted on the y-axis. This is done to check visually

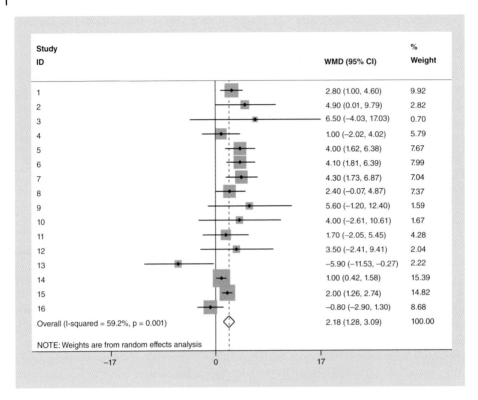

Figure 68.3 Forest plot.

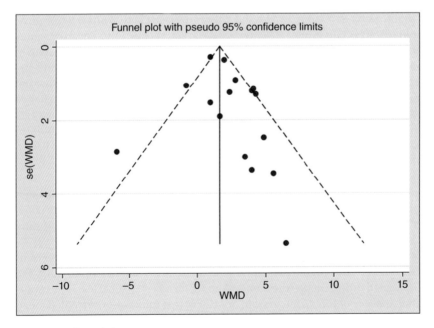

Figure 68.4 Funnel plot.

whether a pattern can emerge such that small or negative studies that are either missed or disproportionately presented in that matrix that would otherwise be present. The mouth of the funnel points downward where studies that are of low power are presented (they are considered to be of low power, as the standard error is quite high and, hence, they are located to the bottom of the *y*-axis), but note as well that because of this, we would expect that the positive and negative studies will be fairly evenly distributed. As the power of the studies will increase, we'd expect that the studies will converge toward the summary estimate that we have seen (the summary estimate line is the dark line). A solid dark line runs through the center and indicates the pooled effect estimate. Studies that have a large sample size and therefore small standard errors are located on the top or tip of the funnel. Studies are arranged according to their effect size but also according to their power or sample size and therefore weights. Note that no studies are located in the left lower quadrant of the funnel plot. This suggests that low-powered studies with negative findings might have been omitted in this meta-analysis, indicating publication bias in this study.

As in this meta-analysis, we noted heterogeneity; therefore, a subgroup analysis of the studies is important. While subgroup analyses can be conducted in different ways and indeed, in the study the authors reported subgroup analyses, one common strategy is to conduct a subgroup analysis by conducting a regression analysis. In the linear regression analysis, referred to as *meta-regression,* the subgroup can be considered on any study characteristic (ideally measured on a continuous scale) and the *y*-variable (or the outcome variable) is the outcome variable or a measure of the outcome variable or effect size. The linear model is used to identify whether specific study characteristics can explain why the results might have been different. A statistically nonsignificant linear model might indicate that the particular explanatory variable did not explain variability in the distribution of the effect size in the studies.

Figure 68.5 shows the meta-regression of the Black and White differences in diastolic blood pressure following ACE inhibitor therapy (Peck et al. 2013), where effect size in terms of reduction in diastolic blood pressure was regressed on the dosage of antihypertensive medication that was administered.

In this meta-analysis, given the heterogeneity of the studies in finding the association between ACE inhibitor dosage and response for diastolic blood pressure, a reanalysis was done. In the reanalysis, the effect size of the studies was regressed on the dose of ACE inhibitors administered. It was found that in the regression, the changes in the diastolic blood pressures decreased as the dosage increased; however, this reduction was not statistically significant (beta coefficient for dose = −0.03, 95% confidence interval: −0.11, 0.06, $p = 0.626$).

Thus, to recapitulate, the basic principles of data analysis in meta-analysis, in the beginning, are that it is important to decide, based on the research question and objective of the meta-analysis, whether the meta-analysis is to answer the larger question of "what is possible" or whether it is reasonable to assume that the studies belong to a larger universe of studies of which these form a sample, and therefore a random effects meta-analysis may be attempted. Alternatively, a formal test of statistical heterogeneity of the studies is conducted, and one of the several measures are estimated to test statistically, but also clinically and methodologically how similar are the studies. If the studies are similar, then both fixed effects and random effects meta-analyses are attempted and the summary estimates are confirmed and discussed. On the other hand, if the studies

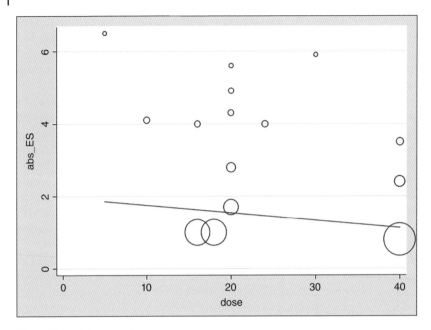

Figure 68.5 Meta regression.

show considerable heterogeneity, then either a meta-analysis is not attempted or several subgroups are analyzed. In all cases, causes of such heterogeneity are explored, using meta-regression and other strategies. A search for publication bias is reported as well, using funnel plots or other visual inspection tools such as L'Abbé plots.

68.8   Conclusion

To conclude, a meta-analysis is a form of review where the analyst conducts a systematic review but where the scope of the review extends beyond narrative synthesis of information and where statistical data analysis involves comparing two alternative forms of treatment or exposure. One of the two treatments can be a novel or one form of intervention, while the other can be an alternative intervention or placebo. In conducting the meta-analysis, the analyst starts with a research team consisting of clinicians or domain experts (domain experts are those individuals or professionals who are knowledgeable and have experience in the matter under study); information specialists and search experts who can conduct robust searches of literature; statisticians; and database experts. The next task for the meta-analyst team is to frame a set of research questions using the PICO method, and search comprehensively all available literature on the topic specific to the question. The next steps are close reading of the retrieved studies, search for gray literature, abstraction of data and organization of the database to prepare for analysis. A set of effect estimates are decided upon, and in general, meta-analyses should answer three related questions:

1. Does evidence exists that the intervention or the exposure is statistically associated with the outcome?

2. Are the results similar across the studies?
3. Were studies omitted, resulting in publication bias that could be uncovered?

At the end of the study, subgroup analyses are conducted to indicate robustness of the analyses. Data are cut in various ways to examine if subgroups were to reveal important insights into the data obtained from the primary studies.

In general, in deciding to conduct meta-analysis, it is important to keep in mind the comprehensiveness of the search process and the heterogeneity of the studies included in the meta-analysis. This is where a decision of whether to conduct meta-analysis at all needs to be made. Assessment of study heterogeneity is particularly important. Moher et al. (2009) proposed preferred formats of reporting meta-analyses (PRISMA) and these emphasize common elements such as clearly mentioning meta-analysis or systematic review on the title, structured abstracts, and clear description of the process of screening articles for the review.

In this chapter, the emphasis was on conducting pairwise meta analyses using a standard approach of identifying studies that compare only two interventions or only two exposure conditions for the same outcome. Newer approaches to meta analyses also include network meta analyses where more than one intervention or one exposure for a range of outcomes (or a matrix of interventions and outcomes are studied). However, a detailed discussion of network meta-analyses is beyond the scope of this review. A place to start while embarking on a meta-analysis is Cochrane Collaboration (www.cochrane .org), which is a rich repository of meta-analyses and systematic reviews on a range of topics. Cochrane Collaboration also provides training and offers a free software package to manage systematic reviews and meta analyses easily – the RevMan package. While RevMan software allows organization and conduct of meta-analysis easy, for additional analyses, statistical software packages such as R and Stata have several packages and routines that enable analysts to conduct meta analyses effectively. It is hoped that this introduction to meta-analysis will enable the reader to embark on reading, interpreting, and thinking about meta-analyses that can be used for their own purposes. When conducted well and appropriately, meta-analyses can provide invaluable information on the comparison of different treatment approaches and exposure-outcome associations for epidemiological studies.

References

Ancoli-Israel, S., Cole, R., and Pollack, C.P. (2003). The role of actigraphy in the study of sleep. *American Academy of Sleep Medicine* 26 (3): 342–392.

Babu, G.R., Jotheeswaran, A.T., and Mahapatra, T. (2013). Is hypertension associated with job strain. *Occupational and Environmental Medicine* 71 (3): 220–227.

Brewer, J.A., Mallik, S., Babuscio, T.A. et al. (2011). Mindfulness training for smoking cessation: results from a randomized controlled trial. *Drug and Alcohol Dependence* 119 (1–2): 72–80.

Evans, D. (2003). Hierarchy of evidence: a framework for ranking evidence evaluating healthcare interventions. *Journal of Clinical Nursing* 12: 77–84.

Guyatt, G.H., Thorlund, K., Oxman, A.D. et al. (2013a). GRADE guidelines: 13. Preparing summary of findings tables and evidence profiles-continuous outcomes. *Journal of Clinical Epidemiology.* 66 (2): 173–183.

Guyatt, G.H., Oxman, A.D., Santesso, N. et al. (2013b). GRADE guidelines: 12. Preparing summary of findings tables-binary outcomes. *Journal of Clinical Epidemiology.* 66 (2): 158–172.

Higgins, J.P.T. and Green, S. (2006). *Cochrane Handbook for Systematic Reviews.* Cochrane Collaboration.

Higgins, J.P. and Thompson, S.G. (2002). Quantifying heterogeneity in a meta-analysis. *Statistics in Medicine.* 21 (11): 1539–1558.

Moher, D., Liberati, A., Tetzlaff, J. et al. (2009). Preferred reporting items for systematic reviews and meta-analyses: the PRISMA statement. *PLoS Medicine* 6 (7): e1000097.

Nissen, S.E., Tuzcu, E.M., Schoenhagen, P. et al. (2004). Effect of intensive compared with moderate lipid-lowering therapy on progression of coronary atherosclerosis. *JAMA* 291 (9): 1071.

Peck, R.N., Smart, L.R., and Beier, R. (2013). Difference in blood pressure response to ACE Inhibitor Monotherapy. *BMC Nephrology* 14 (201).

Petitti, D.B. (1999). *Meta-analysis, Decision Analysis, and Cost-effectiveness Analysis: Methods for Quantitative Synthesis in Medicine.* Oxford University Press.

Moher, D., Shamseer, L., Clarke, M. et al. (2015). Preferred reporting items for systematic review and meta-analysis protocols (PRISMA-P) 2015 statement. *Systematic Reviews* 4 (1).

Wyatt, J.C. (2000). 4. Keeping up: continuing education or lifelong learning? *Journal of the Royal Society of Medicine* 93 (7): 369–372.

Zipkin, D.A., Greenblatt, L., and Kushinka, J.T. (2012). Evidence-based medicine and primary care: keeping up is hard to do. *Mount Sinai Journal of Medicine: A Journal of Translational and Personalized Medicine* 79 (5): 545–554.

69

Missing Values: How to Treat Them Appropriately
David C. Howell

Department of Psychological Sciences, University of Vermont, Burlington, VT, USA

69.1   Introduction

The treatment of missing data has been an issue in statistics for some time, but it has come to the forefront in recent years because of improved techniques for dealing with the problem. The topic covers a range of research designs, including the problem of unequal sample sizes in the analysis of variance and missing data in contingency tables, but this chapter focuses on standard linear models such as multiple linear regression. That discussion will generalize to many other analyses, such as structural equation modeling and linear mixed models. A set of web pages that goes with this chapter is available online.[1] These include a discussion of methods for the analysis of variance and contingency tables, as well as more technical coverage of modern approaches to missing data.

Fortunately, we have come a long way since someone could say that the best treatment for missing data is not to have any. My goal in this chapter is to give the reader an understanding of the issues involved in the treatment of missing data and the ability to be conversant with the approach that is adopted. When it comes to discussing or selecting an approach, it is not necessary to have an in-depth knowledge of the technical issues, but it is necessary to understand the alternatives and to have a grasp of what is involved in each method.

69.2   Types of Missingness

We must begin with the question of why data are missing in the first place. They could be missing for perfectly simple and harmless reasons, such as a participant having an automobile accident and not being able to appear for testing. In such a case, missingness is more of a nuisance than a problem to be overcome. On the other hand, data could be missing on the basis of either the participant's potential score on the dependent variable (Y) or any of the independent variables (X_i). The reasons for missing data plays an important role in how those data will be treated.

---

1 David C. Howell's index of Web pages: https://www.uvm.edu/~dhowell/

A Guide to the Scientific Career: Virtues, Communication, Research, and Academic Writing, First Edition.
Edited by Mohammadali M. Shoja, Anastasia Arynchyna, Marios Loukas, Anthony V. D'Antoni, Sandra M. Buerger, Marion Karl and R. Shane Tubbs.

69.2.1 Missing Completely at Random (MCAR)

Rubin (1976) defined a taxonomy of missingness that has become the standard for any discussion of this topic. This taxonomy depends on the reasons why data are missing. If the fact that data are missing does not depend on any values, or potential values, for any of the variables, then data are said to be *missing completely at random (MCAR)*. Any observation on a variable is as likely to be missing as any other. The example of the careless motorist, who does not appear for testing because of an accident having nothing to do with the study, is a case in point. If you are going to have missing data, this is the ideal case because treatment of the existing data does not lead to bias in the estimated parameters. It may lead to a loss in power, but it will not lead to biased parameter estimates.

Little (1988) has provided a statistical test of the MCAR assumption. His MCAR test is a chi-square test. A significant value indicates that the data are not MCAR. This test is provided in the SPSS Missing Values Analysis (MVA), and should be applied whenever there is some question about MCAR. SAS also includes this test in PROC MI, and in R it can be found in the BaylorEdPsych package.

69.2.2 Missing at Random (MAR)

Data are *missing at random (MAR)* if the probability of missing data on a variable (Y) is *not* a function of its own value *after controlling for other variables in the design*. Allison (2001) uses the example of missingness for data on income being dependent on marital status. Perhaps unmarried couples are less likely to report their income than married ones. Unmarried couples probably have lower incomes than married ones, and it would at first appear that missingness on income is related to the value of income itself. But the data would still be MAR if the conditional probability of missingness were unrelated to the value of income *within each marital category*. Here, the real question is whether the value of the dependent variable determines the probability that it will be reported, or whether there is some other variable (X) where the probability of missingness on Y is conditional on the levels of X.

If data are at least MAR, the mechanism for missingness is called *ignorable*. Thus, we can proceed without worrying about the model for missingness. This is not to say that we can just ignore the problem of missing data, but at least the existence of missing data is not going to seriously distort the results. This helps to explain why Little's test is often ignored. Even if the data are not MCAR, if they are at least MAR, we can pretty much proceed as we usually would.

69.2.3 Missing Not at Random

Data are classed as *missing not at random (MNAR)* if either of the above two classifications are not met. Thus, if the data are not at least MAR, then they are MNAR. When data are MNAR, there is presumably some model that lies behind missingness. If we knew that model we might be able to derive appropriate estimators of the parameters in the model underlying our data. Unfortunately, we rarely know what the missingness model is, and so it is difficult to know how to proceed. In addition, incorporating a model of missingness is often a very difficult task and might be specialized for each application. The following discussion will deal primarily with data that are at least MAR.

69.3   Linear Regression Models

Many of our problems, as well as many of the solutions that have been suggested, refer to designs that can roughly be characterized as linear regression models. The problems – and the solutions – are certainly not restricted to linear regression: they apply to logistic regression, classification analyses, structural equation modeling, and other methods that rely on the linear model. But I will discuss the problem under the heading of linear regression, because that is where it is most easily seen.

Suppose that we have collected data on several variables. One or more of those variables is likely to be considered a dependent variable, and the others are predictor, or independent, variables. We want to fit a model of the general form

$$Y_{ij} = b_0 + b_1 X_{1i} + b_2 X_{2i} + e_{ij}$$

In this model, data could be missing on any variable, and we need to find some way of dealing with that situation. We will assume that the missing data are either MCAR or MAR. A number of approaches to missingness in this kind of situation have been used over the years. I will first discuss the traditional approaches, which are going out of favor if they are not already there. Some of these are important even if only for the fact that they point out problems that later models attempt to address.

69.3.1   Casewise Deletion

The most common approach to missing data in regression analyses is what is called casewise deletion (or listwise deletion or available case analysis). Using this approach, we simply drop from the analysis all cases that include any missing observation. The analysis is then carried out on the data that remain. This is usually the default analysis for most statistical software.

There are definite advantages to casewise deletion. If the missing data are at least MAR, casewise deletion leads to parameter estimates that are unbiased. Casewise deletion also provides a starting point for more appropriate analyses. If the data are MNAR, however, this approach produces biased estimates. The resulting model is difficult to interpret because of confounding with missingness.

69.3.2   Pairwise Deletion

In pairwise deletion, data are kept or deleted on the basis of pairs of scores. In computing the overall covariance or correlation matrix, a pair of scores contributes to the covariance if both scores are present, but does not contribute if one or both of them are missing. Thus, if a participant has data on Y, X_1, X_2, and X_5, but not on X_3 or X_4, that participant would be included in computing r_{YX_1}, r_{YX_2}, and r_{YX_5}, but not in computing r_{YX_3} or r_{YX_4} (and similarly for the rest of the pairs of observations). All available observations would be used in estimating means and standard deviations of the variables.

This method does make use of all available data and thus estimates parameters on the maximum sample size. But that is its only advantage. The major disadvantage is that each correlation, mean, and standard deviation is estimated on a somewhat different dataset. In addition, it is not only possible but also not uncommon that the covariance matrices resulting from this approach and needed for the analysis will not be positive definite. This means that it is impossible to calculate a normal inverse of either matrix, and thus be able to solve the necessary equations.

Pairwise deletion is generally a bad idea, and I can think of no situation in which I would recommend it. As someone once said of stepwise regression, I would characterize pairwise deletion as "unwise" deletion.

69.3.3 Mean Substitution

One approach that is sometimes taken when data on an independent variable are missing is to substitute for the missing scores the mean on that variable for all nonmissing cases. This approach has the dubious advantage of using all the cases, but it has several disadvantages. Mean substitution is now rarely recommended because it can seriously underestimate the standard errors and thus overestimate significance.

69.3.4 Regression Substitution (Imputation by Least Squares)

One additional fairly simple approach to the treatment of missing data is to regress the variable that has missing observations on the other independent variables (even variables not used in the final analysis), thus producing a model for estimating the value of a missing observation. We then use our regression equation to impute (substitute) a value for that variable whenever an observation is missing.

There are advantages to this approach in some situations, but it leads to sometimes serious underestimation of error and should be avoided in favor, at least, of what is called *stochastic* regression approaches, discussed below, where an element of error is deliberately added to the predicted value. Regression substitution is usually not recommended by itself, because it still underestimates the standard errors in the final analysis, but the ideas behind it contribute to the more modern approaches discussed next.

69.4 Modern Approaches to the Problem of Missing Data

I will discuss two approaches to missing data that are currently the preferred methods that overcome many of the problems discussed above. Both were impractical not long ago, but are very practical now that we have high-speed computing. Even your personal laptop can carry them out in a reasonable time. The methods are maximum likelihood, which includes a third method called expectation–maximization (the EM algorithm), and multiple imputation (MI). A discussion of these methods can be found in Little (2005) and Little and Rubin (2002).

69.4.1 Maximum Likelihood

The idea behind maximum likelihood is relatively straightforward, though the computations certainly are not. Suppose that we have a set of data (either complete or missing) and we want to estimate the population means and covariances. (We need those to solve any regression problem.) The maximum likelihood approach effectively "tries out" many possible values for these parameters. For example, the solution selects one set of means and covariances and computes the likelihood of the obtained data given that set of parameters. Then it auditions a different set of values for the parameters and again computes the likelihood of the obtained data. After auditioning many possible sets of means and covariances, it finds those parameters for which the (nonmissing) data are most likely to have occurred. From these means and covariances, it then computes the standard regression solution.

When we have complete data, maximum likelihood reduces to a relatively simple solution that doesn't require much computation. However, when there are missing data, the likelihood values require an iterative solution and require special computer software. Fortunately, most statistical software programs can supply this analysis if requested. The nice thing about the maximum likelihood approach is that it will produce parameter estimates that are unbiased and, assuming that we have computed standard errors appropriately, the standard errors are also unbiased for data that are MCAR or MAR.

69.4.2   Expectation–Maximization (EM)

In maximum likelihood, we "audition" many different possible values for means and variances. But where do those values come from? One common approach is to use the EM algorithm to generate them and evaluate their appropriateness. This algorithm dates from the 1970's (Dempster et al. 1977) and continues to play an important role in missing data analysis.

EM is a maximum likelihood procedure that works with the relationship between the unknown parameters of the data model and the missing data. As Schafer and Olsen (1998) have noted, "If we knew the missing values, then estimating the model parameters would be straightforward. Similarly, if we knew the parameters of the data model, then it would be possible to obtain unbiased predictions for the missing values." (pp. 553–554). This suggests an approach in which we first estimate the parameters, then estimate the missing values, then use the filled-in dataset to re-estimate the parameters, then use the re-estimated parameters to estimate missing values, and so on. When the process finally converges on stable estimates, the iterative process ends.

For many, perhaps even most, situations in which we are likely to use EM, we will assume a multivariate normal model. Under that model it is relatively easy to explain in general terms what the EM algorithm does. Suppose that we have a dataset with five variables $(X_1 - X_5)$, with missing data on each variable. The algorithm first performs a straightforward regression imputation procedure where it imputes values of X_1, for example, from the other four variables, using the parameter estimates of means, variances, and covariances or correlations from the existing data. (It is not important whether we calculate those estimates using casewise or pairwise deletion, because we will ultimately come out in the same place in either event.) After imputing data for every missing observation in the dataset, EM calculates a new set of parameter estimates. The estimated means are simply the means of the variables in the imputed dataset. But recall that when I discussed stochastic regression imputation, I pointed out that the data imputed with that procedure would underestimate the true variability in the data because there is no error associated with the imputed observations. EM corrects that problem by estimating variances and covariances that incorporate the residual variance from the regression. For example, assume that we impute values for missing data on X_1

from data on X_2, X_3, and X_4. To find the estimated mean of X_1 we simply take the mean of that variable. But when we estimate the variance of that variable we replace

$$\Sigma(X_i - \overline{X})^2 \text{ with } \Sigma(X_i - \overline{X})^2 + s^2_{1.234},$$

where $s^2_{1.234}$ is the squared standard error of the regression of X_1 on X_2, X_3, and X_4. Similarly for the covariances. This counteracts the tendency to underestimate variances and covariances in regression imputation. Now that we have a new set of parameter estimates, we will have a new estimated regression equation, and we repeat the imputation process to produce another set of data. From that new set we re-estimate our parameters as above, and then impute yet another set of data.[2] This process continues in an iterative fashion until the estimates converge.

EM has the advantage that it produces unbiased – or nearly unbiased – estimates of means, variances, and covariances. Another nice feature is that even if the assumption of a multivariate normal distribution of observations is in error, the algorithm seems to work remarkably well.

One of the original problems with EM was the lack of statistical software. That is no longer a problem. The statistical literature is filled with papers on the algorithm, and a number of programs exist to do the calculations. A good source, particularly because it is free and easy to use, is an online program by Joseph Schafer called NORM, which is available to run as a stand-alone under the Windows operating system. The paper by Schafer and Olsen (1998) listed in the references is an excellent introduction to the whole procedure. SPSS also includes a missing data procedure (as a separate add-on) that will do EM. The results of that procedure closely match that of NORM, but in my experience the standard errors in the resulting regression are smaller than those produced by data imputed using NORM.

69.4.2.1 An Example

The following example is based on data from a study by Compas (1990, personal communication) on the effect of parental cancer on behavior problems in children. The dependent variable is the total behavior problem T score from the Achenbach Child Behavior Checklist (CBCL). One might expect that the gender of the parent with cancer (SexP) would be a relevant predictor (things fall apart at home faster if mom is sick than if dad is sick). Other likely predictors would be the anxiety and depression scores of the cancer patient (AnxtP and DeptP) and the spouse (AnxtS and DeptS). These five predictors were to be used in a multiple linear regression analysis of behavior problems. Unfortunately, due to the timing of the first round of data collection, many of the observations were missing. Out of 89 cases, only 26 had complete data. The good thing is that it is reasonable to assume that missingness was due almost entirely to the timing of data collection (different families receive a diagnosis of cancer at different times) and not to the potential values of the missing observations. So we can assume that the data are at least MAR without too much concern. The data for this example are available as an ASCII file at the web pages mentioned at the beginning of the chapter.

Using only casewise deletion in SPSS, we obtain the results in Table 69.1. In the variable names, P stands for "patient" and S for "spouse." Notice that the sex of the parent

---

2 Enders (2010) points out that we don't actually insert the estimated missing data into the data set, but the result is the same as if we did, and it is easier to think of it that way.

Table 69.1 Casewise deletion using SPSS.

Coefficients[a]

Model	Unstandardized coefficients		Standardized coefficients	t	Sig.
	B	Std. error	Beta		
1 (Constant)	−2.939	12.003		−0.245	0.809
SexP	−3.769	2.803	−0.183	−1.344	0.194
DeptP	0.888	0.202	0.764	4.393	0.000
AnxtP	−0.064	0.169	−0.062	−0.380	0.708
DeptS	−0.355	0.155	−0.460	−2.282	0.034
AnxtS	0.608	0.166	0.719	3.662	0.002

a) Dependent Variable: Totbpt. $N = 26$, $R^2 = 0.658$

with cancer does not have an effect, which is somewhat surprising, but the patient's level of depression and the depression and anxiety levels of the spouse are all significant predictors. However, as noted above, complete data are available only for 26 of the 89 cases.

We can improve the situation using the EM algorithm. For this analysis, all the variables in the dataset are included. That will be true with MI as well. In other words, we will use variables in the imputation process that we might not use in the subsequent analysis, because those variables might be useful in predicting a participant's score, even if they are not useful in subsequently predicting behavior problems. This is especially important if you have variables that might be predictive of missingness.

The SPSS analysis of the EM-imputed dataset is shown in Table 69.2. The data were imputed using Schafer's NORM program and then read into SPSS. Notice that the regression coefficients are not drastically different from those in the previous analysis with casewise deletion, but the standard errors are considerably smaller. This is due mainly to the large increase in sample size with the imputed data. Interestingly the sex

Table 69.2 SPSS analysis of the EM-imputed dataset.

Coefficients[a]

Model	Unstandardized coefficients		Standardized coefficients	t	Sig.	95% Confidence interval for B	
	B	Std. error	Beta			Lower bound	Upper bound
1 (Constant)	−11.591	6.215		−1.865	0.066	−23.953	0.771
SexP	−3.238	1.749	−0.106	−1.851	0.068	−6.717	0.241
DeptP	0.886	0.094	0.722	9.433	0.000	0.699	1.073
AnxtP	−0.004	0.099	−0.003	−0.039	0.969	−0.202	0.194
DeptS	−0.418	0.097	−0.357	−4.310	0.000	−0.610	−0.225
AnxtS	0.762	0.099	0.631	7.716	0.000	0.565	0.958

a) Dependent variable: TotBPt. $N = 89$ $R^2 = 0.871$

of the patient is much closer to significance at $\alpha = 0.05$. Notice, also, that the squared multiple correlation has increased dramatically, from 0.658 to 0.871. I am much more comfortable with this model than I was with the earlier one, which was based on only 26 cases.

69.4.3 Multiple Imputation

A second method for imputing values for missing observations is known as MI. The original work on this approach was due to Rubin (1987). The interesting thing about MI is that the word *multiple* refers not to the iterative nature of the process involved in imputation but to the fact that we impute multiple complete datasets and run whatever analysis is appropriate on each dataset in turn. We then combine the results of those multiple analyses using fairly simple rules put forth by Rubin (1987). In a way, it is like running multiple replications of an experiment and then combining the results across the multiple analyses. But in the case of MI, the replications are repeated simulations of datasets based on parameter estimates from the original study.

The introduction of new simulation methods, known as Markov Chain Monte Carlo (MCMC), has considerably simplified the task of doing MI, and software is now available to carry out the calculations. Schafer (1997, 1999) has implemented a method of MCMC called data augmentation, and this approach is available in his NORM program referred to earlier. MI is available in SPSS and in SAS as PROC MI and PROC MIANALYZE.

The process of MI, as carried out through data augmentation, involves two random processes. First, the imputed value contains a random component from a standard normal distribution. (I mentioned this in conjunction with regression imputation.) Second, the parameter estimates used in imputing data are a random draw from a posterior probability distribution of the parameters.

The process of MI via data augmentation with a multivariate normal model is relatively straightforward, although I would hate to be the one who had to write the software. The first step involves the imputation of a complete set of data from parameter estimates derived from the incomplete dataset. We could obtain these parameters directly from the incomplete data using casewise or pairwise deletion; or, as suggested by Schafer and Olsen (1998), we could first apply the EM algorithm and take our parameter estimates from the result of that procedure.

Under the multivariate normal model, the imputation of an observation is based on regressing a variable with missing data on the other variables in the dataset. Assume, for simplicity, that X was regressed on only one other variable (Z). Denote the standard error of the regression as s_{XZ}. (In other words, s_{XZ} is the square root of $MS_{\text{residual.}}$). In standard regression imputation, the imputed value of X ($\widehat{X}$) would be obtained as

$$\widehat{X}_i = b_0 + b_1 Z_i$$

But for data augmentation we will add random error to our prediction by setting

$$\widehat{X}_i = b_0 + b_1 Z_i + u_i s_{X.Z}$$

where u_i is a random draw from a standard normal distribution. This introduces the necessary level of uncertainty into the imputed value. Following the imputation procedure just described, the imputed value will contain a random error component. Each time we impute data, we will obtain a slightly different result.

But there is another random step to be considered. The previous process treats the regression coefficients and the standard error of regression as if they were population parameters, when in fact they are sample estimates. But parameter estimates have their own distribution. (If you were to collect multiple datasets from the same population, the different analyses would produce different values of b_1, for example, and these estimates have a distribution.) So our second step will be to make a random draw of these estimates from their Bayesian posterior distributions – the distribution of the estimates given the data, or pseudo-data, at hand.

Having derived imputed values for the missing observations, MI now iterates the solution, imputing values, deriving revised parameter estimates, imputing new values, and so on until the process stabilizes. At that point we have our parameter estimates and can write out the final imputed data file.

But we do not stop yet. Having generated an imputed data file, the procedure continues and generates several more data files (perhaps a new data file after each 200 cycles of the process). We do not need to generate many datasets, because Rubin has shown that in many cases three to five datasets are sufficient. Because of the randomness inherent in the algorithm, these datasets will differ somewhat from one another. In turn, when some standard data analysis procedure (here we are using multiple regression) is applied to each set of data, the results will differ slightly from one analysis to another. At this point, we will derive our final set of estimates (in our case, our final regression equation) by averaging over these estimates following a set of rules provided by Rubin.

For a discussion and example of carrying out the necessary calculations, see the excellent paper by Schafer and Olsen (1998). Another example based on data used in this article is available from the author in the web pages cited at the beginning of this chapter. That site also includes a discussion of using NORM to carry out the calculations.

References

Allison, P.D. (2001). *Missing Data*. Thousand Oaks, CA: Sage Publications.

Dempster, A.P., Laird, N.M., and Rubin, D.B. (1977). Maximum likelihood from incomplete data via the EM algorithm (with discussion). *Journal of the Royal Statistical Society, Series B* 39: 1–38.

Enders, C.K. (2010). *Applied missing data analysis*. New York: The Guilford Press.

Little, R.J.A. (1988). A test of missing completely at random for multivariate data with missing values. *Journal of the American Statistical Association* 83: 1198–1202.

Little, R.J.A. (2005). Missing data. In: *Encyclopedia of Statistics in Behavioral Science* (ed. B.S. Everitt and D.C. Howell), 1234–1238. Chichester, England: Wiley.

Little, R.J.A. and Rubin, D.B. (2002) *Statistical Analysis with Missing Data*. Hoboken: Wiley.

Rubin, D.B. (1976). Inference and missing data. *Biometrika* 63: 581–592.

Rubin, D.B. (1987). *Multiple Imputation for Nonresponse in Surveys*. New York: Wiley.

Schafer, J.L. (1997). *Analysis of Incomplete Multivariate Data*. London: Chapman & Hall (Book No. 72, Chapman & Hall series Monographs on Statistics and Applied Probability).

Schafer, J.L. (1999). Multiple imputation: a primer. *Statistical Methods in Medical Research* 8: 3–15.

Schafer, J.L. and Olsen, M.K. (1998). Multiple imputation for multivariate missing-data problems: a data analyst's perspective. *Multivariate Behavioral Research* 33: 545–571.

Section X

Academic Networking

70

Essentials of Interviewing for Prospective Medical Students and Residents

Frederic J. Bertino[1,2] *and Talal A. Kaiser*[2,3]

[1] Department of Radiology and Imaging Sciences, Emory University School of Medicine, Atlanta, GA, USA
[2] Department of Anatomical Sciences, St. George's University School of Medicine, Grenada, West Indies
[3] Department of Internal Medicine, University of Connecticut School of Medicine, Hartford, CT, USA

70.1 The Importance of a Great Interview

When applying for medical school or residency, the interview is one of the most important steps a future physician must navigate in his or her path to a career in the health professions. Interviews are offered after a medical school or residency program has conducted a careful review of the applicant's credentials. Exam scores, grade-point-averages, and other objective data are often preliminary materials that an admissions committee may use to select candidates for interview. The personal statement – a testament to an applicant's goals and ambitions – may similarly carry weight when being considered for interview after much of the objective criteria has been met.

The interview for medical school and residency is an opportunity for learning; for both the applicant and the program. The applicants have the opportunity to expand on their ambitions stated in the personal statements and evidenced by their academic performance, while also learning about the program. The interview allows the applicant an opportunity for showcasing conversational ability, self-advocacy, and present oneself in the manner that the applicant hopes to be received by the admissions committee. The interview is the most important chance to convey professionalism and an insight into the personality of the applicant.

This chapter highlights the important skills and techniques that every applicant should master before entering an interview for medical school or residency. By unveiling the experience of the typical interview flow and pre-and-post interview agenda, the applicant should feel confident in the scheduling, preparation, conduct, and follow-up of an ideal interview.

70.2 General Interview Skills

The interview is an opportunity for an applicant to showcase his or her achievements and personality otherwise not listed on paper. The interview provides the admissions committee insight into the type of person who may be joining their community in the

A Guide to the Scientific Career: Virtues, Communication, Research, and Academic Writing, First Edition.
Edited by Mohammadali M. Shoja, Anastasia Arynchyna, Marios Loukas, Anthony V. D'Antoni, Sandra M. Buerger, Marion Karl and R. Shane Tubbs.

future. The submitted application often contains achievements, publications, performance in school and on standardized examinations, and letters of recommendation for an admissions committee to review. This application lacks the true essence and personality of the applicant. This shortcoming is overcome by the interview, where the applicant can convey his or her personality and ambition to an admissions committee. When interviewing in health sciences, for either medical school or a residency position, programs expect applicants to be goal-oriented. With the exception of the personal statement, the interview is the only opportunity to personally convey one's future goals and ambitions to a selection committee in the form of a dialogue. This exchange allows both parties (the applicant and the interviewer) to learn each other's desires and, hopefully, find a mutual interest in one another.

Obtaining an interview is an indication that a program selection committee believes a candidate has the potential to succeed in a program based on objective data gathered from an applicant's application. This means that to obtain an interview, the application must have demonstrated exemplary strengths, accomplishments, and great passion for his or her chosen field. The interview is the final opportunity to make a strong impression on the program in an intimate and conversational environment.

70.3 Conversational Ability

Communication skills are imperative for not only replying and accepting invitations to interview but also for the interview itself. Written word and spoken word communication are inherently different. The ability to write well does not imply that one may speak with equal fluency. Additionally, the stress of an interview setting can also make communication challenging. Therefore, practice in public speaking or rehearsing the fluency of important answers is encouraged before the interview. Like giving a speech, rehearsing is the best way to effectively convey one's intents and goals to an admissions committee. Holding mock interviews with friends or experienced colleagues, rehearsing fluidity of speech to oneself in a mirror, and listening to playback of a recording are all ways that the applicant may improve his or her speaking ability.

Arriving at an interview unprepared for commonly asked questions can be detrimental to the applicant's efforts of applying. Later in this chapter, commonly asked questions will be presented (Table 70.1). Having well-thought-out answers for these major questions will help the applicant effectively communicate his or her passions to the admissions committee.

70.4 Advocating for Oneself

Many applicants find it very difficult to talk about themselves. While the qualities of a physician are personified in modesty, selflessness, and empathy, it is important to realize that, during the interview, the applicant must be able to advocate for him or herself with confidence.

The applicant should not understate his or her achievements. Many qualified applicants have been invited for interviews. The interview is an opportunity for the applicant to personally demonstrate his or her accomplishments, speak fluently about goals and successes, and provide proof that the applicant is a great fit for the culture of the residency program or medical school.

It is possible to convey modesty, confidence, and assertiveness simultaneously in an interview. It is usually this attitude that appeals to committees and results in a successful match or acceptance. Speaking confidently and using examples of past achievements and success will serve as proof that the applicant is a good fit for a program offering resources that align with an applicant's needs and goals. Hiding these achievements will not help the applicant. Realize that other equally qualified applicants may have the upper hand at a sought-after position simply by being able to advocate on their own behalf.

70.5 A Note on Honesty

While applicants everywhere strive to be impressive to an admissions committee, it is important to never compromise one's honesty in the process of interviewing. Confidence and conversational ability are enough to demonstrate one's character to an admissions committee. Exaggerating the truth, arrogance, overcrediting oneself, or overtly lying during an interview will be disastrous to the applicant's chances of match or admission and may harm future application efforts for postgraduate work. An interview should be a demonstration of one's strengths by providing objective evidence of one's passions through prior achievements, experiences, and personal statements. These points, when cited honestly from the applicant's curriculum vitae, can convey one's passions and goals sincerely while providing proof for the applicant's continued motivation and work ethic.

70.6 Explaining Weaknesses on an Application

While it might be ideal to have a flawless application, it is not uncommon to have a less-than-desirable mark on an application that may warrant further attention from an interviewer. This attention is not typically an opportunity to criticize an applicant. Instead, it is an opportunity to further examine why such a mark exists and to examine ways the applicant has risen above any shortcomings. The key to explaining weaknesses on an application during the interview is to *introspectively* examine the root cause of the problem. Most importantly, taking steps to fix the problem and acting on them will show dedication to self-improvement and help maintain applicant desirability.

As mentioned before, this is not an opportunity to create excuses without any true result. A failed attempt at a qualifying exam or a poor grade in one's coursework requires a clear explanation for the shortcoming and evidence for subsequent improvement. Failure to improve without any effort placed into correcting the problem will affect the applicant negatively. Explaining the error, and demonstrating that the problem has been fixed, is the best way to convey maturity in an interview. It demonstrates that the applicant is capable of rising from failure and will consistently strive to self-improvement. This resiliency is expected in residency and in medical school, and the steps toward self-improvement will guarantee the applicant a successful career in medicine.

70.7 Preparation: A Means of Reducing Anxiety

It is entirely common and natural for applicants to feel nervous during, or in the time approaching the interview. This is a natural response to a stressful situation. Preparation for an interview is the single best technique to overcome interview anxiety. Preparation involves much of the guidance mentioned in this chapter already, while also keeping a close itinerary of events and travel for the specific interview day.

In the following sections, we hope to reduce the anxiety of the process by providing information about the interview process and giving strategies for success.

70.8 Put Your Best Foot Forward: Attire and Behavior

A large part of preparation stems from proper interview attire. An interview is a professional invitation for the possibility of employment or enrollment. It should be treated with maturity, class, and professionalism. How you present yourself speaks volumes. Professional attire and behavior are paramount for a successful interview.

Dress professionally. For males, this typically means a neutral suit (gray, blue, or black) along with a dress shirt and tie. For females, this typically means a neutral pant suit (gray, blue, or black) or knee-length skirt with a blouse. For both males and females, hygiene and grooming are important. For males with facial hair, make sure it is well groomed. When in doubt about wardrobe, it is always better to err on the side of conservatism. The goal is to be remembered for who you are as an applicant and not for what you wore.[1]

It should go without mention that being on "one's best behavior" is an absolute rule during an interview. Being polite and courteous to everyone you encounter (both in person and electronically via email). Everyone met during the interview process may have a voice in the decision-making process of the applicant's hire. Realize that an ounce of kindness and patience during a hectic season can go a long way.

An applicant should take note of any bad habits so as to avoid them during a conversation with an interviewer. These behaviors are often unconscious. Behaviors that should be avoided include biting one's nails, "picking" at one's teeth or fingers, cracking or popping one's knuckles, and twirling or playing with one's hair. Avoid using filler words in conversation. These are words or phrases like "um," "like," and "you know." These filler words can be distracting from the point the applicant is trying to make. Practicing responses to common questions is a good way to state a point without relying on filler words. In situations where there is an unexpected question, it is always better to take a moment to gather one's thoughts before speaking than to speak aimlessly until an answer forms.

70.9 Interview Expenses

Interviews are expensive – this is an unfortunate fact of the process. Budgeting your expenses (especially while relying on student loans) can be a burden. Remember that

---

1 Preparing for Medical School Interviews, 2015. AAMC Students, Applicants and Residents website: https://students-residents.aamc.org.

while the cost of the interview process may be daunting, an interview is an opportunity for one to begin a career as a physician and should be seen as an investment into a future career. Loans exist for students through the public and private sector to provide additional financial aid during the interview season.[2] Being cost-conscious and saving for the season is advised. This is the time to use your life savings to enter into a career path that will set a level of financial security for the future.

Traveling costs can be expensive. Hotel rooms, food and drink, and transportation costs can quickly add up. However, there are several mechanisms in place to allow you to save money. Many programs offer a "pre-interview dinner" at the expense of the program to allow applicants to meet residents and students. Some programs may even cover the hotel expenses for visiting applicants. There are many services online or through smartphone applications that provide inexpensive rental cars or hotel rooms. Finally, credit card companies offer reward programs for frequent travelers where one may obtain free nights in hotel rooms or discounted airline tickets. Researching these resources ahead of time can often mean differences of hundreds of dollars.

70.10   Scheduling the Interview

Obtaining an interview invitation can be a very exciting experienceand responding to the invitation is the first opportunity to make a positive impression. Respond to interview invitations as soon as possible.[3] Invitations are often sent out to multiple applicants at one time and often more invites are extended than there are interview spots. A prolonged delay in responding may prevent you from scheduling an interview day at all. Be professional and courteous in your communication with the program. Prior to an in-person encounter, the only way a program can assess an applicant is through written words. Choose them wisely.

Do not change a scheduled interview day unless absolutely necessary. Program coordinators have the unenviable task of having to coordinate hundreds of interviews. Frivolous changes only make their jobs more difficult and make it more likely that you will make a poor impression. If you do not plan to attend an interview, be sure to cancel the interview as soon as you know. This allows the interview slot to go to someone else on the program's list and is a demonstration of your professionalism. Never "no-show" to an interview. Programs understand that emergencies, by definition, are unavoidable. Whether it is a missed flight or weather delay, always call the program and explain that you will not make it for your interview. A last-minute phone call is always better than no call at all. The program director community is very small, and stories of unprofessional applicants can very easily travel to other programs.

70.11   Learn about the Program

Prior to the interview, learn as much as you can about the program. Browse the program website, and seek out former classmates from your current program who may be current

---

2 Advice on Applying to Residency Programs, 2012. AAMC website: www.aamc.org.
3 Communicating with Residency Programs, 2015. AAMC Students, Applicants and Residents website: https://students-residents.aamc.org.

students or residents at the program of interest . In addition to having things to ask about on the interview day, knowledge of the program demonstrates interest and will allow you to determine if the program is right for you.

If there is a pre-interview dinner, try your best to attend. This often-informal setting is a great opportunity to get a sense of the personality of the program. More often than not, the residents who attend have no direct say in who is chosen for the program and are more than willing to answer any questions you may have. Additionally, meeting other applicants the night before can reduce anxiety for the following day.

70.12 Pre-Interview Contact

Pre-interview contact with a program can often feel impersonal, as most programs have a general email that is sent out to all interview applicants. Regardless, be sure to respond to all emails courteously. Always address program coordinators as "Mr." or "Ms." It is not common to receive an email from a physician at this stage, so it is not necessary to address the sender as "Dr." unless clearly specified.

A good format to follow when responding to an interview invitation from a program is as follows:

a. Address the sender by the appropriate title.
b. Thank the program for extending an interview invitation.
c. Offer three to four potential days for your interview availability.
d. Demonstrate flexibility in scheduling days.

Once the interview date is confirmed, respond to let the program know you have received the confirmation email. Most programs send out a reminder email a week or so prior to the interview date. Sometimes this is sent through a proxy email address and does not require a reply. If it is sent from the coordinator, it is good form to reply.

70.13 The Interview Day

Get a good night's sleep the night before the interview. While easier said than done, being well rested will help you be more alert and allow you to think clearly. Try and relax the night before the interview. Allow time to wind down before heading to bed to avoid an anxious night's sleep.

On the interview day, be aware of the route from the hotel to the hospital. Print out directions or have them saved to your smartphone or mobile device, arrange for transportation, and give ample time for any delays that may arise. Lay out everything you will need for the interview day (i.e. clothing, notebook, pen, etc.) the night before. Be self-aware and plan ahead for any timing issues. It is important to be punctual.

70.14   Post-Interview Contact

Post-interview communication can be divided into two categories as follows:

70.14.1   Communication from the Applicant to the Program

Always send a thank-you email to the program coordinator and the faculty who interviewed you. It is not usually necessary to send thank-you notes to every person you meet during the interview day and it is not typically expected that you do so. It may be wise to send these emails shortly after the interview to ensure they are not forgotten and to demonstrate to the program your continued interest. Handwritten cards or letters instead of or in addition to email are not expected. However, these have the added benefit of appearing more personal than an email.[4]

For future residents, prior to ranking programs, some suggest that it may be wise to send a "letter of intent" to one's top choice program. Be cautious of this advice. While it is not a match violation to reveal your preferences, it is a violation if you ask the program to reveal the same. Additionally, sending a letter to multiple programs stating that they are your "first choice" is in poor form and will be revealed as deceitful after the match.

70.14.2   Communication from the Program to the Applicant

Many programs have decided to adopt a "No post-interview communication" policy, as they feel it places an unnecessary stress on applicants and programs to demonstrate mutual interest. Additionally, words exchanged prior to a decision being made rarely have the requirement to be true. With this in mind, do not be disheartened if responses are not received. Other communication, such as contact regarding questions about the program, are often encouraged by programs .

Be wary of any communication from the program or faculty. While it is certainly no one's intention to mislead, general courtesies can often be misconstrued as verbal commitments. There is no shortage of stories from applicants who received positive communication from a program, only to end up at a different program. For those applying for residency spots: no communication from a program should influence how a future resident should rank programs.

70.15   Key Questions to Ask and Be Asked

Interview questions can vary in nature. Questions can be career-oriented, in an effort to understand an applicant's plans for the future or person-focused and geared toward getting to know the applicant's personality better. Questions about red flags are fair game. Practicing answers for the common interview day questions is also essential. These answers should by no means be memorized or overly rehearsed, but instead should act as a scaffold for the applicant to build upon and cater to individual programs.[5] A general idea and subset of points should be in the applicant's mind on

---

4 Advice on Applying to Residency Programs, 2012. AAMC website: www.aamc.org.
5 Preparing for Medical School Interviews, 2015. AAMC Students, Applicants and Residents website: https://students-residents.aamc.org.

Table 70.1 Common interview questions.

Why did you choose medicine as a career?
Why did you apply to this specialty/school?
What would you like to specialize in?
Where do you see yourself in 10 years?
What are the greatest challenges facing healthcare today?
Tell us about yourself.
What do you hope to gain by coming to this school/residency?
What qualities can you bring to this program?
Describe a situation in which you encountered a peer acting unprofessionally. How did you handle it?
What is the last book you read for pleasure? Tell me about it.
Describe an interesting case for me.
What are your hobbies?

interview day to ensure he or she conveys a thorough and enthusiastic message of his or her goals. The most commonly asked questions in interviews are listed in Table 70.1.

Every interview typically ends with some form of the question, "Do you have any questions for me?" Sometimes, interviews can even begin with this question! This question is often frustrating for applicants. The importance of this question from an interviewer's perspective cannot be understated. It helps guide the interview, allows one to gauge the interest level of the applicant, and allows the interviewer to cater the interview toward the interests of the applicant. Do not be intimidated by this question and be ready to ask one or two questions. Try and avoid asking questions that can be answered by viewing the program website. The only incorrect answer to this question is, "I have no questions." The common questions to ask an interviewer are listed in Table 70.2.

Table 70.2 Common questions to ask an interviewer.

What do you think are the strengths of the program?
What do you think are the weaknesses?
Where do residents/students end up after they finish the program?
What sort of measures are in place to help residents who are struggling?
Do you foresee any changes to the program administration in the near future?
Do you anticipate any changes to the program structure in the near future?
How do you like the region/city/town?
What is it like to raise a family here?
Are you from this area originally? Was the transition in moving to this area difficult for you/your family?

Further Reading

*Association of American Medical Colleges (AAMC) Interviewing Resources (*2017*). Available at: URL:* https://www.aamc.org/cim/residency/application/interviewing

Iserson, K.V. (2013). *Iserson's Getting into a Residency: A Guide for Medical Students*, 8e. Tucson: Galen Press.

Freedman, J. (2010). *The Residency Interview: How to Make the Best Possible Impression.* MedEdits Publishing.

71

Professional and Academic Societies and Meetings

Philip R. Brauer

Department of Clinical Anatomy, Kansas City University of Medicine & Biosciences, Kansas City, MO, USA

71.1   Academic Societies

An academic society is a learned society promoting a particular discipline or branch of knowledge. Most individuals join an academic society at the local level while in a professional or graduate school. While many initially join an academic society, most do not actively participate and soon drop out or just continue paying annual dues, getting little in return. However, a smaller percentage of members become actively involved and participate and help contribute to the activities and governance of the society. Other members, more advanced in their research careers, provide advice and guidance for its younger members, and it is well documented that their mentoring of students, residents, and early-career researchers has a positive impact on the subsequent success of younger members (Reynolds 2008; Takagishi and Dabrow 2011).

71.2   Professional Societies

Professional societies represent a synergistic group of individuals working to further a particular profession and promote the interests of individuals in that profession. By working as a group, they have a much larger impact than if they worked as individuals. For instance, they promote advocacy, develop policy, cultivate public relations, and impart new knowledge important to the profession. Membership opportunities are available at almost any stage along your career and, in some societies, even into retirement. Almost every professional scientific society offers student, resident, and postdoctoral fellow memberships at a discounted membership fee, as well as regular memberships to faculty, research scientists, and practicing physicians.

71.3   Career Benefits

71.3.1   Research Presentations

Local and regional meetings held by academic and professional societies are a great venue for you to make your first formal research presentation outside your immediate

A Guide to the Scientific Career: Virtues, Communication, Research, and Academic Writing, First Edition.
Edited by Mohammadali M. Shoja, Anastasia Arynchyna, Marios Loukas, Anthony V. D'Antoni, Sandra M. Buerger, Marion Karl and R. Shane Tubbs.

research group. The meeting environment tends to be smaller and less formal, giving you a chance to polish your presentation skills, garner advice from fellow members to improve your presentation, and learn from your mistakes with virtually no consequences. Not only does this experience help improve your presentation and advocacy skills by having to rise to the challenge of preparing and presenting to your peers, but it provides an opportunity to share your findings and discuss your ideas with others interested in your work. Likewise, by presenting your work at national and international meetings, you will garner attention, draw others into a conversation, and initiate new contacts with potential collaborators or individuals connecting you with resources you might need to continue your research. Later as a senior research professional, your continual participation in an academic society can help you recruit staff, faculty, and researcher clinicians as they begin entering more advanced stages of their careers.

71.3.2   Workshops and Training Opportunities

Professional societies can help advance your research training and career development and provide opportunities that may not be available at your current institution. For students, residents, postdoctoral fellows, and new research investigators, professional societies provide valuable online resources, including curriculum vitae (CV) writing, negotiation tactics, technical and grant writing courses, and certifications, that have been developed and implemented by experienced successful researchers.

Many societies hold career-development training sessions during their meetings and align you with more senior members interested in serving as mentors. Depending on the venue, the society may organize specific training workshops in areas of technology and methodologies that are new to you or that you want to know more about. Some programs even offer accredited continuing medical education programs. Furthermore by attending these sessions, you will bring back new information, skills, and insights to share with members at your home institution.

71.3.3   Grants and Travel

Many professional societies offer travel awards and research grants to help jump-start early research careers of its younger members. This includes training grants, "seed money" for research projects, salaries for release time, and supporting travel costs to research symposia.

71.3.4   Recognition

Joining academic and professional societies early in your career has advantages, as your recognition will grow over time, especially if you are actively involved and participating. Societies call for volunteers from the membership to serve on committees, organize meeting platforms sessions, and more. By actively participating, there will be occasions where you will meet and work with established and well-recognized scientists, increasing your profile and recognition in the profession. This greatly enhances the "foot-in the door" opportunities such as establishing new collaborations, participating in career advancing activities, and enhancing your name recognition as more senior level professionals become familiar with you and your work. From personal experience, my early membership participation led to invitations to write manuscripts and textbooks, serve on peer-review committees, and, eventually, to leadership roles in the very professional society I joined as a student. Taking advantage of these opportunities also enhances your CV, brings recognition from your peers and employers, and helps with promotion.

71.3.5 Networking

Professional scientific societies organize annual research meetings at the regional and national level or both. Here you will meet a broad network of researchers including students, residents, educators, practitioners, and clinicians all having different perspectives and with surprisingly similar experiences and challenges. One of the best benefits of attending a professional society's meetings is the opportunity for networking and establishing fruitful collaborations with others in and outside your area of research (see Chapter 14). Where else can you meet journal editors, grant study section members, noted scientists (e.g. Nobel Prize recipients), textbook authors, public policy makers, technology vendors, technical staff, and developers, all in one place sharing and learning from one another?

Make the most of the opportunity to interact with plenary speakers after their presentation or at organized social events held during the conference. Many will be internationally recognized, and this is your chance to question and discuss their research and its relationship to your own. Moreover, many of them will likely be reviewing your manuscripts, serving as editors of the journals where you plan on submitting your manuscripts, or serving on extramural grant peer-review panels.

As you meet, gather contact information and socialize with other investigators. You will, as a collective, share insight, discuss common problems, and brainstorm possible solutions in a unique environment. Not only will you learn about new approaches and solutions to problems, you will likely be introduced to new and emerging technologies and instrumentation from commercial vendors that play an integral part of most scientific meetings. Many companies also bring technical staff interested in providing solutions to your research problems.

Rather than avoiding the sharing of data or resources, society members are generally anxious to share their discoveries and are eager to find opportunities for collaboration. Long-term friendships and fruitful collaborations are established at these meetings, and over time, you will form an extended professional scientific family. These new connections can be important when needing to provide reference letters from outside your current institution.

Social events at these meetings (e.g. ones coupled with a poster session) offer a great occasion to establish collaborations, learn the finer points of research methodologies, and resolve technical problems you may face in your own research. Informal gatherings (e.g. lunch with participants or after-hour events) provide another avenue for networking and strengthening collaborative bonds. You might even meet your next employer or connect with someone aware of upcoming career opportunities not publicly known!

71.4 Getting Started

While this all sounds great, how do you start? First, you will want to find the right group(s) based on your research interest, goals, and possibly your geographic location.

Ask more senior investigators at your current institution which professional societies they belong to and whether they feel this or a particular group might be suitable for you. Then, follow up by investigating the societies' website, their mission, what they offer their members, who represents the leadership, and whether a society has local or regional chapters. When considering joining a professional society, you might want to attend one of its scientific meetings to get a better feel for the group and to meet some of its members.

Larger professional societies may also have local chapters or hold regional meetings that might be more convenient and financially viable for you to attend. These local venues provide a great place to present and fine-tune your research presentations and to garner feedback before submitting your work for publication. By connecting with members in your region, you might find local resources and expertise that you may not have been aware of before. Moreover, it will likely be easier to establish collaborative research efforts with members of the local group because of their proximity. It is becoming harder to find the financial resources, release time, open calendar dates, etc., to attend long-distance, multiday national/international meetings, particularly when starting out in your career. Hence, participation in less costly local and regional meetings sponsored by professional societies offer an alternative that can provide many of the same benefits as national meetings. This local network can also give you an inside track to regional internship and career opportunities.

Participating in activities of local chapters can culminate in leadership opportunities at the national level. By serving on local chapter committees or as a delegate to the national/international society, you enhance your reputation within the society through recognition of your participation at the local/regional level. By participating at the local level, there will be occasions where you meet the national representatives of the organization at sanctioned events, thereby expanding your professional network.

71.5 Conclusions

While there are many tangible benefits to belonging and participating in a professional scientific/medical society (e.g. discounts to meeting registrations, books, free journal subscriptions, etc.), the most important things are intangible. These include making your research known, making the most of important networking opportunities, developing leadership skills, establishing research collaborations, and opening doors for new professional career-building possibilities. Sharing what you have learned with your colleagues upon your return from professional meetings will also have a positive impact on how they and your superiors view your contribution to your institution's mission. Finally, you will return energized and excited by what you have learned and from the new connections and collaborations you developed while at the meeting.

Professional and academic societies can serve as your professional home where you build long-term relationships with an ever-growing sphere of colleagues, sharing common goals in advancing medical knowledge through research as you build a career. You will find that the longer you participate in a society, the more you will want to remain a part of it and see it grow. As you become more engaged with the society, you will find it personally rewarding, and you will become the seasoned professional wanting to work toward benefiting the careers of the new members, as the society did for you!

References

Reynolds, H. (2008). In choosing a research health career, mentoring is essential. *Lung* 186 (1): 1–6.

Takagishi, J. and Dabrow, S. (2011). Mentorship programs for faculty development in academic general pediatric divisions. *International Journal of Pediatrics* 2011: 538616–538621.

72

Getting the Most from Attending a Professional Meeting

Peter J. Ward

West Virginia School of Osteopathic Medicine, Lewisburg, WV, USA

72.1   Introduction

Attending scholarly, professional meetings related to one's field is an invaluable activity in academia. In addition to keeping abreast of new developments in your field, there are opportunities to inform others of your own scholarly work and to build networks in your profession. Meetings are a prime place where students and new faculty are able to connect with peers and to meet the luminaries within their fields. While such meetings or conventions occur frequently and are quite important in academia and other professions, there is not a great deal of literature available that relates to them (Mair and Thompson 2009).

This short chapter will describe the major activities that take place during professional meetings and how a novice can get the most out of attending such a meeting. There is no single way to do this, but I will endeavor to supply some tips that will make attendance intellectually and professionally rewarding. I will assume that the reader is someone who has not yet attended a professional meeting and is looking forward to having the first (or next) meeting make a real impact in his or her professional life.

There are several steps you can take to benefit from attending a professional meeting. First, choose the right meeting to attend. This will require you to be mindful of your budget and find suitable accommodations. Once you are on site, scout out the meeting events and visit vendors and publishers. If possible, prepare to present a topic or a sample of your work. If you are interested, look for ways in which you can become involved in the organization throughout the meeting.

As you become more experienced and attend several meetings, you will naturally find other approaches that work for you. In addition, you will develop a cohort of peers that you will see each meeting. Please take my suggestions as a starting point and elaborate upon them in ways that work for you.

72.2   Choosing a Meeting

If meetings are new to you, the first step is to identify a professional organization that hosts meetings that will be of benefit to you. Consult with colleagues, mentors, and

A Guide to the Scientific Career: Virtues, Communication, Research, and Academic Writing. First Edition.
Edited by Mohammadali M. Shoja, Anastasia Arynchyna, Marios Loukas, Anthony V. D'Antoni, Sandra M. Buerger, Marion Karl and R. Shane Tubbs.

advisors to get a sense of the variety of professional organizations that cater to your profession and have a focus that matches your interests. Once you have a list of professional organizations that you are interested in exploring, start scouting their internet or social media sites. This will give you an idea about what activities occur in the organization and when they meet. Finding the meeting dates and locations will let you know if attending is even feasible, given your schedule and budget. Conference management groups know that ease of attending is a concern for many attendees and will likely try to position meetings in places that people can easily attend (Var et al. 1985; Grant and Weaver 1996; Oppermann and Chon 1997; Zhang et al. 2007; Yoo and Chon 2008; Mair and Thompson 2009).

Some organizations have a large cohort of international members, and their meetings may take place in your country or quite a long way away. While a faraway or exotic locale may be a draw to many members, it also brings additional expenses that may wreck your budget. If there are viable meetings available in your home country, they will be a good place to start. Once you have a meeting or two behind you, you will probably be ready to tackle a meeting in another country. However, if you are traveling with a group of people from the same institution to an international meeting along with an experienced guide, you will probably be just fine. If that is the case, enjoy!

Once you have identified a meeting that you can attend, start perusing the speakers and sessions that will be held. This will let you know if the topics presented have relevance to your interests. At the same time, look for poster or speaking venues where you might be able to present your own work. These two activities, learning about developments in your field and presenting your own work/research, are really the "official" reasons to attend a meeting. If you find a meeting that you can afford, has interesting speakers, and has a venue for you to present your own work, then you have found a meeting worth attending.

Some meetings are small and intimate while others are massive. You will need to attend several meetings to develop a feel for what type of meeting you prefer. There are advantages and disadvantages to either type of meeting. Large meetings have more symposia, more research presentations, a broader array of topics, and more colleagues to meet. Also, if you are hoping to interact with some of the "big names" in your field, it is most likely that they will be attending and presenting their work at larger meetings. However, big meetings can also feel impersonal, and you may experience some difficulty getting your foot in the door if you are hoping to get involved in the association.

Smaller meetings are often more specialized. This is not a problem if the organization's focus corresponds directly with your interests, but it is no fun to find yourself attending an entire meeting that is only peripherally connected to your topics of interest. This is another reason why research ahead of time can help you maximize the benefits of attending a meeting. If you are hoping to get involved in an organization, smaller meetings are often a good place to start because you can become a "known person" simply by engaging in normal, conversational networking. In addition, some professional organizations host large annual meetings and smaller local meetings. While researching the meetings, be sure to see if any of those local meetings might be easy for you to attend.

72.3 Budgeting for Travel and Housing

Not surprisingly, meetings can be expensive, and their cost is a major consideration in whether or not a person chooses to attend (Mair and Thompson 2009). Once you have selected a meeting, start developing a budget that includes travel, housing, food, and other expenses as needed. Once you have an idea of your budget, it is time to see about getting yourself to the conference safely and staying somewhere hygienic while you are there. Consult with your advisor, department chair, and/or secretarial staff since they may be able (or obligated) to make your travel arrangements with approved vendors. Also, they sometimes will be able to find funding to assist in your travel if you are lucky.

Even if someone else will make the purchases for you, it is a good idea to go online and see what flights or trains are available to the conference site. This ensures that your schedule is not too tight and that you will arrive in time for the conference. Be sure to verify that you will be able to reach your hotel from the airport or train station at the time you arrive. Taxi cabs and rideshares are available in most large cities, but they can quickly become expensive. To save a bit of money, ask if your hotel has a shuttle service.

Many conferences occur within a hotel or in a conference center attached to a hotel. If you are able to stay in those locations, you will find that it simplifies your life considerably. However, the counterpoint is that the conference site is often more expensive than other options. Hotels often set aside some rooms at a cheaper conference rate, though – these are first come, first served, so the likelihood of securing one of these rooms is greatest if you try to book it as soon as the conference dates are announced. If the reduced rate is not available, or is still higher than you want to pay, use travel websites or mobile applications to see if other, less expensive hotels are within walking distance of the conference. Be sure to read the reviews associated with each location before making up your mind. In the end, you will have to balance convenience and budgetary concerns as you make your decision.

Students and faculty looking to minimize their expenses will also sometimes find roommates to share the cost of lodging. As you might expect, this works well if you know the other person well already. If you can stay with a friend, the conference can be both an enjoyable social and professional experience. If you are sharing a room with someone you have never met face-to-face, be sure to coordinate with that person early so that you will have enough beds, sheets, pillows, etc. This will also give you some idea about how well you will get along.

72.4 Surveying the Sessions

The speakers and profession-specific educational content associated with a professional meeting are major factors that influence people's desire to attend (Witt et al. 1995; Grant and Weaver 1996; Oppermann and Chon 1997; Rittichainuwat et al. 2001; Yoo and Chon 2008). These sessions keep you abreast of changes in your field, new innovations, and ongoing controversies. In any large meeting, there are probably many sessions scheduled at the same time. Be sure to consult the session catalog, online registry, or their equivalent to identify the sessions that you are most interested in. You can be as meticulous about this as you choose, and it is not uncommon to see people jumping from one talk to another as they try to get the most out of the conference. Regardless of whether you choose to take a relaxed approach or plan your day in five-minute intervals, do survey the sessions and individual speakers to see where your time is best put to use. To get the most from the experience, rank sessions that occur near each other so that you can make up your mind quickly when faced with difficulty moving between each talk. While you can attend sessions focused on your area of interest, be sure to scan the entire catalog to see if there are any talks that may be interesting, even if they are not directly relevant to your research. You may find a new venue or direction for your future work after reaching a bit beyond your comfort zone and taking a broader view.

During the sessions, be sure to write down any questions that arise and, if possible, ask the speaker about it during the Question and Answer (Q&A) session if there is one scheduled. Your question should be well-defined and succinct so that you can articulate it quickly and allow the speaker to craft a coherent response. Avoid compound or nested questions (questions within questions), since they can be difficult for the speaker to follow and they eat into the time for other people's questions. If you are hesitant to ask questions, you can get a sense for how to do it well by watching the reaction of the crowd to questions from other attendees. Emulate those whose questions get nods and thoughtful discussion from others in attendance. Take note of the questions that elicit eye-rolling and try to avoid those errors when you ask questions. Gaining the confidence to speak up and make inquiries is an important skill to develop in academia. These research sessions are an excellent time to get some practice in doing it well.

72.5   Visiting Vendors/Publishers

Meetings are a venue where vendors and publishers aggressively pursue specific segments of their customers. As such, they often bring new and not-yet-released products to meetings to tempt people in the field. It is always worth browsing the vendors to become familiar with their products and services. You may get advance notice of some new developments that will help back at your home institution. Sooner or later, you will probably find yourself tasked with making purchasing decisions in your department or outfitting a laboratory or other research space. At that time, the catalogs and contacts you have gathered from the meeting will come in very handy. Depending on the size of the meeting you attend, there may be some door prizes or little curios offered by the vendors. The days when companies could spend a lot of money on give-away items are gone for now, but this is still a good time to stock up on free pens, highlighters, and pads of paper. You might even score a flash drive or nice insulated grocery bag.

Publishers are also frequently interspersed among the vendors. Publishers are worth consulting for all the same reasons (knowing about updated and revised texts, upcoming books and articles) but also because they are not only there to sell items but to evaluate possible collaborators and authors. If you have any interest in eventually authoring a text or producing a commercial product, the publishers and vendors will be invaluable resources. The people staffing the booths frequently spend a long time with the same company, and you will likely see them frequently as the years pass. By maintaining a line of communication with these people and companies, you are improving your chances of building a solid partnership with them at a future date.

72.6 Presenting Your Work or Research

When you select a meeting, it is a very good idea to ensure that it will be a good venue to present your current and anticipated research or scholarly activity. While publications remain the gold standard of scholarly accomplishment, presenting your work to a group of informed peers is beneficial in several ways. It improves your ability to communicate clearly, allows for immediate feedback/criticism, and gives you the opportunity to meet other researchers with an interest in the same topic.

The most obvious benefit of presenting your work publicly is that you will gain confidence speaking in front of an informed audience. This frequently begins with poster presentations and can progress to a platform speaking session in front of the members. Some graduate programs have courses that help build public speaking skills, but there are many that do not. Either way, gaining confidence in your own ability to convey knowledge concisely will help make you a better academic. When presenting for the first time, be certain to practice multiple times in front of others. This will help the talk flow smoothly and gives you the opportunity to spot typos or other errors and to adapt to any feelings of nervousness that you might experience. Remember that *proper preparation prevents poor performance.*

Another benefit of presenting your work publicly is that you will receive criticism on your work and find ways to improve it. Tactful and constructive criticism is the lifeblood of academia. From peer review to questions after a public seminar, ideas and hypotheses are tested and critiqued to see if they are useful, valid, and accurate. Presenting your work in front of an informed group of people will help you find ways to improve and may even prompt other directions for your work to follow thereafter. This is similar to the process of responding to a peer reviewers' comments in articles that you have submitted for publication, except that it occurs orally and extemporaneously.

If that last sentence fills you with terror, do not panic. When you respond to questions, take your time formulating an answer. If necessary, clarify what is being asked and then respond concisely but conversationally. Have a bottle of water handy so that you can take a small drink if you are confused by a question. This gives you time to formulate a response without appearing flustered. On the other hand, if you thrive off the give-and-take of a Q&A session, this might be your time to shine. Do your best to answer each question succinctly and to expand on your findings, if possible.

A final benefit of presenting your work at a professional meeting is that it provides a venue for networking with established people in your field. Networking sometimes carries a negative connotation as a self-serving, superficial activity. While there are some people who embody this stereotype, becoming capable and comfortable while networking is an essential tool for academics, and it is commonly cited as a major reason to attend conferences (Witt et al. 1995; Grant and Weaver 1996; Oppermann and Chon 1997; Rittichainuwat et al. 2001; Jago and Deery 2005; Severt et al. 2007; Yoo and Chon 2008).

Throughout the entire process, present your work clearly and concisely. Pause to allow others to ask questions and reply as clearly as possible to their queries. To get the most out of interacting with colleagues, remain sincere and be yourself. Ask about their interests and be ready to supply a succinct account of yourself beyond your presentation. Hopefully, this will give you the opportunity to meet new and interesting colleagues that you will be seeing at conferences for years thereafter.

72.7   Getting Involved

Not only do professional meetings provide a venue for networking, education, and presenting your work, but they also open the door for getting involved in their governance, which is itself a way to deepen your professional life. While getting elected to the board of an organization is not immediately feasible for most people, many professional organizations have special interest groups that give members an opportunity to participate on a smaller scale. Getting the most out of your attendance at a conference definitely involves looking into the subgroups that contribute to the overall organization to see if you have any interest in attending their sessions.

While the number and type of groups may vary, there are often education-focused groups, career development groups, and research-specific groups. Check the conference schedule to see if these groups are hosting any seminar sessions. While a new or inexperienced association member would likely have difficulty getting elected to the association's board or governing council, it is often easier to get known and elected as a representative to these special interest groups, but do not be surprised if it takes several attempts. There are frequently many people vying for the same spot. Keep attending the sessions that interest you and putting your work forward. Regardless of the school or program from which you come, the longer you spend in academia, the less *where you are from* matters and *what you can do* becomes more preeminent. If you have a genuine interest in the group's focus, it will be noted eventually.

You might also consider sitting in on business meetings related to the association. Frequently, these are listed alongside the other sessions and are open to membership. These are often a bit dry but will give you an idea about the direction the association is traveling and the endeavors that it is pursuing. In addition to reports about finances and governance, there may also be specific meetings devoted to any journals or publishing projects that are linked to the association.

72.8   Follow-Up and Follow-Through

Inevitably, there will come a time after the meeting or during the day when, despite the best efforts of the organizers, you find yourself with several free hours and no particular interest in the sessions being offered. If this is the case, take a deep breath, relax, and go explore the world outside of the conference. Visit some of the local tourist destinations, travel with friends to a restaurant, go shopping, or some other activity that you enjoy. Different cities vie for the opportunity to host meetings, and there are often many literary or cultural activities available in the host cities. Some large meetings will have tourist information available at the convention center and can arrange for transport if needed.

If you are traveling with your family, be sure to do some research on what events and tourist sites they can visit while you are attending the meeting. Just as balance between the personal and professional must be maintained in everyday life, so too must a balance exist when attending professional meetings. This will hopefully ensure that you return to the meeting for several years and grow as a member.

References

Grant, Y.N.J. and Weaver, P.A. (1996). The meeting selection process: a demographic profile of attendees clustered by criteria utilized in selecting meetings. *Journal of Hospitality and Tourism Research* 20 (1): 57–71.

Jago, L.K. and Deery, M. (2005). Relationships and factors influencing convention decision-making. *Journal of Convention and Event Tourism* 7 (1): 23–42.

Mair, J. and Thompson, K. (2009). The UK association conference attendance decision-making process. *Tourism Management* 30: 400–409.

Oppermann, M. and Chon, K.S. (1997). Convention participation decision-making process. *Annals of Tourism Research* 24 (1): 178–191.

Rittichainuwat, B.N., Beck, J.A., and Lalopa, J. (2001). Understanding motivations, inhibitors, and facilitators of association members in attending international conferences. *Journal of Convention and Exhibition Management* 3 (3): 45–62.

Severt, D., Wang, Y., Chen, P., and Brieter, D. (2007). Examining the motivation, perceived performance, and behavioral intentions of convention attendees: evidence from a regional conference. *Tourism Management* 28: 399–408.

Var, T., Cesario, F., and Mauser, G. (1985). Convention tourism modelling. *Tourism Management* 6: 194–204.

Witt, S.F., Sykes, A.M., and Dartus, M. (1995). Forecasting international conference attending. *Tourism Management* 16 (8): 559–570.

Yoo, J.J.-E. and Chon, K. (2008). Factors affecting convention participation decision-making: developing a measurement scale. *Journal of Travel Research* 47: 113–122.

Zhang, H., Leung, V., and Qu, H. (2007). A refined model of factors affecting convention participation decision-making. *Tourism Management* 28: 1123–1127.

73

Finding Research Opportunities as a Medical Student

Frederic J. Bertino, MD

Department of Radiology and Imaging Sciences, Emory University School of Medicine, Atlanta, GA, USA
Department of Anatomical Sciences, St. George's University School of Medicine, Grenada, West Indies

73.1 Why Pursue Research as a Medical Student?

The publication of evidence-based medicine (EBM), among other forms of medical research, has allowed information to be available to clinicians and laypersons so that the standard of care is always questioned and reassessed. Improvements in medical science are constantly being published; physicians (and physicians-to-be) must remain up to date on these reported practices to be effective clinicians.

Medical students can be valuable members of a research team and contribute to scientific literature. While research offers one avenue of résumé improvement, students are more importantly able to experience in-depth trends in clinical medicine not otherwise learned from a textbook, honing their skills in research and clinical inquiry. This chapter focuses on how medical students can become involved in research at the basic and clinical science levels, and how to be effective team members at various points throughout the research and publication process.

73.2 A Disclaimer: What Benefit Does Research Bring to the Medical Student?

During medical school, students are often stressed to obtain a residency appointment of their choice. In this pursuit, many believe that research is crucial for their professional success. Postgraduate training programs look for well-rounded applicants that in addition to strong US Medical License Examination (USMLE) scores and grades, also demonstrate a curiosity for learning. Involvement in research helps communicate this interest.

Research experience may contribute to a strong postgraduate application, and even subsidize an application with minor academic imperfections. However, examination scores tend to be the factor by which residency programs filter their applicants; unfortunately, a résumé full of quality publications may not be reviewed by admissions staff if the appropriate examination score requirements are lacking.

A Guide to the Scientific Career: Virtues, Communication, Research, and Academic Writing, First Edition.
Edited by Mohammadali M. Shoja, Anastasia Arynchyna, Marios Loukas, Anthony V. D'Antoni, Sandra M. Buerger, Marion Karl and R. Shane Tubbs.
© 2020 John Wiley & Sons, Inc. Published 2020 by John Wiley & Sons, Inc.

In a study by Green et al. (2009), survey results demonstrated that residency program directors considered published medical school research and research experience as the lowest ranked selection criteria, while regarding clerkship grades, USMLE Step 1 scores, and specialty specific clerkship grades as the most important factors for candidate selection (Table 73.1).

Based on this survey, research at the medical school level is not mandatory, and an application without research will not discount a competitive applicant with high grades, examination scores, and excellent recommendations from clinical preceptors. With this in mind, it is important that medical students pursue research opportunities because of genuine interest and a passion for discovery – not because they wish to simply "look good on paper."

Aside from applying to residency programs, involvement of medical students in research is encouraged as it gives them an opportunity to learn about a topic in depth and exposes them to current information about key trends in their chosen specialty. Maintaining a knowledge of cutting-edge literature through reading and active research can position the medical student as a key member of a diagnostic team. This role is extremely important, as the team will look to the student for information on new treatments, diseases, and pathophysiologic mechanisms that are used to guide therapy.

Table 73.1 Rankings of the importance of Academic Selection Criteria from a National Survey of Residency Program Directors, 2006.

Academic criteria	Rank	Statistically different from rank(s)[a]
Required clerkship grades	1	2–14
USMLE Step 1 score	2	5–14
Grades in senior electives (specialty specific)	3	6–14
Number of honors grades	4	6–14
USMLE Step 2 CK[b] score	5	7–14
USMLE Step 2 CS[c] pass	6	8–14
Class rank	7	10–14
Membership in Alpha Omega Alpha	8	10–14
Medical school reputation	9	11–14
Medical school academic awards	10	12–14
Grades in other senior electives	11	14
Grades in preclinical courses	12	14
Published medical school research	13	N/A
Research experience while in medical school	14	N/A

a) To illustrate statistical differences that exist when comparing all other selection criteria, this data column indicates the ranks that are statistically different from the criteria in each row.
b) Clinical knowledge
c) Clinical skills

Source: Reproduced from Green et al. (2009), with permission from Wolters Kluwer Health, Inc.

73.3   Background for the Student-Researcher

The best way to engage in research as a medical student is to demonstrate an interest in contributing to scientific literature. A student may approach a faculty member to act as a mentor and help guide the project. This relationship can be especially beneficial if the faculty member shares the same research interest as the medical student. However, the goal of publishing in a peer-reviewed journal or presenting at a national conference does not always happen due to time limitations, especially for busy medical students. Though publication is desirable, the process of learning how to research properly is the most valuable benefit for the student researcher (Iserson 2013). Any experience in research reflects positively on residency applications or personal statements for post-graduate employment. These experiences are priceless, and are often improve chances of acceptance by admissions committees.

Residents or students of general medicine should be concerned with the production of quality research rather than the topic on which it is centered. If a specific topic or area of specialization is important to the student, then research on that topic is encouraged (Iserson 2013). However, it should be stressed that the fundamentals of research technique and critical thinking be developed as a student so that quality research may be conducted in the future. The essence of research experiences at the student level values learning the proper methods of research in order to provide the foundation for future scientific endeavors.

73.3.1   Have a Research Goal and Time Frame in Mind

Medical students are already pressed for time between studying for examinations and clerkship rotations. It is therefore a reasonable assumption that research will need to be conducted during nights and weekends; taking on a massive project that tests boundaries will unlikely result in a time-efficient publication (Iserson 2013.)

Quality research can be a time-intensive process that requires review, revision, and more review. Likewise, the scope of a student's research project should be simple enough that the medical student or a small team of researchers can complete it in a reasonable amount of time. The student should schedule accordingly and factor in extra time for reviews and revisions. This will make the acceptance of an abstract to a conference or journal an achievable goal without producing a hurried paper to follow it, and will allow the student to meet all expected deadlines (Iserson 2013).

Of course, research timelines are often derailed by unpredictable events, which may delay the publication of a well-engineered project. Fortunately, experts in research are aware of this and accommodate for interruptions in their workflow. While not all deadlines are consistently met, the importance of team communication cannot be understated. A team that communicates openly and honestly can help advance the research, overcoming hurdles, and keeping a project on track for timely publication.

73.3.2   Quality Research Production

Quality research is geared toward a peer review process in a well-respected journal with a high impact factor (IF). High IF journals are publications that hold papers with important, relevant, and known studies, which are cited in subsequent journals and projects. A journal that publishes perfectly conducted, peer-reviewed research reports contributes significantly to the scientific community.

Research may be conducted at the primary level (a study that answers a specific question by performing a specially designed experiment or test) or at the review/analytic level (a study that seeks to draw together correlations from other peer-reviewed studies). The latter type of research aims to simplify and gather other studies for the advancement of primary research or clinical practice.

Unless laboratory time is available and a principal investigator can guide the primary research, review and analytical projects are more accessible for medical students and are very welcome contributions to peer-reviewed journals. Such studies can be produced with greater ease than primary research studies, and when conducted appropriately, they often result in high-quality publications. A common belief among medical students is that analytical or clinical-based research (chart reviews, case analyses, quality improvement) is in fact more straightforward to publish than basic science research. In retrospect, clinical-based projects require less time for data gathering and the correlations that are drawn from the study can achieve high-impact status. Similarly, case reports are relatively easier papers to write as they focus more on gathering accessible data than conducting a primary experiment (Iserson 2013).

There are multiple methods of analytical research that vary in quality, and therefore impact factor. The meta-analysis is arguably the most respected method as it uses extensive statistical techniques to pool, compare, and contrast data among different studies (see Chapters 67 and 68) (Siwek et al. 2002).

A systematic review method without extensive statistical analysis is second to a meta-analysis. Nonsystematic reviews that are based primarily on investigator experience are subject to bias and are unlikely to be conducted at the medical student level. Case-reports or the case-series genre of papers are acceptable as student projects, especially when the project seeks to draw correlations among patient data obtained from charts. These reports, however, have lower IFs unless they report something truly novel or rare.

73.3.3   Student Protection and Ethical Consideration

In the world of research hierarchy, it is unlikely that a medical student will see his or her name as the first author on a publication, especially when many higher-ranking professionals are also contributing to the process (Iserson 2013). The medical student's role in larger team-based projects will typically be relegated to tasks of writing and literature gathering, and will likely appear lower on the list of project contributors.

Order of authorship at the medical student level will likely not make or break a student's status as a potential residency prospect or future researcher. Primary credit for the project should be given to the researcher responsible for the majority of the work. Concerns about authorship must be discussed before the project commences, and the ethical parameters of the medical student's role must be established in writing prior to publication (Iserson 2013). This is critical if, for example, the initial idea for a particular project belongs to the medical student; it will protect the student's rightful claim to authorship (see Chapter 38).

73.4   How to Find Research Opportunities as a Medical Student

There are two types of research one may conduct in medical school: basic science research and clinical research.

73.4.1   Basic Science Research

Professors that teach the first two years of basic medical science in med school often conduct their own primary research when not in the classroom. This practice is evident at larger universities in which primary research is conducted within the medical school's laboratories. Approaching a research professor with a common interest or project idea, the student may be presented an opportunity to work with faculty directly and thereby participate in a quality project.

The downside to primary research on a basic-science level, however, is that these projects are often time-intensive. In order to publish results from a bench research project, a hypothesis must be formulated and tested, and the data recorded and analyzed. This process, though dynamic and exciting, may take more time than the medical student can devote to the project, and the likelihood of publication may be slim in the two-year duration of basic science education.

Research societies exist at some medical schools, which support student endeavors to perform research over the course of a semester or a summer term. For example, St. George's University School of Medicine in Grenada, West Indies, is home to a Clinical Research Society (CRS) that grants students the opportunity to publish quality research. Students in this program apply through the Department of Anatomical Sciences for acceptance into the society; over the course of their second year, the accepted students are instructed in the methods of review and analytical research. During this period, students are able to obtain access and training in the usage of key databases, such as the National Cancer Institute's Surveillance, Epidemiology, and End Results program (SEER) database.

Many medical schools also have journal clubs that provide medical students the necessary tools of literature-reading that are valuable in academic research. It is highly encouraged that a medical student looking to engage in research be a participant in such groups. Inquiring about similar organizations may open up avenues of potential research that medical students are often unaware of.

73.4.2   Clinical Science Research

Performing research as part of a team that includes faculty members of a hospital or clinic is one of the most fruitful paths to publication. Similarly, residents who have research experience are often looking for students to help them with projects. These physicians find great benefit in the aid a student can provide, especially in tasks such as data collection, literature review, and writing. Their expertise in the field will often lead to frequent publication and their willingness to teach may provide the student-researcher with a recurrent opportunity for research experience in the third and fourth year of medical education (Institute of Medicine 2011.)

If the medical student has a particular clinical interest, it is of benefit to seek out a faculty member who specializes in that specific area. Exposure to medical and surgical subspecialties may help to influence a medical student's chosen path for residency. Learning more about the field through research is an excellent method for successful career development. Close professional relationship with these individuals can lead to mentorship, advising the medical student on his or her career path. The research connection and networking among colleagues may lead to future projects with other teams. In fact, the opportunity to write this very chapter came to me as an offer from colleagues of the director of my CRS organization.

A hospital quality-improvement project that enhances management protocol (not simply in-house, but to other institutions) is a relatively underutilized venue for medical student research involvement. It can provide the student-researcher with an ample opportunity to produce high-impact publications and presentations. Furthermore, these types of projects may have significant impact on a local community: they suggest improvements in hospital or patient management that may alleviate problems common in the healthcare industry. Contacting hospital staff and faculty who work in quality-improvement departments or who manage hospital protocol is an ideal way to demonstrate interest. This endeavor may lead to a research opportunity for the student and provide the institution with publishable solutions to pressing problems.

Finally, many hospitals have research departments or are known for being major research centers; their programs may offer an online application system for interested medical students who wish to engage in research.

73.4.3 Choosing a Viable Project

After selecting a field of research, finding a mentor, or joining a project team, the next step is to read extensively on the research topic. It is important that the student has some knowledge of the field before engaging in the actual project (Iserson 2013). Not only does this show initiative to the research team but it also fortifies the medical student with scientific and clinical knowledge that can be used in his or her future practice. Collecting information pertinent to the research topic (and discarding extraneous materials) will be beneficial when the literature review becomes project specific. A background review of the field will further reveal gaps in current research. These gaps often inform the direction of a professional project by seeking to answer a question that has not been asked or thoroughly reviewed previously (Institute of Medicine 2011). Next, if the study is to be retrospective and require data analysis, data set creation is necessary. It is important to discuss the data parameters or variables that are needed to test the hypotheses of interest. The most important task at this stage is to have a clearly defined research question before data collection is begun.

73.4.4 The Literature Review

The literature review is part of a research paper that documents an extensive list of papers, abstracts, journal reviews, and previous research that is relevant to the project in question (Oxman and Guyatt 1988). In many cases, the literature review presents an opportune moment for medical students to contribute to high impact research; faculty members or principle investigators likely have topics of research selected and are in need of eager students to help in the collection of quality literature. The literature review helps the research team become well versed on citable resources for the project as well as identifies weaknesses in the current literature that may point to future projects. Deficiencies in the pursued study can further be recognized by looking at comparable research, providing a stage to modulate the hypothesis and better structure the project for peer review.

Before being used in research, many published reviews deserve careful evaluation of the methodology. The *Journal of the American Dental Association (JADA)* reported that well-documented inclusion criteria were the strongest aspect of high-quality systemic reviews. These reviews were found to explicitly document the methodology used for review, which databases were accessed, the use of multiple databases, secondary searches, terms, and strategy documentation. The reviews that met these criteria were found to diminish potential bias and provide the most thorough review on a particular topic (Major et al. 2006).

Literature review begins with a search through library resources and major research databases, such as PubMed, Scopus, Thomson Reuters Web of Science (formerly the Institute for Scientific Information (ISI) Web of Knowledge), and Cochrane meta-analysis database. Access to full articles are typically granted through the medical school library or within the hospital's medical library (IOM 2011). Establishing targeted and clearly defined questions is crucial to the start of any research project. Consider using the PICO (patient/problem/population, intervention/indicator, comparison, and outcome) method for outlining specific parameters and clinical questions (see Chapter 68 for more details).

The practice of taking notes on pertinent papers and abstracts will help the medical student recall relevant information for inclusion in a study. Keeping organized documentation of previous research and specific citations is important, especially when tasked with the literature review. It is recommended that a medical student wishing to contribute to quality research develop skill in reading papers and abstracts.

73.4.5 Goal Reevaluation

The original hypothesis must be reassessed throughout the research process and should ensure that the literature reviewed is adequate to answer the question. The resources gathered through the literature review may change the direction of the project entirely, which is why it is imperative that adequate time be allotted for research that seeks rapid publication. Focusing the hypothesis and returning to the literature search will help narrow the topic and even inform the type of literature to be included in the published review. The potential for shift in project direction is high; however, a thorough literature review more often than not will provide enough background information to validate quality work.

Figure 73.1 provides a flowchart of the key components to the medical student's role in research. Following this pathway will help keep the medical student focused on his or her task within the research team and allow him or her to lead a very rewarding experience. Students who use these basic guidelines will have an organized understanding of the research process and be able to carry out projects quickly and effectively.

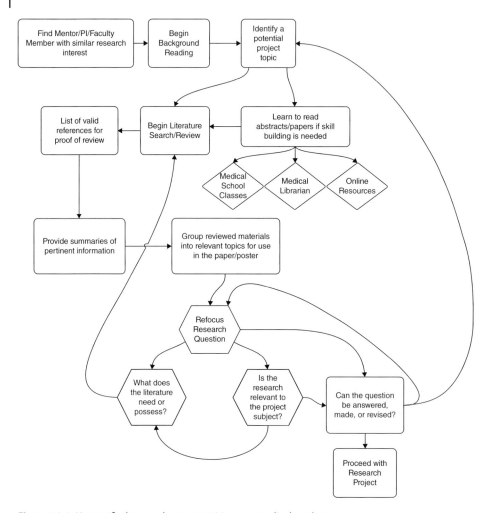

Figure 73.1 How to find research opportunities as a medical student.

73.5 Conclusions

Medical student involvement in research is a wonderful gateway for future clinicians to become involved with academic medicine and medical practice. It provides an avenue for awareness of cutting-edge medical practices and research findings so that even those physicians not looking to publish may stay current in their fields. At the student level, skills that must be developed include the ability to critically read and analyze multiple papers and abstracts; understand where to locate and how to recognize high-impact research; and be able to recognize biases, gaps, or areas for improvement within current scientific literature. These valuable skills will allow the student to develop into a prolific physician-scientist.

Acknowledgment

I would like to extend my personal thanks to Mr. Kell Julliard, director of research at Lutheran Medical Center, Brooklyn, New York, for his advice and assistance on this topic. Mr. Julliard has been very active in including medical students and residents in clinical research for many years and his experience on the subject is very much appreciated.

References

Green, M., Jones, P., and Thomas, J.X. (2009). Selection criteria for residency: results of a National Program Directors Survey. *Academic Medicine* 84: 362–367.

Institute of Medicine (2011). *Finding What Works in Health Care: Standards for Systemic Reviews*. Institute of Medicine of The National Academies.

Iserson, K.V. (2013). *Iserson's Getting into a Residency: A Guide for Medical Students*, 166–170. Tuscon, AZ: Galen Press, Ltd.

Major, M.P., Major, P., Flores-Mir, C. et al. (2006). An evaluation of search and selection methods used in dental systematic reviews published in English. *Journal of the American Dental Association* 137: 1252–1257.

Oxman, A.D. and Guyatt, G. (1988). Guidelines for reading literature reviews. *CMAJ* 138: 697–703.

Siwek, J., Gourlay, M., Slawson, D. et al. (2002). How to write an evidence-based clinical review article. *American Family Physician* 65: 251–258.

74

A Guide to Writing a Curriculum Vitae

Sanjay Patel, Petru Matusz and Marios Loukas

Department of Anatomical Sciences, St. George's University School of Medicine, Grenada, West Indies

74.1 Introduction

A curriculum vitae (CV) summarizes a person's professional career. The term curriculum vitae originates from Latin and means "course of life." It is often used when applying for academic positions, grant funding, and many other professional needs. It is crucial to spend time revising and seeking feedback on a CV until it effectively communicates a person's strengths and potential.

74.2 What Should Be Included in the CV?

The CV has the following headings: personal data, education, professional experience (research and teaching), awards, grants, publications, and references (Leung and Robson 1990). This list is not complete, and every CV will structure the list differently to highlight the applicant's strengths.

74.3 Cover Page

A cover page should be written to support every CV. It gives the applicants an opportunity to introduce themselves, highlight their qualifications, and show interest in the position being offered. It should use full names when possible, for example, "...Dear Samantha Thompson." It should also address the applicant's interest and reasons for applying for the position advertised in the beginning of the letter. The body should include a chronological statement that addresses the applicant's past careers, elaborates on what makes the applicant a good candidate for the position, and his/her future goals. It should end on a positive note without trying to sound overly assertive (Day and Gastel 2006).

A Guide to the Scientific Career: Virtues, Communication, Research, and Academic Writing, First Edition.
Edited by Mohammadali M. Shoja, Anastasia Arynchyna, Marios Loukas, Anthony V. D'Antoni, Sandra B. Buerger, Marion Karl and R. Shane Tubbs.
© 2020 John Wiley & Sons, Inc. Published 2020 by John Wiley & Sons, Inc.

74.4   Personal Data

The personal information included on a CV should include the full name, mailing address, phone number, and email address of the applicant. The applicant's address and academic institution address should be positioned parallel to each other with the academic institution's information aligned to the left margin. Some optional information may also be included such as the applicant's date of birth, permanent citizenship, or visa status. However, social security number, sexual orientation, and religion should never be included the CV.

74.5   Education

An applicant's educational experience is always included in his/her CV and is listed in the subheadings: undergraduate, postgraduate, and continuing education. Each degree earned should be written in initials and include the year that it was awarded, along with the awarding institution's name. In some situations where the institution's location may be ambiguous, it can be useful to specify the city and state.

74.6   Professional Experience

This section includes a list of the professional positions held by the applicant in the past, such as any academic appointments held as visiting professor or clinical researcher. This list should be written in reverse chronological order. At minimum, it should include the institution or company's name, a brief description of duties performed, and the start and end dates the position was held for. By writing a brief description, the applicant has the opportunity to describe his/her level of involvement and duties to other organizations.

Persons applying for teaching positions should include a summary of all their instructional experience. Recent graduates should start with any graduate teaching apprenticeships or graduate teaching experience. Junior faculty should mention classes they have taught and not include graduate teaching assistant work. It is common practice to include the full course title, position held, dates in which the experience took place, and a brief narrative of main duties performed.

The list should be organized either by undergraduate/graduate, specific field/area, or by the institution taught at. If a person has taught at multiple institutions, he/she should organize this list by location. The terms and years taught should be included in parenthesis next to the course title. This allows for readers to see your range of teaching competencies without the list becoming exhausting.

An example would be:

> The Human Body 753 – Teaching Assistant – Spring 2015 – UC School of Medicine, Responsibilities: Lecturer for problem-based learning sessions of 30 students in each; developed syllabus for discussion session, created lesson plans, created visual aids for lectures, created assignments, and evaluated student learning, assigned discussion session grades.

Also a section on research experience in included with the list in reverse chronological order. Every entry includes the institution's name, location, and years involved along with position and advisor's name. Included in each entry would be a brief description of the focus of the research, along with descriptive terms showing level of involvement in the project such as, "developed PCR method to study changes in receptors," or "analyzed changes in synaptic plasticity." To reduce vaunting and keeping the CV professional, do not use the pronoun "I" when describing involvement.

74.7 Awards and Grants

This section is sometimes placed third on the CV; however, regardless of where it is placed, this section shows that the author has been recognized and adds weight to his qualifications. Depending on the author' institution, some may want to separate honors and awards from grants and fellowships. An example of an honor or award that should be included is those that are recognized by an academic organization, placement into an academic competition, or election into an honorary society. List the year the award was received along with the professional group granting the award.

Previously awarded grants, especially those from the NIH, are becoming more emphasized to help gauge the ability of a researcher to be productive and acquire more funding (Svider et al. 2013). Initially, graduates may include academic scholarships, small grants, or department fellowships that support graduate research. Senior researchers will include more competitive funding, such as federal research grants. In this list give the funding organization, institution where the funds were utilized, amount funded, and year.

74.8 Publications

It is important to list all publications completely and uncensored in this section. Tenure committees will focus on this section and specifically look for studies done in referred journals (Hirsch 2005). It is best to include impact factor and rank of journals. This list can be found at 2018 Journal Citation Report, Science Edition (Thomas Reuters 2013).[1] All publications written should be organized into separate categories such as: original articles, case reports, review articles, and books. Give full details of the publication, including all author names, title, journal, and year published. Each publication should be formatted into the American Medical Association citation style. The individual's name should be underlined or bolded when co-authors are included.

Articles that are not yet published should be further clarified as "submitted," or if accepted but awaiting publication, then "in press" (Hailman and Strier 2006). Some committees may look up the publications to become more familiar with your work; therefore, you may want to include a PubMed ID in the citation. If there are many publications, you can subdivide them into categories, depending on subtype. Editorial and ad hoc reviewer work for journals should be cited accordingly but do not include the titles of papers being reviewed, as this should remain anonymous.

---

1 http://jcr.incites.thomsonreuters.com

Recent graduates may want to include talks given at departmental seminars and conference activity. This will show that they have public speaking experience and the history of research that went into a manuscript before it becomes printed in a journal. The talks should be clearly organized by type and include the name of paper presented, name of conference, and date with year only. More senior researchers will include invited talks, keynote addresses, and talks in scientific colloquia. List the talks by giving the title, institution location, and date with year. It is always important to use common sense and keep the list relevant and avoid "padding" your CV (Cleary et al. 2013).

74.9 Additional Categories

Examples of additional categories not mentioned above include:

- Professional services (committee for local institution or academic society)
- Mentor for graduate students of junior faculty (include list of students, years, and if they completed candidacy)
- Professional Society Memberships
- Certifications (i.e. ACLS)

74.10 Formatting/Layout

The completed CV should be clear, concise, and appealing to look at. The format should be consistent throughout the entire CV. It should have 1-in. margins throughout, use 12-point font, and be single-spaced. Only the candidate's name is in 14-point font and center bold at the top of the page. All entries thereafter are left justified with headings in bold and every letter capitalized. Each entry and heading has a blank line space in between. There is no need to use bullet points or italicize (except for book titles).

74.11 Conclusion

The CV is used as a means to organize all of one's professional accomplishments to a prospective employer. It should be updated routinely to ensure that it has the most recent and accurate information on an applicant. Great consideration should be taken to ensure it is clear, concise, and without any errors in grammar. Lastly, it is important to keep the CV factual and follow the principle, "Honesty is the best policy."

References

Cleary, M., Walter, G., and Jackson, D. (2013). Editorial: 'is that for real?': curriculum vitae padding. *Journal of Clinical Nursing* 22: 2363–2365.

Day, R.A. and Gastel, B. (2006). *How to Write and Publish a Scientific Paper*, 6e, 227–233. Westport, CA: Greenwood Press.

Hailman, J.P. and Strier, K.B. (2006). *Planning, Proposing, and Presenting Science Effectively: A Guide for Graduate Students and Researchers in the Behavioral Sciences and Biology*, 2e, 149–164. Cambridge, UK: Cambridge University Press.

Hirsch, J.E. (2005). An index to quantify an individual's scientific research output. *Proceedings of the National Academy of Sciences of the United States of America* 102: 16569–16572.

Leung, A.K. and Robson, W.L. (1990). How to write an effective curriculum vitae: making the most of your assets. *Canadian Family Physician* 36: 2083–2088.

Svider, P.F., Mauro, K.M., Sanghvi, S. et al. (2013). Is NIH funding predictive of greater research productivity and impact among academic otolaryngologists? *Laryngoscope* 123: 118–122.

Index

A Guide to the Scientific Career: Virtues, Communication, Research, and Academic Writing, First Edition.
Edited by Mohammadali M. Shoja, Anastasia Arynchyna, Marios Loukas, Anthony V. D'Antoni, Sandra M. Buerger,
Marion Karl and R. Shane Tubbs.
© 2020 John Wiley & Sons, Inc. Published 2020 by John Wiley & Sons, Inc.